14

D

2 8

DUE

?

Ecological Studies

Analysis and Synthesis

Edited by

J. Jacobs, München · O. Lange, Würzburg
J. S. Olson, Oak Ridge · W. Wieser, Innsbruck

Czechoslovak Academy of Sciences

Scientific Editor: Prof. Dr. Miroslav Penka
Scientific Advisor: Dr. Milena Rychnovská
English Language Editor: Dr. Margaret S. Jarvis

Bohdan Slavík

Methods of Studying
Plant Water Relations

With 181 Figures

Academia
Publishing House of the Czechoslovak Academy of Sciences,
Prague
Chapman & Hall Limited London
Springer - Verlag Berlin · Heidelberg · New York
1974

The picture on the dust cover
closely resembles figure 3.24
on page 198.

Distributors in all western countries:

ISBN 0-412-13230-3 Chapman & Hall Limited London
ISBN 3-540-06686-1 Springer-Verlag Berlin · Heidelberg · New York
ISBN 0-387-06686-1 Springer-Verlag New York · Heidelberg · Berlin

Library of Congress Catalog Card Number 74-3672

Offsetprinting and bookbinding: Brühlsche Universitätsdruckerei Gießen

Preface

To write a handbook of methods is surely to invite criticism, as has already been said several times. On the other hand, there is a great need for methodological manuals in all fields of science. It was therefore decided to compile this book, written in good faith to help scientists, teachers and students who will, it is hoped, use it and judge it good.

To be useful to the reader, such a manual must provide a broad review of the methods available and describe them in sufficient detail to permit preliminary selection and judgement. It has to give — at least for selected methods — a sufficiently detailed description of the equipment and procedure as to be to some extent self-contained. It must assume a critical standpoint as regards the theoretical basis of the methods, the significance of results, and their errors and limitations. It must also furnish examples, pertinent numerical tables, and very complete references. All this and much more is expected of a good manual of methods.

It is difficult to fulfil all these requirements. The result, as always, is a series of compromises. We have tried to make this more than just a list of methods without critical evaluation. Not all techniques are described in the same amount of detail, but theoretical introductions offering a basis for critical evaluation are supplied in all cases. As many methods as possible have been included, with full references, so as to provide an adequate review. Some of the data and some figures, *e.g.* electronic circuits, serve rather to show how difficult and challenging the techniques are than to offer sufficient information for construction of the equipment. Where appropriate, useful technical details concerning devices and measuring procedures have been included for certain selected and particularly promising methods so as to enable the reader to profit by other's experience. Names of commercial suppliers of equipment are given as examples only and do not imply any preferential endorsement.

Simple methods are included as well as highly sophisticated ones, and laboratory methods as well as those suitable for field work. Even methods that yield only approximate results have been included, with the necessary reservations. Tables have been selected on the basis of how often they are needed; there are also some general tables in the appendix, giving information on the symbols and units generally used. Review tables setting out brief critical evaluations and the advantages and disadvantages of the methods may help the reader in a first selection. Methods for measuring leaf area and leaf temperature, isotopic techniques and radiation measurements were not included, as recent descriptions may be found elsewhere. The same is true of methods that fall outside the scope of the manual for obvious reasons, *e.g.* measuring water-vapour exchange on a field scale, large-scale lysimetry, *etc*. Brief or more detailed reviews of methods for determining some ecological factors such as air and soil humidity and dew have, however, been included.

The authors are aware that progress in science to some extent depends on progress in methods, and that this goes on whether people publish manuals of methods, or not. We know this, and our book takes account of the fact.

A great deal of the manual was translated from the Czech by Dr. J. Spížek, and the entire English text was revised by Dr. Margaret S. Jarvis. The authors are greatly indebted to Dr. P. G. Jarvis of Aberdeen for valuable criticism and suggestions. We also thank our other scientific friends and collaborators for their help, also Miss Jarmila Veverková for drawing all the figures and Mrs. Věra Dobrová for help in revision and in preparing the indexes.

January, 1974   *B. Slavík and co-authors*

Contents

List of Authors

List of Frequently Used Symbols, Abbreviations and Units

List of Authors

J. ČATSKÝ, Institute of Experimental Botany, Czechoslovak Academy of Sciences, Flemingovo 2, 160 00 Praha 6, Czechoslovakia.

J. HRBÁČEK, Institute of Basic Agrotechniques, 664 62 Hrušovany near Brno, Czechoslovakia.

P. G. JARVIS and M. S. JARVIS, Department of Botany, University of Aberdeen, St. Machar Drive, Aberdeen AB9 2UD, Great Britain.

V. KOZINKA, Botanical Institute, Slovak Academy of Sciences, Dúbravská cesta 26, 809 00 Bratislava, Czechoslovakia.

H. R. OPPENHEIMER †, Department of Horticulture, Faculty of Agronomy, Hebrew University, Rehovot, Israel.

B. SLAVÍK, Institute of Experimental Botany, Czechoslovak Academy of Sciences, Flemingovo 2, 160 00 Praha 6, Czechoslovakia.

J. SLAVÍKOVÁ, Department of Botany, Science Faculty, Charles University, Benátská 2. 128 01 Praha 2, Czechoslovakia.

J. SOLÁROVÁ, Institute of Experimental Botany, Czechoslovak Academy of Sciences, Flemingovo 2, 160 00 Praha 6, Czechoslovakia.

J. ÚLEHLA, Institute of Basic Agrotechniques, 664 62 Hrušovany near Brno, Czechoslovakia.

---

† Deceased.

List of Frequently Used Symbols, Abbreviations and Units

Every effort has been made to unify symbols and abbreviations used in this book as far as possible. However, in some cases different symbols have been used where the maintaining of symbols originally used by the authors quoted has been preferred in some places. For other symbols and abbreviations see also Tables 1.1, 1.2, 2.4, 6.1, 6.2 and 6.3.

Symbol	Description	Main SI Unit	Units used in this book
A	area	m^2	cm^2
C	concentration	$mol\ m^{-3}$	$g\ cm^{-3}$ or $mol\ l^{-1}$
D	diffusivity, diffusive coefficient	$m^2\ s^{-1}$	$cm^2\ s^{-1}$
E	transpiration or evaporation rate	$kg\ m^{-2}\ s^{-1}$	$g\ cm^{-2}\ s^{-1}$ (see also Section 5.5.8)
G	sensible heat flux by conduction	$J\ m^{-2}\ s^{-1}$	
H	sensible heat flux by convection	$J\ m^{-2}\ s^{-1}$	
I	radiation intensity		
K_Q	thermal conductivity	$J\ m^{-1}\ s^{-1}\ K^{-1}$	$J\ cm^{-1}\ s^{-1}\ {}^{\circ}C^{-1}$
K_w	(absolute) water permeability constant, hydraulic conductivity	s or $m\ s^{-1}$	s or $cm\ s^{-1}$ (see also Section 4.1)
L	latent heat of vaporization	$J\ kg^{-1}$	$J\ g^{-1}$
M	molecular weight		
P	(hydrostatic) pressure	Pa	bar
P	precipitations		mm
Q	heat flux	$J\ m^{-2}\ s^{-1}$	
R	molar gas constant	$8.3143\ J\ K^{-1}\ mol^{-1}$	
R	viscous flow resistance		
R	radiation flux	$J\ m^{-2}\ s^{-1}$	
T	temperature	K	${}^{\circ}C$
V	volume	m^3	cm^3
$\bar{V}$	partial molal volume	$m^3\ mol^{-1}$	$cm^3\ mol^{-1}$

Symbol	Description	Main SI Unit	Units used in this book
W	mass ($\sim$ weight) of water	kg	g
a	absolute humidity of the air	$kg_{H2O}\ m^{-3}$	$mg_{H_2O}\ l^{-1}$
a	chemical energy storage coefficient		$cal\ g^{-1}$
a_w	relative chemical activity of water		
b	atmospheric pressure	bar	bar or torr
c	specific heat	$J\ kg^{-1}\ K^{-1}$	$J\ g^{-1}\ °C^{-1}$
d	distance, thickness of the boundary layer	m	cm
h_w	capillary potential		cm_{H_2O}
l	length	m	cm
m	mass	kg	g
n_w	molar fraction of water		
n_D	refractive index		
p	partial water vapour pressure	bar	torr
q	flux (mass or wolume)	$kg\ s^{-1}$ or $m^3\ s^{-1}$	$g\ s^{-1}$ or $cm^3\ s^{-1}$
r	diffusive resistance	$s\ m^{-1}$	$s\ cm^{-1}$
t	time	s	s
v	linear velocity of mass flow	$m\ s^{-1}$	$cm\ s^{-1}$
w	weight (generally) [$\sim$ mass]	kg	g
z	distance	m	cm
γ	psychrometric constant	$bar\ K^{-1}$	$mbar\ °C^{-1}$
η	dynamic viscosity	$N\ s\ m^{-2}$	$g\ cm^{-1}\ s^{-1}$
μ_w	chemical potential of water	$J\ kg^{-1}$	$J\ kg^{-1}$
π	osmotic pressure	bar	bar
ϱ	density	$kg\ m^{-3}$	$g\ cm^{-3}$
σ	surface tension	$kg\ s^{-2}$	$g\ s^{-2}$
σ	reflection coefficient	fraction	fraction
τ	matric tension	bar	bar
Ψ_w	water potential		
Ψ_s	osmotic potential	bar or $J\ kg^{-1}$	bar
Ψ_p	pressure potential		
Ψ_m	matric potential		

Superscripts

'	uncorrected value, changed value
°	reference value

Main Subscripts

a	air, air boundary layer
c	cuticle
cell	cell
(e)	external
i	intercellular spaces
(i)	internal
l	leaf
m	matrix
mes	mesophyll
n	net
p	pressure
r	root
s	solute
sap	sap
u	wind
v	vacuole
vap	vapour
w	water
(deplasm)	deplasmolysis
(lp)	limiting plasmolysis
(n)	normal state
(p)	plasmolysis
(prot)	protoplast
(sat)	saturation
(sol)	solution

Abbreviations	Description	
c.c.	capillary capacity of soil	
CP	capillary potential	
DPD	diffusion pressure deficit	
e.m.f.	electromotive force	
f.c.	field capacity of soil	
m.w.f.	"mobile" water fraction	
p.p.m.	parts per million	
p.t.l.p.	permanent turgor loss point	
p.w.p.	permanent wilting percentage	
r	albedo	
r.h.	relative humidity of the air	per cent or fraction
r.w.c.	relative water content	per cent or fraction

Abbreviations	Description	
s.d.	saturation deficit	bar or torr
t.s.m.s.	total soil moisture stress	
v.p.m.	volume per million	
w.d.	water deficit, water stress	
w.r.d.	water resaturation deficit	per cent or fraction
$\Delta W_{(sat)}$, w.s.d.	water saturation deficit	per cent or fraction
Δ	difference or increment	

Water in Cells and Tissues

1.1 Introduction and Terminology

Living cells need to be hydrated, *i.e.* more or less saturated with water, in order to function normally. Plant cells are usually incompletely saturated with water. A certain hydration deficiency [determined as the lowered value of the water potential (see below)] often occurs. This represents a driving force for the flux of water through all parts of the plant on the one hand, and a factor controlling its physiological activity on the other.

There has been a long search for suitable means of expression and measurement of the hydration level of plant tissues. One source of the difficulties has been the practical methodological aspects. Here the desire for a relatively simple technique has led — even after most of the theoretical problems have been settled (see later) — to the development of many indirect methods of which none is satisfactory for all purposes. That is why there are so many.

Nevertheless the main problem was the theoretical one of finding an objective expression of the water status of the cell and tissue which would be right and reasonable from the physical point of view and significant as an internal physiological factor.

Chemical activity of water (μ_w) may be expressed by the chemical potential of water. The chemical potential of water in a given system may be measured only relatively as compared with the chemical potential of pure free water, *i.e.* of water containing no solutes and bound by no forces. So it is measured as ($\mu_w - \mu_w^0$), when μ_w^0 is chemical potential of pure free water at the same temperature and pressure. Even the measurement of the relative chemical potential of water with pure free water as reference was not free of problems.

From the thermodynamic point of view, the chemical potential of water is determined unequivocally by the free energy (free enthalpy) per unit mole of water, *i.e.* the energy required for the transfer of that water from one place to another; it is also systematically affected by environmental factors such as pressure, temperature *etc.* However, there is no direct evidence that the physiological activity of water is proportional to the defined chemical activity. The field is thus open for various conjectures about the character of this relationship.

As a result of the free kinetic energy of molecules, the water molecules move from a place where they have a higher mean free energy to a place with a lower one. This effect is called diffusion and its rate (dn/dt) depends on molar concentration gradient

(dC/dx), on the area of the membrane through which the movement occurs (A), and on a diffusion constant (D) depending on the temperature, the size of water molecules, their association *etc.* (first FICK's law of diffusion):

$$\frac{dn}{dt} = -AD\frac{dC}{dx}.\tag{1.1}$$

As the molar concentration of pure water is higher than the molar concentration of water in any aqueous solution, the net diffusive flux of water molecules will always be positive from pure water to a solution, under isothermal conditions. If an aqueous solution and a pure solvent (water) are separated by a membrane completely permeable to water molecules but more or less impermeable to the molecules of the solute, a special type of diffusion, water osmosis, takes place. Diffusion of water from pure water to the solution appears here as an increase in the volume of the solution as it becomes diluted. To prevent osmosis, we must apply a pressure to the solution. The pressure which under isothermal conditions prevents the diffusion of water through a semipermeable membrane to an aqueous solution is termed the osmotic pressure of the solution. This osmotic pressure of the solution is determined by the properties of the dissolved solute.

Osmotic pressure

$$\pi = icRT\,,\tag{1.2}$$

where π is osmotic pressure in units of bars, i is the average number of particles per molecule [osmotic pressure of dissociated compounds is higher ($i > 1$), hence electrolytes exhibit a higher osmotic pressure than undissociated compounds], c is the concentration of the solution (mole litre^{-1}), R is the gas constant (8.314 J mol^{-1} K^{-1}), T is the absolute temperature (in kelvins). Solutions with the same osmotic pressure are mutually isotonic (iso-osmotic, isomysic, isopiestic); a solution with a higher osmotic pressure than that of another solution is hypertonic (hyper-osmotic, hypermysic); a solution with a lower osmotic pressure than that of another solution is hypotonic (hypo-osmotic, hypomysic).

As the free energy of water depends on temperature, diffusion can be brought about by a temperature difference between two cells or tissues even though diffusion would not occur at a uniform temperature. This process is called thermo-osmosis. A local difference of electric potential may also cause the diffusion of water, *viz.* electrosmosis. This is diffusion of water caused by equilibration of the electric potential by one-way movement of the water molecules. A gradient in the free energy of water can also be caused by a pressure gradient, which would result in the mass flow of water along the gradient.

Most mature cells of plant tissues include vacuoles, which are enclosed in protoplast and which usually contain an aqueous solution of inorganic salts, sugars, organic acids and their salts, proteins *etc.* The vacuolated cell is subject to the processes described above. The vacuolar solution and the solution surrounding the cell are the two solutions and the outer layer of cytoplasm acts as a semipermeable membrane, the cell wall being completely permeable to the great majority of compounds. If the osmotic pressure of the vacuolar solution is higher than that of the surrounding solution, osmosis of water into the vacuole occurs. The vacuole and thus also the cell contents as a whole (the whole protoplast) increase in volume. The expansible cell wall is extended until its regressive pressure reaches a value corresponding to the difference between the osmotic pressure of the solutions. A diffusion equilibrium is then reached, osmotic pressure being compensated by the pressure of the cell wall. Turgor pressure, by which the cell wall is distended, then counterbalances the inward pressure of the cell wall.

If the osmotic pressure of the vacuolar solution is lower than that of the surrounding solution, diffusion of water molecules out of the cell into the medium occurs, and the vacuolar solution becomes more concentrated until equilibration of the two osmotic pressures is attained. The vacuolar volume, and hence also the protoplast volume, decreases. Where the cell wall does not contract enough to follow the changes in volume of the protoplast, plasmolysis (detachment of the peripheral cytoplasm from the cell wall) is observed.

The ease of measurement of osmotic pressure of a solution and the described osmotic model of the vacuolated cell made it possible to express water status of the vacuolar sap and of the whole cell in an externally applied solution in terms of pressure (*e.g.* in bars). Thus for a long period many different pressure terms were used for description of water status in plant cell (see Tables 1.1. and 1.2), using a simple equation

$$A = B - C, \tag{1.3}$$

where A denotes the pressure by which the vacuolated cell absorbs water from the medium and is given by the difference between the potential osmotic pressure of the vacuolar sap (B) and the pressure of the cell wall (C). It must be remembered that

Table 1.1 *Equation (1.3) in symbols used by various authors.*

$A \quad = B - C$	(1.3)
$S_z \quad = S_i - W$	(URSPRUNG and BLUM 1916), Saugkraftgleichung, osmotische Zustandsgleichung (HÖFLER 1918)
$DPD = OP - TP$, where $TP = WP$	(MEYER 1954)
$S \quad = O - (E + B)$	(PRÁT 1945)
$NIF = \sum IF - \sum EF$	(BROYER 1947)
$\psi_w \quad = \pi + \tau + P$	(SLATYER and TAYLOR 1960)
$S \quad = W - (P_T + P_A)$	(WALTER 1963)
$\Delta W \quad = \pi - P - \tau$	(WEATHERLEY 1965)
$\psi_w \quad = \psi_s + \psi_m + \psi_p$	(KRAMER *et al.* 1966, SLATYER 1967)

(For explanation of symbols see next Table 1.2.)

Table 1.2 *Terminology and symbols of values of water state.*

Term	Symbols	Numerical value	Authors
Value A in equation (1.3):			
Water potential	ψ, ψ_w	zero to negative	SLATYER and TAYLOR (1960), SLATYER (1967)
Equivalent terms:			
suction force	S	zero to positive	RENNER (1915)
Saugkraft, Saugkraft der Zelle	S_z	zero to positive	URSPRUNG and BLUM (1916)
water absorption power		zero to positive	THODAY (1918)
suction pressure		zero to positive	STILES (1922)
suction tension		zero to positive	BECK (1928)
osmotisches Saugpotential		zero to positive	OPPENHEIMER (1930)
turgor pressure			CURTIS and SCHOFIELD (1933)
net influx specific free energy	NIF	zero to positive	BROYER (1947)
diffusion pressure deficit	DPD	zero to positive	MEYER (1945)
osmotic equivalent of the cell		zero to positive	LEVITT (1951)
hydrature deficit		zero to positive	FUKUDA (1958)
hydropotential	HP	zero to negative	SHMUELI and COHEN (1964)
water potential depression	ΔW	zero to positive	WEATHERLEY (1963/65)
Value B in equation (1.3):			
Osmotic potential (solute potential)	ψ_s	zero to negative	SLATYER (1967) KRAMER *et al.* (1966)
	π	zero to negative	SLATYER and TAYLOR (1960)
Equivalent terms:			
Saugkraft des Zellinhaltes	S_i	zero to positive	URSPRUNG and BLUM (1916)
osmotischer Wert	W or OW	zero to positive	WALTER (1931b, 1965)
osmotic pressure	OP	zero to positive	MEYER (1945)
	π	zero to positive	SLATYER (1967)
osmotic solutes specific free energy			BROYER (1947)

Table 1.2 *(continued)*

Term	Symbols	Numerical value	Authors
Value C in equation (1.3):			
Pressure potential	ψ_p	(exceptionally) negative, (mostly) positive	KRAMER *et al.* (1966)
			SLATYER (1967)
	P	(exceptionally) negative, (mostly) positive	SLATYER and TAYLOR (1960)
Equivalent terms:			
turgor pressure	TP	(negative to) positive	MEYER (1945)
Turgordruck	P_T	positive	WALTER (1963)
wall pressure	WP	(positive to) negative	MEYER (1945)
Wanddruck	W	(positive to) negative	URSPRUNG and BLUM (1917)
opposite pressure of the surrounding cells	B	positive	PRÁT (1945)
Gegendruck der Nachbar-zellen	P_A	positive	WALTER (1963)
hydrostatic specific free energy			BROYER (1947)
Further components of ψ_w:			
Matric potential	ψ_m	zero to negative	KRAMER *et al.* (1966) SLATYER (1967)
	τ	zero to negative	SLATYER and TAYLOR (1960)
Equivalent terms:			
interfacial tension			EDLEFSEN (1941)
non metabolic specific free energy			BROYER (1947)
matric pressure (matric suction)	τ	zero to positive	SLATYER (1967)

with the exception of the turgor pressure (pressure exerted on the cell wall) only potential pressures are involved.

Figure 1.1 shows the changes in these values (*A*, *B*, *C*), as well as changes in relative volume of the cell (cell wall), of the protoplast (prot), and of the vacuole (v), which take place if the cell is immersed in solutions of varying potential osmotic pressure. [These changes are found in cases where the cytoplasmic layers (plasmalemma, tonoplast) are completely impermeable to the solute and completely permeable to water.] However, the same changes occur during any water intake (towards the left) or water loss (towards the right) by wilting or, conversely, by saturation, for example.

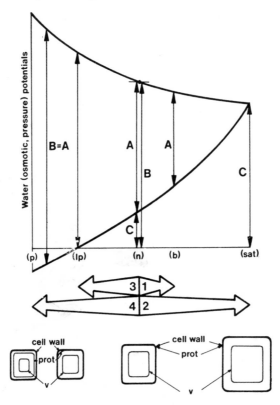

Fig. 1.1 A diagrammatic representation of changes of values *A*, *B*, and *C* of equation (1.2) during loss of water from the cell (dehydration, wilting) [changes towards the left] or during uptake of water by the cell [changes towards the right]. Abscissa: relative cell volume, ordinate: bar. On the abscissa, the hydration changes are plotted and designated with the same symbols as in Table 1.1. (*n*) is the normal state of the cell, (*sat*) full saturation — full turgor, (*lp*) limiting plasmolysis, (*p*) plasmolysed state — zero turgor. — On the ordinate, the values of the individual components of water potential are plotted (*A*, *B*, *C*). The numbered arrows show the course of changes after immersion of the cell at a normal state (*n*) (1) in a solution with an osmotic pressure equal to the *A* value at the (b) point; (2) in pure water; (3) in a solution causing limiting plasmolysis; (4) in a very hypertonic solution. prot — protoplast, v — vacuole.

The values of equation (1.3) have varied terminology. This is summarized in Tables 1.1 and 1.2. Value A has been called suction force (RENNER 1915, Ursprung and BLUM 1916), absorption power (THODAY 1918), suction pressure (STILES 1922), suction tension (BECK 1928), effective osmotic pressure (SHULL 1930), turgor pressure (CURTIS and SCHOFIELD 1933), osmotic cell equivalent (LEVITT 1951). The term diffusion pressure deficit (DPD) (MEYER 1938, 1945), has been criticized from the theoretical point of view by RAY (1960). FUKUDA (1958) proposed the term hydrature deficit, SHMUELI and COHEN (1964) hydropotential (HP), WEATHERLEY (1965a, 1965b) water potential depression (ΔW).

Various expressions are also used for term B of the equation 1.3: suction force of the water content (URSPRUNG and BLUM 1916), osmotic value of the cell (WALTER 1931b), osmotic pressure of the cell sap (MEYER 1945). Expressions such as wall pressure and turgor pressure are commonly used by most authors for the value C.

In the English literature the symbols and terminology of MEYER (1945) were most commonly used. In the German literature the terminology of URSPRUNG and BLUM (1916) (a review by URSPRUNG 1939, and BLUM 1958) is still used for values A and C, whereas WALTER's (1931b) expression osmotic value is used for the value B.

All these are expressed in terms of "pressure" since the values of equation 1.3 are in this case homologous with osmotic or diffusion water pressure. As can be seen, the terminology used varies considerably. Without exaggeration, it can be said that practically every author has suggested a new term.

Nevertheless, new concepts and terminology based on thermodynamic ideas were still sought for expressing the general situation common to all the described states and processes, suitable for use in the study of water relations between plant and soil and plant and atmosphere. Indeed, many methods used when measuring "pressure" constants are thermodynamic (tensiometric measurements of diffusion pressure deficit, cryoscopic estimation of the osmotic pressure of the cell sap etc.). Analogously with the term "partial chemical potential" (μ) expressing free energy (free enthalpy) per mole, EDLEFSEN (1941) used the term "specific free energy of water" to express free energy per unit of mass of water. This free energy is thus the work necessary to transfer a unit of water mass from one place to another. The total specific free energy of the system includes several components, caused by (1) hydrostatic pressure (hydrostatic specific free energy), (2) osmotic properties of the solution (specific free energy of osmotic solutions), (3) gravitational field and (4) interfacial tension. BROYER (1947) suggested the term "net influx specific free energy" (NIF) which is the difference between the algebraic sum of specific free energies causing water movement into the system (IF) and the algebraic sum of specific free energies causing water efflux out of the system (EF) (*i.e.* out of the cell):

$$\text{NIF} = \text{IF} - \text{EF} . \qquad (1.4)$$

In addition to the components (1) and (2), BROYER also uses non-metabolic specific free energy caused by adsorption and imbibition more or less homologous with the component (4) and metabolic free energy. Terminology based on thermo-dynamics has also been suggested by other authors, for soil conditions in particular.

TAYLOR and SLATYER, at the Madrid Symposium of UNESCO, 1959, (SLATYER and TAYLOR 1960, TAYLOR and SLATYER 1962) successfully attempted to unify the concepts and terminology. They defined individual components of the total spe-cific free energy of the system. This should make it possible both to analogize with the terms used so far and with methodological procedures used in studying the water relations of the plant and its environment (soil, atmosphere). Some of the symbols in their review are not used in the more recent monograph by SLATYER (1967a).* In order to avoid further complications, the symbols used in SLATYER's monograph are used in this handbook.

The total water potential of the system concerned (*i.e.* subcellular structures, cell, tissue, soil layer) − symbol Ψ_w, is the total specific free energy of water in the system. Water potential is proportional to the relative chemical potential of water, *i.e.* to the difference between chemical potential of pure water (μ_w^0) and chemical potential of water in the system (μ_w):

$$\Psi_w = \frac{\mu_w - \mu_w^0}{\overline{V}_w} = \frac{\Delta\mu_w}{\overline{V}_w}, \tag{1.5}$$

where $\overline{V}_w$ is the partial molal volume of water [$cm^3\ mol^{-1}$]. Here the dimension of water potential is energy per unit volume, [$J\ mol^{-1}$].

When expressed in terms of pressure the value of water potential is analogous with other pressure terms like diffusion pressure deficit or suction pressure (see Table 1.1 and Section 1.2). In that case the following relationship must be noted:

$$\Psi_w = -\ (DPD) = -\ S\ . \tag{1.6}$$

(Abbreviations see again Table 1.1.)

Diffusion pressure deficit (DPD), suction force, or similar terms are normally zero (in pure water) or positive, whereas water potential is normally zero or nega-tive. The higher the DPD (or suction force), the lower (the more negative) the water potential, *e.g.* DPD of 5 bar corresponds to water potential of − 5 bar.

---

* Symbols π and τ originally suggested by SLATYER and TAYLOR (1960) for osmotic and matric potentials (osmotic and matric components of the total water potential — see later) which usually have a negative numerical value are now designated by SLATYER (1967a) as osmotic pressure (π) and matric tension (τ), which have positive numerical values. Instead symbols ψ_s and ψ_m are used for osmotic potential (water potential due to solutes) and matric potential, respectively. Similarly the symbol P for pressure potential has been abandoned and substituted by the analogous symbol ψ_p.

According to its thermodynamic origin water potential may be expressed in units of energy, thus preferably in joules per g water (see Section 1.2).

The relationship between water potential and relative activity of water (a_w) and thus also the relative water vapour pressure p/p^0 (p is actual water vapour pressure in the system, p^0 vapour pressure of pure water) is given by the following equation (see also Fig. 1.2):

$$\Psi_w = \frac{RT}{\overline{V}_w} \ln a_w = \frac{RT}{\overline{V}_w} \ln \frac{p}{p^0}, \qquad (1.7)$$

where R is gas constant [J mol^{-1} K^{-1}], T is absolute temperature [K] and $\overline{V}_w$ partial molal volume of water [1 mol^{-1}]. For the relationship between water potential and relative air humidity see Table 1.3.

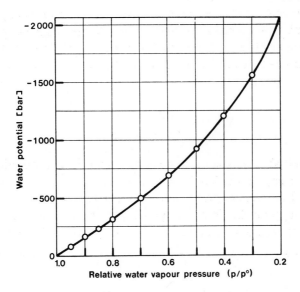

Fig. 1.2 The relationship between water potential (ordinate) and relative water vapour pressure (abscissa) at 20°C. (See Table 1.3).

Water potential consists of at least three mutually independent components:

$$\Psi_w = \Psi_s + \Psi_p + \Psi_m, \qquad (1.8)$$

where Ψ_s is the osmotic potential, Ψ_p the pressure potential and Ψ_m the matric potential.

Osmotic potential (Ψ_s) is the decrease in the total water potential due to compounds dissolved in water, *i.e.* due to the osmotic pressure (π) of aqueous solutions (*e.g.* cell sap, soil solution). Equation 1.2 expresses the dependence of osmotic pressure on concentration of the solute. The osmotic potential of a solution is negative, and is thus lower than the water potential of pure water, which is zero.

Pressure potential (Ψ_p) is the increase in total water potential due to hydrostatic pressure (P). It is positive in most plant cells, as the cell wall pressure is normally higher than atmospheric pressure, and negative in xylem vessels and soil tensiometers.

Matric potential (Ψ_m) represents the component of specific free energy of water which is associated with water status on interfacial borders (*i.e.* in the matrix): *e.g.* in the colloidal structures of cytoplasm, in micellar structures in cell walls, in capillary and ultracapillary systems in soil *etc.* The matric potential normally has negative numerical values, being lower than the water potential of pure free water. In non-vacuolated cells the matric potential of cytoplasmic structures − along with pressure potential − plays a decisive role in determining the total water potential of the cell.

As the physiological activity of cell structures is mainly dependent on the hydration level, expressed as chemical potential of water, it depends on the water potential in these structures.

It follows from the equation (1.7)

$$\Psi_w = \frac{\varrho}{M} RT \ln \frac{p}{p_0} \tag{1.9}$$

where ϱ is water density [g cm^{-3}]; M − molecular weight of water; R − the gas constant; T − absolute temperature [K]; p − actual water vapour pressure; p_0 − saturated water vapour pressure at the temperature T, that the water potential decreases proportionally with the decrease in relative water vapour tension. On the basis of the relationship given in equation (1.7), vapour tension corresponds to the relative humidity of the air. WALTER (1931b) suggested a very illustrative term "hydrature" (hy) for expressing the hydration level of cells or tissues. Hydrature is then the relative pressure of water vapour, *i.e.* the partial water pressure expressed as a percentage of the saturated water vapour pressure at a given temperature. The hydrature of pure free water is 100%; a solution of potential osmotic pressure of 20 bar has a hydrature of about 98.5% at 20 °C; and a hydrature of 90% corresponds to an osmotic pressure as high as 140 bar (see Fig. 1.2). The relation between water potential and relative tension of water vapour, shown in Fig. 1.2, is useful for expressing the relationship between water status in plants and air humidity (Table 1.3).

WALTER (1931b, 1955, 1963, 1965, 1966, 1967) KREEB 1967, KREEB and BORCHARD (1967), and WALTER and KREEB (1970) provide evidence that the hydrature of the cytoplasm of any vacuolated cell (in a turgid or non-turgid state) corresponds to the hydrature of the vacuolar sap and that the hydration level of the cytoplasm is hence always determined by the chemical potential (*i.e.* osmotic potential) of the water in the cell sap. On the other hand, STOCKER (1960), SHMUELI and COHEN (1964 and 1967) and SLATYER (1966a, 1967a, 1967b) demonstrated that the water potential in the cell structures is given by the value of the water potential of the cell, because the water potential of cytoplasm (given by the relative pressure of water vapour) is higher in a cell with a positive pressure potential (*i.e.* with turgor) than its osmotic potential. Thus only after loss of turgor, when the cell wall pressure is zero, is the chemical activity determined by the osmotic potential (osmotic pressure) of the vacuolar sap alone, since the water potential is then equal to the osmotic potential of the cell sap. The total water potential (Ψ_w) must be the same at any point in the cell at equilibrium, but the component potentials (like osmotic potential in the vacuole), of course, vary from point to point.

This introduction is oversimplified at many points; it is intented only as a simple explanation of the basic terminology used in this manual in describing the principles

Table 1.3 *The relationship between water potential* (bar *at 20°C) and relative air humidity* (*r. h. in percent p/p°*). *From* WALTER *1931b. All values of water potential are negative.*

Graded by 0.1% r. h.									
%	.9	.8	.7	.6	.5	4.	.3	.2	.1	.0
99	1.36	2.70	4.06	5.42	6.78	8.13	9.48	11.21	12.2	13.6
98	14.9	16.3	17.6	19.0	20.4	21.7	23.1	24.5	25.9	27.2
97	28.7	30.0	31.4	32.8	34.2	35.7	37.0	38.3	39.7	41.1
96	42.5	43.9	45.3	46.7	48.1	49.5	51.0	52.4	53.8	55.2
95	56.5	57.9	59.4	60.8	62.2	63.6	65.0	66.5	67.9	69.3
94	70.7	72.1	73.5	75.0	76.4	77.8	79.2	80.6	82.1	83.5
93	84.9	86.4	87.9	89.3	90.8	92.2	93.6	95.0	96.4	98.0
92	99.4	101	102	103	105	106	108	109	111	112
91	114	115	116	118	120	121	123	124	126	127
90	129	130	132	133	135	136	138	139	141	142

Graded by 1% r. h.									
	9.	8.	7.	6.	5.	4.	3.	2.	1.	0.
8	157	173	188	204	220	236	252	268	285	302
7	319	336	354	371	389	406	424	444	463	482
6	501	521	540	560	581	602	623	644	667	689

Graded by 10% r. h.

50% — 934	30% — 1575	10% — 2928
40% — 1205	20% — 2082	0% — ∞

and techniques of different methods used for estimating the hydration level of plant cells and tissues. For more details see the cited original papers concerned with the theoretical background, particularly the monographs by SLATYER (1967a) and KRAMER (1969).

1.2 Units of Water Potential and Examples of Absolute Values of Water Potential and its Components

Values of Ψ_w, Ψ_s. Ψ_p and Ψ_m are expressed most frequently in units of pressure: bars, atmospheres or cm of water column. However, they may also be expressed in units of free energy, thus for instance in J g^{-1} or erg g^{-1}.

Tables 1.4 and 1.5 present basic mutual conversions of units used to express values of water potential. The given approximate conversions are valid only for standard conditions. It is necessary to bear in mind when converting pressure units (bar, atm, dyn cm^{-2} and cm_{H_2O}) to units of free energy that they are equal only numerically, and that their dimensions are different.

It follows from the equation (1.3) that the absolute value of the water potential of a cell or tissue depends primarily on the osmotic potential. Osmotic potential varies mostly within the range from -4 to -30 bar (*i.e.* osmotic pressure between $+4$ to $+30$ bar). However, higher and also considerably lower values may be found. Average values of osmotic potential of the so-called cell sap of leaves (see Section 1.4.2) are given by WALTER (1960), for example: succulents -3 to -7 bar, aquatic plants about -8 bar, swamp plants -9 to -14 bar, shade plants -7 to -9 bar, cultivated plants of the temperate region -10 to -20 bar, sun plants -12 to -20 bar, tree leaves -15 to -19 bar, conifers -16 to -22 bar, evergreen plants -15 to -23 bar, plants of semiarid regions -20 to -40 bar, halophytes -20 to -60 bar (-200) bar. Tubers and bulbs have values from -7 to -21 bar, fruits from -9 (tomato) to -38 (grapes) bar.

At full saturation, the negative osmotic potential is balanced by the positive pressure potential (pressure of the cell wall and surrounding cells), so that the water potential of the whole cell is zero. The water potential decreases (becomes more negative, while the value of the diffusion pressure deficit increases) as a water deficit develops. The water potential reaches the value of the prevailing osmotic potential (osmotic pressure of vacuolar sap) when turgor is lost (*i.e.* when turgor pressure equals zero). In extreme cases (*e.g.* sclerophyllous species), when negative Ψ_w values are expected (negative turgor, see Section 1.5.1), the cell water potential may be lower than the osmotic potential, *i.e.* the diffusion pressure deficit may be higher than the osmotic pressure.

The pressure potential reaches its maximum at full saturation with maximum turgor, when it is numerically equal to the negative value of the osmotic potential at that time. It becomes zero with the loss of turgor, when the water potential in

Table 1.4 *Conversion of various units of water potential.*

Energy per volume (pressure)			Energy per weight	Energy per mass		
bar	atm	dyn cm^{-2}	cm$_{H_2O}$	erg g^{-1}	J g^{-1}	J kg^{-1}
1	= 0.987	= 1.000×10^6	≈ 1019.7	≈ 1.000×10^6	= 0.10	= 100.0
1.013	= 1	= 1.013×10^6	≈ 1033.2	≈ 1.013×10^6	= 0.1013	= 101.3
1.000×10^{-6}	= 0.987×10^{-6}	= 1	≈ 1.02×10^{-3}	≈ 1	= 1.000×10^{-7}	= 1.00×10^{-4}
9.807×10^{-4}	= 9.679×10^{-4}	= 9.8×10^2	≈ 1	≈ 9.8×10^2	= 9.8×10^{-5}	= 9.8×10^{-2}
10.00	= 9.87	= 1.000×10^7	≈ 10197.0	≈ 1.000×10^7	= 1	= 1000.0
0.01	= 0.00987	= 1.000×10^4	≈ 10.2	≈ 1.000×10^4	= 0.001	= 1

The above mentioned approximate conversions hold for standard conditions and pure free water. It should be kept in mind when converting pressure units of free energy and *vice versa* that they are equivalent only numerically but expressed in basically different units.

Table 1.5 *The mutual relationship between various units of water potential (or its components) in pressure (tension or suction) units* (atm, bar, cm_{H_2O}) *and free energy of water* ($J\ g^{-1}$, $J\ kg^{-1}$, $erg\ g^{-1}$). *All values of water potential are negative.*

Pressure units			Equivalent free energy of water		
atm	bar	cm_{H_2O}	$J\ g^{-1}$	$J\ kg^{-1}$	$erg\ g^{-1}$
1	1.013	1 033.2	0.1013	101.3	1.013×10^6
2	2.026	2 066.4	0.2026	202.6	2.026×10^6
3	3.039	3 099.6	0.3039	303.9	3.039×10^6
4	4.052	4 132.8	0.4052	405.2	4.052×10^6
5	5.065	5 166.0	0.5065	506.5	5.065×10^6
6	6.078	6 199.2	0.6078	607.8	6.078×10^6
7	7.091	7 232.4	0.7091	709.1	7.091×10^6
8	8.104	8 265.6	0.8104	810.4	8.104×10^6
9	9.117	9 298.8	0.9117	911.7	9.117×10^6
10	10.13	10 332	1.013	1 013	1.013×10^7
11	11.14	11 365	1.114	1 114	1.114×10^7
12	12.16	12 398	1.216	1 216	1.216×10^7
13	13.17	13 432	1.317	1 317	1.317×10^7
14	14.18	14 465	1.418	1 418	1.418×10^7
15	15.195	15 498	1.519	1 519	1.519×10^7
16	16.21	16 531	1.612	1 621	1.621×10^7
17	17.22	17 564	1.722	1 722	1.722×10^7
18	18.23	18 598	1.823	1 823	1.823×10^7
19	19.25	19 631	1.925	1 925	1.925×10^7
20	20.26	20 664	2.026	2 026	2.026×10^7
21	21.27	21 697	2.127	2 127	2.127×10^7
22	22.29	22 730	2.229	2 229	2.229×10^7
23	23.30	23 764	2.330	2 330	2.330×10^7
24	24.31	24 797	2.431	2 431	2.431×10^7
25	25.33	25 830	2.533	2 533	2.533×10^7
26	26.34	26 863	2.634	2 634	2.634×10^7
27	27.35	27 896	2.735	2 735	2.735×10^7
28	28.36	28 930	2.836	2 836	2.836×10^7
29	29.38	29 963	2.938	2 938	2.938×10^7
30	30.39	30 996	3.039	3 039	3.039×10^7
40	40.52	41 328	4.052	4 052	4.052×10^7
50	50.65	51 660	5.065	5 065	5.065×10^7
100	101.3	1.0332×10^5	10.13	10 130	1.013×10^8
1000	1013	1.0332×10^6	101.3	*ca.* 101 300	1.013×10^9

Table 1.6 *Changes of values* ψ_w, ψ_s, ψ_p *and volume of cell and protoplast in cells with various hydration level.*

Hydration degree of cell	Plasmolysed cell	Limiting plasmolysis	"Normal" state	Complete water saturation
Subscripts	(p)	(lp)	(n)	(sat)
Water potential ψ_w	minimum $\psi_{w(p)} = \psi_{s(p)}$	$\psi_{w(lp)} = \psi_{s(lp)}$	$\psi_{w(n)}$	maximum $\psi_{w(sat)} = 0$
Osmotic potential ψ_s	minimum $\psi_{s(p)} = \psi_{w(p)}$	$\psi_{s(lp)}$	$\psi_{s(n)}$	maximum $\psi_{s(sat)} = \psi_{p(sat)}$
Pressure potential ψ_p Pressure of cell walls	$\psi_{p(p)} = 0$ or negative	$\psi_{p(lp)} = 0$	$\psi_{p(n)}$	maximum $\psi_{p(sat)} = \psi_s$
Cell volume V_{cell}	$V_{cell(p)} = = V_{cell(lp)} = 1$	$V_{cell(lp)} = 1$	$V_{cell(n)} > 1$	$V_{cell(sat)} > 1$
Protoplast volume V_{prot}	$V_{prot(p)} < < V_{prot(lp)} < 1$	$V_{prot(lp)} = 1$	$V_{prot(n)} > 1$	$V_{prot(sat)} > 1$

vacuolated cells is equal to the osmotic potential of the vacuolar sap. In most crop plants the loss of turgor occurs at water potentials of about -20 bar.

The matric potential of cellular structures corresponds to the osmotic potential of the vacuolar sap when no hydrostatic pressure differences between the sap and the "matrix" occur.

1.3 The Determination of Water Potential (Ψ_w) of Cells and Tissues

1.3.1 Introduction

A relatively large number of methods for the determination of water potential have been described. This shows that it has been and still is very difficult to work out a perfect and more or less universal method, and that only experience will show which methods are the most suitable for a particular purpose.

Methods for measuring water potential are basically of three kinds: (1) compensation methods, (2) direct methods measuring water vapour pressure above the tissue and (3) pressure chamber method.

(1) In compensation methods we usually look for the solution of a known osmotic potential which is equal to the water potential of the tissue (or — in some cases — of the individual cells) to be determined. That solution is isotonic (isopiestic methods).

The general principle of compensation methods is as follows: The net water transfer between the test solution and the tissue sample as a result of the equilibration of the difference of water potential is measured in a set of uniform parallel tissue samples in a set of graded test solutions. Thus (Fig. 1.3) having a graded set of test solutions of known concentrations *i.e.* of known osmotic potentials,

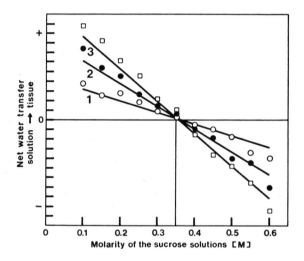

Fig. 1.3. An example of the relationship between net water transfer between the tissue (+ means water uptake by the tissue) and the reference test solution (ordinate) and molarity M of the sucrose solution (abscissa) used in compensation methods. 1,2,3 are lines obtained after three different (increasing) time intervals of compensation. The measured water potential ψ_w in this case corresponds to 0.36M sucrose, *i.e.* −9.6 bar.

each of them being equilibrated with one of the set of parallel uniform samples of tissues, we plot the rate and sign (+ or −) of the net water transfer against osmotic potential of each test solution. We will get a more or less straight line crossing the zero change line (horizontal line in Fig. 1.3) at a point which corresponds to an osmotic potential value equal to the water potential of the tissue. Complete equilibration is not necessary, only sufficient time is required to elapse for measurable water transfer, to ensure a distinct intercept. The result similar to that in Fig. 1.3 is obtained when similar conditions for net water transfer are ensured in each sample plus test solution. Only measurements falling on a straight line are significant. Incomplete series comprising only plus or only minus values, as well as sporadic determination of zero water transfer, are useless.

Compensation between the water potential of the sample of the tissue and the osmotic potential of the test solution can take place either in liquid (Section 1.3.2) or in vapour (Section 1.3.3) phase. When compensating in the liquid phase, parallel samples of tissues are immersed directly into test solutions of a graded concentration series. When compensating in the vapour phase, the parallel samples are placed above test solutions in closed vessels and the movement of water vapour occurs along the gradient of water vapour pressure. In both cases the direction and rate of the net transfer of water depend on the gradient of water, or of osmotic potentials, between the sample and the test solution.

All parallel samples of tissues of one set of samples must be uniform and comparable in origin and size (surface). The uniformity is a serious factor limiting the accuracy of the results.

(2) In direct methods the water vapour pressure in a closed vessel containing the sample is measured. Thermocouple or thermistor psychrometers (Section 1.3.4), thermocouple dew-point hygrometer (Section 1.3.5) or hanging drop method (Section 1.3.6) are used.

(3) In the pressure chamber method (Section 1.3.7) the water potential of leaf cells is compensated by external pressure exerted from outside on the leafy shoot or single leaf. The point of compensation is indicated when water begins to flow from the leaf mesophyll cells into the xylem vessels against the osmotic gradient of the vessel sap. This flow is detected by the outflow of sap from the vessels on the cut surface of the shoot or leaf petiole which is outside the pressure chamber.

A summary Table 1.9 of methods for the determination of water potential may be found on pages 72 and 73 and used for preliminary selection of the method suitable for the purpose.

In addition to the methods determining the mean water potential of whole tissues more or less exactly, determinations of freezing point have also been used for estimating water status in plant tissues (see Section 1.3.8.)

1.3.2 Compensation Methods in the Liquid Phase

In these methods, rate and direction of net water transfer between graded test solutions and samples of a set of parallel probes of the tissue (cells) is determined when the tissue is immersed directly in the test solutions. The water transfer can be detected by measuring (a) changes in concentration of the test solution (refractometric and smear methods), (b) changes in size (volume) of the tissue sample (cell and segment methods), (c) changes in the volume of the test solutions (potometric method).

Cells damaged at the cut surfaces of the tissue samples influence the water potential of neighbouring intact cells. There is no turgor pressure in damaged cells, so their water potential is equal to the osmotic potential, being thus lower (more negative) than before damage. The damaged cells withdraw water from neighbouring cells, so the water potential of these cells also decreases. As a result, slightly lower values of water potential are obtained by this method. V. ÚLEHLA (1926) points out the effect of changes in swelling of the cell wall and cytoplasm and changes in cytoplasmic permeability of cells of the tissue placed in the test solution. SPURNÝ (1951) compared the auxanographic (measurement of thickness), volumetric and gravimetric methods, using tissues of *Opuntia phaeacantha*, and found that there were considerable differences in the estimates of water potential and its decreases in the test solution, even in this very suitable material. CRAFTS *et al.* (1949) and MEYER and WALLACE (1941) found that potato tissue continued to shrink even after cells were plasmolysed.

1.3.2.1 Test Solutions

A solution which is in direct contact with living cells or tissues should have several important properties:

(1) It must not be harmful to living cells and must not cause undesirable structural changes. Solutions of potassium nitrate, often used in the past, are therefore not suitable.

(2) The cytoplasmic membranes such as the plasmalemma and tonoplast should be practically impermeable to the solute. Their semipermeability can be expressed as the reflection (selectivity) coefficient, which is the ratio of the apparent osmotic pressure to the potential osmotic pressure. A membrane which is completely impermeable to a solute has a reflection coefficient $\sigma = 1$; a completely permeable one has a reflection coefficient $\sigma = 0$. In most cases the impermeability is not complete. All liquid phase methods are based on the assumption that the reflection coefficient is one. The penetration of the solute into leaf tissue is probably the most common source of error in techniques based on liquid phase compensation, as pointed out by SLATYER (1966b).

(3) The solute (osmoticum) should not be metabolized by plants nor subject to change caused by microorganisms during prolonged immersion of the tissue in the solution, if necessary.

Sucrose, mannitol and polyethylene glycol are the most commonly used osmotica at present. It should be pointed out that none of them completely fulfils the three requirements above.

It has been demonstrated that s u c r o s e does penetrate into intact cells and tissues (*e.g.* WEATHERLEY 1954, ORDIN *et al.* 1956, GOODE and HEGARTHY 1965 *etc.*). The reflection coefficient for sucrose is 0.6 to 0.7 (SLATYER 1966b). This uptake of sucrose causes errors in methods using liquid phase compensation. It decreases the potential of the tissue during the equilibration, so that the estimates of water potential are too low (too negative). The sucrose taken up is very readily metabolized. Sucrose may also be hydrolyzed at tissue surfaces (*e.g.* on the surface of roots, BURSTRÖM 1957) and it is readily fermented in an unsterile solution at room temperature. In both cases the osmotic potential of the solution decreases and so does the reflection coefficient. This can be prevented by adding a grain of thymol. However, thymol is toxic to plant cells. Sealed solutions may be stored unchanged in a refrigerator for a few days.

There is evidence that mannitol is absorbed by higher plants (BURSTRÖM 1953, THIMAN *et al.* 1963, TRIP *et al.* 1964, GOODE and HEGARTHY 1965, FRANKOVÁ and KOLEK 1967, MANOHAR 1966a, LAWLOR 1970). Mannitol is detected in the guttation sap relatively soon after adding it to the nutrient solution (KOZINKA and KLENOVSKÁ 1965). SLATYER (1966b) found the reflection coefficient for mannitol to be 0.8 to 0.9. It is not yet known to what extent mannitol can be metabolized by higher plants (BURSTRÖM 1953, ORDIN *et al.* 1956, FERGUSON and STREET 1958, THIMAN *et al.* 1963 and others); it is also suspected of being toxic (GROENEWEGEN and MILLS 1960, STROGONOV and LAPINA 1964, TAYLOR 1965). Mannitol is less useful than sucrose because of its limited solubility in water: only solutions up to 0.8M can be prepared at 20 °C, corresponding to an osmotic potential of −25 bar.

Polyethylene glycol (PEG) with a molecular weight 400 to 20 000, seems to be a satisfactory osmoticum (JACKSON 1962, WALTER 1963, MANOHAR and HEYDECKER 1965, CHEESMAN *et al.* 1965, JARVIS and JARVIS 1965, JANES 1966, LAWLOR 1969 and others). It is usually known by its commercial names, Carbowax and

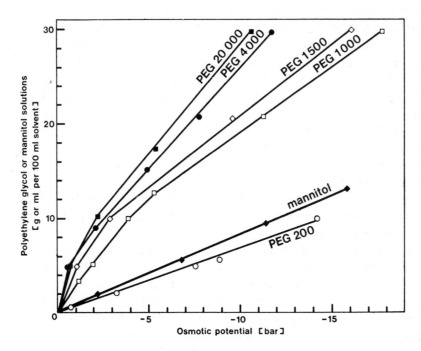

Fig. 1.4 Osmotic potential of the solutions of polyethylene glycol (different molecular weight fractions) and of mannitol. (LAWLOR 1970)

Lutrol. On the other hand LAGERWERFF *et al.* (1961) found that polyethylene glycol with a molecular weight of 6 000 was toxic, although a preparation with a molecular weight of 20 000 was acceptable as an osmoticum. According to these authors, polyethylene should be dialyzed or passed through a ion exchange column to remove toxic heavy metal impurities. LESHEM (1966) found polyethylene glycol to be toxic for *Pinus halepensis* seedlings.

LAWLOR (1970) found that impurities were not responsible for the toxic effects. Using randomly labelled ^{14}C polyethylene glycol (mol. weight approx. 4 000) and determining PEG analytically, LAWLOR (1970) found that PEG 200 (like mannitol) entered plants with undamaged roots and did not cause any damage to plants, whereas PEG 1 000, 4 000 and 20 000 reduced water absorption by roots and caused desiccation of the plant, probably by blocking the pathways of water movement.

In Fig.1.4 examples of the relationship between concentration and osmotic potential for water solutions of mannitol and polyethylene glycol of different molecular weight are found (from LAWLOR 1970).

Substances of the highest available purity and distilled water, should be used to prepare test solutions. Solutions of graded concentration, corresponding to various osmotic potentials, are prepared by diluting a more concentrated solution (1M or 1.5M). This concentrated solution should be prepared in such a way that 1 or 1.5 moles of a compound are weighed with a minimum accuracy of 1/1000 molecular weight (*e.g.* maximum error of 1 per thousand), dissolved and the volume made up to 1 000 ml with distilled water. This is a molar (or 1.5 molar) solution. For preparing a molal (or 1.5 molal) solution, 1 000 ml distilled water are added to the same amount of substance. A 0.95M solution, for example, is then prepared by mixing 95 ml of the concentrated solution (1 mole per litre) with 5 ml of water. Large volumes are preferred, as they can be measured with the greatest accuracy. Grading of the solutions in units of 0.05M is usually sufficient.

Since the relation between the osmotic potential of aqueous solutions of different osmotica and their concentration is different, tables (*e.g.* Table 1.7 for sucrose) and/or direct determinations of osmotic pressure must always be used (see Section 1.4.2).

MICHEL (1972) recently criticaly revised tabular data of osmotic potential of sucrose solutions from classical sources, in particular those of MORSE (1914), and ROBINSON and STOKES (1949 and 1959).

1.3.2.2 Cell and Segment Methods

The original method of URSPRUNG and BLUM (1916) gives the Ψ_w value of one cell; a change in the cell volume is the criterion of water transfer between test solution and cell. It is hence necessary to measure as accurately as possible the volume of the same cell before and after exposure to the test solution. The cell

Table 1.7 *Osmotic potential* (ψ_s[bar]) *and refractive index* (n_D^{20}) *of sucrose solutions of various molar concentrations* (M — *moles per 1 litre of the solution*) *at 20°C. The table is based on table data of* URSPRUNG (*1939*) *constructed on the basis of empirical findings of* MORSE, BERKELEY *and* HARLEY *and on the table of* GUSEV (*1960*).

M	bar	n_D^{20}	M	bar	n_D^{20}
0.000	− 0.00	1.3503	0.200	− 5.36	1.3603
05	13	06	05	50	06
10	26	08	10	64	08
15	40	11	15	79	10
20	54	13	20	94	13
25	67	15	25	− 6.08	15
30	80	18	30	22	18
35	94	21	35	36	21
40	− 1.07	24	40	50	23
45	21	26	45	65	25
50	34	29	50	79	27
55	47	31	55	93	30
60	61	34	60	− 7.07	32
65	74	36	65	22	35
70	87	38	70	36	37
75	− 2.01	41	75	51	39
80	14	43	80	65	42
85	27	46	85	80	44
90	41	48	90	94	47
95	54	51	95	− 8.09	49
0.100	67	53	0.300	24	52
05	81	55	05	38	54
10	95	58	10	53	57
15	− 3.08	60	15	68	59
20	21	63	20	82	62
25	34	65	25	97	64
30	47	67	30	− 9.12	67
35	61	70	35	26	69
40	75	72	40	41	71
45	88	75	45	56	73
50	− 4.01	77	50	70	77
55	14	80	55	86	79
60	27	82	60	−10.00	82
65	41	85	65	15	84
70	54	87	70	30	87
75	67	90	75	46	90
80	81	92	80	62	92
85	94	95	85	78	94
90	− 5.08	97	90	94	97
95	22	1.3601	95	−11.10	99

Table 1.7 *(continued)*

M	bar	n_D^{20}	M	bar	n_D^{20}
0.400	−11.25	1.3702	15	−18.56	1.3807
05	42	04	20	74	10
10	58	07	25	93	12
15	74	09	30	−19.12	15
20	89	12	35	31	17
25	−12.05	14	40	49	19
30	22	16	45	68	22
35	38	18	50	86	24
40	53	21	55	−20.06	27
45	69	23	60	24	29
50	85	25	65	43	33
55	−13.02	28	70	61	35
60	18	30	75	81	38
65	34	33	80	99	40
70	50	35	85	−21.18	43
75	68	37	90	37	45
80	84	40	95	58	48
85	−14.01	42	0.700	77	50
90	17	45	05	97	53
95	33	47	10	−22.16	55
0.500	50	50	15	37	57
05	67	54	20	56	60
10	83	56	25	76	62
15	99	59	30	95	64
20	−15.15	61	35	−23.16	67
25	33	63	40	35	69
30	49	66	45	55	72
35	67	69	50	74	74
40	84	71	55	95	76
45	−16.03	74	60	−24.15	79
50	20	76	65	37	81
55	38	79	70	59	84
60	56	81	75	80	86
65	74	84	80	−25.01	88
70	93	86	85	22	91
75	−17.11	89	90	44	93
80	28	91	95	66	96
85	46	93	0.800	87	98
90	65	95	05	−26.08	1.3901
95	83	98	10	30	03
0.600	−18.00	1.3800	15	51	05
05	18	02	20	72	08
10	37	05	25	94	11

Table 1.7 *(continued)*

M	bar	n_D^{20}	M	bar	n_D^{20}
0.830	−27.15	1.3914	1.045	−37.43	1.4018
35	35	16	50	68	21
40	55	18	55	94	23
45	76	21	60	−38.19	26
50	96	23	65	44	28
55	−28.16	26	70	70	31
60	36	28	75	−39.00	33
65	57	31	80	30	36
70	77	33	85	56	38
75	97	36	90	81	41
80	−29.17	38	95	−40.06	43
85	43	41	1.100	32	45
90	68	43	10	93	50
95	88	45	20	−41.43	54
0.900	−30.09	48	30	−42.04	59
05	34	50	40	55	63
10	59	52	50	−43.05	69
15	85	54	60	66	74
20	−31.10	56	70	−44.27	79
25	30	59	80	77	84
30	50	61	90	−45.38	88
35	76	64	1.200	99	93
40	−32.01	67	10	−46.60	97
45	26	69	20	−47.21	1.4102
50	52	72	30	81	07
55	77	74	40	−48.42	12
60	−33.02	76	50	−49.03	17
65	28	79	60	64	22
70	53	81	70	−50.24	27
75	78	83	80	95	31
80	−34.04	86	90	−51.56	36
85	29	89	1.300	−52.27	41
90	54	91			
95	80	94			
1.000	−35.05	96			
05	30	99			
10 -	56	1.4001			
15	86	04			
20	−36.16	07			
25	42	09			
30	67	11			
35	92	14			
40	−37.18	16			

volume is calculated from individual cell sizes measured under the microscope. This calculation is correct only if the cell shape resembles a rotation body (sphere, cylinder or cone) or a parallelepiped. A critical contour drawing of the cell is made, using special drawing equipment, and the cell volume is calculated. The preparation is made in liquid paraffin to prevent water loss by evaporation. A test solution is added to the section directly on a microscope slide, under the microscope, so that the section is not moved and the cell is still visible. A new section must be prepared and a new cell must be chosen for each solution concentration (ERNEST 1931, 1934a, b). The technique is therefore very laborious and time-consuming. In addition, the water potential of only one cell is measured. The accuracy of the method is not high because the cell volume has to be calculated from the cell contour map. The authors themselves suggested using the vertically projected surface area instead of calculating the volume. This projected area may be measured from the drawing by planimetry. However, this is only a slight simplification.

URSPRUNG and BLUM (1924) and MOLZ (1926) described a so-called simplified segment method for measurement of water potential of whole tissues. The method takes advantage of the fact that changes in the volume of tissue segments may be deduced from changes in their length: A set of trapezoidal segments (5 to 15 mm by 2 to 5 mm), the size chosen to enable reproducible length measurements, is cut from the sample (e.g. leaf) under liquid paraffin with a razor blade, avoiding the leaf veins. Each segment is transferred with a drop of liquid paraffin onto a microscope slide and its length is measured by means of an objective scale or a calibrated eyepiece scale. Each segment is then put in one of a series of sucrose solutions for about 30 minutes. Then the length is measured again in a drop of the same solution.

Sclerophyllous leaves of evergreen species or some xerophytes, and needles of coniferous species, do not change in length in osmotically varied solutions. The thickness of sections may then be measured instead (Hartlaubmethode: URSPRUNG and BLUM 1927). The sections are placed vertically into a slit between two pieces of glass; one fixed, one movable. These pieces can be made of the two halves of a microscopic slide cut with a diamond. One half is stuck to a glass base; the other can slide up and down between two fixed parallel bars of glass. When working with conifer needles it is necessary to ensure sufficient contact with the test solutions, perhaps by preparing large cut surfaces. Accurate measurements of changes of thickness may be obtained by using the so-called "lever method", based on recording the change by means of a lever functioning as a scale pointer, as in an auxanometer (Hebelmethode: URSPRUNG and BLUM 1930).

The method has numerous shortcomings; the principal ones have been summarized by V. ÚLEHLA (1926), OPPENHEIMER (1930) and ARCICHOVSKIJ (1931). Equilibration is through volume changes rather than the length changes which are actually measured; change in length may be negligible in relation to change in volume and thus may scarcely be detectable. Changes in the dimensions of individual cells and thus also of tissues are not necessarily related to changes in the total volume.

The segment method has been used successfully to measure the water potential of parenchymatous tissue taken from potato tubers (PRINGSHEIM 1931, LYON 1936, 1940, MEYER and WALLACE 1941). MEYER and WALLACE (1941) also used the gravimetric method with the same material, the result of osmotic equilibration being indicated by the weight of the tissue. The gravimetric method has the inadequacies described above, and also has other sources of error. For example, surface wetting during the second weighing may cause an error. Comparing the gravimetric method with refractometry, ASHBY and WOLF (1947) found additional errors caused by infiltration of the solution into intercellular spaces as a result of a decrease in surface tension caused by substances released from the tissues. Gravimetric methods involving direct exposure of leaf discs on the surface of test solutions are therefore not recommended.

1.3.2.3 Refractometric and Smear (SHARDAKOV) Methods

Some of the main shortcomings of the methods described above do not apply to the following methods, which estimate the water potential of all kinds of tissue on the basis of water transfer between test solutions and the tissue, as indicated by changes in the concentration of the test solutions. A sufficiently large ratio of volume and exchange surface of the tissue to volume of the surrounding test solution is required, so that the changes in concentration are as large as possible, Smear (densitometric) and refractometric methods differ in the means of detection of the concentration changes.

The smear method, which is ingenious but rather subjective, was originally designed by ARCICHOVSKIJ and OSSIPOV (1931) and extended by SHARDAKOV (1938, 1948). Water is removed from the tissues in a hypertonic solution, so that the solution surrounding the tissues is diluted, becomes less dense and moves upward. As the dilution is associated with a change in refractive index, visible ascending bands may be observed. A hypotonic solution around the tissue becomes more concentrated, resulting in descending smears or bands. To increase visibility of the smears, ARCICHOVSKIJ and OSSIPOV constructed special equipment for illumination, including magnifying apparatus. SHARDAKOV (1938, 1948) used the same principle but his technical procedure was slightly different. The tissue samples are immersed in small volumes of test solutions of sucrose lightly coloured with methyl orange or methyl blue. After exposure, drops of the coloured solutions are transferred with a micropipette to colourless solutions of the same initial concentration. If a coloured drop rises, the test solution was hypertonic relative to the water potential of the tissue, if it sinks the opposite was true and if it remains still, the test solution had the same water potential as the tissue sample, so there was no change in its concentration brought about by immersion of the sample. This simple procedure requires no special equipment and is not sensitive to temperature changes. It is suitable for field conditions. However, it is sometimes rather difficult and subjective to estimate whether the drop remains steady when working with solutions of only slightly different densities (fine grading of test solutions), so the accuracy of the method is limited.

A more objective measure of a change in concentration of the test solutions is the change in refractive index after contact with the tissue. The method was originally worked out by MAXIMOV and PETINOV, first published by ASHBY and WOLF (1947), and only subsequently published by MAXIMOV and PETINOV (1948). The same method was described by LEMÉE and LAISNÉ (1951) and discussed by GAFF and CARR (1964).

Test solutions of sucrose are prepared, *e.g.* with 0.1M or 0.05M gradations in concentration (see Section 1.3.2.1). Mannitol solutions can also be used (KNIPLING and KRAMER 1967). Small test tubes of about 10 mm diameter and 45 mm height are filled with 0.5 ml aliquots of the test solutions. MAXIMOV and PETINOV (1948) recommended 1 to 2 ml for leaf discs of diameter 7 to 8 mm; this is rather too much in our experience. For the smear method, REHDER (1959) used 0.5 ml for 20 to 30 pieces of leaf blade of about 10 to 20 mm². SLATYER and MCILROY (1961) recommended 0.5 g fresh weight of sample and 2 to 3 ml of solution. KRAMER and BRIX (1965) used test tubes 13 by 10 mm in size, containing 3.5 ml of the solution. Generally it is recommended (SLAVÍKOVÁ 1963a, b) that the volume of test solution should be the minimum required to wet the pieces of tissue. However, the volume used should be sufficient to allow two drops to be taken for refractometric measurements. One test tube of each concentration of solution is needed for each estimate.

Reports of the compensation time are variable. MAXIMOV and PETINOV (1948) recommended 30 to 45 min for discs of leaf tissues at a temperature of about 30 °C and 1 to 2 hours or even 4 hours at normal or low temperatures, respectively. KRAMER and BRIX (1965) used 30 to 60 min exposure. SLAVÍKOVÁ (1963a, b) found an optimum of 3 to 4 hours for roots and REHDER (1959, 1961) and REHDER and KREEB (1961) also recommended an exposure of 3 hours. WALTER and ELLENBERG (1958) found that exposure for several hours is necessary for pine needles.

Prolonged exposure should not have any effect on the estimates of water potential but should only increase the measured differences and hence the slope of the lines in Fig. 1.3. However, it has been found that the estimates of water potential obtained decrease with increasing duration of contact of tissue with the test solution (V. ÚLEHLA 1926, 1928, KOVÁŘ 1930, STILES 1930, SPURNÝ 1951). Using SHARDAKOV's method, REHDER (1959, 1961) also found that the water potential decreased with increasing duration of exposure. The same was found by LEMÉE and GONZALES (1965), who repeatedly obtained decrease followed by increase when repeatedly cutting marginal annuli from the leaf discs used, *i.e.* decreasing the diameter of the discs during compensation. These changes are thus probably associated with irreversible changes of osmotic conditions at the cut surfaces. V. ÚLEH-LA (1926, 1928) explained the decrease in water potential of the tissue in terms of changes in imbibition and permeability of cytoplasm. STILES (1930) suggested that the decrease in osmotic potential is caused by the release of osmotically active electrolytes from the vacuoles into the test solution. Whatever the explanation may be, reproducible relative results obtained using standard exposure times are comparable with those obtained by the gravimetric method of SLATYER (LEMÉE and GONZALES 1965). HELMUTH and GRIEVE (1971) used leaf discs of 2 mm diameter only which resulted in less negative values of water potential in sclerenchymatous to mezophytic material.

In order to eliminate the error due to the cutting of the plant material, TYURINA (1972) applied the osmotic solutions directly to the surface of the leaf *in situ* in isolated compartments of a special leaf chamber.

SLAVÍKOVÁ (1963), using primary roots which are readily wetted and have a rhizodermis which is easily permeable to water, in contrast to leaf epidermis, found no decrease in the estimate of water potential with increase in exposure time. Good compensation, *i.e.* good slope as in Fig. 1.3 had already occurred after about 30 min. However, longer exposure is advantageous because differences from the initial concentrations of the test solutions are larger.

KNIPLING and KRAMER (1967) found differences of 1 to 5 bar in water potentials as determined by psychrometric and dye methods in the same material (tree leaves, tomato, sunflower, tobacco, cotton) in the range 0 to 30 bar. Errors in measuring leaf water potential in some woody plants were discussed by BRIX (1966).

In addition to the systematic errors of the method the numerical accuracy of the determination depends on (1) how closely the concentrations of the test solutions are graded, (2) how rapid the water transfer rate is in relation to the compensation time.

Indication of Changes in Concentration of Test Solutions by the Refractometric Method

The refractive index of drop samples is measured in an Abbe refractometer, which makes it possible to estimate the refractive index to the fourth decimal place. Manipulation of the refractometer is described in the instrument manuals. Accuracy in estimating the last decimal place may be increased by adjusting the boundary between the dark and light fields of the refractometer several times and taking the average reading. Table 1.6 on pages 21 to 23 shows the relationship between the refractive index of sucrose solutions, their molarity, and osmotic potential.

The difference in the temperature of the refractometer prisms between the first and the last reading should not exceed 2 °C. If necessary, in the laboratory the temperature of the prisms can be controlled by flowing water coming from an ultrathermostat.

It is recommended that the refractive index of a drop of the test solution taken immediately after introducing the sample and stirring should be taken as the initial value. This is important as concentration changes may be brought about after immersion of the sample by bleeding of cell sap from cut surfaces, dissolution of salt sediments on leaf surfaces or soil impurities on roots or by water on wet leaf surfaces. It is advantageous to stir the solution containing the sample by blowing bubbles through the pipette which is later used to take the sample drop. Moisture from the breath should be avoided. On the other hand, a different technique may be used, in which the first volume of the solution is replaced by a new solution after the first has rinsed the material and been discarded. This procedure avoids, to some extent, the cut edge problem previously described. The final value of the refractive index of the test solution after the compensation time is measured again after thorough mixing.

Indication of the Concentration Changes of Test Solutions by the Densitometric Method (Smear, "Schlieren", dye or SHARDAKOV method)

SHARDAKOV (1938, 1948) and later REHDER (1959), SLATYER and McILROY (1961), SCHLÄFLI (1964, 1966), KNIPLING (1965, 1967a), SCHRETZENMAYR (1966, 1967) and others added a water-soluble dye (e.g. methyl orange) to the test solutions in which the samples are immersed. The dye is added to the solution either directly in solid form after exposure to the samples, in order not to change the solution concentration, or to a larger volume of the test solutions before transfer to the test tubes and addition of the samples. On the other hand, MOURAVIEFF (1959), BRIX and KRAMER (1962) and KOZLOWSKI (1964) added methylene blue, methyl orange or Gentian violet (in powdered form) to the control solution of each initial concentration and transfered a drop of this dyed solution to each of the test solutions from which the samples had been removed.

After exposure of the tissue in the test solution, the solution is carefully mixed and a sample drop of the coloured solutions is taken with a new, carefully washed and dried pipette of an inner diameter of 2 mm, about 18 cm long and drawn into a capillary. The tip of the pipette is then immersed 1.5 cm below the level of an undyed solution containing the same initial concentration of sucrose (about 10 ml in a test tube). The pipette is positioned in such a way that its tip does not move in the solution and a "cloud" of the coloured measured solution is allowed to flow out of the pipette. The use of a pipette with its tip bent at a right angle is recommended, so that the drop comes out horizontally. If the "cloud" remains within the range of 5 mm below to 2 mm above the mouth of the pipette, both solutions have the same density, i.e. the same concentration. The drop moves up or down when the solution is more or less dense.

The dye method does not show quantitative concentration differences but only whether the difference is zero, positive or negative. Interpolation (as in Fig. 1.3) cannot therefore be used to obtain values of water potential. However, if one of two adjacent concentrations is plus and the second minus, the resulting value must lie approximately in between. The accuracy of the method depends on the gradations of the range of concentrations of the test solutions used. The concentration differences between adjacent standard solutions should not be greater than 0.005M, because the accuracy cannot then exceed 1 bar.

A modification of the refractometric method was used by STOCKING (1945) to measure the water potential in intact tissues of the petioles of water melon leaves (*Cucurbita pepo* L.). One of the graded series of sucrose solutions was injected into the hollow petioles and samples were taken at intervals with a syringe. Changes in concentration were measured by the refractometric method.

1.3.2.4 Potometric (Volumetric) Method

In addition to the compensation methods already described, a potometric method was suggested by ARCICHOVSKIJ *et al.* (1931) to measure the water potential of tissues of stems and branches. Small potometers are filled with graded sucrose solutions and fixed to stems or branches after removal of the cambium.The authors used this method to determine the gradient of water potential with height, and periodic changes in water potential. A similar method had already been described by NORDHAUSEN (1921).

1.3.3 Compensation Methods in the Gaseous Phase

The direction and relative rate of the water vapour transfer between each of a series of parallel tissue samples and each of a set of graded test solutions of known osmotic potentials in closed vessels is determined either (a) by changes in fresh weight of the tissue samples (gravimetric method) or (b) by changes in the volume of the osmotic solution (capillary method).

1.3.3.1 Gravimetric Method

Accurately weighed replicate samples of tissue are placed in a series of closed vessel. The relative humidity of the air in the space around the sample in each vessels is adjusted by a solution (usually NaCl see Table 1.8 page 34) of a known osmotic potential. The vessels are placed for several hours in a reliable constant-temperature bath and the tissue samples are reweighed after exposure in order to ascertain the changes in fresh weight and hence the relative rate and direction of the water vapour transfer during compensation time.

The gravimetric method was first designed by ARCICHOVSKIJ and ARCICHOV-SKAJA (1931), further developed by SLATYER (1958) and modified by KREEB (1960) and KREEB and ÖNAL (1961). There must be no temperature difference between the tissue and the solution, as any such difference would immediately result in undesirable changes in the water vapour pressure difference. Respiration of the tissue may therefore decrease the reliability of the results. In order to shorten the compensation time it is also important that the dead space around the sample is as small as possible and that the shortest possible diffusion path is ensured by having only a short distance between the sample and the solution. The surface area of the solution must be as large as possible so that the relative humidity of the air corresponds to the osmotic pressure of the solution after closing the vessels. However, the

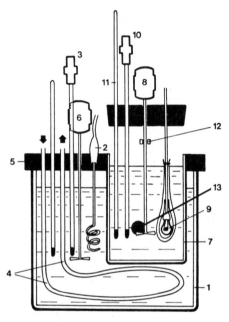

Fig. 1.5 A schematic section of the water bath. 1 — outer bath, 2 — heater, 3 — contact thermometer in the outer bath, 4 — cooling coil, 5 — polystyrene sheet, 6 — stirrer in the outer bath, 7 — inner bath, 8 — stirrer in the inner bath, 9 — heating bulb, 10 — contact thermometer in the inner bath, 11 — calorimetric thermometer, 12 — ball bearing, 13 — position of the measuring chambers. Screening by aluminium foil is not shown.

sample must be placed in the vessel in such a way that it cannot be contaminated by the solution (e.g. by capillary rise). A relatively large sample surface is also advantageous. There must be no losses by evaporation, which would affect the water potential of the solution, during adjusting and placing the sample. Examples of suitable equipment will be given.

Water Bath

A good thermostatic water bath is necessary. Long term constancy of the inner temperature of $\pm$ 0.05 or better should be maintained. Short term variations of the temperature which may result in impairing perfect isothermal conditions in the measuring vessels, in particular between test solution, air and tissue sample, must

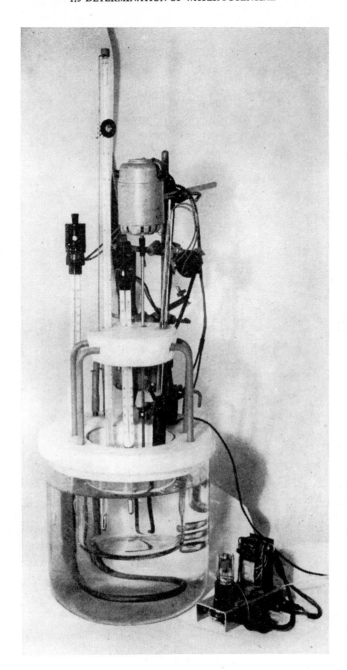

Fig. 1.6 A view of the water bath with a thyrathron relay.

be avoided. These anisothermal conditions may cause convection of air and distillation resulting in rather large errors. The following equipment may be put together
from usually available material and is practically the same as the equipment of
SLATYER and KREEB.

The water bath is double (Fig. 1.5 and 1.6). Its size is chosen according to the
internal volume required. Thorough mixing is difficult in baths that are too large
(above 50 − 100 l), and thermal capacity is low in baths which are too small (less
than 10 l). The vessel used as the outer bath (1) may be a glass jar or the lower part

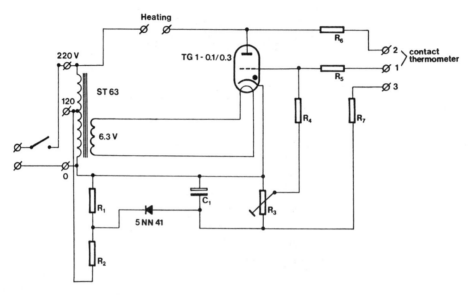

Fig. 1.7 A diagram of the thyrathron relay. $R_1 = 30$ kΩ, $R_2 = 20$ kΩ, $R_3 = 50$ kΩ, $R_4 = $
$= 0.1$M Ω, $R_5 = 2.2$M Ω, $R_6 = 2.2$M Ω, $R_7 = 2.2$M Ω, $C_1 = 20$ μF, 1,2,3 connections to the
heater.

of a thick-walled 20 l bottle. It is heated by a 500 W heater (2) connected to a contact thermometer (3) by electronic and mercury relays. A temperature of 24.5 °C
is maintained so that at room temperature the bath need not be cooled. In other
cases, the outer bath is cooled by tap water passing through a copper tube (4) forming a simple loop. The outer bath can be covered with a polystyrene sheet (5)
with the necessary holes cut into it. The inner bath (7) can also be made of a small
jar or the lower part of a horizontally cut bottle, somewhat excentrically placed in
the polystyrene sheet. The round shape of the vessels facilitates thorough mixing of
the whole water volume. The inner bath (or the whole bath) should be insulated
from any heat and light radiation from outside in order to prevent local increase
in temperature and hence anisothermal conditions and direct effect of light on the
tissue. A contact thermometer with a mercury bulb (3) is inserted through the
polystyrene sheet covering the outer bath so that the mercury bulb and the sub-

merged heater (2) are at the same depth, together with the propeller of the stirrer (6), the speed of which can be regulated.

Water in the inner bath is also mixed by a propeller (8) or by a pump stirrer which ensures vigorous and thorough mixing. A darkened 25 W bulb (9) immersed in the bath in a polyethylene bag is used for additional heating. The bulb is turned on by a reliable closing relay (a diagram of a thyrathron relay is shown in Fig. 1.7) which is operated directly by a contact thermometer (10) set at 25.0 °C. The temperature of the inner bath is monitored by a calorimetric thermometer (11) with

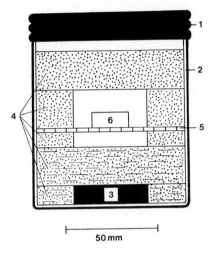

50 mm

Fig. 1.8 The measuring chamber, according to KREEB (1960). See the description in the text.

a suitable scale (graduation of 0.01 °C or less) or a suitably adjusted Beckmann thermometer. The heating bulb switches on at about 30 to 60 s intervals in the arrangement described with a small temperature gradient between the warmer inner bath (25 °C) and the colder outer one (24.5 °C). The inner bath is not covered, as the evaporation from the surface of the liquid cools it slightly. For field work, KREEB (1961) used an 80 l plastic vessel for a simple water bath maintained at a temperature of 30 °C by means of a 100 W bulb connected through a relay to an accurate thermometer. The arrangement was found to be useful, although the accuracy was only about ± 1 to 2 bar because of the dispersion of the measured changes in weight.

Measuring Vessels and Solutions

A suitable arrangement was described by KREEB (1960 — Fig. 1.8): Screw-topped (1), rubber-washered, water-tight glass vessels (2) of about 6 cm in diameter were weighed with pieces of lead (3) placed in the polyurethane foam rings (4) lining the vessels. The foam was wetted with the appropriate solution in such a way that solutions (%). (Compiled from different sources).*

Molarity	Osmotic potential of NaCl solutions at 20 °C [bar]	Relative humidity above NaCl solutions at 20 °C [%]	Molality	Osmotic potential of NaCl solutions [bar]		Osmotic potential of KCl solutions [bar]	
				at 20 °C	at 25 °C	at 20 °C	at 25 °C
0.05	— 2.34	99.83	0.05	— 2.30	— 2.34	— 2.29	— 2.33
0.10	— 4.39	99.68	0.10	— 4.54	— 4.62	— 4.52	— 4.59
0.15	— 6.44	99.53	0.15	— 6.77	— 6.88	— 6.71	— 6.83
0.20	— 8.50	99.37	0.20	— 9.00	— 9.15	— 8.90	— 9.05
0.25	— 10.57	99.22	0.25	—11.22	—11.41	—11.07	—11.26
0.30	— 12.65	99.07	0.30	—13.44	—13.68	—13.24	—13.47
0.35	— 14.77	98.92	0.35	—15.67	—15.95	—15.40	—15.68
0.40	— 16.89	98.77	0.40	—17.91	—18.23	—17.57	—17.88
0.45	— 19.05	98.61	0.45	—20.16	—20.52	—19.73	—20.09
0.50	— 21.22	98.44	0.50	—22.41	—22.81	—21.90	—22.30
0.55	— 23.41	98.27	0.55	—24.67	—25.12	—24.06	—24.51
0.60	— 25.60	98.10	0.60	—26.94	—27.44	—26.23	—26.72
0.65	— 27.82	97.94	0.65	—29.22	—29.77	—28.40	—28.94
0.70	— 30.04	97.79	0.70	—31.51	—32.10	—30.57	—31.16
0.75	— 32.30	97.63	0.75	—33.81	—34.46	—32.74	—33.38
0.80	— 34.56	97.47	0.80	—36.12	—36.82	—34.92	—35.61
0.85	— 36.86	97.31	0.85	—38.45	—39.20	—37.10	—37.84
0.90	— 39.16	97.14	0.90	—40.79	—41.58	—39.28	—40.07
0.95	— 41.48	96.97	0.95	—43.14	—43.99	—41.47	—42.31
1.0	— 43.80	96.80	1.0	—45.50	—46.40	—43.66	—44.55
2.0	— 97.7	93.0	2.0	—95.70	—97.80		
3.0	—168	88.3					
4.0	—252	83.0					
5.0	—327	78.4					
Saturated solution	—374	75.8					

* Osmotic potentials of NaCl solutions expressed in joules per kg water are tabulated for a wide range of temperatures in LANG (1967).

just a few drops could be squeezed out by lightly pressing the foam. A stainless-steel gauze (5) was placed between the polyurethane rings half way up the vessel and the sample (6) was placed in a small central space (diameter 3 cm, height 3 cm). The bottom polyurethane discs were wet with the solution. SLATYER filled similar vessels up to 1.2 cm below the top edge with pure silica sand saturated with NaCl solutions. A nylon gauze bearing the sample was supported just above the surface of the sand. Aqueous solutions of NaCl made up to correspond to osmotic potentials of −5, −10, −15, −20, −25 ... −35 bar (interpolated according to Table 1.8) are the test solutions. It is important that the lid can be closed without raising the pressure in the vessel.

Samples of Leaf Tissue, Weight Determination, Exposure and Evaluation of Results

SLATYER (1958) used discs of about 7 mm in diameter, cut with a cork borer. KREEB (1960) cut leaf samples roughly 3 × 8 mm in size. Total samples size depends on the accuracy of the balance used. The samples are prepared and put into vessels situated in a humid chamber (*e.g.* inoculation "Hansen" chamber) lined with wet filter paper. The vessels should first be thermally equilibrated (24 hours) in the water bath.

The exposure time does not appreciably affect the estimated water potential. An exposure time of 4 to 8 hours, preferably 8, is sufficient (BARRS and SLATYER 1965). KREEB (1960) exposed the samples for 20 to 24 hours.

Samples are weighed with an accuracy of ± 0.1 mg, if possible, with an accurate torsion or automatic analytical balance also lined with wet filter paper before weighing. Small, closed, previously tared weighing bottles may also be used. They should be kept at constant temperature and humidity and never handled directly.

Errors caused by weight loss through respiration naturally increase with exposure time. KREEB and ÖNAL (1961) found that the error caused by respiration was equivalent to 0.5 to 3 bar in potato leaves exposed for 48 hours. Correction of this error is therefore recommended as follows. A replicate sample in each series is killed* immediately and its dry weight determined. The experimental samples, the water potential of which has been measured, are also fixed immediately after weighing and their dry weight determined. The differences between the two corresponding dry weights are added to the weights of the experimental samples after exposure. The ratio of the weight after exposure to that before exposure is calculated from these corrected weights (Fig. 1.9). For exposure periods of 24 hours or less, this correction is not necessary with many kinds of leaves.

The method is suitable for measuring medium to high water potentials. SLATYER (1958) supposed that, theoretically, an accuracy of ± 0.2 bar may be obtained;

---

* For field work, it is useful to fix the leaf discs to the bottom of a cork with entomological pins and then to place the cork over a vessel containing boiling water (an alcohol burner is sufficient). This kills the samples, which are taken to the laboratory for drying and weighing.

later SLATYER and McILROY (1961) estimated the practical error (without correction for respiration) to be $\pm$ 3 bar. These figures represent extremely high values of error.

Results of BARRS and SLATYER (1965) seem to indicate that during exchange of water vapour between the plant material and the test solution, the loss of water from plant tissues to a hypertonic solution is faster than its uptake from a hypotonic solution. This is probably responsible for the fact that estimates of water potential

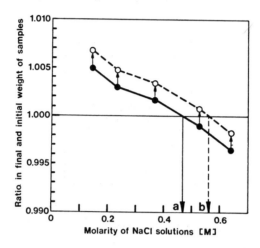

Fig. 1.9 An example of the relationship between the ratio of final and initial weight (ordinate) of samples exposed during gravimetric determination of water potential above NaCl solutions of different molarities (abscissa). The vertical arrows show the shifting of the results after the correction for weight loss due to respiration (KREEB and ÖNAL 1961): **a** is the estimate of water potential with no correction for respiration (about 0.45M NaCl, *i.e.* −19.1 bar), **b** water potential corrected for respiration (about 0.55M NaCl, *i.e.* −23.4 bar).

tend to be too high rather than too low, so that the error of the method is within the range of $\pm$ 1 to 3 bar.

A simplified gravimetric method has recently been designed by J. ÚLEHLA *et al.* (1965). Weighed samples (discs, segments of leaf blades) are placed between two strips of filter paper, through which one of a series of graded test solutions is drawn, protected from evaporation by polyethylene sheets. The samples are separated from the solutions by perforated PVC film, so that equilibration takes place in the gaseous phase.

1.3.3.2 Capillary Method

In the capillary method, only one tissue sample is placed in a vessel exposed as above, together with a set of short, thin capillaries open at both ends, containing the graded series of test solutions. The distance between the two menisci of the solution in each capillary is measured with a microscope eyepiece micrometer before exposure. The distance is measured again after an exposure of several hours and the differences plotted against the values of osmotic potentials of the test solutions used and evaluated as described in Section 1.3.1.

The capillary method suggested originally for osmotic pressure measurement by HALKET (1913) and developed by URSPRUNG and BLUM (1930), was designed primarily for the determination of the water potential of soil samples and solutions. The principle on which it is based allows it to be used for measuring the water potential of plant samples.

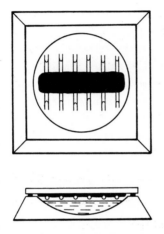

Fig. 1.10 The capillary method (URSPRUNG and BLUM 1930). See the description of the method in the text.

A series of thin-walled capillaries filled with a graded series of solutions of known osmotic potentials is fixed to a strip of plasticine. The set is then placed on a slide covering the so-called "embryo" dish, *i.e.* a glass truncated pyramid with a shallow concave hollow filled with the sample to be measured (see Fig. 1.10). The vessel is closed by a glass cover and the changes of distance between the two menisci in each capillary measured immediately and after 1 to 6 hours by means of a microscope with an eyepiece scale. Capillaries made from inert glass, inner diameter 0.3 mm, carefully washed (in chromic acid, ethanol, ether and hot water in turn) and dried before filling, are broken into pieces at least 5 mm long and filled by touching the surface of the test solutions with one end. The ends are then dried with filter paper, excess solution is drawn off, and the distance between the menisci is adjusted.

There must be no temperature differences between the capillaries and the sample (as in previous methods). Reasonably constant temperature must be maintained to prevent changes in volume of the solution in the capillaries, in particular during the measurement of the distance between the menisci under the microscope (!).

The "dead" air space should be kept as small as possible to facilitate water vapour exchange. Unlike the previous method, the water vapour pressure in this space is determined predominantly by water vapour pressure at the surface of the sample and not by the test solution(s), as was the case in the gravimetric method. Accurate fine gradation of the series of the solutions is also essential. Interpolation is not possible, because of different conditions for diffusive water vapour transfer in different capillaries.

1.3.4 Psychrometric Methods

After recent development, psychometric techniques have become almost standard for precise laboratory determination of water potential in plant tissues and in other samples. For a general review and recent advances see also WIEBE et al. (1971) and BROWN and VAN HAVEREN (1973 in press).

These methods are used to measure water potential by determining the wet bulb depression in a closed gaseous system which is in equilibrium with the sample. This wet bulb depression is calibrated empirically against the wet bulb depression given by salt solutions of known osmotic potential. Isothermic conditions must be maintained i.e. the sample, enclosed atmosphere and thermocouple junction(s) (or thermistor) must be at the same temperature, as water vapour pressure depends to a great extent on the temperature. The wet bulb depression must be measured with an accuracy of 0.001 °C. For example, a water potential of -43.8 bar (equal to the osmotic potential of 1M NaCl solution) corresponds to 96.8% relative humidity, for which the empirically determined wet bulb depression of a non-ventilated thermocouple by MONTEITH and OWEN (1958) was found to be 0.21 °C at 25 °C. To measure water potential with an accuracy of 0.25 bar or 0.5 bar, we therefore have to be able to measure the wet bulb depression with an accuracy of ± 0.0012 °C or ± 0.0025 °C, respectively.

1.3.4.1 Principal Types of Psychrometers

There are three main ways of measuring the wet bulb depression. In the first (type A), a thermocouple junction is used alternately as wet and dry: the output of the thermocouple is red first when the junction is dry, then condensation of a fine film of liquid water on the surface of the junction is brought about by Peltier cooling. The same thermojunction then functions as a wet junction as long as water remains on its surface (SPANNER 1951, MONTEITH and OWEN 1958, KORVEN and TAYLOR 1959). The difference between the two readings is taken as equivalent to the wet bulb depression and calibrated using solutions of known osmotic potential as samples.

In the second type (type B) the output of a thermocouple with the thermojunction permanently wetted by a small drop of pure water is measured (RICHARDS and OGATA 1958, EHLIG 1962). The difference between this temperature and the constant temperature in the measuring vessel is then again plotted (in term of e.m.f) against osmotic potentials of the solutions used for calibration. Thus in this so-called wet-loop or droplet psychrometer a diffusive flux of water from the water droplet (the source) to the tissue sample serving as the sink is measured. With this type a long-term thermal stability of the vessel, *i.e.* of the bath ($\pm$ 0.001 °C), is required in order to take its temperature as a constant reference temperature, or an additional similar thermocouple without a water droplet is included in every sample vessel and measured as reference.

For measurement with a psychrometer of type B, a special compensation technique (isopiestic technique) has been suggested using sucrose solutions of different known osmotic potentials instead of pure water for the droplet on the wet junction of the thermocouple (BOYER and KNIPLING 1965, BOYER 1966) (see later).

KREEB (1965b) developed a psychrometric method (type C) using a thermistor. The dry bulb temperature is measured with a thermistor to which a miniature chamber filled with wetted filter paper can be tightly attached. The thermistor may also be temporarily immersed in a similar chamber containing water. The wet bulb temperature is then measured by the same thermistor after temporarily removing the chamber (or after taking the thermistor out of a similar chamber containing water). The method differs from the others in the type of temperature sensor and in the alternate measurements of wet bulb and dry bulb temperatures.

1.3.4.2 Constant-temperature Water Bath

The water bath used for the psychrometric measurements is similar to that described in Section 1.3.3.1 for gravimetric measurements of water potential. A double bath with a small temperature gradient (0.5 °C in our case) between the two parts, with the continuous cooling balanced by an immersed, low wattage heater, was found to be very useful. The sensitivity of normal contact thermometers is sufficient if they are connected to an electronic relay. However, it is useful to connect the 25 W heating bulb of the inner bath directly by means of a thyrathron relay (Fig. 1.7), which operates efficiently, reliably and immediately.

The temperature difference between the two baths is chosen so that the heating bulb is switched on and off at a suitable frequency. This depends primarily on the heat capacity of the inner bath and the power input of the bulb. The arrangement is optimal when the intervals between switching on and switching off are similar. If a temperature sensor with a negligible delayed action (*e.g.* thermocouple, small resistance thermometer) is used to measure the water temperature in the inner bath, a variation of water temperature of almost 0.05 °C, may be found even if the water is thoroughly stirred. However, it is the temperatures of the sample, the thermocouple and the enclosed air, in the measuring chamber, which must remain constant and this is achieved by the relatively large thermal capacity of the measuring chamber. The position of the heating bulb is important. It should be immersed in such a way that the circulating water stream (which is visible because of impurities moving in the water) driven by the stirrer moves from the bulb directly to the contact thermometer (LØVLIE and ZEUTHEN 1962). However, there must be no direct thermal radiation from the bulb. The water in the inner and outer baths must be electrically earthed. The inner bath has to be large enough to allow thorough and rapid mixing even when the measuring chambers are immersed in it.

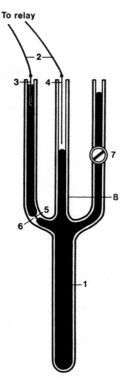

Fig. 1.11 The thermoregulator (BRIX and KRAMER 1962). The regulator is made of glass (1) filled with mercury, the lower part of a size of 200 × 25 mm; (2) are copper inlets to the relay; (3) and (4) are tungsten contacts; (5) is a platinum wire connecting mercury in the left arm with mercury in the other parts of the device across a glass septum (6). The meniscus of mercury in the capillary (8) [0.5 mm inner diameter] rises with increasing temperature up to the contact with the tungsten wire (4), which closes the tap (7).

LØVLIE and ZEUTHEN (1962) suggest the use of a toluene thermometer to meassure the temperature variation of the bath. It is made of a thin-walled tube about 250 cm long, coiled into a spiral. The tube is immersed in the bath and filled with pure toluene (about 15 ml) so that the upper meniscus is in a thin capillary set into the upper end of the spiral and projecting above the surface of the bath. Details may be found in the author's paper. A calorimetric thermometer graduated in 0.01 °C, or a Beckmann thermometer set at the required range, may also be used.

BRIX and KRAMER (1962) used a mercury thermoregulator, shown in Fig. 1.11. It is useful because of its high switching sensitivity but the large temperature inertia of the big mercury bulb as compared with a normal contact thermometer is a disadvantage.

It is always necessary to find out whether it is useful to cover the inner bath with a thermally insulating cover (*e.g.* a suitably perforated polystyrene sheet) to prevent evaporation and change in the cooling system caused by change in the relative humidity in the laboratory. Constant room temperature is not necessary when using a double bath. Some authors working with single baths maintain constant room temperature (variation of $\pm 0.2\ °C$).

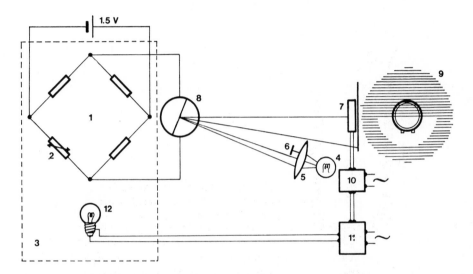

Fig. 1.12 Temperature control (Løvlie and Zeuthen 1962). A Wheatstone bridge (1), where (2) is a thermistor, is situated in the bath (3). A projection lamp (4) illuminates through a lens (5) a photocell (7) by the boundary light/darkness by means of a razor blade (6). The front view of the photocell with a crescent light index thrown by the mirror galvanometer (8) is designated (9). The photocell current amplified by the amplifier (10) turns on the heating bulb (12) of the bath through a relay (11).

Løvlie and Zeuthen (1962) used (Fig. 1.12) a resistance thermometer connected in a Wheatstone bridge, instead of a contact thermometer, to actuate the heating bulb. The Wheatstone bridge was located directly in the bath. The out-of-balance signal was displayed on a mirror galvanometer, the moving spot from which impinged on a photo-electric cell. The current was then amplified and through a relay switched on the heating bulb. Korven and Taylor (1959) used a similar device, with a thermistor detector in the bath forming one arm of a Wheatstone bridge.

Barrs and Slatyer (1965) used very successfully an Aminco 4−204 supersensitive thermoregulator operating through an Aminco 4−5300 electronic relay or a Sunvic toluene-mercury thermoregulator operating through a proportioning head and EA3T electronic relay.

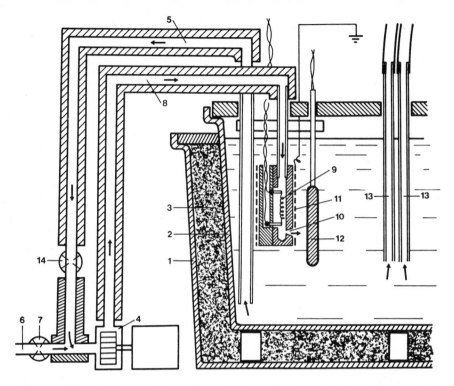

Fig. 1.13 Diagram of a multiple water bath (MILLAR 1971c). Two polyethylene tanks (1) and (2) — one within the other — were separated by 8 cm mineral fiber (3). Water pump (4) [4 l min^{-1}] drove water from the inner tank (30 l) by pipe (5) insulated with plastic foam and from a refrigerated water storage tank by (6) in proportion adjusted by needle valve (7). The mixture was pumped into the bath through lagged pipe (8) with control heater (9) [operated by a relay] through a 3 mm orifice (10). The heater (80 W, 9.4 cm length nichrome resistance wire 0.13 mm diameter) in acrylic block was enclosed in a Faraday cage of stainless steel mesh (11). The temperature controller (12) was situated in front of the orifice slightly to one side. The heating and cooling cycles were balanced manually and are 5 to 8 s. Siphons (13) connected to the chilled water tank prevented an increase of bath water level. Needle valve (14) was normally open and permitted rapid change of bath temperature.

SPOMER (1972b) described a simple inexpensive water bath with maximum short time fluctuations of $\pm 5 \times 10^{-4}$ °C and drift of $+2 \times 10^{-3}$ °C d^{-1} for continuous periods up to 25 days. A large volume mercury-toluene thermoregulator with a mechanical heat proportioner was used.

More complicated baths have been designed by SPANNER (1951), KIJNE and TAYLOR (1963) and HOFFMAN and SPLINTER (1968). Refrigerated baths are necessary for work below room temperature (RICHARDS and CAMPBELL 1948, KLUTE and RICHARDS 1962). MILLAR (1971c) described a sophisticated multiple water bath with temperature control within 0.001 °C at an ambient temperature 30 ± 1 °C (Fig. 1.13).

1.3.4.3 Construction of Psychrometers

Type A (Peltier psychrometer)

The construction of type A psychrometers is relatively simple and the material is readily available.

The thermocouple designed by MONTEITH and OWEN (1958) (see Fig. 1.14) — which may serve us for the description of the type — was fixed on a rubber stopper fitting a test tube (psychrometric chamber) containing the sample. The wires (1) mounted in drilled copper plugs (2) pass through the rubber stopper (3). The plugs were inserted into the stopper and sealed with a sealing material (*e.g.* araldite) on the upper side. Copper wires, about 1 mm thick, project 5 mm into the psychrometric chamber (4). Two wires, one constantan (6), one p-chromel (5), 0.10 to 0.15 mm thick, were soldered to the main wires (reference junction). They are crossed and soldered 8 mm from the reference junction and the ends were cut off with a razor-blade (sensing junction). The soldered junction was immersed for about 15 min in boiling distilled water, to remove traces of $ZnCl_2$ in the flux.

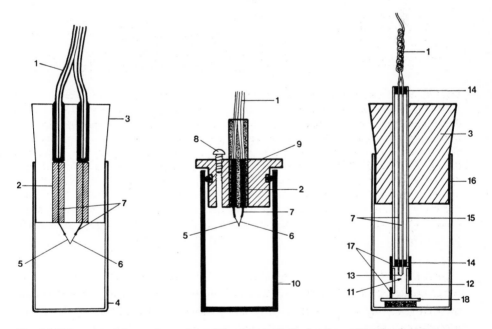

Fig. 1.14 Thermocouple psychrometers of the A type. Left: simple modification by MONTEITH and OWEN (1958). Centre: modification with a brass chamber. Right: miniature modification by MANOHAR (1966a,b,c): 1 — insulated wires, 2 — copper rods, 3 — rubber stopper, 4 — glass measuring chamber, 5 — chromel and 6 — constantan wires, 7 — copper wires, 8 — screw for pressure equilibration after closing the chamber, 9 — brass stopper with O-rings, 10 — brass chamber, 11 — psychrometric chamber, 12 — silver cylinder, 13 — thermo-couple, 14 — araldite seals, 15 — glass tube, 16 — glass specimen tube, 17 — polyethylene sleeves, 18 — leaf disc.

MANOHAR (1966a) used a neutral flux of aniline phosphate in ethylene glycol, with an alloy of tin and lead (60 : 40) as solder for the reference junction, the sensing junction being welded. WAISTER (1963) recommended arc welding under oil in order to prevent oxidation. BOX (1965b) soldered the free junction of a robust A-type psychrometer with silver. LANG and TRICKETT (1965) and CAMPBELL *et al.* (1968) preferred welding to soldering.

Welding may be achieved by a thermocouple arc-welder using inert gas (argon) *e.g.* Model 125 SRL thermocouple welder by Spembly Technical Products Ltd., England, Welding Machine Model 1-06502 Unitek Corp., Monrovia California U.S.A. A simple circuit diagram of an arc-welder unit using graphite rod was designed by BROWN (1970). Welding may be done either in an inlet gas flux (nitrogen, LOPUSHINSKY and KLOCK 1970a, b) or in a small pool of light-weight optically clear oil in a Petri dish. Traces of the oil are removed by dipping the thermocouple into acetone.

MONTEITH and OWEN (1958) achieved a sensitivity of 60 µV per degree C with their thermocouple. The junction of p-chromel with constantan forms a wet thermojunction (after wetting — see below), whereas the junctions with the copper wires have the same temperature as the copper plugs and the water bath and serve as reference junctions. During measurement, the galvanometer connected to the outlets of the thermocouple measures the temperature differences, if any, between the constantan/chromel free junction and the reference junctions *i.e.* constantan/ /copper and chromel/copper, which are at bath temperature. Prior to measurement of the psychrometric difference, current is allowed to flow into the thermocouple in the opposite direction (about 30 mA for about 30 s) at 1.5 V. The constantan/ /chromel junction is cooled by the Peltier effect (by about 1.5 °C) to below the dew point as the relative humidity of the surrounding atmosphere is very high, so that a water film condenses on the surface. At the same time, the reference junction between these wires and the copper wires warms up as a result of Joule heat, which is rapidly dissipated from the copper wires into the copper plugs. As soon as the constantan/chromel junction (sensing junction) is wet, the cooling current is turned off and the thermoelement is switched over to the microvoltmeter, to measure the temperature difference between the wet junction and the reference junctions, *i.e.* the wet bulb-depression.

The theory of type A psychrometer was thoroughly discussed by RAWLINS (1966), PECK (1968, 1969) and DALTON and RAWLINS (1968).

RAWLINS (1966) derived a theoretical equation for the electromotive force of both types (A and B) of thermocouple psychrometer, based on net heat exchange of the wet thermocouple junction, taking into account the construction, size and material of the thermocouple and the geometry of the chamber. He came close to the empirical results and suggested criteria for construction of both types. Maximum psychrometer sensitivity is achieved when the heat flux to the wet thermojunction through the wires is negligible. This heat flow to the wet junction is more dependent on the area of the wires near the wet junction than on their length. Thus the sensitivity of existing psychrometers should not be affected by shortening the wires. Any difference in either temperature or vapour pressure between the sample and the atmosphere in the measuring vessel will cause error in estimation of water potential. Sinks for water vapour in the vessels, resulting from condensation even on thoroughly cleaned vessels, may cause significant errors.

PECK (1968, 1969) emphasised the significance of heat sinks and calculated their optimum size for sufficient elimination of heating of the reference junctions by Peltier cooling of the sensing junction; this was unusually large.

Since the time when first psychrometers of the A type were constructed, there have been many modifications, using different materials, different geometry and different construction of the thermocouples.

BOX (1965a, b) designed a modification for measuring water potential of leaf discs with big copper heat sinks at the reference junctions.

A miniature thermocouple psychrometer with an extremely small psychrometric chamber (volume 0.06 cm^3) was designed by MANOHAR (1966a, b) (Fig. 1.14 right). The walls of the psychrometric chamber (1) consist of a silver cylinder (2) of internal diameter 4 mm, 8 mm high, connected by a polyethylene sleeve (10) to a glass tube (6) holding the micropsychrometer. The thermojunction constantan chromel p (3) is soldered together under oil using the solder mentioned above. At a distance of 0.5 mm from this junction the chromel p and constantan wires are soldered to copper wires (4) fixed by araldite seals (5) in a glass tube (6) (3 mm inner diameter). The wires are coiled for a length of 25 cm to produce a constant-temperature heat sink (7). The psychrometric device with its miniature psychrometric chamber (1) is mounted in a rubber stopper (9) inserted into a glass specimen tube (8). The miniature psychrometric chamber (1) can be lowered by sliding the glass tube (6) in the rubber stopper downwards, so that the chamber can be gently pressed with a second polyethylene sleeve (10) against either filter paper (when calibrating) or the surface of the sample to be measured. The distance of the thermocouple from the measured surface, and the geometry, affect the calibration considerably (MANOHAR (1966b). The psychrometer was originally designed for measuring the water potential of germinating pea seeds.

MILLAR (1971a) constructed a rather complicated A type psychrometer (Fig. 1.15 left) providing almost doubled range of measurements (0 to −130 bar) and minimizing main sources of errors (adsorption of water vapour on the walls of the chamber, leaf resistance and respiration, see Section 1.3.4.8).

A thermojunction consisting of bismuth/bismuth plus 5% tin was found to be very useful for effective cooling by the Peltier effect and e.m.f. of the connection (SPANNER 1951). This junction may be cooled by the Peltier effect by up to 4 °C and yields 126 µV per °C. A thermocouple with bismuth wires was also used by KORVEN and TAYLOR (1959), who described its extreme fragility. Spraying the thermo-

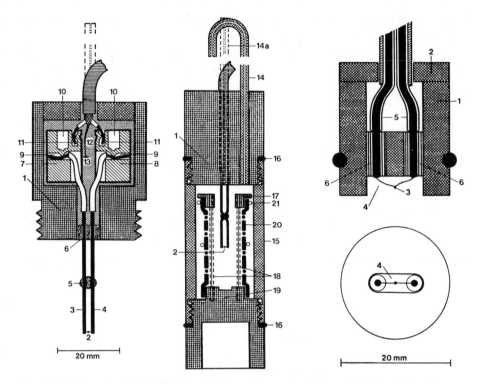

Fig. 1.15 Left and middle: Thermocouple psychrometer of type A designed by MILLAR (1971a,b).

Left: The thermocouple assembly: 1 — polished stainless steel housing. 2 — sensing thermojunction of 1 mm long chromel wire (25 µm diameter) and 1 mm long constantan wire (26 µm diameter) welded to the ends of constantan (3) and chromel (4) wires, respectively (254 µm diameter), connected by a spacer (5) of resolidified Apiezon wax "W", 6 — plug of the same material, 7 and 8 — two copper block heat sinks (each 0.8 cm^3) with holes (9) where constantan and chromel wires (3) and (4) were crimpsed with a punch inserted into holes (10). Similarly copper wires (11) are connected with the heat sinks. The shielding of the cable (12) is connected to (1) at (13). The inner parts of the assembly are insulated by epoxy resin.

Middle: Arrangement of MILLAR's psychrometer for plant samples. Capillary tube (14) maintains the chamber at atmospheric pressure. It is in fact situated not at right side [as shown in heavy lines] but behind the axis of the thermocouple assembly [in broken lines (14a)]. 15 — cylinder, 16 — neoprene gaskets, 17 — flanged disc, 18 — three rods (spaced at 120° around the disc) connecting the disc to the boss (19), 20 — plant sample, 21 — spring clips holding the disc.

Right: Thermocouple psychrometer probe (A type) for measuring leaf and soil samples (HOFFMAN and SPLINTER 1968a,b). 1 — Teflon body, 2 — Teflon cover, 3 — sensing thermojunction chromel/constantan (both wires 25 µm diameter), 4 — oblong chamber 9 mm long machined out of Teflon body, 5 — copper lead wires from a shielded cable, 6 — copper splice connectors.

couple with a dissolved plastic makes it more rigid and decreases the "drift", *i.e.* gradual change in sensitivity, but equilibration is then much slower.

HOFFMAN and SPLINTER (1968a, b) designed a thermocouple psychrometer probe (Fig. 1.15 right) which is similar to the type constructed by RAWLINS and DALTON (1967) for measurements of soil water potential *in situ* (described in Section 3.1.3.1) but the measuring junction is situated not in a ceramic bulb but in the centre of a small oblong equilibrium chamber machined out of the Teflon body. This Teflon body may be attached with epoxy resin to the leaf surface.

HSIEH and HUNGATE (1970) designed a temperature compensated A type psychrometer enabling either conventional measurement or direct compensation of the fluctuations of "dry" temperature (Fig. 1.16).

A four-terminal Peltier type psychrometer (MILLAR *et al.* 1970) is a special design of an A-type thermocouple psychrometer suitable especially for nonisothermal conditions for *in situ* measurements. In the normal two-terminal design of A type psychrometers, the main error in measured value of wet bulb depression

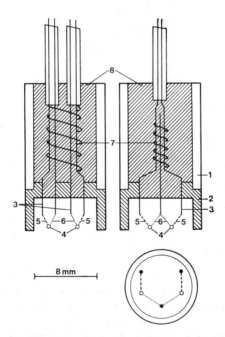

Fig. 1.16 Temperature compensated psychrometer of type A (HSIEH and HUNGATE 1970). Left: Type which can be used either as a conventional psychrometer or as a thermally compensated one depending on the connection.

Right: A simplified temperature compensated thermocouple psychrometer. 1 — Pyrex glass tube, 2 — Teflon block with a chamber cavity 8 mm in diameter (could be attached with grafting wax to the leaf surface), 3 — copper wires, 4 — sensing thermojunctions, 5 — constantan wire (25 μm diameter), 6 — chromel wire (25 μm diameter), 7 — copper wire winding in solder, 8 — epoxy resin.

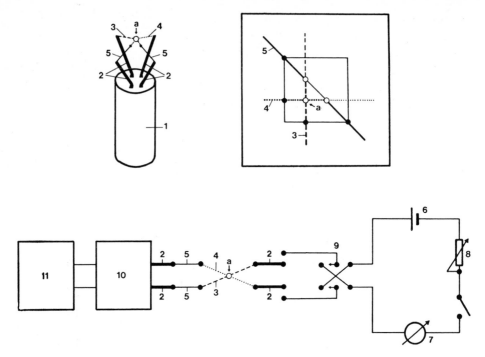

Fig. 1.17 Diagram of the four-terminal A-type thermocouple psychrometer (MILLAR *et al.* 1970).
Top left: View of the thermocouple assembly. 1 — Teflon rod, 2 — copper posts, 3 — constantan wires (3 mm long, 25 µm diameter), 4 — chromel wires (3 mm long, 25 µm diameter), 5 — Karma wire (8 mm long, 57 µm diameter). The junctions were welded while the wires were placed in position shown right over a hole in a piece of cardboard. 4 — welding points, a — sensing junction, 6 — battery 1.35 V.
Bottom: Diagram of the circuit. 7 — microampermeter, 8 — potentiometer 2.5 kΩ, 9 — switch, 10 — microvoltmeter, 11 — recorder.

is caused by the difference in temperature between the reference junctions (in the neighbourhood of the heat sinks) and the sensing junction chromel/constantan, caused by non-equilibration of the temperature after heating the reference junction when cooling the sensing junction, both by Peltier effect. This is because the thermal paths to the sensing (wet) junction and to the reference junction are different: the sensing junction is heated or cooled mostly by the air in the vessel, the reference junction mainly by conduction through the copper leads from the bath. So the temperature of sensing junctions in non-isothermal conditions is influenced in a different way from the temperature of reference junctions. In four-terminal Peltier type psychrometer, improved thermal insulation of the reference junctions was achieved: the reference junctions (see (1) in Fig. 1.17) are not heated while passing the Peltier cooling current while junctions (3) are heated and of course sensing junction (4) is cooled. In addition to this the reference junctions are thermally insulated by wires made of a material with extremely low thermal conductivity and — this is necessary — with almost zero thermal e.m.f. against copper (wires like Evanohm, Karma, Minalpha were used). The cooling current used was 4 mA for 30 s, the sensitivity was about 0.4 μV bar^{-1} at 25 °C. The four-terminal psychrometer showed an output of 0.5 to 1 μV per °C of air temperature change, while the two-terminal psychrometers had a variation of about 13 μV per °C.

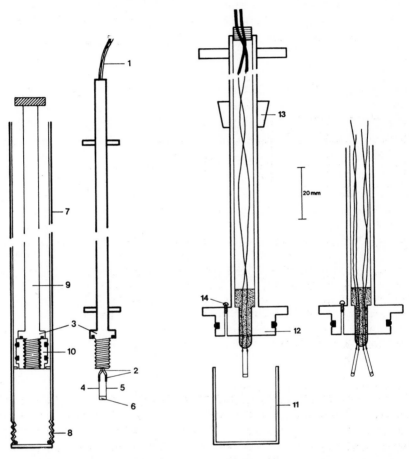

Fig. 1.18 Droplet psychrometer [type B] (RICHARDS and OGATA 1958) modified by BARRS and SLATYER (1965). 1 — PVC coated twin flex, 2 — single wire, 3 — brass flange, 4 — chrom-l wire, 5 — constantan wire, 6 — silver ring, 7 — brass tube 25 mm in diameter, 300 to 400 mm long, 8 — stainless steel cap with neoprene washer, 9 — brass rod closing the measuring chamber during equilibration, 10 — brass piston with O-rings.

Fig. 1.19 Another modification of the droplet psychrometer [type B] (KRAMER 1967). Left: a normal assembly with a single thermocouple. Right: a type with two thermocouples for the elimination of the metabolic heat error (see Section 1.3.4.8). 11 — brass chamber. 12 — brass cover with O-rings and normal thermocouple with silver ring, 13 — stopper used to fix the assembly in the bath, 14 — screw for pressure equilibration.

Type B (Droplet or Wet-loop Psychrometer)

In contrast to thermocouple devices where the wet junction is formed temporarily by a condensed water film by cooling the junction below the dew point by the Peltier effect, there are arrangements where the wet junction is made by direct wetting, as described by RICHARDS and OGATA (1958). Its slightly improved modifications are shown in Figs. 1.18 and 1.19. The outlets are formed by two copper leads 7×0.12 (1), insulated with softened PVC. Only one wire (2) comes always from both leads 6 mm under the flange (3). Chromel p (4) on one side and constantan (5) (diameter 0.025 mm, length 6 mm) on the other are soldered to the wire 10 mm under the cover (reference junction). The lower ends of the chromel and constantan wires, which are parallel and vertical are soldered to a silver ring (6) (inner diameter 2.5 mm, thickness 0.25 mm, height 1.0 mm). The ring holds the standard drop of distilled water used to wet the constantan/chromel junction. A solder consisting of Cd and Sn (75 : 25) (BARRS and SLATYER 1965) was found suitable. This thermocouple has an internal resistance of 20 ohm and its sensitivity corresponds (according to EHLIG 1962) to 0.5 µV per bar of water potential. KLUTE and RICHARDS (1962) and RAWLINS (1966) suggested reducing the length

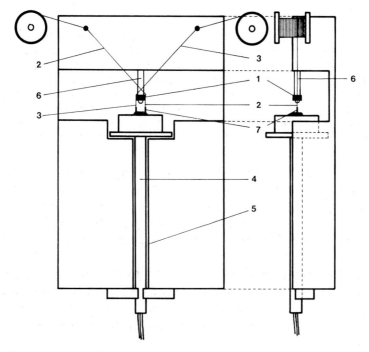

Fig. 1.20 A device for soldering the silver ring (1) to constantan (2) and chromel p (3) wires (KRAMER 1967). The brass tube (4) is fixed firmly in a vice (5) and the silver ring is held in position by a peg (6). The wires (2 and 3), previously soldered to the copper wires (7), are arranged over the ring and then soldered to it.

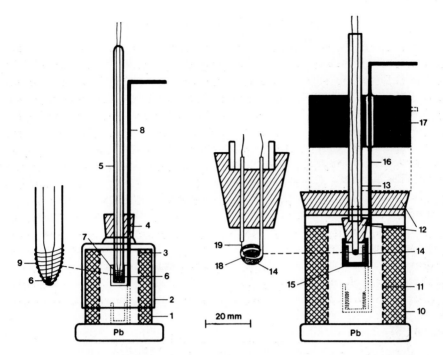

Fig. 1.21 Thermistor psychrometer assembly (KREEB 1965c): Left: a modification using a thermistor sealed in a glass tube. Right: a modification of the droplet psychrometer with a miniature thermistor. 1 — glass vessel, 2 — rubber cap, 3 — copper net cylinders holding plastic foam insulation, 4 — rubber stopper, 5 — glass tube, 6 — thermistor, 7 — plastic "spoon" filled with water, 8 — holder of the plastic cylinder "spoon", 9 — wetting thread, 10 — aluminium vessel, 11 — copper net holding plastic foam insulation, 12 — rubber stoppers, 13 — Plexiglass tube, 14 — water drop, 15 — copper vessel with wet filter paper, 16 — copper holder, 17 — copper plate Pb-lead weigh, 18 — miniature thermistor with inlets (19).

of the thin wires to 4 mm or even to 1 mm in order to improve mechanical strength without changing the output. LANG and TRICKETT (1965) substituted a wire loop for the silver ring, since according to RAWLINS (1966), this may be a better wet junction as it is wetted on its whole surface, in contrast to the silver ring. A device facilitating the soldering of silver ring to the wires was described by KRAMER (1967, Fig. 1.20).

For the isopiestic technique (BOYER and KNIPLING 1965, BOYER 1966), a psychrometer was designed enabling rapid exchange of the inner plunger without causing appreciable changes in the relative humidity and temperature of the air inside, which might affect the water potential of the sample. See page 58 and Figs. 1.26 and 1.27.

Type C (Thermistor psychrometer)

Two modifications of thermistor psychrometers (KREEB 1965b) have been described (see Fig. 1.21). In the first modification a thermistor (6) sealed in a glass tube (5) was used (*e.g.* NTCG, type Valvo EB 8 32005 P/1K5). A thread (9) was wrapped round the end of the tube to ensure good surface wetting. This sensor measures the "dry" temperature when a plastic shoe (7) filled with water is tightly attached to it so that evaporation from the thermistor surface is prevented. The thermistor then measures the temperature of the water in the shoe, which is equilibrated with the temperature of the whole measuring vessel immersed in the constant-temperature bath.

A miniature thermistor (*e.g.* Valvo E 209 AE/P 1K5) is used in the second modification of the thermistor psychrometer, preferred by KREEB (1965b), (Fig. 1.21 right). The thermistor (18) is fixed to a small copper wire hook in which a drop of distilled water (4) is held, keeping the thermistor permanently wet. The inlets of the thermistor (19) are soldered to copper wires, one of which forms the hook. The thermistor is fixed to the rubber stopper by the wires. A small copper vessel (15) filled with wet filter paper, may be attached to the rubber stopper from below [by means of an externally operated holder (16)]. The thermistor measures the dry-bulb temperature when surrounded by the attached vessel (15), *i.e.* when placed in an atmosphere saturated with water vapour so there can be no evaporation from the water drop in the copper loop. It measures the wet-bulb temperature when the vessel (15) is lowered and water from the drop around the thermistor freely evaporates, depending on the water vapour pressure in the measuring chamber, which is in equilibrium with the water potential of the plant material.

1.3.4.4 Measuring Chambers and Arrangement of Samples

The design of the measuring chambers is very important. The chambers should preferably be constructed so that the air space between the sample and the thermocouple (or thermistor) is as small as possible. The relative humidity corresponding to the water potential of the sample then develops very rapidly, especially if the surface of the sample is as large as possible. Practical experience shows that the amount of plant material should be such as just to cover the inner walls of the chamber. Any further increase in the amount of material does not increase the rate of equilibration of water potential and decreases the effectivity of temperature equilibration. It follows that the chamber should be small, its size being determined by the size of the thermocouple device and minimum surface of the sample material (*e.g.* leaves) required.

Some authors recommend cutting the plant material into small pieces (about 5 × 10 mm). BARRS and KRAMER (1969) and MANOHAR (1971) confirmed previous observations (*e.g.* MANOHAR 1966b, d) that water potential measured by a psychrometer (generally by a vapour phase determination) is higher (less negative) in sliced tissues than in unsliced samples. The remaining intact cells actively accumulate solutes released from the damaged cells. Since pressure potential changes more rapidly with cell volume than does the osmotic potential, the net result is an increase in measured water potential of sliced tissues, the difference being as high as 20 to 50%. When measuring the water potential of a soil sample, it is convenient to fill the whole tube with soil and remove a cylinder of soil with a cork borer. The thermocouple device is inserted in the resulting space. It is also convenient to fill the tube with filter paper to absorb the cell sap or solution when measuring the osmotic potential of a solution or of dead tissue (EHLIG 1962). WAISTER (1963a, b)used glass

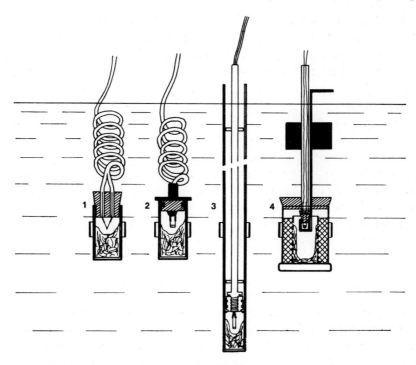

Fig. 1.22 Arrangement of different types of measuring chamber in water bath during equilibration and measurement. 1 — thermocouple of the A type (MONTEITH and OWEN 1958), 2 — thermocouple of the B type (KRAMER (1967), 3 — thermocouple of the B type, modification by BARRS and SLATYER (1965), 4 — miniature thermistor psychrometer (KREEB 1965b).

chambers 17×24 mm in size, the useful volume being 4 cm^3. The chambers should be wrapped in aluminium foil to prevent heating by radiation (*e.g.* from the heating bulb in the bath). Spiral winding of the immersed part of the insulated inlets is useful as their immersed parts can then be up to about 30 cm long and prevent heat dissipation from the meassuring chamber.

The internal geometry and the arrangement of the sample, especially the length of the diffusive pathways between the thermocouple junction and the sample, the area of the exposed surface of the sample and the volume of the aii in the chamber, have an appreciable effect on the results. In all cases the same arrangement must be ensured for calibration and measurement.

Chambers of various shapes, and their position in the bath, are shown in Fig. 1.22. Chamber (1) is a simple glass tube 25 × 75 mm (MONTEITH and OWEN 1958). Chamber (2), for the Type B psychrometer, is made from thin-walled brass tubing, the lower end of which is closed and sealed. Metal is to be preferred to glass as surface water vapour adsorption by metal is much less. KREEB (1965c) used aluminium vessels (4) weighted with lead (Pb) for the thermistor type psychrometer. Further details may be seen in Fig. 1.21. A copper plate (2.5 × 6 cm) is soldered vertically onto the copper holder (17) of the removable vessel passing through the rubber stopper (12) of the vessel. The copper parts facilitate rapid temperature equilibration between the water bath and the contents of the vessel.

The measuring chambers in the baths should be wrapped in alluminium foil (not shown in Fig. 1.22) in order to prevent heating by direct visible radiation and (if the chambers are made of translucent material) direct effect of light on the sample (*cf.* ROBERTS 1969 for the gravimetric method).

The sample is put into the measuring chamber in the humid box lined with wet filter paper (transfer box), in which the sampling vessels are stored. Sampling vessels which prevent any changes in water potential by evaporation should be used.

Enclosure of the plant material in the chamber may theoretically cause changes in its water potential through water vapour equilibration itself (SHMUELI and COHEN 1964) and through anaerobic conditions (BARRS 1965b). Salt exudates or dust on leaf surfaces may cause serious errors, mostly decreasing the measured values of water potential (EHLIG 1962, KREEB 1960, EHLIG and GARDNER 1964, LANG and BARRS 1965, BOX 1965a, b, BARRS 1968a, KLEPPER and BARRS 1968).

The inner surface of the chambers should be thoroughly clean and kept free of any contamination (*e.g.* by residues from solutions used in previous calibration procedures) and from any moisture. KRAMER (1967) suggested scouring the inner surface of brass chambers with fine steel wire wool mounted on a shaft connected to an electric motor. The chambers and the caps with the thermocouple device must then be washed with distilled water and dried carefully with a jet of dry air.

If a rubber stopper is used in the A type psychrometer, it is covered with vaseline and one end of a piece of fine string is closed in the vessel with the stopper. The string is then pulled out, so that there is no overpressure as a result of inserting the stopper. If a brass vessel with the psychrometer device sealed into a brass lid is used, there is a hole in the lid, closed by a small screw with a neoprene O-ring, to prevent any change in atmospheric pressure in the vessel by closure. The brass lid is sealed with a rubber O-ring (Fig. 1.14).

Before closing the chamber, the thermocouple is heated for 5 min above a 100 W bulb to make it warmer than the bath into which the vessel is to be placed, as there must be no condensation of water on the stopper or the thermoelement. All the vessels in the series to be measured are then immersed in the water bath.

1.3.4.5 Measuring Procedures

Temperature equlibration in an accurate water bath ensures isothermic conditions inside the measuring chamber. *i.e.* the sample, air and psychrometer have the same temperature, and the water vapour pressure of the sample and the air surrounding the psychrometer are in equilibrium. It is necessary to determine empirically the time required for temperature equilibration of each type of measuring chamber, bath and material. The wet bulb depression obtained should not change after further exposure, in our experience. Four to six hours are sufficient for most plant material. Fig. 1.19 shows an example of the determination of the equilibration time required for a thermocouple psychrometer (MONTEITH and OWEN 1958). KRAMER and BRIX (1965) used 6 and 6 to 8 hours for the same type of psychrometer. EHLIG (1962) used 2.5 to 6 hours when working with cotton, peppers and clover. WAISTER (1963) recommended 4 to 5 hours (leaves of *Salvia patens*), BARRS and SLATYER (1965) found that 12 to 24 hours was adequate for another modification of the large measuring chamber. The shortest equilibration time needed to calibrate a thermistor psychrometer was 45 min (KREEB 1965b). The time must not be too long, as irreversible changes in the plant material might result in changes in water potential. Commercial thermocouple units are obtainable with equilibration in minutes.

Equilibration and subsequent measurement of a large number of samples is very time-consuming. In order to minimize the working time, LAMBERT and VAN SCHILFGAARDE (1965), using a cam timer and a millivolt recorder, developed a device for reading one psychrometer at 20-minutes intervals. LANG and TRICKETT (1965) designed an automatic scanning system for both A and B types of psychrometers with output up to 30 μV. BARRS (1968) designed an automatic scanning and recording system for B type psychrometers with two thermocouples each (dry and wet), using commercially available parts. ROWSE and MONTEITH (1969) described a fifty channel apparatus in which outputs from A type psychrometers were recorded by converting the movement of light spots of a reflecting galvanometer into pulsed signals, which were printed. HOFFMAN et al. (1969) designed a system which automatically reads and stores initial and final e.m.f. and cools the measuring junctions of A type psychrometers. It is built from inexpensive commercial components. By adding a stepping relay, 150 psychrometers can be read in an hour. CONDON et al. (1971) designed a semi-automatic measuring system for operating and scanning 23 psychrometers of the A type within 8 minutes. LAWLOR (1972) described an automatic multichannel thermocouple psychrometer based on an operational amplifier.

Meters for Measuring Thermocouple Output

Either a reflecting galvanometer can be used, measuring the current directly, or some type of a microvolt potentiometer, measuring the resulting electrical potential.

(1) The reflecting galvanometer should have a sensitivity of about 6×10^{-9} A at an inner resistance of up to 20 ohms and a short deflection period. The galvanometer should measure with a minimum accuracy of 16.6 cm per °C if it can be read with an accuracy of 0.5 mm per m distance and if water potential is to be measured with an accuracy of 0.5 bar.

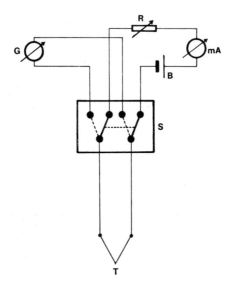

Fig. 1.23 A scheme of connection of the thermocouple device (T), mirror galvanometer (or microvoltmeter) (G) and a source of current (B) for Peltier cooling. mA — milliamperemeter, R — potentiometer (0—2000 ohms), S — switch, B — 2 V batery. See also Fig. 3.4.

MONTEITH and OWEN (1958) used a sensitive lamp and scale galvanometer with a sensitivity of 167 mm per μA at 1 m scale distance with 10 Ω internal resistance (Tinsley type 4500). BARRS and SLATYER (1965) used a Cambridge short-period reflecting galvanometer (Cat. No. 41127/1) giving a deflection of 300 mm per μA at 1 m. WAISTER (1963, 1964) also used a Tinsley type galvanometer 4500 with an internal resistance of 4.5 Ω and a sensitivity of 95 mm per μA at 1 m. KRAMER (1967) used a lamp and scale galvanometer (Leeds and Northrup Model 2285) with a period of 7.5 s and an external damping resistance of 25 Ω. Tinsley & Co Ltd. offer a photocell galvanometer amplifier (Type 5214) with maximum sensitivity 5000 mm/pV in a 10 Ω circuit.

(2) A potentiometer (recorder) with a stable DC preamplifier may be used. RICHARDS and OGATA (1958) and KLUTE and RICHARDS (1962) used a microvolt potentiometer according to TEELE and SCHUHMAN (1939). In most cases electronic amplifiers and voltmeters have proved to be suitable, *e.g.* Microvolt Ammeter Model 150B (and models 148 and 155) by Keithley Instrumentation Inc., Hewlett-Packard 419A DC Null Voltmeter, Leeds and Northrup No. 9834 guarded d.c.

null detector, MJ55 Microvoltmeter with build-in cooling current supply and switches by Wescor Inc. (459 South Main Logan, Utah 84311, USA). More sophisticated equipment for recording the results has already been mentioned.

Measurement with Psychrometers of the A Type

The method using the Peltier effect is as follows (see the diagram in Fig. 1.23): The thermocouple (T) may be switched by means of a fast and reliable switch (S) either to a galvanometer (G) or microvoltmeter or through a regulated resistance and an ammeter (mA) to a 1.5 V battery (B) so that the constantan side is connected to the negative pole of the battery. First the current (regulated by a resistance) and time for best cooling must be found experimentally (Fig. 1.24). A current of about 30 mA for 30 s is usually used with the constantan/chromel thermocouple.

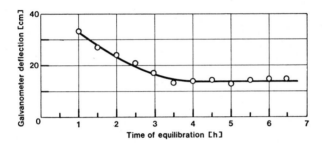

Fig. 1.24 An example of determination of the time required for equilibration in a water bath of the thermocouple chamber of the A type as modified by MONTEITH and OWEN (1958).

During measurement, the thermocouple is first switched over to galvanometer or microvoltmeter. The meter should show practically no deflection from zero. If it does, there must be undesirable e.m.f.'s from the assembly other than the thermocouple device. It is then necessary to check all connections and contacts, which all have to be low-thermal ones, preferably pure copper to pure copper, electrical earthing of the water baths, the shielding of the conducting wires and the reversing switch. When the deflection is minimal (of the order of 10^{-7} V), it is possible to proceed further by switching over to the battery to cool the thermojunction by the Peltier effect. After the previously determined optimum cooling time with an empirically determined current (see Fig. 1.25), the thermocouple is rapidly switched back to the meter. The maximum deflection, reached after several seconds, is read. The response of galvanometers is often too slow, microvoltmeters are faster and may reach a real plateau.

Sometimes it is not possible to wait for stabilization of the deflection as the water film condensed on the thermocouple evaporates relatively rapidly, so the so-called ballistic method, based on the determination of the maximum deflection, is used. This deflection is proportional to the wet bulb depression.

Measurement with Psychrometers of the B Type

Sometimes the measuring chamber is first equilibrated for several hours in a water bath. A brass rod (Fig. 1.18, 9) is screwed in, in place of the thermocouple assembly. After equilibration, the rod is replaced by the thermocouple assembly. However, the thermojunction is usually wetted at the start of equilibration by placing a standard drop of pure water (4 μl) on the silver ring with a microsyringe. The vessel is then placed in the bath and allowed to reach constant temperature. The galvanometer or microvoltmeter is read at intervals after about 30 min until

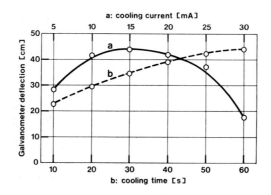

Fig. 1.25 Empirical determination of the maximum galvanometer deflection using different cooling currents (a) and different cooling times (b) for the Peltier effect. (0.5N NaCl solution used for the determination.)

a steady output corresponding to less than ± 0.1 bar per hour is reached (1 to 1.5 h). There is an output of 0.5 μV even with pure water instead of the sample. This is probably because of adsorption on the walls of the vessel.

Calibration is with filter paper soaked in NaCl solutions lining the bottom and the walls of the vessel. The calibration curve is linear down to −50 bar (Fig. 1.29). The measured steady output is proportional to the rate of evaporation from the water drop, which is proportional to the gradient of water potential between the surface of the drop and the surface of the sample, through the atmosphere. It therefore also depends on diffusion of water within the plant sample, since the water potential of the sample is lower than that of pure water (RAWLINS 1964). This causes an error in the direction of measured water potential values which are too high. In order to avoid this error, BOYER (1966) suggested the so-called isopiestic technique.

Isopiestic Technique

It was shown by RAWLINS (1964), BOYER and KNIPLING (1965) and BOYER (1966) that the rate of evaporation from the water droplet of the RICHARDS and OGATA (1958) psychrometer depends on the resistance to water vapour diffusion into the leaf sample. Evaporation is proportional to the difference between the zero water potential at the water droplet surface and the water potential of the leaf sample, so there is no evaporation when this difference is zero. Only then is the measured output of the thermocouple not affected by the rate of water vapour transfer, which is dependent on the resistance to diffusion in the leaf tissue. BOYER and KNIPLING

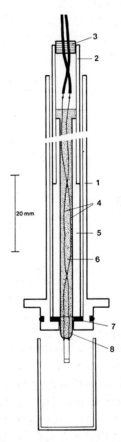

Fig. 1.26 Thermocouple psychrometer for the isopiestic technique (BOYER and KNIPLING 1965; modification according to KRAMER 1967). 1 — brass tube 8 mm in diameter, 2 — Plexiglass tube, 3 — cork, 4 — copper wires, 5 — brass plunger, 6 — epoxy resin, 7 — rubber washer, 8 — wax.

(1965) therefore suggested the use of droplets of solutions of different osmotic potentials instead of pure water. One thermocouple device was substituted for another containing a droplet of a different test solution in the silver ring, without changing the water potential of the atmosphere in the chamber or of the sample. Using the device of BOYER and KNIPLING (1965) shown in Fig. 1.26, it was possible to get the same reading one hour after changing the plunger with the thermocouple device, using the same test solution.

The measurement procedure is as follows. Constant outputs are reached using two or three test solutions and pure water, in turn. It is advantageous to include a solution with an osmotic potential close to the expected water potential of the sample. The outputs are plotted against the osmotic potential of the test solutions (see Fig. 1.27) and the intersection of the line obtained with the line of zero deflection, on the abcissa, is the isopiestic point (A), corresponding to the water potential of the sample. The intersection of the ordinate, at zero potential (B), corresponds to the deflection when there is pure water in the ring of the thermocouple. This isopiestic technique provides very accurate determination of the water potential. However, it is slightly more laborious than other psychrometric method.

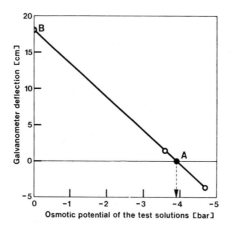

Fig. 1.27 Isopiestic technique (BOYER and KNIPLING 1965). Three measurements were made (empty circles), the first (point B) with pure water, the other two with droplets of test solutions on the thermocouple silver ring (section A). Intersection A: isopiestic point corresponding to $\Psi_w = 3.9$ bar on the abscissa.

Measurement with a Thermistor Psychrometer

In the thermistor device of the first type (Fig. 1.21 left) the "spoon" (7) is filled with distilled water and fitted on to the lower end of a glass tube containing the thermistor. In the second type B a drop of distilled water is placed in the loop of copper wire around the thermistor and the filter paper lining the copper vessel (15) is wetted with distilled water. A rubber stopper with the thermistor device is inserted in the shoe. The sample is then placed in the measuring chamber, which is closed and put in a 25.00 ± 0.01 °C water bath.

Dry-bulb temperature with the "spoon" attached closely to the thermistor is measured first after temperature equilibration. The bridge and an amplifier described in Fig. 1.45, designed also for thermistor cryoscopy (KREEB 1965b), are used. A low bridge voltage of 0.5 V prevents heating of the thermistor. Relatively constant measured voltage reflects constant temperature inside the measuring chamber. The copper "spoon" is then carefully removed from the thermistor sensor by means of a copper holder (Fig. 1.21, 16), and after about 5 min an equilibrium deflection corresponding to the wet-bulb temperature is reached. The amplifier zero is checked (in order to detect drift), the "spoon" is put on again and the stabilized deflection corresponding to the dry-bulb temperature is read again (after about 10 min). Average values of at least four repetitions of this procedure are taken. Good agreement between values shows that equilibration of the water potential had taken place. If the difference between dry-bulb and wet-bulb temperature decreases during the series of measurements, equilibration has not been attained and the exposure in the bath must continue. When the chamber is opened after the measurements, it is necessary to check that the water drop still remains on the thermistor, as only then are the measurements correct.

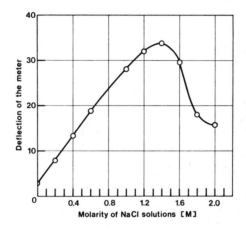

Fig. 1.28 An example of a calibration curve for the A type psychrometer as modified by MONTEITH and OWEN (1958).

1.3.4.6 Calibration

For an unventilated psychrometer of type A, the relationship between the psychrometric difference and the high values of relative air humidity which correspond to the normal range of water potentials is not linear. Each psychrometer must be calibrated and a calibration curve constructed (as in Fig. 1.28). This is done by measuring the output of the thermocouple enclosed in a chamber lined with filter paper soaked in one of a range of known NaCl solutions (see Table 1.8, page 34) in the same way as during the measurement. The water potentials corresponding to the concentrations of the NaCl solutions are plotted against the corresponding galvanometer (potentiometer) deflections. Fig. 1.29 shows calibration curves for psychrometers of the A and B types and a theoretical line for fully ventilated psychrometers. KRAMER (1967) found a practically linear relationship down to -50 bar for the B type. KREEB (1965c) also showed a practically linear relationship between galvanometer deflection and osmotic potential of solutions between zero and -45 bar for the thermistor psychrometer.

Calibration of the psychrometer is valid only for the same geometric conditions as obtained during calibration, as already mentioned. The calibration does not change for several weeks or even months if the psychrometer is kept clean and intact.

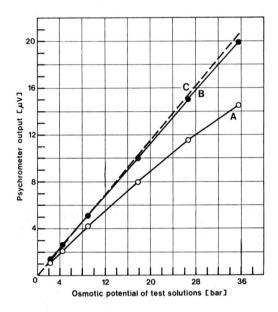

Fig. 1.29 An example of calibration curves for psychrometers of the A and B types. C is a theoretical line for fully ventilated psychrometers (BARRS 1965a).

In order to allow measurements to be made even if calibration of each psychrometer would be too time-consuming, CAMPBELL *et al.* (1966) suggested a sample changer allowing sequential measurement for six simultaneously equilibrated sample chambers using only one calibrated psychrometer assembly. A compact sample chamber C-52 made by Wescor Inc. Logan, Utah, U.S.A. enables to measure exchangeable samples by one calibrated psychrometer without requiring a constant temperature bath.

1.3.4.7 Psychrometric Measurements of the Water Potential of Intact Plants

Recently, several modifications of psychrometric devices have been described, enabling direct measurements of water potential of intact leaves, seeds, tree trunks or soil to be made. The main technical requirement of such measurements is the elimination of rapid temperature changes either by a sophisticated temperature control system (as is the case with leaves) or by the assumption that the temperature of the sample is not subject to short-term fluctuations. This problem was thoroughly discussed by RAWLINS and DALTON (1967).

From the biological point of view, the main problem in measuring water potential of intact leaves using a psychrometric method is the necessary enclosure of the leaf or at least of a part of it in the measuring chamber. If the whole leaf is enclosed, as by LANG and BARRS (1965), its water potential equilibrates with that of the enclosed space and, because transpiration stops, with the water potential of the conducting elements of the xylem. If only a part of the leaf is enclosed, as by BOYER (1968), it must be assumed that its water potential is in equlibrium with the more or less unchanged water potential of the rest of the leaf.

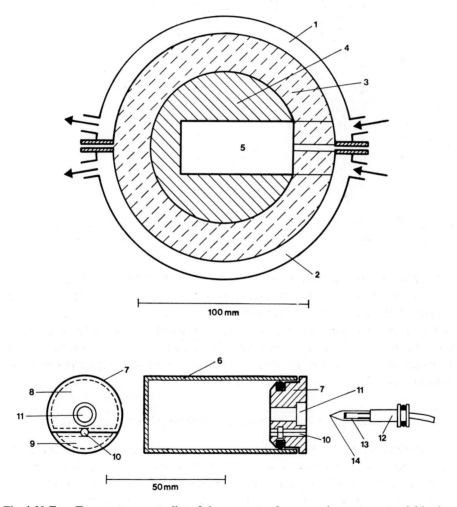

Fig. 1.30 Top: Temperature controller of the apparatus for measuring water potential in the xylem of intact plants (LANG and BARRS 1965). See description in the text.
Bottom: Cross-section of the sample chamber with the thermocouple assembly in the apparatus shown in Fig. 1.26. See the description in the text.

LANG and BARRS (1965) constructed a spherical thermostatic device (shown in Fig. 1.30, top). The outer envelope is formed by a water jacket consisting of two hemispheres (1.2) into which water from an ultrathermostat is introduced; the inner envelope (3) is made of polyurethane. A solid copper sphere (4), 9 cm diameter, had a cylindrical hole (5) bored for the sample chamber (3 cm diameter). The sample chamber (Fig. 1.30, bottom) was made of stainless steel (6) and was closed with a special plug (7) made up of two dissimilar parts (8, 9) between which a space was left for the leaf petiole (10). The plug also had a central hole (11) for insertion of the psychrometer (12). This was a shortened thermocouple psychro-

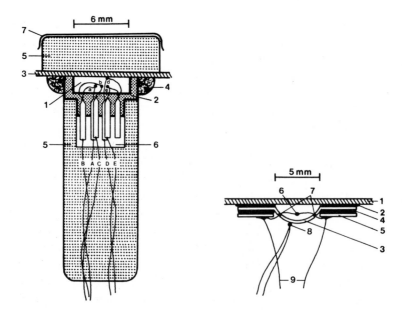

Fig. 1.31 Top: Double compensated thermocouple psychrometer designed by CALISSENDORFF (1970) for *in situ* measurement of leaf water potential (From WIEBE *et al.* 1971). 1 — measuring chamber (6 mm diameter, 2 mm depth) machined in a Teflon body (2), which is attached to the leaf (3) and sealed by a pliable adhesive seal (4), 5 — foam insulators, 6 — epoxy resin, 7 — reflecting aluminium foil; a and b: compensated three lead wire psychrometer [leads A, B, C] (see also HSIEH and HUNGATE 1970, Fig. 1.16), d and e: differential thermocouple for measuring the difference between leaf (d) and air (e) temperature [leads D, E].
Bottom: A miniature A type silver-foil thermocouple psychrometer for measuring leaf water potential *in situ* (HOFFMAN and RAWLINS 1972) 1 — leaf, 2 — silver impregnated water-based conductive coating, 3 — silver-foil disk with central indentation (psychrometer chamber), 4 — laminated material electrically insulating adhesive film, 5 — two half-washers of silver foil (50 μm thick) to which the chromel and constantan wires (25 μm thick) of the thermocouple junction (6) and two copper lead wires (9) (50 μm) are soldered, 7 — holes for chromel and constantan wires insulated with epoxy resin droplets, 8 — additional copper/constantan thermocouple cemented to the outer side of the psychrometer silver-foil chamber. The whole psychrometer is insulated with 2.5 cm plastic foam (not shown).

meter (13) of the RICHARDS and OGATA (1958) type with a thermocouple junction of (14) chromel p/alumel (0.25 mm). The output of the psychrometer corresponds to the wet bulb depression. The junction chromel p/alumel, together with a reference junction in an ice bath, measures the dry-bulb temperature inside the chamber. The thermostatic device makes it possible to maintain constant temperature with a maximum drift of $\pm$ 0.01 °C per hour. As it is very difficult to maintain a particular temperature, it is necessary to calibrate the psychrometers for the whole range of temperatures observed.

BOYER (1968) enclosed only a relatively small part of the leaf in the measuring chamber, the constant temperature of which was maintained by a water jacket with water circulated from an ultrathermostat.

The thermocouple psychrometer (type A) designed by HOFFMAN and SPLINTER (1968a, b — Fig. 1.15) and HOFFMAN and HERKELRATH (1968) can also be used for measuring leaf water potential *in situ*.

CALISSENDORFF (1970) described a double compensated thermocouple psychrometer for measurement of leaf water potential *in situ* with an additional differential thermocouple for measuring the temperature difference between leaf and air in the chamber (Fig. 1.31 top.).

A miniature A type psychrometer for measuring leaf water potential *in situ* was designed by HOFFMAN and RAWLINS (1972, Fig. 1.31 bottom). The chamber with a single-junction psychrometer is formed by an indentation (5 mm in diameter) in a silver-foil disk (50 μm thick, 13 mm in diameter) which is attached to the leaf with a silver-impregnated heat conducting coating. The thermojunction (25 μm chromel and constantan wires, junction approximately 0.2 mm in diameter) is inserted into the chamber through two holes (and sealed in them) in the foil indentation, the wires are soldered on the outer side on two silver-foil half washers to 50 μm copper wires. The psychrometer is insulated on the back side with 2.5 cm of foamed plastic. Fluctuations of the thermocouple temperature were found to be less than 0.025 °C when the temperature of a simulated leaf was changing at a rate of 1 °C per minute. The medium error of the measured leaf water potential was not higher than ± 1 bar.

Thermocouple psychrometers enclosed in small porous ceramic cups have been used for soil water potential measurements (*e.g.* RAWLINS and DALTON 1967, see Section 3.1.3.1 and Figs. 3.2, 3.3 and 3.4) and, inserted in holes drilled into the sapwood of tree trunks, for measuring water potential in trees (SPOMER 1964 and WIEBE *et al.* 1970); they are produced commercially for general use *e.g.* by Wescor Inc. (Logan, Utah, U.S.A.).

1.3.4.8 Accuracy, Sources of Error, and Comparison of Psychrometric Methods

If properly calibrated, the psychrometric methods yield results the accuracy of which depends primarily on the sensitivity of the meter used and — for a galvanometer — on its deflection period in ballistic measurements. The accuracy of the thermistor psychrometer depends on the performance of the amplifier. It is easy to repeat the measurement several times when using an A type psychrometer, which utilizes the Peltier effect for cooling, or a thermistor psychrometer.

For a modified psychrometer of the B type, EHLIG (1962) found the mean error $s_{\bar{x}}$ to be 0.1 bar with a sensitivity of 0.51 to 0.58 μV per bar. Corresponding errors in leaf water potential within the range of 0 to −20 bar and −2.3 to −12.4 bar were found to be ± 0.25 bar and ± 0.17 bar, respectively.

BARRS (1964, 1965a, 1968) pointed out that heat of respiration of the plant sample in the chamber may cause errors by heating the chromel p/constantan junctions more than the reference junctions with copper wires, as the temperature of the reference junctions is more closely coupled to the bath temperature. As a result, a current is obtained when measuring dry-bulb temperature using the A type psychrometer. The deflection is opposite to that obtained when measuring the wet-bulb temperature. Hence for the B type psychrometer, in which dry-bulb temperature is not measured, the apparent wet-bulb depression is increased. The error caused by heat from the plant sample is eliminated in the A type psychrometer by measuring both the dry-bulb and wet-bulb temperatures, which are influenced in the same way. In the B type psychrometer the error results in estimates of water potential which are 10 to 20 per cent too low (BARRS 1965a); the error can be eliminated by using two thermojunctions, one permanently wet and the other dry (see Fig. 1.19, right). Both thermocouples must of course be calibrated and measured simultaneously. The output of the dry thermocouple must be added to the output of the wet thermocouple. This correction is reported to be very difficult in material with a changing rate of respiration resulting probably from tissue deterioration (KRAMER 1967), e.g. tobacco leaves.

RAWLINS (1964) showed a systematic error during measurement with a permanently wet thermojunction of a type B psychrometer. In this type of psychrometer the rate of cooling (and so the output) depends on the flux of water vapour from the water drop through the surrounding atmosphere and into the sample, along the gradient of water potentials. If there is resistance to diffusion at the surface of the plant sample, water vapour flux resulting from evaporation at the wet thermojunction is slowed down. This will result in decreased evaporation and decreased cooling, i.e. in higher wet-bulb temperature. The measured values of the water potential would thus be too high. This error is usually about 8 per cent (4 to 12 per cent) (KRAMER 1967). It can be eliminated by using the isopiestic technique (see Section 1.3.4.5). However, BARRS (1965a) could not confirm that the resistance to water diffusion was the cause of the error.

Adsorption of water vapour on the walls of the measuring chamber may be another source of error. It can be eliminated by completely covering the inner surface with the sample or, during calibration, with filter paper soaked with test solutions or pure water; or by coating the walls of the chamber with vaseline (BOYER 1967a).

It is necessary to maintain the geometry of the measuring chamber unchanged during calibration and measurement (*i.e.*, the distance between the surface of the sample and the thermocouple, the volume of enclosed air *etc.*).

A very good, detailed discussion of the possible sources of error in psychrometric measurements is that by BARRS (1968, pp. 302−310). General evaluation of precision was also published by WAISTER (1964).

It would be premature to try to make a conclusive evaluation of the psychrometric methods for determination of the water potential at the present state of their development. These techniques will certainly be developed further and considerable improvements may be expected. At the present time we can say that type A, using the Peltier effect, is reliable for determining the wet-bulb depression, as it includes measurement of the dry-bulb temperature. The rapid evaporation of the water film condensed after passage of the Peltier cooling current requires the ballistic type of measurement, *i.e.* determination of the maximum value of a rapidly changing galvanometer deviation, which might not be accurate. The e.m.f. of the thermocouple is the same as for type B, but the internal resistance is up to 20 times lower, and the resulting current higher than in type B. Calibration of the type A thermocouple is not linear within the range used. The thermocouple construction is relatively simple and robust.

The type B psychrometer (the so-called wet-loop or droplet psychrometer) measures only the wet-bulb temperature; the dry-bulb temperature is assumed to be identical with the bath temperature, and therefore constant. In types with two parallel identical thermocouples, one permanently wet, the other permanently dry (KRAMER 1967, Fig. 1.19 right), the thermocouple output corresponding to the dry temperature can be measured and used as reference for calculating the wet-bulb depression. The calibration curve, *i.e.* the relationship between galvanometer deflection and water potential, is linear. The galvanometer deflection after equilibration is stable and can be read very accurately, so that irregularities indicating errors in the measurement can easily be discovered. Construction of the type B psychrometer is rather difficult (*e.g.* soldering the silver ring) and it may not be easy to obtain thin chromel p and constantan wire commercially. The galvanometer used may have a higher internal resistance than the galvanometer for the thermocouple of the type A.

The method of measurement of the dry-bulb temperature by a permanently wetted thermistor (thermistor psychrometer), according to KREEB (1965c), is very useful. The principle could probably also be usefully applied to thermocouple methods. The construction of the thermistor psychrometer is not technically difficult, but a high quality amplifier (Fig. 1.45) is required.

The isopiestic technique eliminates possible errors caused by the resistance of the tissue to water diffusion, but it is more time-consuming and laborious than other methods, as a minimum of three measurements is required for each determination.

There are several ways of making psychrometric measurements less laborious:

by automation and recording the measurements (see also Section 1.3.4.5); or by using one calibrated thermocouple for several chambers simultaneously equilibrated with samples (see *e.g.* CAMPBELL *et al.* 1966, Section 1.3.4.6). These modifications may be applied to both A and B type psychrometers. Another possibility is to avoid equilibration in thermostatic bath using special construction of the chamber (see Section 1.3.4.6).

As already mentioned, psychrometers of type B give slightly higher (*i.e.* less negative) estimates of water potential than psychrometers of type A (BARRS 1964, SLATYER and BARRS 1965, see Section 1.3.4.8).

A detailed discussion of the construction, application, error sources, calibration and operation of the type A psychrometer can also be found in the papers of MILLAR (1971a, b).

1.3.5. Dew-point Thermocouple Hygrometry

Quite recently NEUMANN and THURTELL (1973 in press) introduced a completely new method using a thermocouple hygrometer for measuring water potential of samples or *in situ* by means of determining dew point depression. Two thermocouples in a miniature chamber have a common measuring junction which enables simultaneous cooling of the junction by Peltier effect and measuring the resulting temperature depression. Calibration of the hygrometer is made as follows: First in dry air the characteristic parabolic relationship between the cooling current and the temperature depression is plotted using a $x-y$ plotter. Then after having introduced water vapour into the chamber, part of the electrical energy is used to condense water on the cooled junction if this is below dew point. Then the cooling current is decreased manually until the pen of the $x-y$ recorder touches the curve obtained in dry air: at that point the microvoltmeter deflection is proportional to the dew point and may be calibrated in terms of water potential.

CAMPBELL *et al.* (1973) discussed the theory of the procedure and suggested electronics automatically maintaining the temperature of the measuring junction at dew point. A theoretical sensitivity of 0.75 μV bar^{-1} and a change of the sensitivity of 0.45% °C^{-1} was calculated so that the error of $\pm 6\%$ due to temperature changes of the sensor between 20 to 50 °C may be expected. The dew point meter is commercially available (HR-33T Dew Point Microvoltmeter by Wescor Inc., Logan, Utah).

The method seems to be very promising.

1.3.6 Volumetric Tensiometric Method (Hanging Drop Method)

The method originally worked out by WEATHERLEY (1960) for microosmometric measurement of osmotic potential of solutions and cell sap can also be used to measure the water potential of tissues and soil (MACKLON and WEATHERLEY 1965,

TINKLIN 1967). It is described in Section 1.4.2.3. The authors recommend the use of a pile of leaf discs cut out with a cork-borer from which smaller discs are cut out from the centres to leave a hollow cylinder made up of a pile of leaf annuli. The end of the volumetric pipette reaches the centre of the column of annuli (Fig. 1.32). When measuring the water potential by this method, it is necessary to allow a longer equilibration period for samples in the water bath than for determination of osmotic potential. There is a transfer of water vapour between the surface of the

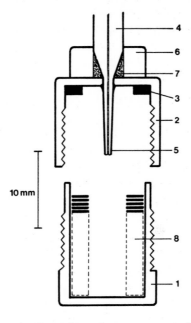

Fig. 1.32 Chamber for volumetric tensiometric measurement of water potential in leaf discs (MACKLON and WEATHERLEY 1965). 1 — chamber base, 2 — cap, 3 — rubber washer, 4 — pipette, 5 — jet, 6 — collar, 7 — epoxy resin, 8 — leaf annuli.

tissue and the drop of the test solution, depending on the steepness and direction of the gradient of partial water vapour pressure. This tissue surface is the cut surfaces of the leaf annuli, and therefore is made up mainly of injured cells without turgor. It is therefore not clear whether the results obtained correspond to the water potential of the intact cells or are changed by equilibration with the content of the cells damaged by cutting.

Calibration is described in Section 1.4.2.3. The difference between the water potential of the tissue sample and the known osmotic potential of the test solution in the pipette must not be greater than $\pm$ 4 bar. Preliminary pilot measurements using various test solutions are therefore required. The same tissue samples can be used for the final determination and for the pilot experiments, provided that there is no water loss by evaporation during manipulation.

1.3.7 Pressure Chamber Method

The atmospheric pressure around an enclosed leafy shoot is increased until a pressure is reached at which xylem sap just appears at the cut end of the shoot. The xylem sap is then at atmospheric pressure (DIXON 1914, SCHOLANDER *et al.* 1964, 1965, 1966). It has been suggested by these authors that the pressure measured in this way compensates the original negative pressure in intact xylem vessels.

At the point where the sap just appears at the cut end of the shoot, the applied pressure is just sufficient to force water into the xylem vessels from the surrounding intact cells and thus return the meniscus of the xylem sap to the cut surface. As these intact cells have semipermeable cytoplasmic membranes the expressed sap has a very high water potential, close to zero. Thus the pressure applied just increases the water potential of the leaf cells to a value equal to the osmotic potential of the xylem sap at atmospheric pressure. Therefore the pressure exerted

$$P = \Psi_{w(1)} - \Psi_{s(sap)} \, , \tag{1.10}$$

where $\Psi_{w(1)}$ is the water potential of the leaf cells (at least of the tissues surrounding the vessels) and $\Psi_{s(sap)}$ is the osmotic potential of the xylem sap, which is usually very high (close to zero).

The pressure chamber technique is very simple in principle. The shoot (the whole above-ground part or an intact leaf) is detached as quickly as possible and enclosed in a steel pressure chamber so that the cut end of the stem or petiole protrudes from the chamber in which it is sealed. The chamber is then hermetically sealed and the pressure inside is gradually increased by compressed air or nitrogen from a cylinder until small sap droplets appear at the position of the xylem vessels on the cut surface. This can be detected with a hand lens or binocular microscope. Fig. 1.33 shows the whole assembly, with two possible ways of hermetically sealing the whole cylinder and sealing in the stem or petiole. O-rings can be used for both. The example on the right shows a very simple way of sealing in the stem by means of a conical rubber stopper (8) with a hole of a diameter which can be adjusted to fit the stem (PIERPOINT 1967). Silicon rubber polymerized *in situ* makes it easy to prepare a specially shaped hole for stems or petioles with a non-circular cross section. Four screws (6) serve to tighten the washer without damaging the stem or petiole.

It is sometimes useful when measuring and increasing the pressure step by step to cut a thin section from the bottom of the stem or petiole from time to time, to prevent drying of the ends of the xylem vessels. It is essential to prevent transpiration loss from the shoot during measurement. This can be done (1) by humidifying the air or nitrogen entering the pressure cylinder and/or (2) by covering the inner walls of the cylinder with wet filter paper. The errors caused by transpiration losses in the chamber may otherwise be considerable (SLAVÍK, unpublished results). WARING and CLEARY (1967) modified the pressure chamber technique for field use by de-

signing equipment weighing less than 18 kg. Inexpensive equipment was described
by KNIPLING (1968c), TOBIESSEN (1969), BLUM *et al.* (1973) and others. A miniature
pressure chamber was designed for measuring individual needles of conifers (personal communication of R. B. WALKER 1973). Pressure chambers are also commercially available *e.g.* from PMS Instruments, Corvallis, Oregon 97330, U.S.A.,
Charles W. Cook & Sons Ltd. 97 Walsall Road, Birmingham B 42 ITT, England.

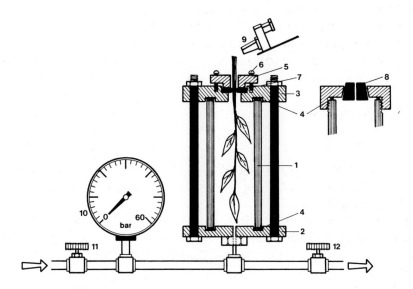

Fig. 1.33 A pressure chamber. 1 — steel pressure chamber, 2 — lower cover, 3 — upper
cover, 4 — O-rings, 5 — insertion held with four screws (6) and used to seal around the stem
of the plant by means of an O-ring (7), 8 — rubber stopper, 9 — binocular microscope, 10 —
pressure gauge, 11 — inlet valve, 12 — outlet valve. Upper right: Another method of closing
the pressure chamber and sealing the stem.

The accuracy of the results depends on (1) accuracy of the pressure gauge, (2)
the rate at which the pressure is gradually increased and (3) primarily, as almost
always, on the way in which the samples are handled, particularly on prevention of
water loss from the plant material while it is being enclosed in the pressure chamber.

The rate of increasing the pressure influences the reading. KAUFMANN used a rate
of one bar every 4 to 5 s. WARING and CLEARY suggested a maximum rate of about
0.7 bar s^{-1}, starting at the pressure which is about 4 to 5 bar lower than the expected
value of *P*. This was found to be too rapid for *Picea, Pinus* and *Sequoiadendron* by
RICHTER *et al.* (1972) and may yield water potential measurements which are 2 to
4 bar too negative. RICHTER *et al.* (1972) used a rate of 1 bar s^{-1}.

The rate of pressure increase also influences the temperature in the chamber.
PURITCH and TURNER (1973) found experimentally at the beginning of the measurement a rapid increase of temperature in excess of 8 °C above ambient tempera-

Table 1.9 Review of methods used for determination of water potential.

Method		Suitable material	Mean accuracy	Reliability [1]	Equipment [2]	Degree of laboriousness [3]
Cell method (URSPRUNG and BLUM 1916)	laboratory	one cell	rather low	*	* microscope	***
Segments method and its modifications (URSPRUNG and BLUM 1926, 1929, 1930)	laboratory possibly even field	leaves without sclerenchymatic tissues	rather low	*	* microscope	**
Refractometric method (ASHBY and WOLF 1947, MAXIMOV and PETINOV 1948)	laboratory and field	leaves, very suitable for roots	±0.5 to 1 bar	**	* refractometer	**
Smear method (ARCICHOVSKIJ and OSIPOV 1031, SHARDAKOV 1938)	field and laboratory	leaves	± > 1 bar	*	*	**
Potometric method (ARCICHOVSKIJ et al., 1931)	exceptionally	cambium in woody plants	low	*	**	***

Method	Application	Material	Accuracy	[1]	[2]	[3]
Gravimetric method (ARCICHOVSKIJ and ARCICHOVSKAYA 1931, SLATYER 1958, KREEB 1960)	laboratory and field	most materials	+1 to −3 bar	** correction for respiration	***	**
Capillary method (URSPRUNG and BLUM 1931, 1930)	laboratory	most materials	±1 bar	*	* microscope	***
Volumetric method (WEATHERLEY 1960, MACKLON and WEATHERLEY 1965)	laboratory	most materials	±0.5 to 1 bar	**	**	**
Psychrometric method (SPANNER 1951, MONTEITH and OWEN 1958, RICHARDS and OGATA 1958, KREEB 1965, BOYER and KNIPLING 1965, and others)	laboratory possibly field	most materials	up to ±0.1 bar	***	***	* to **
Dew point hygrometry (NEUMANN and THURTELL 1973, CAMPBELL et al. 1973)						
Pressure chamber method (SCHOLANDER et al., 1964, 1965, 1966, BOYER 1967)	laboratory and field	cut shoots and leaves	±0.2 bar	**	**	*

1) * low — *** high. 2) * simple — *** complex. 3) * low — *** high.

ture, proportional to the rate of pressure increase. Then the temperature declined during continued pressure increase until equilibration with ambient was reached after the pressure ceased. When the inside pressure was annulled the inside temperature fell to low subzero values. These temperature changes may affect both the water potential (which is temperature dependent) and the plant itself especially in case when measurements with the same plant are repeated.

In coniferous trees the appearence of resin bubbles, which usually start at a lower pressure than sap, may be mistaken for sap and interfere. RICHTER and ROTTEN-BURG (1971) suggested detecting the endpoint by looking for a sharp increase in conductivity between a set of parallel miniature electrodes inserted (about 1 mm deep) into the cut surface of the stem as soon as sap wets the surface. The simple device had a 9 V battery, 100 µA meter and a parallel potentiometer of $1M\Omega$ for controlling the sensitivity.

BOYER (1967a) found that in most cases the sum of the pressure necessary to bring about sap outflow from cut xylem vessels in the procedure described, and the osmotic potential of this sap, were similar ($\pm$ 2 bar) to the water potential of the leaves on the shoot, as measured by the isopiestic psychrometric technique. KAUFMANN (1968a) compared the water potentials of forest trees as measured by psychrometric and pressure chamber methods and found poor agreement in oaks. BOYER (1967a), KAUFMANN (1968a) and RITCHIE and HINCKLEY (1971) denote the resistance to the movement of water through the xylem and the infiltration of intercellular spaces and (in woody species) of the xylem vessels which are no longer active and by absorption of water by nonvascular xylem as possible sources of errors resulting in measured values of water potential which are too low (too negative) in comparison with values determined with psychrometric method.

KLEPPER (1968), KLEPPER and CECCATO (1969), DE ROO (1969a, b), BEGG and TURNER (1970), DETLING and KLIKOFF (1971), DUNIWAY (1971), BOYER and GHORASHY (1971) CARR (1972), SPOMER and LANGHANS (1972) and BLUM et al. (1973) found satisfactory correlation enabling reliable application of the method. KAUFMANN (1968b) as well as BARRS et al. (1970) and FRANK and HARRIS (1973) recommend drawing calibration curves for each species and age of the plant.

1.3.8 Determination of Freezing Point of Tissues

Cryoscopy of aqueous solutions used for measuring the osmotic potential of saps in plants is described in detail in Section 1.4.2.2. Freezing point of the tissues may theoretically also be used for estimating the free energy of water in the tissues (cf. e.g. MARSHALL 1961, ABELE 1963 and subsequent authors). The main practical difficulty in using cryoscopy for whole plant tissues is the establishment of a steady state equlibrium between the heat loss due to external cooling and the latent heat gain during freezing of water in the tissue, where heat transfer cannot be facilitated by mixing as it is in cryoscopy of solutions.

Nevertheless, attempts have been made to suggest procedures for estimating freezing point of the tissues, in particular of leaf tissues, by CARY and FISHER (1969, 1971) and FISHER (1972).

Recent development of solid state cooling devices enabled CARRY and FISHER (1969, 1971) to design a portable device for measuring the freezing-point depression in living plant samples. Two Wheatstone bridges were used in the electronic scheme, one containing the two thermistors which are in close contact with the sample to be measured (leaf segment 4 to 5 cm^2 folded several times), the other included the thermistor used to monitor the temperature of the cold block. The cooling block (using a Peltier cell) contained a copper tube chamber 19 mm in diameter for the sample pack which was inserted between the thermistors. During cooling the temperature plateau of the sample was measured; this occurred after ice inoculation, taken as the freezing point. Calibration was made by KCl and sucrose solutions soaked up on a piece of filter paper inserted between the thermistors. The authors found the average difference between results obtained by this freezing-point method and measurements by A type thermocouple psychrometers to be 2.6 bar in several crop plants.

OGURA (1971) used a laboratory device with double cooling for quick and for slow cooling. He transferred the sample from the first to the second slowly cooling chamber in order to get a more pronounced and stable freezing-point plateau. This is otherwise strongly dependent on sample volume and structure, which caused different heat exchange equilibrium in different samples and in solution samples used for calibration. OGURA used the same device for measuring osmotic potential of expressed cell sap.

1.4 Determination of Osmotic Potential

Osmotic potential of the vacuolar sap of cells or tissues may be determined basically in two ways: *in situ* in the cells, by the so-called cell methods (Section 1.4.1), or by first obtaining the sap from the tissues and then determining its osmotic potential by a direct or indirect method (Section 1.4.2).

A summary Table 1.12 of methods for the determination of osmotic potential may be found on pages 106 and 107 and used for preliminary selection of the method suitable for the purpose.

1.4.1 Cell Methods

Both cell methods can be used for highly vacuolated cells in tissues which can be isolated *in vivo* or be observed in isolation under the microscope (*e.g.* leaf epidermis), and also for algae.

1.4.1.1 Method of Limiting Plasmolysis

The method of limiting plasmolysis involves finding the solution in a series of previously prepared test solutions (of a solute for which the semipermeable plasmatic membranes of the cell have a selectivity coefficient (reflection coefficient σ near to 1) which brings about incipient plasmolysis (see page 6) in about half the cells of a living tissue, *e.g.* the epidermis. At limiting plasmolysis the point of zero turgor pressure (zero pressure potential) has been reached. The osmotic potential of the cell vacuolar sap is then equal to that of the solution. This is the osmotic potential at limiting plasmolysis ($\Psi_{s(1p)}$), which is as much higher than the original normal osmotic potential ($\Psi_{s(n)}$), as the volume at incipient plasmolysis is lower than the original normal volume:

$$\Psi_{s(n)} = \Psi_{s(1p)} \frac{V_{(1p)}}{V_{(n)}} . \tag{1.11}$$

[This relationship holds for simplified conditions and non-electrolyte test solutions (BRIGGS 1957)]. To calculate $\Psi_{s(n)}$ (see Table 1.6), it is therefore necessary to find the ratio of the cell volumes before and after plasmolysis ($V_{(1p)}/V_{(n)}$). As it is generally impossible to measure cell volume, measurement is simplified by measuring only two dimensions of rectangular cells. The calculation is then based on the assumption that the third dimension changes in the same way as the measured dimension nearest in size.

$$\frac{V_{(1p)}}{V_{(n)}} = \frac{l_{(1p)}}{l_{(n)}} \frac{w^2_{(1p)}}{w^2_{(n)}} , \tag{1.12}$$

where l and w are length and width, respectively. If the cell walls of turgid cells are only slightly expandable, the difference between $V_{(n)}$ and $V_{(1p)}$ and thus between $\Psi_{s(n)}$ and $\Psi_{s(1p)}$ is only small.

Plasmolysis is an artifact which does not occur normally in natural conditions and there is some evidence that it may partially damage the cell.

The epidermis is stripped from a leaf, or a section of living tissue (one to three cell layers thick) cut with a razor blade or hand microtome under a drop of liquid paraffin to prevent evaporation. A set of parallel samples are placed in a series of test solutions in wide-mouth bottles arranged in order of decreasing osmotic potential. Preliminary measurements determine the required period of exposure, the so-called plasmolytic time. This depends on various factors, *e.g.* the permeability of the cytoplasm to water. It may be from five minutes to several hours, depending on the material. The time is determined most rapidly by studying plasmolysis of the experimental tissue in a very hypertonic solution. It is then certain that plasmolysis will occur, and the degree of plasmolysis is determined after suitable times (10, 20, 40, 80 min). During the measurement proper the tissue is first left in test solutions for the appropriate time (usually for 10 to 30 min) and then transferred in a drop of the solution to a microscope slide. The occurrence of plasmolysed cells is then determined under the microscope. If there is plasmolysis in some cells, about 100 cells are examined and the number of cells in which plasmolysis has just started is counted. The point at which the protoplast just separates at one place from the cell-wall is called limiting (incipient) plasmolysis. Plasmolysis might not be detected in cells with small vacuoles, and the plasmolytic method cannot therefore be used.

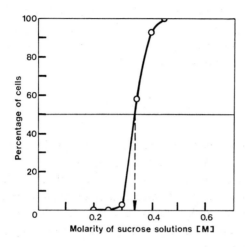

Fig. 1.34 Percentage of cells exhibiting limiting plasmolysis (ordinate) in tissues immersed in solutions of sucrose of varying molarity (abscissa). The point at which just 50% of cells exhibited limiting plasmolysis corresponds to 0.34M sucrose solution, *i.e.* an osmotic potential of −9.4 bar.

The samples in all the test solutions are then examined in the same way. The time of exposure in the solutions does not affect the results, provided that it is longer than the time required for the plasmolysis. The results are plotted in terms of the relationship between the percentage of clearly plasmolysed cells and molarity (osmotic pressure or osmotic potential) of the test solutions. It can be seen in Fig. 1.34, that in this example the concentration of sucrose solution in which 50 per cent of the cells were in a state of limiting plasmolysis was 0.34M, which corresponds to an osmotic potential of −9.41 bar at 20 °C (see Table 1.7, page 21).

The accuracy of the measurement depends to some extent on how closely the test solutions are graded, although the result may be obtained from interpolation as in Fig. 1.3. Because of the adherence of protoplasm to the cell wall the results may differ by several bars from those obtained by the method of so-called limiting deplasmolysis, in which the solution is found in which limiting deplasmolysis of a completely plasmolysed tissue occurs. The $\Psi_{s(lp)}$ values are lower (more negative) than $\Psi_{(deplasm)}$.

1.4.1.2 Plasmometric Method

The plasmometric method worked out by HÖFLER (1917, 1918a) may only be used in some cases, where the volume of the protoplast and of the cell can be calculated from the measurements made under the microscope. The volume of the protoplast (V_{prot}) and the volume of the whole plasmolyzed cell (V_{cell}) is measured and calculated in the measured cell after osmotic equilibrium between the cell and a known test solution has been reached. The osmotic potential is then expressed by the equation

$$\Psi_{s(lp)} = \Psi_{s(sol)} \frac{V_{prot}}{V_{cell}}.$$ (1.13)

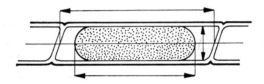

Fig. 1.35 A regularly shaped plasmolysed cell which may be used to calculate the ratio V_{prot}/V_{cell} during plasmometric determination of osmotic potential.

The tissue is left in a hypertonic solution long enough for complete equilibrium between the cells and the solution to be reached. The tissue is then taken out of the solution and suitable cells are found under the microscope. The volume of cells and protoplasts can only be measured with sufficient accuracy if the cells have a more or less regular geometric shape and if very regular convex plasmolysis occurs (see Fig. 1.35). An eye-piece micrometer is used to measure the cell dimensions required for sufficiently accurate calculation of cell volumes, *i.e.* length and diameter of cylindrical cells, radius of spherical surfaces, length of protoplasts *etc*. Dimensions common to the protoplast and to the whole cell may be used to calculate V_{prot}/V_{cell} if the required dimensions cannot be determined in some cells. It is possible to use the areas of vertical projections instead of the volume ratio, as in the planimetric method of PRÁT (1922) and TOTH and FELFÖLDY (1955). A similar simplified procedure was suggested by HÄRTEL (1963). The outlines of several cells and of their plasmolysed protoplasts are drawn; the areas are measured with a planimeter and used to calculate V_{prot}/V_{cell}.

The osmotic potential of the hypertonic solution used is then multiplied by V_{prot}/V_{cell} to obtain $\Psi_{s(lp)}$, as the volume of the plasmolysed cell (not the protoplast volume) is the volume at zero turgor.

The plasmometric method is extremely valuable for suitable cells (but there are few such cells), as a single measurement in one plasmolytic solution is sufficient. It can be used to determine the osmotic potential of a single cell, whereas the method of limiting plasmolysis can be used only to determine the average value of the whole homogeneous tissue. Only one tissue and one hypertonic solution are required. However, there must again be minimum permeability of the semipermeable cytoplasmic membranes to the solute used for the test solution (reflexion coefficient σ near to one). Vacuoles must be large in comparison with the volume of the protoplast. There must be no osmoregulation, *i.e.* no physiological metabolic changes in the osmotic potential of the vacuolar solution under the influence of the external test solution. The accuracy of the plasmometric method is greater than that of the method of limiting plasmolysis, provided that direct size measurement and calculation of volumes is possible.

1.4.1.3 Zero Turgor Methods

At zero turgor (zero pressure potential) the water potential of the cell (tissue) is equal to the osmotic potential of the cell sap $\Psi_{s(lp)}$). Limiting plasmolysis is not the only as indication of zero turgor. This point can be also detected by measuring changes in elasticity of cells immersed in different test solutions.

For the detection of the point of zero turgor (zero pressure potential) of cells equilibrated in an isotonic solution TAZAWA (1957) used a sensitive "turgor balance". This chain balance exerts a measured pressure on the cells in different solutions. TAZAWA extrapolated the relation between the pressure necessary to maintain the same cell deformation (*i.e.* the same position of a pointer mounted on the arm in contact with the cells), and the osmotic potential of the solution, to the point of zero pressure (zero weight on the balance). The osmotic potential found by this procedure corresponds to the point of zero turgor even when there is no visible plasmolysis.

The resonance method (Section 1.5.4) suggested for the estimation of pressure potential of tissues may also be used for the detection of the point of zero turgor; and so may the pressure chamber technique (Section 1.3.7).

1.4.1.4 Visual Melting Point Method

BEARCE and KOHL (1970) measured osmotic potential of vacuolar sap *in situ* by measuring the temperature at which thin slices of tissue (*e.g.* epidermis) previously rapidly frozen thawed under a polarizing microscope.

1.4.2 Methods Used for Determination of the Osmotic Potential of Expressed Cell Sap

In most cases the cell sap can be released from tissues by pressure after the semipermeability of the cytoplasmic membranes has been destroyed. The expressed cell sap is assumed to have the same osmotic potential as the vacuolar solution which was *in situ* in a dynamic equilibrium with the cytoplasmic structures, all of them being under the same cell wall pressure.

However, some measurements seem to indicate that some of the osmotically active cations remain bound to immobile anions in the cytoplasm, so that the measured osmotic potential of the expressed cellular sap is slightly higher (the osmotic pressure slightly lower) than the osmotic potential of the vacuolar sap in the cell. During the pressing, the cell sap may be contaminated by other solutions in the tissue (*e.g.* xylem and phloem saps) and filtered by passing through the tissue structures.

In addition, these methods are suitable for accurate measurement of the osmotic potential of sap released *e.g.* by root bleeding, free solution from plant bodies (*e.g.* coconut milk, the content of sieve tubes and xylem vessels *etc.*).

The advantage of these methods is that they give $\Psi_{s(n)}$ values for cells in a normal state of turgor. The average values for whole tissues or organs are obtained.

1.4.2.1 Preparation of Cell Sap

The determination of osmotic pressure of the expressed cell sap is so accurate that the source of the most important errors is the process of preparation of the cell sap from a plant sample.

Only in special cases can the sap be obtained directly by a puncture technique, *e.g.* by means of a microsyringe or micropipette from *Nitella* and *Vallonia* cells or from *Amoeba*. Phloem sap can be obtained in relatively large quantities by the aphid-stylet technique. KENNEDY and MITTLER (1953) and MITTLER (1958) developed this procedure into a valuable method. A rapidly feeding aphid (*i.e.* an aphid with a good honeydew production) is anaesthetized with a gentle stream of CO_2 and severed from its mouth part, the style remaining inserted in a single phloem sieve tube. Exudation continues undisturbed, frequently for hours, probably with unchanged velocity during the first time period. The velocity was found to be up to 5 μl per hour.

The tissue has to be fixed (killed) in order to destroy the semipermeability of the living cytoplasmic membranes, which would prevent release of many osmotically active compounds from the vacuoles of living tissues subjected to pressure.

Fixation by low temperature is advantageous, as there are then no chemical changes such as hydrolysis or condensation. However, changes resulting from expression of the sap may occur if samples are stored for a time. These changes may affect the quantitative composition of the cell sap more than those caused by high temperature. High temperature causes irreversible denaturation of proteins and consequent inactivation of enzymes. The fixed sample can thus be maintained at a low temperature for a very long time and then subjected to pressure at a normal temperature without change in properties, providing that microbial contamination is prevented. Fixation should be as rapid as possible. Killing by low temperature occurs mostly during thawing of the samples. A vessel containing the sample can be immersed in liquid nitrogen (temperature − 197.7 °C) several times and then left to thaw. Cooling to −20 °C is sufficient for some material (tubers), so that a deep freeze may be used. The tubes containing the samples can be placed in crushed solid carbon dioxide (temperature −78.5 °C). The frozen sample is pressed as quickly as possible after thawing, or stored in a deep freeze until required.

The use of high temperatures is simpler (WALTER 1928, 1931a, b). Closed vessels containing the samples are immersed in a boiling water bath. The resulting pressure inside the closed vessels might burst them, so vessels with a screw cap or a similar rigid closure or glass vessels with rubber stoppers, inside aluminium screw-cap vessels are used. Another useful procedure is to use glass tubes (size 25 × 70 mm) closed with rubber stoppers. The stoppers are replaced prior to boiling, by stoppers pierced by syringe needles. The needles allow equilibration of the pressures inside and outside the tubes, while the loss of vapour from the samples (through the syringes) is negligible. It is advantageous to place the samples in a wire basket (in marked positions) immersed in a boiling water bath in such a way that the contents of the tubes are completely under water. The boiling time is 10 to 20 min, depending on the material: the samples must not be boiled too long as mash-like solutions would be obtained on pressing.

KREEB (1965a) drew attention to the fact that the samples need not actually be boiled to kill them but that 15 min at 90 °C is sufficient. He collects the material directly in 20cm^3 all-glass syringes (Fig. 1.38 A), the needle tips of which are closed from the outside by rubber stoppers. The material is then killed simply by placing the syringes for 15 min in a 90 °C water bath.

Plant material containing high concentrations of easily hydrolysable glucosides cannot be fixed by boiling. Thus in *Laurocerasus officinalis* ROEM., the leaves of which contain prunulaurasine, the osmotic potentials of sap from boiled material are lower by 10 or even 20% than those of sap from material fixed by low temperature (WALTER 1963). Mash-like samples, the osmotic potential of which is directly proportional to the boiling time (KOZINKA 1960), are obtained from leaves of apricot trees (*Prunus armeniaca* L.) even after boiling for only a short time. We have never been able to press the required amount of cell sap from lime tree leaves (*Tilia cordata* HILL, and *Tilia platyphyllos* SCOP.), because of the presence of foaming substances.

Samples may also be fixed by poisonous vapours. This method was found to be suitable and very effective, though inactivation of enzymes cannot be excluded. Chloroform vapour, which rapidly destroys the semipermeability of a tissue, is often used. The loss of semipermeability is so rapid that the cell sap runs out of the tissue even before the application of pressure (NĚMEC 1899, KOZINKA 1960). KOZINKA (1960) used saturated chloroform vapour in a desiccator, or added small quantities (0.5−4 ml) of liquid chloroform immediately after sampling, to 150 ml tubes containing the samples. A correction of $\Delta T = 0.102$ °C, based on the maximum solubility of chloroform in water (7.10 g per 1 litre at 20 °C), should be deducted from the cryoscopically determined melting point when using samples fixed with chloroform (WALTER 1931b). However, as found by KOZINKA (1960), a solution of chloroform in water is not cryoscopically analogous with an aqueous solution of a solid substance, as the chloroform vapour pressure has some effect.

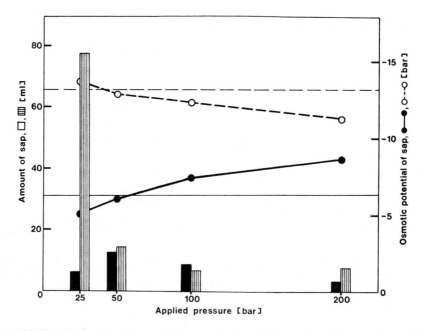

Fig. 1.36 The relationship between the amount of sap (columns) and its osmotic potential (lines) [ordinate] and increasing pressure during pressing [abscissa] in killed (dotted line and dashed columns) and living (full line and black colums) leaves of sunflower (*Helianthus annuus*). Horizontal lines represent the mean osmotic potentials of cell sap at a pressure of 200 atm. (WALTER 1963).

It is therefore necessary to take values for the depression of freezing point of distilled water exposed to saturated chloroform vapour under the same temperature conditions for the same period of time as the tissue sample when correcting the depression of freezing point. This correction is often considerably larger after prolonged chloroform treatment than that recommended by WALTER (1931b). A procedure based on adding 0.5 ml chloroform to 10 -- 15 g fresh weight of the samples and keeping them closed for 24 hours at 0 °C is also useful (KOZINKA 1960). It was found that ΔT differences were small, even smaller than the theoretical value of 0.102° C in most samples.

An oil hydraulic press should be used for pressing. It is advisable to check the pressure, as the same pressure can then always be used. WALTER (1931b) recommends a standard pressure of 100 bar. Fig. 1.36 shows the relationship between the amount and osmotic potential of individual fractions of the cell sap from fixed and non-fixed samples of sunflower leaves (*Helianthus annuus* L.) expressed by gradually increasing pressure. KREEB (1965a) found similar results with leaves of various *Eucalyptus* species and of *Ilex aquifolium*. The osmotic potential of the sap expressed by pressures in the range of 10 to 100 bar changed only very little. Fig. 1.36 also shows that there is a difference between the osmotic potentials of cell sap samples pressed from fixed and fresh tissues. It is apparent that the results depend on the pressure used (WALTER 1963). In addition, it is evident that more

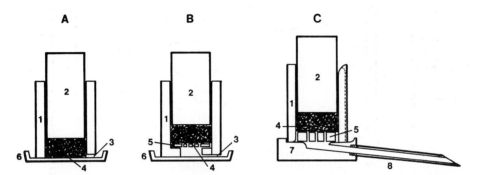

Fig. 1.37 Three types of pressure devices used to press sap from tissues. 1 — steel cylinder, 2 — full piston, 3 — outlet slot in the mantle, 4 — killed tissue, 5 — perforated detachable bottom, 6 — flat tin dish used to collect the sap, 7 — stand with the outlet tube (8). A and B: SLAVÍK (1954), C: CURRIER (1944) as modified by BROYER and FURNSTAL (1941).

than 76% of the sap from the killed tissue was obtained using pressures of less than 25 bar and that its osmotic potential differed from the mean for the cell sap obtained using 200 bar by only 2.8%.

Medium to large samples (10 – 50 g fresh weight) are put into special pressing devices, some types of which are shown in Fig. 1.37 (BROYER and FURNSTAL 1941, SLAVÍK 1954). Other types of press have been described by SCHEUMANN (1964, for small samples), SUSPLUGAS *et al.* (1965), SREENIVASAYA and SRIMATHI (1967). They enable sap to be obtained in sufficient quantity and of good purity. The fixed tissue is placed in the press as it is or wrapped in a piece of fine, closely woven nylon cloth to prevent large amounts of undesirable tissue fragments in the sap.

Very small quantities of cell sap can be obtained from small samples by fixing the samples in glass tubes (about 10 × 45 mm) closed with rubber stoppers pierced with syringe needles and then pressing out the sap with hand pincers (of the type usually supplied with a hand refractometer set used to determine the refractive index in the food industry).

KREEB (1965b) recommends the use of a metal hypodermic syringe (20 cm³) as a press (Fig. 1.38 B). The tip is soldered and the piston of the syringe used for pressing (*e.g.* by means of a hand press) until no more sap is expressed. The expressed sap flows up between the plunger and the syringe wall and accumulates in the hollow plunger.

The expressed sap is most easily stored in the original sample tubes. Concentration change resulting from evaporation should be avoided during transfer and further manipulation. The sap may be stored for a relatively long period in a deep freeze, the time of storage depending on the material.

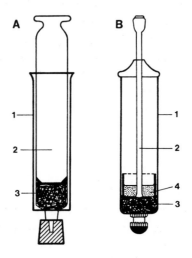

Fig. 1.38 Hypodermic syringes used both for collecting small samples and for killing and pressing the samples collected (KREEB 1965b). A — glass syringe, B — metal syringe. 1 — cylinder, 2 — piston, 3 — plant material, 4 — pressed cell sap.

If it is impossible to treat the samples immediately, the following procedure, tested and described by WALTER, may be used as a last resort. The weight of the fresh samples is determined. The samples are then air-dried as quickly as possible and are stored for the required period in the air-dry state. The sample is crushed before use and distilled water is added to make up the original fresh weight. The samples are enclosed in vessels and shaken at a high temperature for 24 hours, so that osmotically active compounds come into solution. The sample is then pressed in the usual way. According to THREN (1934), the osmotic potential of the sap obtained in this way is not very different from the osmotic potential of the sap from the fresh fixed tissue.

A sample of normal size may not always yield sufficient sap to determine the osmotic potential. The EHLIG procedure (1962) may then be used, the osmotic potential of a fixed sample being determined by the psychrometric method, normally used for estimation of the water potential (see Section 1.3.4). It is necessary to remove any salt exudates from the surface of the plant before fixation. However, this is rather difficult.

There is still discussion about the extent to which the measured osmotic potential of the expressed cell sap corresponds to the osmotic potential of the vacuolar sap *in situ* in the cells. There are differences in opinions about the cellular origin of the so-called cell sap which is expressed. It is evident that the major part comes from the vacuoles. It remains to be seen to what extent this portion can be altered by possible filtration by fixed cytoplasmic membranes during expression. In addition, the expressed cell sap certainly contains solutions from conducting tissues, vessels and sieve tubes, the osmotic potential of which may of course be different from that of the vacuolar sap. It also contains the solutions normally present in the free space *i.e.* in cell walls and occasionally in those parts of protoplasts located outside semipermeable barriers. It must be kept in mind that these barriers are not completely impermeable to some osmotically active substances, even *in vivo*.

The xylem sap from herbaceous or woody plant shoots, or from leaves, can be extracted using the pressure chamber technique (Section 1.3.6). Sap from the xylem of tree trunks can be obtained by centrifugation (LEVITT 1964) or by suction, using a pump (SCHOLANDER *et al*. 1966).

1.4.2.2 Cryoscopy

Cryoscopy is a very accurate physical method of measuring the freezing point of solutions. According to Raoult's law, solutions have a lower freezing point than the pure solvent (pure water). The decrease ΔT is caused by decrease of the relative vapour pressure of the solvent $(p^0 - p)/p^0$ where p^0 is vapour pressure of the pure solvent, p vapour pressure of the solution. The depression of freezing point ΔT is also proportional to the concentration of a dissolved solute

$$\Delta T = Kic,$$ (1.14)

where K is the so-called cryoscopic constant ($K = 1.86$ for water), c is the molecular concentration of the dissolved solute, i is the degree of dissociation.

Osmotic potential

$$\Psi_s = \frac{RT}{\bar{V}_w} \ln \frac{p}{p^0},$$ (1.15)

where R is the gas constant (see Eq. 1.7), T absolute temperature, p/p^0 is the relative water vapour pressure of the solution and $\bar{V}_w$ is the partial molal volume of water. The following equation may be written:

$$\Psi_s = 0.021(\Delta T)^2 - 12.06 \, \Delta T.$$ (1.16)

Various modifications of cryoscopy for small (1.5 ml) to negligible (10^{-6} ml) amounts of sap have been worked out. The accuracy of cryoscopic determination depends on the technique (stirring of the sample, degree of supercooling) and the

accuracy of temperature measurement. Most modifications of cryoscopy are sufficiently accurate, the accuracy exceeding differences caused by heterogeneity of
the material. A relatively simple device is needed to measure samples of medium
size (1.5 ml and more).

A standard cryoscopic device consists basically of a vessel containing a freezing
mixture, *e.g.* a 0.5 to 0.75 l vacuum flask (Fig. 1.39) filled with a mixture of crushed
ice and a solution of sodium chloride, mixed so that its temperature is -7 to $-8\,°C$
(2). The shallow cork closing the flask has holes for a wire stirrer (brass) (3) for the

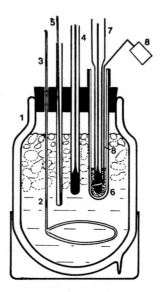

Fig. 1.39 A schematic section of the cryoscope (SLAVÍK 1954).
See the description in the text.

freezing mixture and for a thermometer (4), preferably with 0.1 °C graduation. Another hole is used to insert a narrow tube sealed at its lower end for freezing inoculation capillaries (5), and there is a further hole for placing the vessel which contains
the cell sap (6), the mercury bulb of the thermometer (7) and a stirrer for the sap,
made from a light wire loaded with a rubber stopper (8). The dimensions of this
vessel are chosen according to the size of the mercury bulb of the cryoscopic thermometer, in such a way that the amount of sap is as small as possible and the stirrer
mixes the whole sap column well. The sap should reach about 3 mm above the top
of the mercury bulb of the thermometer. About 1.5 ml are usually used. The thermometer is usually attached to the vessel by a collar which also holds the stirrer,
driven by a motor, the speed of which can be regulated. A vertical lever is eccentri-
cally located on its axle. The stirrer can of course be hand-operated, but a motor
gives a uniform speed of stirring (about two revolutions per second) and is easier to
work with (Fig. 1.40).

Accurate physical determination is carried out by first cooling the vessel by immersion in a $\pm0.5\,°C$ bath. The vessel is then taken out, dried on the outside, and rapidly put into a large tube immersed in the cryoscope cooling bath, so that further cooling takes place in a so-called air jacket. This cooling is substantially slower than direct cooling, but the air jacket ensures that the freezing point is not

Fig. 1.40 A picture of the complete cryoscopic device with a motor driving the stirrer.

decreased as a result of direct contact of the vessel with the freezing mixture when the temperature increases after the beginning of crystallization of ice (see below). On the other hand, the necessary transfer of the sample vessel to the air jacket causes difficulties and the supercooling in the jacket is very time-consuming.

As an accuracy of determination of the freezing point depression ΔT of 0.01 °C (corresponding to 0.12 bar) is sufficient for biological purposes, a modification has been suggested in which the vessel is immersed directly in a stirred cooling mixture. SLAVÍK (1952) found that the spread of individual determinations decreases with increasing temperature of the freezing mixture, so that it is already less than 0.01 °C at −8 °C (Fig. 1.41). This temperature of the cooling mixture still allows sufficiently rapid supercooling, which would otherwise be the slowest stage of the determination. CURRIER (1944b), who used the procedure with an air jacket, as well as WALTER (1931b), drew attention to the fact that any device should be calibrated, as the ΔT values obtained are in all cases complicated by errors associated

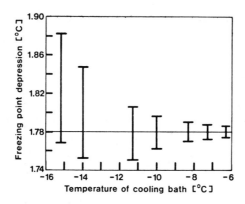

Fig. 1.41 An example of the relationship between scattering freezing point depression ΔT values and temperature of the cooling bath using cryoscopy without air jacket (SLAVÍK 1952).

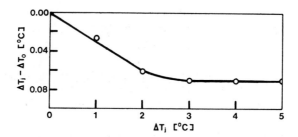

Fig. 1.42 The relationship between differences of freezing point depression ΔT using cryoscopy with (ΔT_j) and without the air jacket (ΔT_0) and the absolute magnitude of ΔT_j found by cryoscopy with the jacket (KREEB 1958).

with supercooling (see below) and release of heat during freezing. Standardization of all instruments, by measuring solutions of known molarities and comparison of the measured ΔT values with tabulated for correction, is therefore recommended. However, this procedure is still not perfect as it is not easy to make up solutions of accurate molarities. In fact, the accuracy of molar solutions is often checked cryoscopically. Analytically pure substances are to be preferred for the most accurate preparation of solutions of electrolytes. KREEB (1957) found the correction for cryoscopy without the air jacket empirically as suggested by SLAVÍK (1952). The difference between the values obtained with and without the jacket naturally depends on the absolute magnitude of ΔT, as both are dependent on the difference between the temperature of the cooling bath and the freezing point of the solution (sap). As the difference also depends on the heat capacity of the vessel containing the solution, Fig. 1.42 (according to KREEB, 1958) is shown only as an example. It is necessary to calibrate any instrument modification individually.

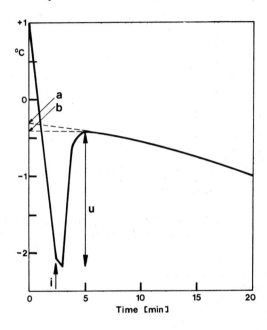

Fig. 1.43 Example of the time course of the temperature of stirred sap during cryoscopy. See the text. i — point of inoculation with an ice crystal, a — a real (extrapolated) value of freezing point, b — read freezing point, u — supercooling.

Fig. 1.43 shows the time course of the state of the sample of sap during cryoscopy. The solution began to freeze at point (i) as the freezing was started (seeded) by ice crystals formed in the capillary immersed into the solution at that point of time. The difference between the lowest temperature and the freezing point represents the degree of supercooling (u). The highest temperature reached subsequently (b) is slightly lower than the extrapolated temperature (a) which is the real measured freezing point (melting point), although (b) can in practice be taken as the freezing point. The difference between (a) or (b) and the freezing point of pure ice (pure water) is the required depression of freezing point ΔT. Supercooling of 1 °C is always used in order to make it possible to use Tables 1.10 and 1.11. During the first measurement the sample is supercooled to about ± 1 °C below the assumed freezing point. During the second measurement, seeding inoculation is at the time when the temperature has decreased to 1 °C below the freezing point indicated by the first pilot measurement.

Cryoscopic thermometer. The size of the bulb of a mercury thermometer determines the necessary sample volume. The smaller the bulb, the smaller the amount of the sap or solution required. The thermometer should have 1/50 °C graduations in the range of about +0.2 to −4 or −5 °C. A Beckmann adjustable thermometer may also be used. However, this has too large a mercury bulb, so that 4 ml or more of the sample sap are needed. Thermometers with 0.1 °C graduations can be used for less accurate estimations.

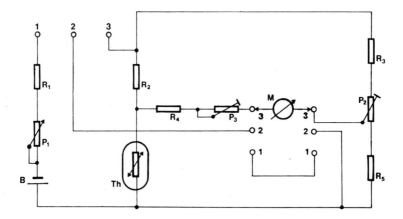

Fig. 1.44 A diagram of the circuit of the thermistor used for cryoscopy (ZEMÁNEK and VEPŘEK 1963 unpublished). $R_1 = 10$ kΩ, $R_2 = 16$ kΩ, $R_3 = 16$ kΩ, $R_4 = 33$ kΩ, $R_5 = 20$ kΩ, P_1 and P_3 linear potentiometer 20 kΩ, P_2 linear potentiometer 5 kΩ, Th — thermistor 12 NR 15, B — battery, M — meter

Thermistors, which are of negligible size, may be useful. The thermistor cryoscope used by ZEMÁNEK and VEPŘEK (1963, unpublished) is shown in Fig. 1.44. A thermistor with a resistance R_{20}, 8.6 kΩ serves as sensor. A symmetrical bridge supplied by a 4.5 V battery source is used. A thermistor current of 50 µA is set by a potentiometer P_1. The bridge is equilibrated at 0 °C by the potentiometer P_2. This sets the zero value of the measuring range at 0 °C. The other end of the range is set by the potentiometer P_3, at −5 °C in this instrument. This range corresponds to about 100 scale divisions of the mirror galvanometer. The bridge is furnished with a tripole, three position switch, which switches the instrument off when in position (1) and simultaneously short-circuits the galvanometer. The thermistor current is set by the potentiometer P_1 in position (2) to 10 scale divisions. The measurement proper is carried out in position (2) with full sensitivity which makes it possible to read the temperature with an accuracy of 0.025 °C. The necessary volume of sap or solution for the thermistor cryoscope is 0.2 to 0.3 ml. The sample volume cannot be any less, although the dimensions of the thermistor detector would permit it to be, as the error caused by cooling of the small vessel immersed directly into the freezing mixture would be considerable. It would be necessary to use an air jacket, which would prolong the measurement. The position of the thermistor immersed in a sample of sap or solution (which is usually in a measuring

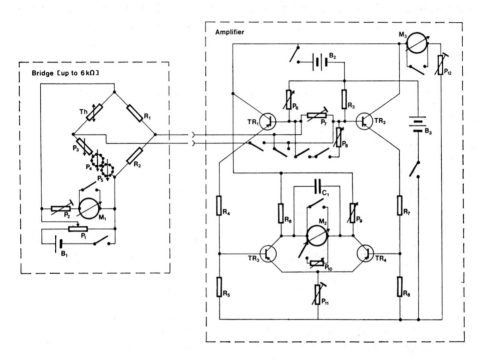

Fig. 1.45 A diagram of the bridge and amplifier used for microcryoscopic measurements (KREEB 1965b).
Bridge: Th — thermistor $1-3$ kΩ, $R_1 = 300\ \Omega$, $R_2 = 50\ \Omega$, $P_1 = 100\ \Omega$, $P_2 = 5$ kΩ, $P_3 = 10\ \Omega$, $P_4 = 10 \times 10\ \Omega$, $P_5 = 10 \times 100\ \Omega$, B_1 — battery 1.5 V, M_1 — microampermeter.
Amplifier: $R_3 = 10$ kΩ, $R_4 = 100\ \Omega$, $R_5 = 10$ kΩ, $R_6 = 10$ kΩ, $R_7 = 100\ \Omega$, $R_8 = 10$ kΩ, $P_6 = 10$ kΩ, $P_7 = 10$ kΩ, $P_8 = 50$ kΩ, $P_9 = 10$ kΩ, $P_{10} = 50$ kΩ, $P_{11} = 10$ kΩ, $P_{12} = 25$ kΩ, TR_{1-4}: transistors OC 304/2, $C_1 = 100\ \mu$F, $M_2 = 50\ \mu$A, M_3 — milliammeter, B_2 and B_3 — batteries 4.5 V.

tube narrowed at its bottom) is important, primarily because the distance from the walls must be kept constant during all measurements and during calibration.

KREEB (1965c) described a field thermistor cryoscope with a battery-driven transistorized amplifier of small dimensions. The plan of the device is shown in Fig. 1.45. For constructional details see the original paper. The author uses a $-8\ °$C bath, supercools without an air jacket first to $-4\ °$C and then to 1 °C below the value of the freezing point found in preliminary experiments and inoculates with capillaries or syringe needles of frozen water. His procedure is similar to that with a mercury thermometer, which will be described later, except that the sap (0.15 to 0.3 cm^2) is not stirred. The accuracy of the measurement is ± 0.5 bar, each measurement takes about 2 minutes.

Use of thermocouples, *e.g.* copper/constantan, needs precise solid state micro-

Table 1.10. *Osmotic potential* ($-$bar) *at the freezing point temperature corresponding to the freezing point depression* (ΔT, °C) *within the range 0 to* -5.99°C. (*Compiled from* HARRIS *and* GORTNER *1914 and* HARRIS *1915 by* WALTER *1931b and converted to bars.*) *All values of water potential are negative!*

	0	1	2	3	4	5	6	7	8	9
0.0	0.000	0.123	0.244	0.367	0.488	0.611	0.733	0.855	0.978	1.099
0.1	1.222	1.344	1.466	1.588	1.710	1.833	1.955	2.077	2.199	2.321
0.2	2.443	2.565	2.686	2.808	2.931	3.053	3.175	3.297	3.419	3.541
0.3	3.663	3.786	3.907	4.030	4.151	4.274	4.395	4.517	4.640	4.761
0.4	4.884	5.005	5.128	5.249	5.371	5.493	5.615	5.738	5.859	5.981
0.5	6.103	6.225	6.347	6.469	6.591	6.713	6.835	6.956	7.079	7.200
0.6	7.323	7.445	7.566	7.689	7.810	7.932	8.054	8.176	8.297	8.420
0.7	8.542	8.663	8.785	8.907	9.029	9.150	9.273	9.395	9.516	9.638
0.8	9.760	9.882	10.00	10.12	10.25	10.37	10.49	10.62	10.74	10.86
0.9	10.98	11.10	11.22	11.35	11.47	11.59	11.71	11.83	11.95	12.07
1.0	12.20	12.32	12.44	12.56	12.68	12.80	12.93	13.05	13.17	13.29
1.1	13.41	13.53	13.66	13.78	13.90	14.02	14.14	14.26	14.38	14.51
1.2	14.63	14.75	14.87	14.99	15.11	15.24	15.36	15.48	15.60	15.72
1.3	15.84	15.96	16.09	16.21	16.33	16.45	16.57	16.69	16.82	16.94
1.4	17.06	17.18	17.30	17.42	17.55	17.67	17.79	17.91	18.03	18.15
1.5	18.27	18.40	18.52	18.64	18.76	18.88	19.00	19.13	19.25	19.37
1.6	19.49	19.61	19.73	19.85	19.98	20.10	20.22	20.34	20.46	20.58
1.7	20.71	20.83	20.95	21.07	21.19	21.31	21.44	21.56	21.68	21.80
1.8	21.92	22.04	22.16	22.29	22.41	22.53	22.65	22.77	22.89	23.02
1.9	23.14	23.26	23.38	23.50	23.62	23.74	23.87	23.99	24.11	24.23
2.0	24.35	24.47	24.60	24.72	24.84	24.95	25.07	25.19	25.31	25.44
2.1	25.56	25.68	25.80	25.92	26.04	26.17	26.29	26.41	26.53	26.65
2.2	26.77	26.90	27.02	27.14	27.26	27.38	27.50	27.62	27.75	27.87
2.3	27.99	28.11	28.23	28.35	28.48	28.60	28.71	28.83	28.95	29.07
2.4	29.19	29.32	29.44	29.56	29.68	29.80	29.92	30.05	30.17	30.29
2.5	30.41	30.53	30.65	30.77	30.90	31.02	31.14	31.26	31.38	31.49
2.6	31.62	31.74	31.86	31.98	32.10	32.22	32.35	32.47	32.59	32.71
2.7	32.83	32.95	33.07	33.20	33.32	33.43	33.56	33.68	33.79	33.92
2.8	34.04	34.16	34.28	34.40	34.52	34.64	34.76	34.88	35.01	35.13
2.9	35.24	35.36	35.49	35.61	35.73	35.85	35.97	36.09	36.21	36.34

voltmeters or a sensitive mirror galvanometer (with a low internal resistance and a very short deflection period). The measured e.m.f. is proportional to the difference in temperature between the cold junction kept at 0.00 °C (distilled water with ice in a vacuum flask) and the warm junction immersed in the sap in the vessel (DIXON 1911, DIXON and ATKINS 1910, HERRICK 1934, BODMAN and DAY 1943, YODA 1961).

A number of commercially available cryoscopes, with rapid and convenient

Table 1.10 *(continued)*

	0	1	2	3	4	5	6	7	8	9
3.0	36.46	36.58	36.70	36.82	36.94	37.07	37.19	37.31	37.43	37.54
3.1	37.66	37.78	37.91	38.03	38.15	38.27	38.39	38.51	38.64	38.76
3.2	38.88	39.00	39.12	39.23	39.36	39.48	39.60	39.72	39.84	39.96
3.3	40.08	40.21	40.33	40.45	40.57	40.69	40.80	40.93	41.05	41.17
3.4	41.29	41.41	41.53	41.65	41.78	41.90	42.02	42.14	42.25	42.37
3.5	42.50	42.62	42.74	42.86	42.98	43.10	43.22	43.35	43.47	43.58
3.6	43.70	43.82	43.94	44.07	44.19	44.31	44.43	44.55	44.67	44.79
3.7	44.91	45.03	45.15	45.27	45.39	45.51	45.64	45.76	45.88	46.00
3.8	46.11	46.23	46.35	46.48	46.60	46.72	46.84	46.96	47.08	47.21
3.9	47.32	47.44	47.56	47.68	47.80	47.93	48.05	48.17	48.29	48.41
4.0	48.52	48.64	48.77	48.89	49.01	49.13	49.25	49.37	49.50	49.61
4.1	49.73	49.85	49.97	50.09	50.21	50.34	50.46	50.58	50.69	50.81
4.2	50.93	51.06	51.18	51.30	51.42	51.54	51.65	51.77	51.90	52.02
4.3	52.14	52.26	52.38	52.50	52.62	52.74	52.86	52.98	53.10	53.22
4.4	53.34	53.47	53.58	53.70	53.82	53.94	54.06	54.19	54.31	54.43
4.5	54.54	54.66	54.78	54.90	55.03	55.15	55.27	55.39	55.50	55.62
4.6	55.75	55.87	55.99	56.11	56.23	56.34	56.46	56.59	56.71	56.83
4,7	56.95	57.07	57.19	57.31	57.43	57.55	57.67	57.79	57.91	58.03
4.8	58.15	58.27	58.39	58.51	58.63	58.75	58.88	58.99	59.11	59.23
4.9	59.35	59.47	59.59	59.72	59.83	59.95	60.07	60.19	60.31	60.44
5.0	60.56	60.67	60.79	60.91	61.03	61.15	61.28	61.39	61.51	61.63
5.1	61.75	61.87	62.00	62.12	62.23	62.35	62.47	62.59	62.71	62.84
5.2	62.95	63.07	63.19	63.31	63.43	63.56	63.67	63.79	63.91	64.03
5.3	64.15	64.27	64.39	64.51	64.63	64.75	64.87	64.99	65.11	65.23
5.4	65.35	65.47	65.59	65.71	65.82	65.95	66.07	66.19	66.31	66.43
5.5	66.54	66.67	66.79	66.91	67.03	67.15	67.26	67.38	67.51	67.63
5.6	67.75	67.87	67.98	68.10	68.23	68.35	68.47	68.59	68.70	68.82
5.7	68.94	69.07	69.19	69.30	69.42	69.54	69.66	69.79	69.91	70.02
5.8	70.14	70.26	70.38	70.50	70.62	70.74	70.86	70.98	71.10	71.21
5.9	71.34	71.46	71.58	71.70	71.82	71.93	72.05	72.18	72.30	72.42

operation, have appeared in recent years, *e.g.* Drucker - Burian - Kreeb type made by W. Schulze, Heppenheim a.d.B., West Germany, Advanced Osmometer made by Advanced Instruments Inc., Newton Highlands, Massachusetts, USA, Aminco - Bowman Cat. No 5-2050 made by American Instrument Co., Silver Spring, Maryland, USA.

The measuring procedure. The sap in the measuring tube is immersed directly in the freezing mixture at −7 to −8 °C and supercooled, while it is stirred continuously, to about 1 °C below the expected freezing point. Then a capillary containing frozen distilled water is inserted in the supercooled sap while the vessel is still in the mixture. The mixture is rapidly and continuously stirred. The temperature of the cryoscopic thermometer starts to rise immediately after freezing begins and goes up to temperature (b) (see Fig. 1.43). The temperature subsequently decreases slowly. The ice forms flakes or possibly a thin "slurry" in the solution. Measurements under conditions when a crust of ice forms are not reliable and must be repeated. The two values determined should differ by less than 0.01 °C, otherwise the measurement must be repeated. The vessel, thermometer and stirrer are carefully washed and dried after each measurement. The freezing point of distilled water is subtracted from the freezing point determined in this way. It is difficult to determine the freezing point of pure water (as it is hard to supercool water so that it does not freeze in

Table 1.11. *The correction of K_Δ for the supercooling by 1 °C for standard procedures used to determine freezing point depression (Δt) by microcryoscope (WALTER and THREN 1934, WALTER 1936) for various values of the freezing point depression (°C).*

°C	.00	.01	.02	.03	.04	.05	.06	.07	.08	.09
0.0	.000	.002	.004	.006	.008	.009	.011	.013	.015	.016
0.1	.018	.020	.021	.023	.025	.026	.028	.029	.030	.032
0.2	.033	.035	.036	.038	.039	.041	.042	.044	.045	.046
0.3	.047	.048	.050	.051	.052	.053	.054	.055	.056	.057
0.4	.058	.058	.059	.059	.060	.060	.061	.061	.062	.062

°C	.0	.1	.2	.3	.4	.5	.6	.7	.8	.9
0.						.063	.067	.071	.074	.078
1.	.082	.086	.090	.093	.097	.101	.105	.109	.112	.116
2.	.120	.124	.128	.132	.135	.139	.143	.147	.150	.154
3.	.158	.162	.166	.169	.173	.177	.181	.185	.188	.192
4.	.196	.200	.204	.207	.211	.215	.219	.223	.227	.230
5.	.234	.238	.242	.246	.249	.253	.257	.260	.264	.268

the form of an ice crust), so accurate NaCl solutions may be used instead. Their freezing points can be found in tables (Table 1.8 on page 34 for NaCl).

The calculation of the osmotic potential based on ΔT. Tables 1.10 and 1.11 serve for the calculation. Table 1.10 (according to WALTER 1931b) gives osmotic potential (in bar) at the freezing point of the solution. Table 1.11 shows the correction (K_Δ) for supercooling by 1 °C (WALTER 1936). These corections were calculated according to empirical data for 1.5 ml of solution and a Burian - Drucker cryoscope with an air jacket, and are to be subtracted from the freezing point depression ($K_\Delta = 0.038 \, \Delta T' + 0.044 \, [°C]$). The real value of the freezing point depression ΔT at various supercooling temperatures can be calculated from the measured value $\Delta T'$, according to the Eq. 1.17:

$$\Delta T = 0.968 \, \Delta T' - 0.044s , \qquad (1.17)$$

where s is supercooling in °C (WALTER and THREN 1934).

The real ΔT value and its corresponding water potential is thus calculated as follows; the correction K_Δ (listed in Table 1.11) is subtracted from the empirically found $\Delta T'$. The osmotic potential in bar at the freezing point temperature, Ψ_s, is then found for the corrected value of ΔT (Table 1.10).

Calibration of the cryoscope. Tables 1.10 and 1.11 allow calculation of the osmotic potential corresponding to the measured ΔT. However, the correction Table 1.11 is only empirical and holds only for the so-called standard cryoscope made by Burian - Drucker. Different cryoscopic devices have different properties, mainly affecting the course of the temperature changes during freezing of the cell sap or solution. Differences in measuring procedure and in temperature of the freezing mixture may also cause considerable differences in the measured $\Delta T'$ values, even though the osmotic potential of the solution is the same. In addition, inaccuracy in the calibration of the thermometer itself, and its various inertia, must also be considered when using the cryoscope with a mercury thermometer. This is even more important for thermocouple or thermistor cryoscopes.

Therefore, any cryoscope and measuring procedure (amount of sap, stirring, seeding *etc.*) must be first calibrated using pure water and solutions of known osmotic potential. Fresh, carefully prepared NaCl solutions, the osmotic potentials of which are shown in Table 1.8 (page 34), are most frequently used. The calibration curve for a cryoscope with a mercury thermometer will give the relationship between the osmotic potential and the measured freezing point directly. The galvanometers of thermistor or thermocouple cryoscopes may be directly calibrated in values of osmotic potential calculated for the required temperature.

The correction for supercooling, and recalculation for a temperature of 20 °C, may be neglected in standard measurements when only relative differences in water potential are required. However, the results must be carefully corrected and recalculated if absolute values of osmotic potential are required, *e.g.* for comparison with estimates of water potential obtained by other methods.

MOSEBACH (1940) designed an apparatus used to determine the melting point of the sap in the same way as in organic chemistry by means of Kofler's block. A glycerol bath, cooled by a cooled salt solution passing through a spiral tube, is used. The temperature is then increased continuously and the temperature at which a frozen sample, sealed in a capillary attached with a rubber ring to the mercury vessel of a Beckmann thermometer immersed in the bath, just melts is determined. The author uses a special optical device for simultaneous observation of the sample and the thermometer. The method makes it possible to determine the melting point of very small quantities of sap. However, the device is rather complex.

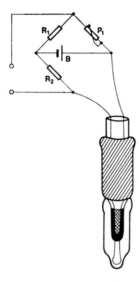

Fig. 1.46 A diagram of the device used to determine melting point (MARR and VAADIA 1961). $R_1 = 2$ kΩ, $R_2 = 2$ kΩ, $P_1 = 10$ kΩ, B — battery 1.5 V.

HARGITAY et al. (1951) described a procedure in which a polarizing microscope is used to detect the appearance of ice crystals in the sap (0.1 to 10 µl) in a capillary. The device suggested by PRAGER and BOWMAN (1963) is a very sophisticated procedure for measuring the freezing point of samples of millimicroliter size with high accuracy (2%).

A rapid cryoscopic method for drops of sap was described by MARR and VAADIA (1961). The instrument is illustrated diagrammatically in Fig. 1.46. The lower part of a tube containing, at an accurately marked position, an isolated thermistor immersed in a drop of sap, is put for a moment into an ethanol-solid CO_2 mixture. The sample freezes immediately, the tube is taken out, carefully dried and placed above a water-ice mixture (0 °C) in a vacuum flask. The sample warms up slowly and the course of change in thermistor temperature is followed with a galvanometer or recorded by a recorder. Results of calibration with a series of sucrose solutions of known molarity are shown in Fig. 1.47; the galvanometer deflection for individual molarities corresponding to the melting points of the solutions is found by constructed tangent lines. The sensitivity of the method is greater than 0.1 bar. However, as seen in Fig. 1.47, the procedure may be used only up to 0.25M sucrose, corresponding to an osmotic potential of -6.7 bar. This range is not usually sufficient so that this is not a useful standard procedure.

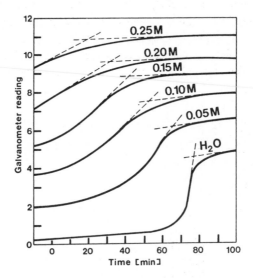

Fig. 1.47 The temperature course of melting sucrose solutions during calibrating for the determination of melting point (MARR and VAADIA 1961).

RAMSAY and BROWN (1955) constructed apparatus for the determination of the melting point of very small amounts of sap (10^{-3} to 10^{-4} mm^3). It is based on the microscopic determination of the melting point of a sample sealed in a capillary, using an ethanol bath cooled with ethanol/solid CO_2 mixture and heated electrically.

1.4.2.3 Tensiometric Methods

These methods involve measuring the water vapour pressure at the surface of a solution. Compensation methods are based on comparison of water vapour pressure above the solution (sap) of unknown osmotic potential and solution of known osmotic potential. The shift of water vapour from one to the other is detected either by volume changes of the solutions (capillary method and hanging drop method) or indirectly by temperature differences between drops of sample and known solution caused by condensation or evaporation of water and measured by a thermocouple (thermoelectric methods).

Capillary Methods. These rather laborious methods belong to the first group of tensiometric methods. Measurements are made of the changes in the known volumes of solutions with graded water vapour pressure, sealed in a small air space with a solution of unknown vapour pressure. BARGER (1904) and RAST (1924) filled capillaries alternately with an unknown solution and with one of a series of known solutions, separated by air spaces (Fig. 1.48). On the basis of the volume changes, shown by change in the distance between the two menisci, they found the solution with the same water pressure as the unknown solution, *i.e.* with a volume which did not either increase (as it does in the case of hypertonic reference solutions) or decrease (as in the case of hypotonic reference solutions). As the solutions compared were always in the same capillary, they were at the same temperature. However, as different solutions were drawn into the same capillary through the same opening,

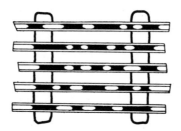

Fig. 1.48 Capillary method (BARGER 1904 and RAST 1924).
See the description in the text.

there was always some mutual contamination. YASTREBOV (1954) therefore described a technique which makes it possible to fill the same capillary with pairs of solutions (known and unknown) without mutual contamination. He used capillaries with an inner diameter of 0.8 to 1.0 mm. An open side arm was drawn out of one side wall of the capillary by means of the gas microburner on the device shown in Fig. 1.49, so that it was possible to draw one of the solutions from each end of the capillary. The air in between leaked or was sucked out through the open side arm as the capillary was filled with the second solution. This side arm was then sealed. The sealed capillary was placed on a slide with two fine marks made with a diamond at a distance corresponding roughly to the distance between the central menisci of the two solutions. The distance between the menisci and these marks was measured as accurately as possible, using a micrometer eyepiece. The whole preparation was then placed in a thermostat at a fairly high temperature (35 °C is best when the material is heat-resistant, otherwise 25 °C is convenient) and the distances were measured after 4−6 hours and after 8−10 hours. When differences between the Ψ_s values of the two solutions are large (of the order of 0.1M NaCl), the volume of the hypertonic solution decreases, and the volume of the hypotonic solution increases. When the differences between the two solutions are smaller, the volumes of both solutions decrease. This is because there is some condensation of water vapour in the colder side arm; however, the volume of the solution with a lower water vapour pressure (the hypertonic solution) always decreases less than the other. It is thus possible, by selection of suitable known concentrations of reference solutions, to determine Ψ_s of the solution or sap with an accuracy of up to 0.000004 bar, as even such small Ψ_s differences are manifested by different evaporation or condensation. The method is very accurate, the temperature does not have to be strictly constant (the compared solutions are always at the same temperature) and only small amounts of the liquid are required; on the other hand, the preparation of the capillaries is rather laborious and difficult.

THÖNI (1965) suggested the use of capillaries (70 mm long, 1 mm in diameter) which are silicone-coated inside (by means of 2% silicone in tetrachlormethane) preventing mutual contamination of the drops of the solutions. The capillaries are closed at one end and the drops are put in with capillary (Pasteur) pipettes.

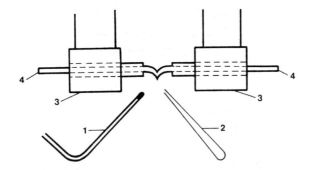

Fig. 1.49 The device for the preparation of capillaries with side arms for the YASTREBOV'S modification of the capillary method (1954). 1 — gas microburner, 2 — forceps, 3 — cork holder, 4 — capillary.

Thermoelectric Methods. An interesting tensiometric thermoelectric method used to determine the osmotic potential of a solution or sap was described by HILL (1930), BALDES (1939), BALDES and JOHNSON (1939) and modified by ROEPKE and BALDES (1938), ROEPKE (1942), and VAN ANDEL (1952). If drops of two solutions are placed in a space saturated with water vapour, one of known, the other of unknown water vapour pressure (and hence of Ψ_s), then the two drops have different temperatures because of different condensation during equilibration of water vapour pressure. The difference is measured with a galvanometer connected to the thermocouples on which the drops are placed.

This device is shown in Fig. 1.50. The thermocouples are sheets of manganin and constantan forming, close to their junction, two loops into which drops of reference and sample solutions are placed. The procedure is as follows: A drop of distilled water is placed in one loop with a pipette, a drop of the sample solution (e.g. 0.06 ml) in the other. The size of the drops is not critical. The chamber containing the thermocouples is lined with wet filter paper, or has a wet paper disc at the bottom (RYCZKOWSKI 1960). The whole apparatus is immersed in a water bath at

25 ± 0.01 °C and condensation on the drop of the sample solution begins and its temperature increases. Steady state is reached after about 30 – 35 minutes, when the amount of heat released by condensation equals that lost by dissipation and radiation, so that the temperature of the drop no longer changes. The galvanometer reading is recorded, the polarity of the thermocouple changed and the galvanometer read again. The thermocouples are taken out, carefully washed and dried. The location of the drops in the rings of the two thermocouples is then reversed and the whole procedure repeated. The average of the two readings is taken. The

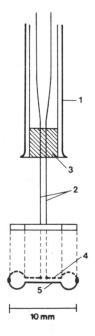

Fig. 1.50 The thermocouple device according to VAN ANDEL (1953). 1 — glass tube, 2 — copper wires, 3 — sealing, 4 — manganin foil strip, 5 — constantan foil strip. Both connections of the thermocouples are on the inside of rings holding drops of the solutions compared.

location of the thermocouple device in the wet chamber is very critical. Every thermocouple device must be calibrated with a series of solutions of various Ψ_s to determine the linear relationship between galvanometer deflection and molar concentration of the solutions. VAN ANDEL (1952) used solutions of boric acid and gives the accuracy as $\pm 0.002 M$ H_3BO_3. He used the method up to a difference of $0.20 M$ H_3BO_3. If Ψ_s is above this, it is necessary to use a solution for the reference drop, instead of water, so that the difference does not exceed $0.2 M$ H_3BO_3, *i.e.* about $0.03 M$ NaCl. It is also necessary to wet the lining paper in the chamber with the solution instead of water, to prevent any substantial dilution of drops with high Ψ_s values.

One determination takes about 1 hour, but several may be carried out simultaneously with separate thermocouple devices and wet chambers. 0.01 ml of sap or solution is required for each measurement. The author did not publish readings of the mirror galvanometer used, but his calibration graph shows that the difference of 0.2M H_3BO_3 corresponds to a deflection of more than 200 mm for a distance of 250 cm.

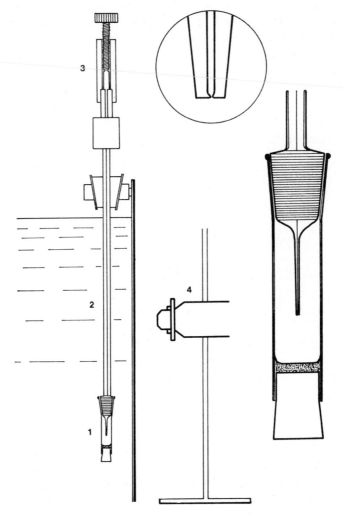

Fig. 1.51 Arrangement of the microosmometer (WEATHERLEY 1960). 1 — glass tube with ground inner walls (see details on the right side of the figure), 2 — micropipette, 3 — screw plunger, 4 — horizontal microscope.

Several types of tensiometric osmometers are commercially available (e.g. Vapour Pressure Osmometer by Mechrolab, Mountain View, California, USA, by Hewlett-Packard 1101 Embarcadero Rd., Palo Alto, California 94303 USA and by

Wescor Inc., 459 South Main, Logan, Utah 84311, USA). Their design allows very rapid measurement of small volumes of solution.

Hanging Drop Method. The volumetric method suggested by WEATHERLEY (1960) seems very promising. The rate of distillation (evaporation and condensation) between an unknown solution and a known reference solution is measured by a microosmometer (Fig. 1.51). The ground-glass inner walls and the bottom of the tube (1) (which is always washed with chromic acid) are wetted with 0.1 ml of the unknown solution. The tube is attached with a ground-glass joint to a micropipette (2). The pipette is about 20 cm long, and the inner and outer diameters are 0.1 and 4.5 mm, respectively. Its tip is specially shaped, as can be seen in the inset in the diagram. A reference solution of known osmotic potential is drawn into the pipette by a screw plunger (3) sealed to its upper end. The solution is held by its surface tension in the same position in the opening of the pipette tip and reaches approximately to the mark in the middle of the tip. The pipette and the tube are immersed vertically in a water bath, the temperature of which does not vary by more than ± 0.01 °C. The temperature of the pipette must first be sufficiently equilibrated with the bath temperature. The solution is then pushed out of the pipette with the plunger so that a hanging drop is formed on the tip. After 10 minutes the reference solution is drawn back into the pipette with the screw plunger, the exact position of the upper meniscus is measured with a horizontal microscope (4) with an eyepiece scale, and the solution is pushed out again into the hanging drop. The equilibration of vapour pressure between the surface of the drop and the surface of the solution on the walls and the bottom of the tube takes place, resulting in either evaporation from or condensation on the hanging drop of reference solution. After 20 and 40 minutes the drop is drawn back into the pipette in such a way that its lower meniscus is again in the pipette opening, and the position of the upper meniscus is measured.

Changes in the position of the upper meniscus result from changes in volume of the reference solution. In calibration, these changes are plotted against the difference between Ψ_s for the solution in the pipette and the solution in the tube. The relationship is basically linear, but sometimes differs in the two parts of the curve (plus and minus) particularly for large differences of osmotic potential. As the difference between the osmotic potentials of the sample and reference solutions should not be more than ± 4 bar, it is necessary to choose a suitable reference solution for various ranges. Examples of calibration curves are given in Fig. 1.52. The mean error is ± 0.06 bar, according to the author.

The reverse procedure may also be used (the unknown solution is drawn into the pipette and the walls of the tubes are wetted with the reference solution). Only 0.001 ml of the sample is then required. However, the surface tension of the cell sap is frequently not sufficient for this.

The method does not require any costly equipment (*e.g.* sensitive galvanometer

etc.) nor, in comparison with other tensiometric methods, is very accurate maintenance of the bath temperature necessary. A single determination takes about
1 hour. Several solutions may be measured in one bath at the same time. A similar
procedure can be used for measuring the water potential of plant samples, see
Section 1.3.5.

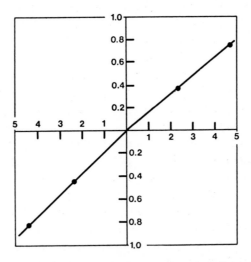

Fig. 1.52 Example of a calibration graph for the microosmometer (WEATHERLEY 1960).
Abscissa: differences of water potential between reference and sample solution [bar], ordinate:
mean rate of distillation, *i.e.* evaporation from or condensation on the hanging drop of
a reference solution, expressed as scale units of the horizontal microscope after 10 min exposure.

1.4.2.4 Psychrometric Methods

Psychrometric methods (described in Section 1.3.4) may be used for the measurement of osmotic potential, in the same way as for the measurement of water potential. It is possible to measure not only the osmotic potential of expressed cell sap
but also of fixed tissue, by this method. Thus EHLIG (1962) used the method of
RICHARDS and OGATA (thermocouple psychrometer of type B) to measure the
water potential of living leaf tissue, which was then killed by low temperature
(mixture of acetone and dry ice), and the osmotic potential of the fixed tissue after
thawing. It is necessary to assume that the water vapour pressure in the space containing the sample and the thermocouple device is determined primarily by the
solution released from the fixed tissue. In fact, the sum of osmotic and matric potentials is measured. It is possible to measure, relatively simply, the total water
potential and its components − osmotic potential and matric potential − in the
same sample of tissue. The psychrometric procedure is extremely valuable in this
respect. Discs of filter paper soaked in the sap pressed from killed tissue may also
be used.

1.4.2.5 Refractometric Method

The refractive index (n_D) of solutions depends on the particular dissolved compounds and their concentration. The relationship between the n_D and osmotic potential allows this method to be used to determine osmotic potential, particularly for solutions in which the osmotic potential increases slowly with increasing molar concentration so that a small change in osmotic potential corresponds to a considerable change in n_D. The opposite is true of electrolytes. Fig. 1.53 shows

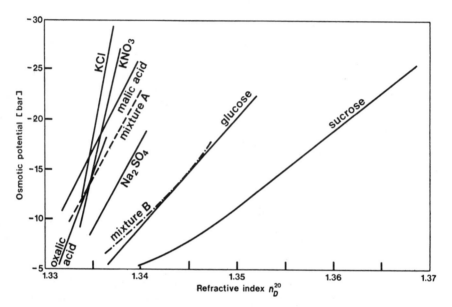

Fig. 1.53 The relationship between osmotic potential (ordinate in bar) and refractive index (abscissa: n_D^{20}) of solutions of mixtures of various compounds. Mixture A: 0.35M sucrose + + 0.8M oxalic acid + 0.67M KCl (1 : 1 : 1), mixture B: 0.4M sucrose + 0.8M oxalic acid + + 0.4M KCl (1 : 1 : 1) (SLAVÍK 1969).

examples of the relationship between the osmotic potential and n_D of aqueous solutions of some substances representative of the osmotically active components of the cell sap. Any change in the concentration of the cell sap caused by a change in its composition changes the empirical regression curve between Ψ_s and n_D. However, any changes in concentration caused only by dilution or concentration, *i.e.* by changes of water balance of a tissue or organ, result in a simultaneous change in Ψ_s and n_D. It is then possible to use n_D for the determination of the osmotic potential of the cell sap. The relative quantitative composition of the cell sap would not be expected to change over a short period (7 − 14 days) of the ontogenetic development of an organ or a whole plant, so over such periods, changes of the n_D of the cell sap of a particular material grown under identical ecological con ditions, will correspond fairly closely to changes in osmotic potential. The regression of the relationship between n_D and osmotic potential determined cryoscopically is found empirically, for instance by cryoscopy of a sufficient number of identical samples, if possible over the whole measured range (SLAVÍK 1959, DAVIS 1963, KOZINKA and NIŽNÁNSKY 1963, KREEB 1965d). An example is shown in Fig. 1.54.

Determination of n_D itself is very simple and rapid. It takes 30 to 40 s, to do twice. It is carried out in a universal refractometer (Abbe refractometer). It is ad-

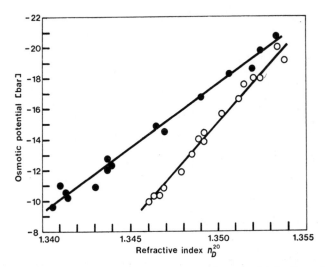

Fig. 1.54 Example of the relationship between osmotic potential (ordinate in bar) and refractive index of the cell sap of sugar beet leaves (solid circles) and oak leaves (*Quercus petraea* LIEBL) (hollow circles) taken during the first half of August (SLAVÍK 1959).

visable to maintain constant temperature of both measuring and illumination prisms by means of water from a flow-through thermostat when working in an environment with fluctuating temperature. One drop of sap (0.02 ml) is sufficient for a measurement. The accuracy of the reading is increased by changing the focus of the instrument after the first setting of the boundary light and dark, re-focusing, and taking the mean value of the two readings. It is important to wash and dry the prisms carefully.

Methods for calculating the osmotic potential directly from n_D by using tables of the relationship between the concentration of sucrose solutions and their n_D, are not well-founded. Nevertheless, this procedure is sometimes used and has been recommended in the literature.

The contribution of electrolytes to the refractive index of cell sap is very small compared to that of metabolites (SLAVÍK 1959, Fig. 1.53). SHIMSHI and LIVNE (1967) proposed combined refractometric and conductimetric measurements,

Table 1.12 *Summary of methods used for the determination of osmotic potential.*

Method	Suitable material	Measured value	Sample size	Mean accuracy	Reliability [2]	Equipment [3]	Degree of laboriousness [4]
Limiting plasmolysis	leaf epidermis, moss leaflets, algae	$\psi_s^{(lp)}$	n × tissue[1]	0.01 M ~ 0.5 bar	*(*)	* microscope	**
Plasmometry	cells of suitable shape	$\psi_s^{(lp)}$	individual cells in one preparation	0.01 M ~ 0.5 bar and higher	*	* microscope	***
	even for irregular cells (PRÁT 1922)			lower		* microscope	
Zero turgor method	*Nitella* interno-dial cells and similar material	$\psi_s^{(lp)}$	individual cells	0.01 M ~ 5 bar	*	***	***
Cryoscopy with mercury thermometer	average cell sap from tissue	$\psi_s^{(n)}$	1.5 to 4 ml	±0.12 bar	***	* special thermometer	**
with thermistor	average cell sap from tissue	$\psi_s^{(n)}$	0.2 ml	±0.12 bar	***	**	**
MARR and VAADIA (1961)	average cell sap from tissue	$\psi_s^{(n)}$ down to −6.7 bar only		±0.12 bar	**	**	**

Method / Author	Object	Symbol	Amount	Accuracy			
MOSEBACH (1940)	average cell sap from tissue	$\psi_s^{(n)}$	0.02 ml	±0.12 bar	***	***	**
RAMSAY and BROWN (1953)	average cell sap from tissue	$\psi_s^{(n)}$	10^{-3} mm^3 = 10^{-6} ml	±0.12 bar	***	***	**
Tensiometry BARGER (1904), RAST (1924)	average cell sap from tissue	$\psi_s^{(n)}$	n × 0.002	±0.2% to ±0.0002%	*	*	**
YASTREBOV (1954)	average cell sap from tissue	$\psi_s^{(n)}$	n × 0.002		**	*	***
WEATHERLEY (1960)	average cell sap from tissue	$\psi_s^{(n)}$	0.06 to 0.1 ml	±0.5 to 1%	**	*(*)	**
VAN ANDEL (1952)	average cell sap from tissue	$\psi_s^{(n)}$	2 × 0.01 ml	±0.3%	**	***	**
Refractometric method	average cell sap from tissue	$\psi_s^{(n)}$	0.02 ml	0.5 bar	*	* refractometer	!* calibration "ad hoc" necessary
Deformation method LOCKHART (1959)	coleoptile, stem with primary structure	$\psi_s^{(1p)}$	n × 10 segments	0.01 M ~ 0.3 to 0.5 bar	*	*	**

[1] n = number of osmotic concentrations used. [2] * lower — *** high. [3] * simple — *** complex. [4] * low — *** high

making the following simplifying assumptions: (1) The n_D is due to sucrose only. (2) There is a unique value dependence of osmotic potential on conductivity of the cell sap. (3) The osmotic potential of a cell sap is the sum of the osmotic potentials determined by n_D and conductivity measurements.

1.4.3 Deformation Method for Measuring Osmotic Potential of Tissues

This method (LOCKHART 1959) is completely different in principle. The author found that the deformability (*i.e.* elasticity) of a tissue incubated in a hypertonic mannitol is a linear function of the difference between the known osmotic potential

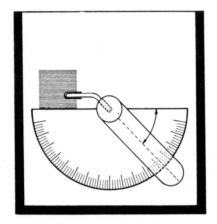

Fig. 1.55 The front view of the device for measuring the deformability of (stem) tissues when using the deformation method for determination of osmotic potential (LOCKHART 1959).

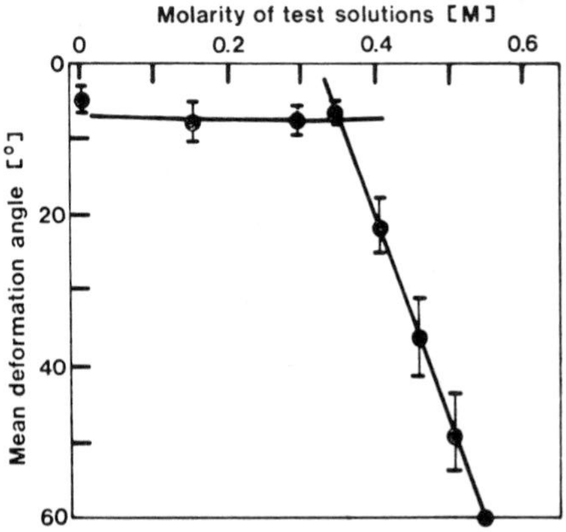

Fig. 1.56 The relationship between the mean deformation angle of stem tissues exposed to test solutions of different molarity. The intercepts of both linear portions with the abscissa indicate the molar concentration corresponding to the osmotic potential at limiting plasmolysis (LOCKHART 1959).

of the solution and the osmotic potential of the tissue ($\Psi_{s(lp)}$). So the osmotic potential of the tissue can be measured *in situ*. The method can be used generally with tissues in which mechanical strength depends on turgor, such as segments of coleoptiles of grasses, stem of young seedlings *etc.*

Approximately 20 mm segments of the material are prepared and placed in small groups (10) into a series of test solutions, *e.g.* with gradations of 0.1M. It is advisable to use mannitol, as the incubation is relatively long, so that a non-metabolizable solute is required.

After incubation for about 1.5 to 3 h, (depending on the material, time not being critical), the sections are taken out of the solution and their deformability measured in a device made from plexite (Fig. 1.55). A hole 5 mm deep is made in the side of a horizontal rotating axle. Its diameter should be such as to make it just possible to insert one end of the segment. A glass tube of a suitable diameter (about 2 mm) is attached by 5 mm of its length to the other end of the segment. This acts as ballast. The weight of the tube is chosen as to bend the plasmolysed tissue on the device by a maximum of 90°. An arm is attached to the axis opposite the hole. The arm is turned until the glass weight is exactly in the horizontal position, as indicated by the horizontal lines on the plate. The angle of the arm axis to the horizontal is then read on the scale of the rotating arm. The average of the angles measured for segments taken from the same test solution is plotted against concentration of the solution. An example is shown in Fig. 1.56. As the tissues are plasmolysed, values of deformation of a segment with zero turgor are linearly dependent on the concentration of the hypertonic solution used. The extrapolation of the line to the point of minimum deformability, in a segment with positive turgor, then gives the Ψ_s value which is taken to represent the $\Psi_{s(lp)}$ during just limiting plasmolysis.

The weight of the ballast affects the slope of the line, but not the position of its intersection with the horizontal line, *i.e.* of minimum deformation. Osmotic equilibrium does not have to be reached, so the time of incubation in the test solution is not important. The method does not require previous calibration.

1.5 Determination of Turgor Pressure (Pressure Potential)

1.5.1 Introduction

Pressure potential (Ψ_p) — usually called turgor pressure* — is the component of water potential which is due to hydrostatic pressure. The higher the pressure, the higher (less negative) the water potential.

At the point of limiting plasmolysis (lp), the pressure potential (*i.e.* the pressure of the cell wall exerted on the protoplast, the so-called wall pressure) is assumed to be zero. In a fully turgid cell, turgor pressure reaches its maximum if there is no other external pressure (*e.g.* pressure exerted by the surrounding cells). A cell saturated with water in a tissue normally has zero water potential. The inward pressure is called wall pressure, the pressure in the opposite direction but of the same magnitude is called turgor pressure. A cell is said to be fully turgid when the wall pressure and the (equal but opposite) turgor pressure are at a maximum.

There is a negative hydrostatic pressure (*i.e.* less than normal atmospheric pressure) in the xylem vessels of transpiring plants, thus lowering the water potential of

---

* The term turgor is also associated with the term relative turgidity. This is a rather illogical expression for the relative water content frequently used in English and American literature. It is expressed in per cent of relative water saturation, *i.e.* the relative turgidity is 100% when the cells are fully saturated. Therefore, relative turgidity = 100 — water saturation deficit. If a tissue has a water saturation deficit ($\Delta W_{(sat)}$) = 18%, its relative turgidity is 82%. It would be advisable to abandon the term (see Section 2.5.1).

the cell sap inside the vessels. It is related to the cohesive tension of water columns.

The measurement of the pressure component of the water potential of cells or tissues is the most difficult. There are only very few methods which can be used for direct determination of turgor pressure and the corresponding opposite pressure of the cell walls, and for measurement of the pressure exerted by neighbouring cells. The pressure potential is therefore most frequently found by calculation.

1.5.2 Calculations of Pressure Potential

Using equation 1.8, and assuming that the osmotic potential of the sap in vacuolated cells is — at least in a steady state equilibrium — in equilibrium with the sum of matric and osmotic potentials in the cytoplasmic structure and in the cell wall, if both are under the same pressure (turgor pressure in this case), it is simple to calculate the pressure potential Ψ_p inside the cell or tissue if the total water potential at the surface of the cell or tissue $\Psi_{w(cell)}$ and the osmotic potential of the vacuolar sap $\Psi_{s(v)}$ are known:

$$\Psi_p = \Psi_{w(cell)} - \Psi_{s(v)} . \tag{1.18}$$

The water potential at the surface of the cell and the osmotic potential of the cell sap may be measured by the various methods described in the appropriate sections of this manual.

URSPRUNG and BLUM (1924) and URSPRUNG (1939) described three ways of calculating wall pressure, i.e. pressure potential of the cell, using cell methods for measuring water potential of the cell and osmotic potential of the vacuolar sap:

(1) Assuming that the increase in wall pressure is proportional to the increase in cell volume, the following equation is valid:

$$\frac{\Psi_{p(n)} - \Psi_{p(lp)}}{\Psi_{p(s)} - \Psi_{p(lp)}} = \frac{V_{(n)} - V_{(lp)}}{V_{(s)} - V_{(lp)}} , \tag{1.19}$$

where $\Psi_{p(n)}$, $\Psi_{p(lp)}$, and $\Psi_{p(s)}$ are pressure potentials in the normal (measured) state, at limiting plasmolysis and at full water saturation, respectively; and $V_{(n)}$, $V_{(lp)}$, and $V_{(s)}$ are corresponding cell volumes.

(2) The simplest calculation of the pressure potential is as follows:

$$\Psi_{p(n)} = \Psi_{s(n)} - \Psi_{w(n)} , \tag{1.20}$$

$$\Psi_{p(n)} = \Psi_{s(lp)} \frac{V_{(lp)}}{V_{(n)}} - \Psi_{w(n)} , \tag{1.21}$$

where $\Psi_{s(n)}$ and $\Psi_{s(lp)}$ are osmotic potentials in the normal (measured) state or at limiting plasmolysis, respectively, and $\Psi_{w(n)}$ is water potential in the normal state.

(3) The following equation may be used if the cells are immersed in a test solution with an osmotic potential $\Psi_{s(sol)}$ numerically close to the water potential of the measured cells:

$$\Psi_{p(n)} = \frac{V_{(n)} - V_{(1p)}}{V_{(sol)} - V_{(1p)}} \left\{ \Psi_{s(sol)} \frac{V_{(1p)}}{V_{(s)}} - \Psi_{s(sol)} \right\}. \tag{1.22}$$

Using similar calculations, many authors have found negative pressure potentials in stressed tissues, down to as low as -150 bar in leaves of xerophytes, and averaging about -2 bar in mesophytes. The reality of such results, and their possible explanation, is still controversial.

1.5.3 Direct Measurements of Pressure Potential

It is only very rarely possible to measure directly the hydrostatic pressure related to the turgor pressure of living cells and tissues. BOURDEAU and SCHOPMEYER (1958) measured the pressure in oleoresin ducts of *Pinus eliotti*, and BUTTERY and BOATMAN (1966) the pressure in laticiferous phloem of *Hevea brasiliensis*. GREEN (1966) and GREEN and STANTON (1967) used a similar procedure to measure pressure potential in *Nitella* cells. Capillaries (*e.g.* 100 mm long and 0.7 mm inner diameter), closed at one end and connected at the other to a hypodermic needle (fine bore steel tubing), were inserted into the cell of *Nitella*. The capillaries, which originally contained air at atmospheric pressure, were after insertion partly filled with cell sap. The decreasse in volume of the air compressed by the hydrostatic turgor pressure in the cell was measured. The ratio of the volume of air before and after insertion corresponds to the pressure potential of the cell. In order to prevent the air dissolving in the sap, GREEN (1966) maintained a slow reduction in the ambient atmospheric pressure during the experiment.

Positive pressures in xylem vessels, brought about by so-called root pressure in non-transpiring and detopped plants, can be measured by similar methods. However, there is usually negative pressure in the xylem of transpiring plants. DIXON (1914), SCHOLANDER *et al.* (1964, 1965, 1966) used a pressure chamber technique (see Section 1.3.6) to measure this tension. A leafy shoot or a leaf is enclosed in a pressure chamber, so that the cut stem or petiole sealed in the cover of the pressure chamber protrudes into the surrounding atmosphere. The menisci of the sap columns in the xylem vessels are brought back to the cut surface by increasing the pressure of air or nitrogen inside the pressure chamber. This is detected by the beginning of sap outflow in the form of small droplets on the cut area. The lowest pressure necessary for this was assumed by the authors to be a direct measure of the tension (negative pressure) in the intact xylem vessels before cutting the stem. Nevertheless, a more detailed analysis, even an incomplete one, shows that the appearance of sap at the cut surface is related to water movement from the surrounding cells into the vessels. The increased water potential of these cells, brought about by increased pressure in the pressure chamber, causes the movement. The mechanism of the water movement is osmosis, *i.e.* outflow of water across the semipermeable cytoplasmic membranes of these cells, against the gradient of osmotic potential. However, the pressure chamber method may be used for the determination of negative pressure potential in the xylem vessels of transpiring plants, if the osmotic potential of the sap in the vessels is taken into account.

The theory of pressure-bomb measurement of s.c. pressure-volume curve was recently advanced by TYREE and HAMMEL (1972), TYREE *et al.* (1973a, b) and TYREE and DAINTY (1973).

HAMMEL (1968) described a method for the direct measurement of turgor pressure in the phloem of tree stems. He used stainless steel syringe needles (0.25 mm inner diameter) with ground and then squeezed and highly polished points and with the other end attached to a glass capillary. The device was filled with dyed water, then the free end of the capillary with the remaining air column was sealed. The device was attached to a source of external pressure and the length of the air column in the capillary was calibrated in terms of external pressure by plotting the ratio $a_{(actual)} b_{(actual)}/a_p b_{(ref.)}$ against P (where $a_{(actual)}$ and a_p are the lengths of the air column at actual barometric pressure and at the applied pressure P, respectively, P is the external pressure applied [corresponding to the turgor pressure] $b_{(actual)}$ is the actual barometric pressure and $b_{(ref.)}$ is the reference barometric pressure). When measuring, the syringe needles were inserted through the bark until they stopped and an abrupt compression of the air column occurred. Then P was assumed to be equal to the turgor pressure in the phloem elements.

STEPANOV (1969) designed a method and device for measuring turgor pressure mechanically: the leaf to be measured is pressed between two tensiometric resistors, the resistance of which changes with changes in the leaf turgor pressure and is measured with a Wheatstone bridge. The method yields rather comparative values and should be carefully calibrated with another method.

1.5.4 Indirect Methods for Determination of Pressure Potential

A very special technique was used by KAMYIA *et al.* (1963) for measuring cell volume as a function of internal pressure: they cut the cylindrical cell of *Nitella* at one end and squeezed out the protoplast, then filled the empty wall with mercury and measured the pressure necessary to regain the original cell volume.

The resonance method of VIRGIN (1955), used by FALK *et al.* (1958), measures turgor directly but in relative and not absolute values. The principle is that any solid body caused to vibrate, vibrates at a certain resonance frequency, which is characteristic of the mechanical properties of the body, *e.g.*, if one end of a steel wire several cm long is fixed and an electromagnet supplied with a current of variable frequency is brought nearer to the other end, the free end of the wire begins to vibrate only if the frequency of the current corresponds to its resonance frequency.

In this way, the resonance frequency of various materials can be measured, for example a piece of plant tissue with a piece of magnetic metal attached to the free end. The resonance frequency of the tissue changes if there is any change in its mechanical properties, which are partly determined by the pressure of the cell walls and hence are proportional to its turgor.

Resonance frequency depends on homogeneity of the resonating material. Heterogeneous bodies, including plant tissues, also resonate at several other, mostly higher, frequencies, which are not sought during measurement of turgor and can lead to errors in measurement.

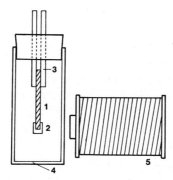

Fig. 1.57 The device for the determination of turgor by the resonance method. 1 — sample strip of tissue, 2 — iron, 3 — glass holder, 4 — vessel, 5 — bobbin (VIRGIN 1955).

A device designed for the indirect determination of turgor is schematically illustrated in Fig. 1.57. A strip or a small block, 1 × 1 × 10 mm (1) is fixed at one end to a glass holder (3) (a tube or two flat slides). A narrow piece of thin (0.025 mm) non-hardened steel sheet (2) is attached to the free end of the strip and the whole holder is inserted in a glass chamber (4). An electromagnet (5) with an impedance of 600 Ω is placed opposite the tissue strip with the attached metal sheet.

The electromagnet from an ordinary electric bell may be used but more powerful electromagnets give better results. The electromagnet is supplied with alternating current of variable frequency from a laboratory oscillator or tone generator. The output of the generator may be increased by a simple amplifier. Vibration of the end of the tissue strip is observed by means of a horizontal microscope with a bulb.

The current frequency bringing about maximum resonance of the tissue strip is determined; this frequency corresponds to "relative turgor"*. VIRGIN (1955) showed that best results are obtained when the resonance frequency of the tissue is in the range 70 to 80 cycles per s.

---

* Do not confuse with relative turgidity!

The resonance method was found useful for continuous assay of turgor of tissue after immersion in solutions of an osmotically active compound (Fig. 1.58). However, a completely satisfactory method of measurement of turgor of strips of leaf tissues in air has not yet been worked out. Ψ_s of the first point on the bottom horizontal part of the curves (Fig. 1.58) corresponds to zero Ψ_p and hence, in practice, to limiting plasmolysis. Making the simplified assumption that the resonance of the tissue is linearly proportional to the pressure potential (turgor), the resonance

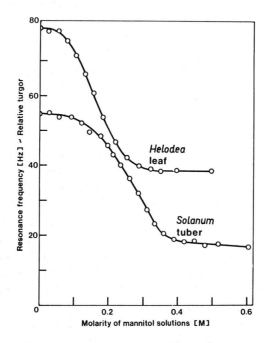

Fig. 1.58 The relationship between the resonance frequency of the tissue and molarity of the mannitol solution which is in osmotic equilibrium with the tissue (Virgin 1955).

method can be used for the calculation of the pressure potential in the following way. The resonance frequency in air, and the osmotic potential of fully saturated tissue and of the tissue at the point of limiting plasmolysis is determined. A calibration curve of the relationship between resonance frequency and pressure potential (turgor) is constructed from these values. This would apply only for the assumed linear relationship between resonance frequency and pressure potential. However, it is evident from Fig. 1.58 that this relationship is not linear; and it would be possible also to determine the resonance frequencies and osmotic potentials of tissue sample in which various water potentials had been artificially induced.

1.6 Matric Potential, "Bound", "Free" and "Mobile" Water

1.6.1 Introduction

Matric potential represents the decrease in water potential caused by the linkage of water at interfacial boundaries in the matrix. Matric potential is an important component of water potential in protoplasm (*e.g.* in its colloidal ultrastructures, by dipoles of water molecules forming hydrogen bonds on hydrophilic sites of organic molecules, proteins in particular), in cell wall structures (colloidal micellar structures), in intercellular spaces, and in capillaries of vascular elements (*incl.* decrease of water vapour pressure on concave menisci).

Matric potential may be determined primarily by plotting the water potential against water content at constant pressure and constant osmotic potential. Matric potential (Ψ_m):

$$\Psi_m = \int_1^{n_w} \tau \, dn_w, \tag{1.23}$$

where τ is the slope of a curve of water release and n_w is water content expressed as molar fraction of water (*i.e.* relative molar fraction of water in the system).

It can be assumed that the water potential is the same in all cell structures or parts in an energetically equilibrated state. However, such complete intracellular equilibration of the hydration level is seldom attained, as changes in water balance of tissues and of individual cells take place continuously. Not only the type of linkage, but also the amounts of the individual water fractions bound in different ways, as well as the amounts at the various sites in the cells, are then important.

Generally, "bound" water is the name given to that fraction which behaves in some respects differently from an aqueous solution, *e.g.* it does not freeze at a particular sub-zero temperature, or does not move into a hypertonic solution during immersion, or does not evaporate from a tissue placed in air with a relatively low water vapour pressure *etc.*

It should be emphasized that although there are different types of linkage of water there is a practically continuous transition between them; the boundaries are only conventional and based on more or less accidentally chosen criteria resulting from the methods used for determination. Estimates obtained by different procedures are not comparable. Comparison of estimates for the same material by the same method is also questionable. Consideration of bound water has become fashionable from time to time in scientific work but much confusion has arisen from inadequate definition of the concept. Many complex properties of plants (*e.g.* resistance to some unfavourable factor) are also often said to be related to the quantity of bound water. Such conclusions are logical but the importance of results along these lines is often overestimated.

In addition to the linkage of water, its localization in the cell is also important, as this determines availability during changes of cell water balance. The so-called mobile fraction of cell water is the fraction which is first diluted when the cell is taking up water or first concentrated, *i.e.* its water potential decreases, when the cell is losing water (SLAVÍK 1955, 1963a).

1.6.2 Determination of Matric Potential

Theoretically, it is possible to determine the matric potential by measuring changes of the total water potential (Ψ_w) during a period of changing water content and constant osmotic potential (Ψ_s) and pressure potential (Ψ_p).

WIEBE (1966) used a commercial pressure membrane apparatus (see Section 3.1.3.2) with a dialysis membrane LSg 60 for direct measurement of the matric potential in several artificial systems (agar, gelatine, paper) as well as for measurement of frozen and thawed tissues (potato tuber, root of mangold [*Beta vulgaris*], stem sections of *Asparagus officinalis*) with different water contents. The material was placed on the membrane supported by nylon cloth instead of a metal gauze (to provide drainage), covered with a layer of sheet rubber and then with sponge rubber to facilitate perfect contact between the membrane and the material. Selected pressures ranging from 0.1 to 15 bar were used each for 48 hours, 24 hours having been found sufficient for the equilibration. The water content of the material was assumed to correspond to the matric potential equal to the pressure used.

Using a pressure chamber technique (see Section 1.3.6), BOYER (1967b) applied increasing pressure (0 to 25 bar) to frozen and thawed shoots of sunflower, yew (*Taxus cuspidata* SIEB. and ZUCC.) and *Rhododendron roseum* REHD. Collecting every drop of sap expressed at the cut surface until equilibrium was reached, he calculated directly the changes of water content at the different pressures applied, *i.e.* at different pressure potentials.

1.6.3 Determinations of "Free" and "Bound" Water

There is a large number of methods used to determine "bound" water; these methods cannot all be presented here and a review by KRAMER (1955) is recommended. See also OKUNTSOV and TARASOVA (1952). Only five of the methods are described below.

Recently the use of nuclear magnetic resonance spin echo and electronic paramagnetic resonance for the determination of water status in plant cells and tissues is discussed *e.g.* by FEDOTOV et al. (1969), SEDYKH and ISHMUKHAMETOVA (1970), GUSEV and SEDYKH (1971).

1.6.3.1 Dilatometric Method

This method was described by CHESHEVA (1934) and modified by GUSEV (1960). It involves measuring increase in tissue volume during freezing at a selected temperature below zero. About 3 to 5 g of accurately weighed fresh sample material (leaves *etc.*) are placed in a 40 to 50 ml dilatometer (Fig. 1.59) which is carefully filled with gasoline (petrol) in such a way that air bubbles are completely expelled after closing the ground glass stopper with a thermometer. Gasoline is then slowly added until it reaches a calibrated side tube. The dilatometer is immersed in a Dewar vessel containing a cryohydrate solution maintaining the temperature which is to be used during the measurement, usually from -3 to $-10\,°C$. GUSEV worked at $-6\,°C$. The gasoline meniscus first goes down on cooling (decrease in volume resulting from decreased temperature). However, the meniscus moves up again

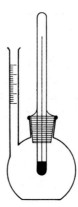

Fig. 1.59 Dilatometer (GUSEV 1960).

when free water begins to freeze, as ice has a greater volume than water. When the temperature inside the dilatometer reaches the selected temperature of the cryo-hydrate solution, the dilatometer is left to stand for 5 min, then shaken several times, so that all the "free" water freezes, including the supercooled water in the sample. It is necessary to wait again for 5 min, after which the further upward movement of the meniscus (increase of volume) is recorded. The temperature of the cryohydrate solution must not change during this time. The amount of frozen, *i.e.* "free" water, is then calculated from the increase in volume (after correction for the decrease of gasoline volume brought about by cooling).

1.6.3.2 Refractometric Method

The refractometric method (DUMANSKIĬ and VOĬTSEKHOVSKIĬ 1948) measures the percentage of water which is taken from the tissue by a very hypertonic (30%) sucrose solution during a 2 h exposure. Exactly 1.5 ml solution, the refractive index (n_D) of which is first found by means of an Abbe refractometer, is pipetted into a small flask. The flask is weighed and 20 to 40 discs about 7 mm in diameter, cut by a cork-borer from the leaf blade, are immersed in the solution. The flask is closed and weighed again. The n_D of the solution is determined after 2 h. Calculation from the original volume and the decrease of the n_D gives the dilution, which is the amount of water taken from the tissue, *i.e.* the free water. The initial dry weight and water content were determined in a replicate sample, so the free water may be expressed as a percentage of all the water contained in the tissues. The difference between the total water content in the tissue, found by determining the dry weight of the measured sample (or the replicate sample), and the determined content of free water represents the amount of "bound" water.

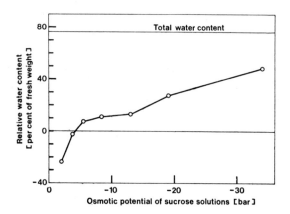

Fig. 1.60 An example of the relationship between the relative water content removed from the tissue by hypertonic solutions and taken up by the tissues from hypotonic sucrose solutions of various osmotic potentials (GUSEV 1960).

As there is a whole series of fractions of bound water, GUSEV (1962) recommends its more detailed analysis by determining by the refractometric method the amount of water taken from replicate tissue samples by test solutions from a series with decreasing osmotic potential. He uses sucrose solutions with osmotic potentials of -2, -4, -6, -9, -13, -18, -24 to -32 bar. The result then demonstrates the relationship between the amount of water taken up by or released from the tissue and the osmotic potential of the solution (see an example in Fig. 1.60).

1.6.3.3 Cryoscopic Method

"Bound" water in living or dead tissues has also been determined cryoscopically (*e.g.* NEWTON and MARTIN 1930, GORTNER and GORTNER 1933, HERRICK 1934) or calorimetrically, assuming that only free water freezes so that the heat released on melting is proportional to the amount of frozen water. Cryoscopic measurements are subject to considerable error, because sufficiently rapid equlibration is hardly possible in the tissue, so that the temperature which is measured [with an inserted thermocouple (HATAKEYAMA 1961) or a mercury thermometer] does not necessarily

correspond to the temperature throughout the tissue. It is interesting that WALTER
and WEISMANN (1935) found that the amount of heat released during freezing (*i.e.*
the area under the curve of the course of temperature with time) was the same in
both living and killed tissues. See also Section 1.3.8.

1.6.3.4 Cobalt Chloride Method

OHYAGI (in: HIGASHI 1955) and TAKAOKI (1962) improved the cobalt chloride
method for the determination of the "bound" water content of tissue slices. The
method is based on the original suggestions of HATSCHECK (1936). Weighed tissue
slices (*e.g.* $1 \times 1 \times 0.1$ cm) are immersed in 20% cobalt chloride ($CoCl_2 . 6 H_2O$)
solution for 24 hours. This time can be shortened by infiltration by a vacuum pump at
40 mm Hg. The slices are then slowly dried at 40 °C until their original blue colour
changes to a standard colour shade, obtained by dissolving 50 mg of $CoCl_2 . 6 H_2O$
in 5 ml of 92% ethanol. This violet shade corresponds to the water potential taken
as the conventional limit of "bound" water. The slice is weighed immediately after
reaching the violet colour and its water content at that point is taken as "bound"
water content.

1.6.4 Determination of "Mobile" Water

It has been observed that the decrease in osmotic potential of the cell sap during
loss of water from wilting excised leaves is more rapid than would be expected
from the theoretical increase of concentration. This suggests that water is lost un-
equally from the various fractions, with a more pronounced water loss from one
of them, the so-called "mobile" fraction (SLAVÍK 1963a).
The procedure used to determine the "mobile" water fraction is as follows:
Replicate samples of leaf segments (leaflets *etc.* − avoid heterogeneity of the leaf
blade!) saturated with water are used. In the first, the osmotic potential of the cell
sap is determined after killing; the second is allowed to wilt to a selected water
deficit, found by weighing, before its osmotic potential is determined. Such pairs
may be used over a suitable range of water deficit, from full water saturation to
20 to 30% water saturation, for example.
Then

$$\frac{\Psi_s[\Delta W_{(sat)}]}{\Psi_{s(sat)}} = \frac{(m.w.f.) \dfrac{\Psi_s[\Delta W_{(sat)}]}{\Psi_{s(sat)}}}{(m.w.f.) - (1 - \Delta W_{(sat)})}, \qquad (1.24)$$

where $\Psi_s[\Delta W_{(sat)}]$ is osmotic potential with known $\Delta W_{(sat)}$, $\Psi_{s(sat)}$ is osmotic potential at full saturation, (m.w.f.) is mobile water expressed as the fraction of the
water content at full saturation, and $\Delta W_{(sat)}$ is the relative water content expressed
as a fraction (*e.g.* water saturation deficit 21% corresponds to $\Delta W_{(sat)} = 0.79$).

It follows that

$$\text{m.w.f.} = \frac{\dfrac{\Psi_s[\Delta W_{(sat)}]}{\Psi_{s(sat)}}(1 - \Delta W_{(sat)})}{\dfrac{\Psi_s[\Delta W_{(sat)}]}{\Psi_{s(sat)}} - 1}. \tag{1.25}$$

Mobile water fractions of 0.9 to 0.7 were found, depending on the rate of water loss (SLAVÍK 1963a).

WARREN WILSON (1967) suggested a similar procedure for calculating of the matric potential, assuming that the bound water [*i.e.* 1 − (m.w.f.)] is due to the matric potential. The dynamic aspect of the m.w.f. determination, and the necessary equilibration between osmotic and matric potential in a steady state, were not taken into consideration.

2

Water Content

2.1 Direct Methods of Water Content Determination

2.1.1 Determination of the Water Content by Drying

The water content of plants is most commonly determined by drying the tissue, organ or plant at a high temperature (usually 60 to 105 °C). The material is weighed immediately after sampling and placed in a preheated oven to be killed as quickly as possible, in order to prevent dry weight losses resulting from rapid respiration at higher temperatures. If it is not possible to weigh the material immediately after sampling, the samples are placed in weighed glass or metal containers, hermetically sealed and weighed as soon as possible. They are then dried in the same containers, after opening. It has been found useful when drying leaf discs to pierce the samples with stainless steel pins and fix them, 5 to 10 at a time, to a flat plate. Samples are then easily weighed on a torsion balance, without the dead weight of the container, and there are not dry weight losses through the samples sticking to the container walls, *etc.* The drying time depends on the size and compactness of the sample; the samples are generally dried to "constant" weight, *i.e.* so-called oven-dry weight, which is usually reached after 12 h with medium size samples. Drying for 2 h at 70 °C is usually sufficient for discs (8 mm in diameter) from leaves of common cultivated plants. No differences in the rate of drying, nor in the final value, were found when higher temperature (105 °C) were used. Three hours are required for drying at 60 °C. Some authors point out that a fraction of the water remains bound in the tissue even after drying at 105 °C; however, this error is negligible in routine determinations.

Reproducibility and accuracy of results is determined by the size and homogeneity of the sample. STREBEYKO and KARWOWSKA (1958) showed that dry weights of samples of about 8 g leaves or organs of common crop plants are measured with a maximum of 1% error.

This method of determining the water content has one disadvantage, *i.e.* the sample usually also loses some or all the volatile compounds, some of the protein nitrogen *etc.* However, this error is not often important.

2.1.2 Chemical Determination of Water by Titration

It is sometimes advantageous (*e.g.* with material containing a large amount of volatile compounds) to determine the water content chemically by titration with the so-called Fischer reagent (FISCHER 1935). The method is based on oxidation of sulphur dioxide by iodine in the presence of water. The reagent is prepared as follows (MITCHELL and SMITH 1948): 762 g of iodine are dissolved in 2 420 ml of anhydrous pyridine in a nine litre flask with a ground glass stopper, and 6 000 ml of anhydrous methanol are then added. Of this solution 3 000 ml are transferred to a four litre flask immersed in crushed ice about one to two days before use, and 135 ml of liquid SO_2 added very carefully after cooling. The flask is closed and the mixture stirred until is completely homogeneous. Sulphur dioxide may also be added in the gasous state (FISCHER 1935) but one then has to be very careful and cool the system very effectively as the warms up considerably. The reagent is standardized in the following way (FISCHER 1935): 10 ml of anhydrous methanol are first titrated in an Erlenmeyer flask until the solution changes colour from chromic yellow to brown. According to FIECHTER and VETSCH (1957), the colour shade reached must remain the same for 5 min: this situation is ensured by titrating to the first detectable brown colour of the solution and then adding a few more drops of the reagent. Ten ml of methanol containing 1% water (*i.e.* 0.1 ml of water) are titrated in the same way. The difference shows the volume of reagent required for the titration of 0.1 ml of water. According to GREENE and MARVIN (1958), it is more suitable to use sodium tartrate (dihydrate) as the standard, because methanol absorbs moisture from the air to a considerable extent.

The determination of the water content of plant material is as follows (GREENE and MARVIN 1958): 100 to 200 mg of tissue cut in pieces is immersed in a tube containing 10 ml anhydrous methanol and the tube is sealed. After 4 h a sample of the methanol (1 to 2 ml) is taken out for titration. A sample of pure methanol is titrated as blank in each series.

2.1.3 Other Methods

Determination of Water in Microscopic Sections

A method for determining water in small sections of wheat grain tissue was proposed by RUCH et al. (1963). The method is based on changes in the refractive index of the tissue brought about by changes in water content. The section is immersed in indane, a liquid with a refractive index ($n_D = 1.537$) similar to that of the dry tissue ($n_D = 1.53$) and photographed in an interference microscope adjusted to give fringes in the field. Afterwards, the preparation is completely dried under reduced pressure (0.13 mbar, 96 °C, 3 h), re-immersed in indane and photographed again. From the shift in the interference fringes between the two photographs it is possible to calculate the water content of the tissue. The method could be modified for other material.

Determination of Water by Distillation with Toluene

Distillation with toluene (MILLER 1942, 1943) is another method which can be used for material with a high content of volatile compounds. The sample is distilled with toluene in a special device and the amount of water distilled off is determined with a micropipette (MILLER 1942) or by weighing (MILLER 1943).

2.2 Indirect Methods of Water Content Determination

2.2.1 Measurement of the Absorption of Radioactive Radiation

The measurement of the absorption of radioactive radiation during its penetration through part of the plant, or occasionally through the whole stand (see later), is a method which can be used for continuous study of changes in plant water content. These methods are used when the content of non-aqueous compounds of the tissue does not change during the determination, so that the decrease of radiation intensity during penetration through the measured organ is proportional to the hydration level of the organ.

Determination of the Water Content of Leaves

Non-destructive beta radiation is usually used for the determination of the water content of leaves (e.g. BOLGARINA and EĬDUS 1956, YAMADA et al. 1958, MEDERSKI 1961, 1964, GARDNER and NIEMAN 1964, NAKAYAMA and EHRLER 1964, MEDERSKI and ALLES 1968, NESTEROV and NIKOLAEV 1968, 1969, BUSCHBOM 1970, JONES 1973).

The interaction of radiation with the absorber (leaf) is given by an exponential equation (SIRRI 1949, NAKAYAMA and EHRLER 1964):

$$I = I^0 \, e^{-\mu x}, \tag{2.1}$$

where I is the intensity of transmitted radiation [counts s^{-1}], I^0 is the intensity of incident radiation [counts s^{-1}], μ is the mass absorption coefficient [cm^2 mg^{-1}] and x is the absorber thickness [mg (f.w.) cm^{-2}].

x is determined from the fresh weight per unit area of the leaf. As water represents a considerable fraction of the fresh weight, changes in x depend on changes in the water content (if there is no change in dry weight). The sensitivity of the method, i.e. the magnitude of the change in count rate during unit change of the absorber thickness x, is then expressed by the equation

$$\frac{dI}{dx} = -I^0 \mu e^{-\mu x}. \tag{2.2}$$

It follows that the sensitivity of the method may be increased by increasing both the intensity of the incident radiation I^0 and the mass absorption coefficient μ. The relative error of the measurement of changes in x for a given absorber thickness $\Delta x/x$ can be found from equation (2.3)

$$\frac{\Delta x}{x} = -\frac{e^{\mu x} I}{\mu x \, I^0}. \tag{2.3}$$

Table 2.1 *Radiation sources used in beta gauge (modified after* WEAST *1968/69 and* BUSCHBOM *1970).*

Isotope	Half-life	Particles (β^-) energy [MeV]	Source activity [μCi]	Distance between source and detector [mm]	References
Carbon-14	5 770 a	0.156	*	3.175	MEDERSKI (1961) WHITEMAN and WILSON (1963) NAKAYAMA and EHRLER (1964) SKIDMORE and STONE (1964) MEDERSKI and ALLES (1968)
			5.8	3.5	ROLSTON and HORTON (1968) TAKECHI et al. (1970)
Phosphorus-32	14.3 d**	1.71		140	BOLGARINA and EÏDUS (1956) ANTOSZEWSKI and LIS (1973)
Sulphur-35	86.7 d	0.167			YAMADA et al. (1958)
Calcium-45	152 d	0.255	10	80	HOPMANS (1971)
Krypton-85	10.4 a	0.67	350	50	BUSCHBOM (1970)
Technetium-99	2.1×10^5 a	0.29			GARDNER and NIEMAN (1964) NAKAYAMA and EHRLER (1964) LANG et al. (1969)
Promethium-147	2.5 a	0.225	2.16	1.59	NAKAYAMA and EHRLER (1964) EHRLER et al. (1965, 1966)
			$3; 6; 30 \times 10^2$ 1 000	1.57	JARVIS and SLATYER (1966) BIELORAI (1968)
			2.0	3.0	PEYNADO and YOUNG (1968)
			5.13		ROLSTON and HORTON (1968)
			0.47	0.4	YANG and DE JONG (1968)
			600	30	JONES (1973)
Thallium-204	3.9 a	0.77			NAKAYAMA and EHRLER (1964)
			1 000		JARVIS and SLATYER (1966)
				234	TAKECHI (1970), TAKECHI et al. (1970)

* Commercially prepared ^{14}C disc in an aluminium holder.
** A correction must be taken for desintegration.

It is evident from this equation that the error is minimum when $x = 1/\mu$. If the instrument error is negligible in respect to the statistical fluctuation od source radiation, the minimum error is observed when $x = 2/\mu$ (NAKAYAMA and EHRLER 1964).

Carbon-14, promethium-147, technetium-99, thallium-204 and krypton-85 are useful sources of beta radiation for leaves of usual thickness (mostly from 10 to 40 mg (f.w.) cm^{-2}). In practice, different sources have been used in different devices; they are summarized in Table 2.1. JARVIS and SLATYER (1966a) used several different, very small, radiation sources for leaves of various thickness during continuous beta-particle gauging in their leaf chamber for studies of gas exchange. They used 0.3, 0.6 and 3 mCi of ^{147}Pm, for leaves $10-30$ mg cm^{-2} thick; and 1 mCi of ^{204}Tl for thicker leaves. Leaf thickness x at which the minimum error is observed are given in Table 2.2.

Table 2.2 *Suitable absorber thickness (i.e. fresh weight divided by area of leaf sampled) for various radiation sources in beta gauge (after* NAKAYAMA *and* EHRLER *1964).*

Radiation source	$2/\mu$	Suitable range of the leaf thickness [mg f.w. cm^{-2}]
carbon-14	7.6	5 to 20
promethium-147	12.3	10 to 40
technetium-99	17.5	10 to 40
thallium-204	64.5	15 to at least 70

A Geiger-Müller tube, crystal scintillator or ion chamber may serve as detector of the radiation passing through the plant material. A thin, mica, end-window GM tube (*e.g.* MEDERSKI 1961, NAKAYAMA and EHRLER 1964, JARVIS and SLATYER 1966a), which is readily available and may be used in the field, is usually used. On the other hand, an ion chamber detects high intensity radiation and hence permits the sensitive use of sources of high activity.

During the measurements, the leaf, usually attached, is placed between the source of radiation and the GM tube (the distance between the two devices varies, *e.g.* 1.59 mm – NAKAYAMA and EHRLER 1964, 8 mm – JARVIS and SLATYER 1966a, 234 mm – TAKECHI 1970), the same part of the leaf always being exposed. The amount of radiation passing through the tissue is measured for 1 or 2 min for a single measurement or is continuously recorded using a rate meter. Depending on the count rate (*i.e.* source strength, leaf thickness) various periods of integration may be used. For a fairly low count rate, say 50 counts s^{-1}, a period of integration of about 1 min is desirable. Using this arrangement, MEDERSKI (1961) found a linear relationship between the logarithm of the count rate and the water saturation deficit (Fig. 2.1). A similar but non-linear relationship was found by JARVIS and SLATYER (1966a, b). However, these authors observed a linear relationship between relative water content (water saturation deficit) and "effective leaf thickness"

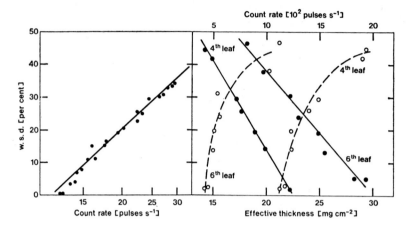

Fig. 2.1 The intensity of beta-radiation after passing through leaves with different water saturation deficits (MEDERSKI 1961).

Fig. 2.2 The relation between w.s.d. and effective thickness (solid lines) and between w.s.d.
and count rate (dashed lines) for a young, thin cotton leaf (4th from the apex) and an older,
thicker leaf (6th from the apex) (JARVIS and SLATYER 1966b).

(Fig. 2.2), which corresponds to absorber thickness x in Eq. 2.1 and in the following
equations. These and other authors (*e.g.* YANG and DE JONG 1968, MEDERSKI and
ALLES (1968) draw attention to the considerable changes in the relation between
count rate and water potential as the leaves age. Leaves also differ from one another
according to species, position on the plant and environmental conditions during
development. MEDERSKI and ALLES (1968) pointed out that very young leaves
should not be monitored for more than a few hours before removing them for
calibration. Individual variability among different plants of the same species, age
and type of cultivation is also relatively high (YANG and DE JONG 1968). This variability can be reduced by subtracting the counts measured on the dried sample
from those for the fresh sample (WHITEMAN and WILSON 1963).

Calibration of beta-ray gauges for measuring leaf water content is tedious and
cumbersome, involving repeated simultaneous determinations of beta-ray absorption and leaf water content, or water potential (JARVIS and SLATYER 1966b).
According to these authors, the calibration procedure can be much simplified, as
long as the beta-ray absorption at a single known relative water content (or water
saturation deficit, see Section 2.5) is measured, together with the saturated and dry
weights of a sample of leaf discs. The authors explained the above conclusion,
which serves as the basis for their calibration procedure as follows. Water saturation
deficit (w.s.d.) can be related to effective thickness x [mg (f.w.) cm^{-2}], in the following way (weight in mg):

$$\text{w.s.d.} = \frac{\text{saturated weight} - \text{initial (fresh) weight}}{\text{saturated weight} - \text{dry weight}} \times 100 \, [\%] \, . \qquad (2.4)$$

The effective thickness x is given by

$$x = \frac{\text{initial (fresh) weight}}{A} \; [\text{mg cm}^{-2}], \qquad (2.5)$$

where A is the area of the leaf sample [cm^2].
Thus

$$\text{w.s.d.} = \frac{\text{saturated weight} - Ax}{\text{saturated weight} - \text{dry weight}} \times 100 \; [\%]. \qquad (2.6)$$

The slope of the relationship between w.s.d. ($\Delta W_{(\text{sat})}$) and x, $d\Delta W_{(\text{sat})}/dx$ is then obtained by differentiation, giving

$$\frac{d\Delta W_{(\text{sat})}}{dx} = \frac{-100 \, A}{\text{saturated weight} - \text{dry weight}}. \qquad (2.7)$$

Thus the slope is determined by the water content at zero water saturation deficit (*i.e.* at relative water content $= 100\%$) and, knowing the slope, the intercept may be obtained from one estimate of x at a known value of w.s.d. Since, for the particular source-detector geometry and characteristics used in a beta gauge, x can be uniquely related to count rate, calibration can simply be achieved by determining only:

a) the count rate/effective thickness relationship for the system used,
b) the saturated weight, *i.e.* the weight of a disc sample of a known area at zero water saturation deficit,
c) the oven dry weight of the sample.

The relationship between count rate and effective thickness, which is independent of the absorber characteristics, is first established by using leaf discs of varying water content or aluminium foil of varying thickness. For this purpose, discs punched out of foil are mounted in small Plexiglass holders with transparent windows (*e.g.* of Mylar), and each holder arranged, in turn, in the beta gauge, so that the foil discs are in the same geometrical relationship as a leaf used in the experiments. An example of the relationship is shown in Fig. 2.3 (JARVIS and SLATYER 1966b).

A series of measurements can then be made with a leaf in position, and no further calibrations between count rate and effective thickness are required, as long as the source-detector geometry and characteristics remain unchanged.

At the end of each experimental run a disc sample of a known area is punched out of the leaf being monitored and fully saturated using the procedure described in Section 2.5.3. Each disc is then placed in one of the standard holders and a count rate obtained. The sample is then weighed to obtain the saturated weight, and oven-dried and weighed to obtain the dry weight. The slope $d\Delta W_{(\text{sat})}/dx$ is cal-

culated from Eq. 2.7 and the intercept from Eq. 2.5, saturated weight being substituted for initial weight. A range of typical calibration curves for a series of
leaves of different ages growing on a single plant, and the dependence of their
slopes and intercepts on leaf age, are shown in Figs. 2.4 and 2.5 (JARVIS and SLA-
TYER 1966b, MEDERSKI and ALLES 1968). However, as pointed out by MEDERSKI
and ALLES (1968), the procedure of JARVIS and SLATYER is based on the assumption
that the tissue does not shrink in area during the loss of water. It is necessary to
use another calibration method, or to determine the error experimentally, when
using material in which there is appreciable shrinkage (*cf.* JONES 1973).

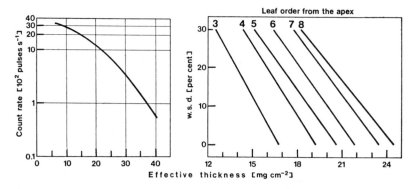

Fig. 2.3 Relationship between count rate and effective thickness, with the same source-
detector characteristics and geometry as for experimental measurements (JARVIS and SLATYER
1966b).

Fig. 2.4 Beta-gauge calibration curves, for a series of cotton leaves numbered from the
apex (JARVIS and SLATYER 1966b).

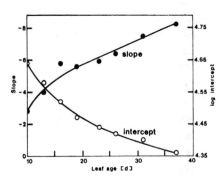

Fig. 2.5 The log of slope and log of *y*-intercept of the calibration lines as a function of leaf
age. Slope is the difference between the log count at zero w.s.d. and log count at 20 % w. s. d.
divided by 20. For convenience, the slope values on the ordinate have been multiplied by 10^3
(JARVIS and SLATYER 1966b).

Determination of the Water Content of Thicker and Denser Organs

More penetrating gamma radiation must be used to measure the water content of thicker and denser organs (*e.g.* in trunks of trees). KLEMM (1959, 1965, 1966; see also KLEMM and KLEMM 1964) placed the radiation source (1 mCi of ^{137}Cs) in a 28 mm thick lead container with a 5 mm aperture, through which a narrow radiation beam emerged. The beam penetrates the trunk and its intensity after passage is measured with a constant-temperature GM tube. KŰHN (1961; see also HINSCH and NIEMANN 1964) measured the water content of tree trunks by the deceleration of fast neutrons by hydrogen atoms and the measurement of the intensity of the decelerated neutrons. The radiation source (50 mCi Ra/Be) was placed in a hole drilled in the trunk and the intensity of decelerated neutrons measured by a detector at the surface of the trunk. The author compared this method with others, *e.g.* with the measurement of scattering of gamma radiation.

Determination of the Water Content of a Plant Community

In principle, the same device used for the measurement of the water content of leaves and other organs may also be used, suitably enlarged of course, to follow the water content of a whole vegetative cover (UNGER 1958, 1959, 1963; compare also with GLUBRECHT *et al.* 1959, KAINDL and HUBMER 1960). This measurement is also based on the assumption of negligible changes in the content of compounds other than water in the material. UNGER used the method both for measuring the water content of the stand and for recording the increase in organic matter during the whole vegetation period.

The main part of the device is a measuring unit bridging the stand and moved on tracks above it. A source of gamma radiation is fixed to one supporting pillar at the side of the stand and a detector is attached to the pillar at the other side. Both elements can be moved simultaneously in 10 cm steps, to various heights. It is thus possible to measure the decrease in the radiation intensity, *i.e.* the water content (or better the content of matter) during the vegetation period. ^{137}Cs and ^{60}Co are suitable sources for absorber thickness of 12 to 70 g (f.w.) cm^{-2} and 20 to 120 g cm^{-2}, respectively. Any type of radiation detector, *i.e.* ion chamber, scintillation chamber, or occasionally even a GM tube, may be used. (For a detailed technical description of various modifications to this device see UNGER 1963).

2.2.2 Measurement of Dielectric Constant (Capacitance) and Related Procedures

Some methods for the determination of plant water content utilize the dependence of the dielectric constant on the presence of water. APPERLOO (1956) extracted water from a sample of plant material with dioxane, the dielectric constant of which was then measured. It is also possible to measure the dielectric constant of the leaf tissue directly. BATYUK and BRATACHEVSKIĬ (1958) enclosed discs of leaf tissue in a Plexiglass cylinder (10 mm high, 15 mm in diameter) with an aluminium cap (which formed one electrode). The capacitance of the condenser so formed depends on the water content of the tissue sample. A small transistorized device for measuring the dielectric constant of leaves *in situ* was constructed by GEZALYAN (1966, 1968, KAZARYAN and GEZALYAN 1968). Similar procedures have been used by MURR (1965), ISHMUKHAMETOVA and KHOKHLOVA (1971) and ISHMUKHAMETOVA and STARTSEVA (1971). ISHMUKHAMETOVA and RYBKINA (1972) have used this method for studying the water status in isolated chloroplasts.

Methods of measuring changes in capacitance of a suitable condenser when air as a dielectric is replaced by the crop have been used by many authors (FLETCHER and ROBINSON 1956, CAMPBELL *et al.* 1962, JONES 1963, 1964, 1965, 1966, ALCOCK 1964, HYDE and LAWRENCE 1964, JONES and HAYDOCK 1970, and others), for the estimation of changes in the water content of vegetation. Basically the apparatus is a radio frequency (R.F.) tuned circuit in which the probes are a condenser. The frequency of oscillation of the R.F. circuit is related to the capacitance between the electrodes of the condenser. Insertion of material between the electrodes causes an increase in the capacitance and hence a reduction in the frequency of the oscillation, which can be measured.

Selective absorption of centimetre wave radiation (hyperfrequency) by water has also been used for water content assessment (KOSMIN and FILIPPOV 1968; for theory see also CHARBONNIÈRE 1970). GOSSE and DE PARCEVAUX (1970) found a good correlation between the water content of a leaf or a canopy and the absorption of 3 cm (10 000 MHz) radiation. They used a commercially available generator (klystron) and receptor of centimetre radiation (Philips). The accuracy of this new method is not yet sufficient for precise measurements.

2.2.3 Measurement of Electrical Conductivity of the Tissue

BOX and LEMON (1958), NAMKEN and LEMON (1960), BRACH and MASON (1965), and KREEB (1963, 1966) have estimated changes in the water content of leaves from changes in the electrical conductivity of leaf tissue (*cf.* also NIKITIN 1966 for other tissues). The conductivity of a tissue is determined by the concentration of electrolytes and the water content. Changes in conductivity are proportional to changes in water content and the concentration of electrolytes, which changes only slightly over a short time.

The conductivity of the tissue has usually been measured by inserting two platinum wire electrodes into the mesophyll or by fixing them with a forceps-like holder to the upper side of the leaf from which the outer layer of cells has been shaved away with a razor blade, so that the electrodes make good contact with the mesophyll (*e.g.* KREEB 1963). The conductivity is measured with an A.C. bridge with a transistor multivibrator. Similar methods have formerly been used for determining the permeability of protoplasm, for the assessment of tissue vitality and of temperature damage, for control of watering, *etc.* (see WALTER and KREEB 1970, p. 163, for literature references).

This kind of electrode does considerable damage to the tissue. A useful modification of the method involves the use of the electrically conductive silver solution "Argentol" for sealing the electrodes to the leaf surface (KREEB 1966). However, "Argentol" contains organic solvents which infiltrate into and damage the mesophyll tissue. In a further recent modification, the electrodes were made directly on the leaf surface using a conductive silver paste (*e.g.* a mixture of very fine silver powder and a dispersion glue or paint). A suitable paste is made of 2 parts of "Mowilith DM 21" (Farbwerke Hoechst AG.) and 1 part of "Silberbronce D 12" (Degussa) (WALTER and KREEB 1970, pp. 164–5).

As conductivity is temperature-dependent, leaf temperature should be measured simultaneously, and the results corrected before data are compared.

2.2.4 Other Methods

MILBURN (1966) and MILBURN and JOHNSON (1966) have constructed a sensitive vibration detector to make audible the vibrations which might be produced by water cavitations in the tissue during water loss or uptake.

Nuclear magnetic resonance (NMR) has been used for measurement of the water content in grain, and the procedure could be modified for other plant material. KUZNETSOV (1961) has designed a device for this purpose, consisting of a permanent magnet with a tension of 2 400 ersteds, a radio-frequency generator, low-frequency amplifier, oscillograph and audio-generator.

Similar procedure has been used by SAMUILOV *et al.* (1971) for determining the state of water in plant tissues.

Regression equations expressing reflectance of incident radiation from the adaxial leaf surface as a function of the water saturation deficit and water content were developed by THOMAS *et al.* (1971). Monochromatic reflectance at the water absorption bands at 1.45 and 1.93 µm, measured with a spectrophotometer with a reflectance attachment, was significantly related ($P = 0.01$) to w.s.d. in leaves of cotton, maize and citrus. However, because of variations in internal leaf structure during leaf development, the predictability of leaf water status from reflectance measurements was poor. Within the usual w.s.d. range (up to 30% w.s.d.), reflectance changes in cotton were small and might not be predictable because of variation among leaves.

2.3 Determination of the Water Content of Cell Walls

Cell walls are considered to be the probable main path of water movement through parenchymatous tissue and to be part of the so-called "apparent free space" (BRIGGS and ROBERTSON 1957; KRAMER 1957), *i.e.* that part of the volume of plant tissues into which dissolved compounds from outside can freely enter by diffusion. The volume of this free space is considered to vary between 3 and 50 % of the total tissue volume. As the cell walls also serve as an important reservoir of water in plant tissues, the cell wall water content has recently became a subject of interest to plant physiologists.

HÄRTEL (1951) determined imbibition curves for cell walls of mesophytes and found them similar in whole leaves and pulverized preparations.

GAFF and CARR (1961) used a preparation of cell walls of leaves of *Eucalyptus globulus* LABILL., from which soluble compounds and residues of protoplasm had been removed. The cellulose preparation was then equilibrated above sucrose solutions (0 to -26 bar) giving water contents ranging from a maximum of 150 to 95% of the dry matter. The fraction of dry cell walls in leaves was then estimated in three ways to be 13%, 27% or 16% of the fresh weight of leaf blades. The maximum water content of the cell walls was therefore equivalent to roughly 40% of the water content of fully saturated leaves.

For whole, dried (above $CaCl_2$. 6 H_2O) and washed leaves of *Eucalyptus aggregata, E. niphophilla* and *Ilex aquifolium*, KREEB (1965a) found that the water content of the cell walls at full saturation was equivalent to 53 to 85% of the dry weight, *i.e.* 18 to 64% of the total water content.

2.4 Root Bridge Method

The normal morphological structure of plants makes direct determination of the components of the water balance by weighing impossible. Any separation of absorptive and transpiring parts of the plant disturbs their normal relations, annuls natural gradients of water potential and disturbs other correlations. It is thus impossible to make measurements of water balance of the same plant over a long period of time or during its whole ontogenesis.

Partial separation of the above-ground and underground parts of the plant (to make it possible to measure both the components of water balance simultaneously) without destroying the connection between them can be achieved by the use of the so-called root bridges method proposed by WERNER (1931, 1933, 1948).

A root bridge between the active part of the root system and the shoot may be formed in practically all plants in such a way that the growing root of a seedling is adjusted by degrees to a dry atmospheric environment, so that one long root is joined with a woody suberized, rhizodermal layer, which enables it to tolerate dry conditions. This above-ground root part may be considerably extended and a plant with a root bridge, or root thread, up to 200 cm long may be obtained. As a result it is possible to weigh separately the above-ground plant part and the roots with soil, while preserving the connection between them. The root thread forms root hairs in a humid environment and is able to recommence absorptive activity.

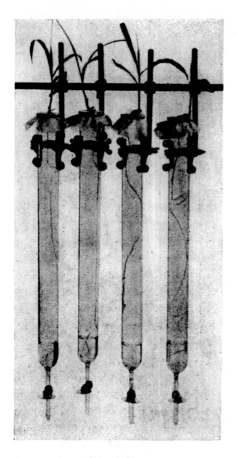

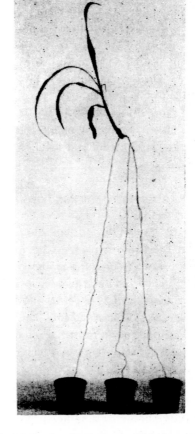

Fig. 2.6 Cultivation of plants with root bridges by decreasing the level of the nutrient solution in glass cylinders.

Fig. 2.7 Young plant of maize with three root bridges.

Hence the root and shoot environments can be controlled separately as required.

The first method of growing plants with root bridges was described by WERNER (1931). His simple device was formed by five interconnected T-pieces connected with a float chamber maintaining a constant level of the nutrient solution. Plants fixed in sheets of cork moved along wooden rods behind each T-piece. The root of a germinated seed passed through the hole made in the cork and reached the nutrient solution contained in the T-piece. Care was taken to ensure that only 2 to 4 cm of root remained immersed in the nutrient solution, by drawing up the root each day by the amount of the daily increase in root length. Alternatively, and somewhat more advisably, the level of the nutrient solution can be lowered while the position of the plant remains fixed. Glass cylinders about 60 cm long and 3 to 5 cm in diameter are useful for this type of cultivation (Fig. 2.6). Seeds are fixed to polyamide nets on the upper edge of the cylinder. The surface of the solution is covered with liquid paraffin, to prevent increase in the humidity of the air in the cylinder above. The germinated seed is protected with polyethylene film and slightly moist filter paper for the first two days after transfer to the cylinder.

This procedure is useful for obtaining a fairly small number of experimental plants. The low percentage of losses during the extension of the root thread is one of its advantages. It is easy, particularly with cereals, to grow plants with two or even more bridges (Fig. 2.7).

A larger number of plants with a root bridge may be successfully grown in a large vessel of nutrient solution which is aerated, or changed every second or third day.

The roots of seedlings (3 to 4 cm long) are inserted into the nutrient solution through the corresponding holes in two PVC or aluminium plates covering the vessel so that the seeds, and later the young plants, remain on the upper plate. At first, the level of the nutrient solution is lowered as the roots grow down. The upper plate, carrying the plants, is then raised when the level of the solution is about 5 cm below the plates. The high air humidity between the surface of the nutrient solution and the lower plate helps to decrease the percentage of losses. There are more losses during the extension of the root threads with this procedure. However, it requires less space and care than the preceding method.

When the root threads are sufficiently long, the plants are raised no further (or the level of the nutrient solution is lowered no further) and the major portion of the root system is left to grow in the nutrient medium so that there is adequate absorption of water and nutrients during the experiment proper. The main advantage of growing the plants in a nutrient solution is the more rapid extension of the root threads, but there are subsequent difficulties in transferring the plants to soil. Considerable care should be taken during transfer and manipulation of the root thread to prevent its damage, which would result in an additional contraction of the conducting elements in the root thread. If weight changes of the soil as well as of the plant are to be followed (water uptake and transpiration), a vessel made from a material which does not absorb water nor change its weight for any other reason (*e.g.* PVC or metal sheet; Fig. 2.8), with a closely fitting cover preventing evaporation of water from the soil, should be used. The root thread passes through a hole in the cover with smooth round edges. A suitable support should be used for the above-ground part of the plant when transplanting. A wire hook is recommended as it makes the transfer of the plant easier and can be used for hanging the plant on a balance.

Cultivation in the soil-filled paper cylinders used by Sítař (1957) was found to be very suitable for growing root bridges in soil. As the roots of the plant in the cylinder (about 8 cm high) reach the bottom, another cylinder is put below and the two are connected by a collar made of parchment or other tough paper. When the roots reach the lower cylinder, the upper is left to dry slowly and the dry soil is removed

FIG. 2.8 Young plants of kale with root bridges, rooted in PVC containers. The thin root bridge (a) allows sufficiently precise weighing; the root bridge (b) is too thick for this method.

with a brush. This procedure is repeated until the root bridge is sufficiently long. Since frequent watering would damage the paper, they are placed on several cm of fine, well-watered sand from which they take up water. Water cannot penetrate in a sufficient amount to the roots in the upper cylinder when a further cylinder is added to the base of the series, so the bottom two cylinders are then carefully watered for two to three days.

The root thread extend more slowly during cultivation in soil than in nutrient solution, but the development of the above-ground part of the plant is faster and the plants are more resistant to drying out.

Prior to the beginning of the measurements, *i.e.* before weighing, the root bridges should be adjusted to bend easily, as their flexibility increases the accuracy of weighing. This is particularly important with dicotyledonous plants, in which the root threads exhibit secondary growth. The accuracy of weighing the two parts of the plant is determined by several factors. The errors caused by the connecting bridge may be minimized by taking the following precautions: (a) the bridge is long and thin, (b) the center of the bridge is fixed somewhat higher than the weighed parts and (c) the position of the weighed parts does not change, *i.e.* the plant remains on the balance during the experiment. In monocotyledonous plants minimum errors are $\pm$ 0.02 to 0.05 g, in dicotyledonous plants $\pm$ 0.1 to 1 g.

Studies of water relations using the root bridge method are relatively recent. This method has so far been used mainly by WERNER's coworkers. It has already been possible to develop a root thread in most cultivated plants, including woody plants. In spite of numerous attempts, the method has failed in onion and asparagus plants.

2.5 Water Saturation Deficit (Relative Water Content)

2.5.1 Characteristics of These and Analogous Terms

The water saturation deficit (w.s.d., $\Delta W_{(sat)}$) is a useful indicator of the state of water balance of a plant mainly because it expresses the absolute amount of water which the plant requires to reach full saturation. The w.s.d. is hence very important which the plant requires to reach full saturation. The w.s.d. is hence very important, for example for the analysis of the water balance of a whole stand, for quantitative calculations of the overall water requirement in a certain area of the stand, and for comparison of the water relations of a plant or a stand with soil moisture content in the root zone, *etc.* On the other hand, the water saturation deficit is often correlated with leaf water potential, so that experimental values of w.s.d., $\Delta W_{(sat)}$, which is more easily measured than Ψ_w, serve as a measure of water potential, (*e.g.* WEATHERLEY and SLATYER 1957, CORNEJO and VAADIA 1960, WHITEMAN 1961, CARR and GAFF 1962, JARVIS and JARVIS 1963, 1965, WHITEMAN and WILSON 1963, BEGG *et al.* 1964, SHEPHERD 1964, TAYLOR 1964, GARDNER and EHLIG 1965, LEMÉE and GONZALEZ 1965, MACKLON and WEATHERLEY 1965, WEATHERLEY 1965b). However, the relationship between $\Delta W_{(sat)}$ and Ψ_w varies considerably according to type and age of material (*cf.* SLATYER 1960). A calibration curve must be found experimentally for each type of material (*e.g.* KNIPLING 1967, YANG and DE JONG 1968). The relationships between $\Delta W_{(sat)}$ and Ψ_w has been discussed in detail by BARRS (1968), who gives further references (*cf.* also MILLAR *et al.* 1968, NOY-MEIR and GINZBURG 1969, GLINKA and KATCHANSKY 1970, JORDAN and RITCHIE 1971, KLEPPER *et al.* 1971, NAMKEN *et al.* 1971, POSPÍŠILOVÁ 1973, 1974, SÁNCHEZ-DÍAZ and KRAMER 1973).

The term water saturation deficit (Wassersättigungsdefizit) was first introduced by STOCKER (1928), on the basis of work by ILJIN (1923) (for later modifications of the term and for references see HEWLETT and KRAMER 1963). It is the amount of water which the plant, or part of the plant, requires to reach full saturation; it is the difference between the water content of the plant at a given time and the water content at maximum saturation, *i.e.* at full turgor* expressed as per cent of maxi-

---

* Maximum (100%) saturation during the determination of water saturation deficit (relative water content) is an artificially produced saturation which does not correspond exactly to the maximum possible saturation of the plant or tissue in a natural situation (*cf.* also BURSTRÖM 1966). Both terms, water saturation deficit and relative water content, are thus methodically defined technical terms (*cf.* also temperature effect, p. 148).

mum water content:

$$\Delta W_{(sat)} = \frac{\text{saturated weight} - \text{initial (fresh) weight}}{\text{saturated weight} - \text{dry weight}} \times 100 \, [\%]. \qquad (2.8)$$

The term water saturation deficit is analogous to the term relative water content, r.w.c. (water content at full turgor being the basis of both, *cf.* SLATYER and BARRS 1965, BARRS 1968). The terms are identical as far as their determination is concerned, differing only in their numerical expression. The relative water content (in this narrower sense of the term, in which it is usually used) expresses the water content in per cent at a given time as related to the water content at full turgor:

$$\text{Relative water content} = \frac{\text{initial weight} - \text{dry weight}}{\text{saturated weight} - \text{dry weight}} \times 100 \, [\%], \qquad (2.9)$$

$$= 100 - \text{water saturation deficit.}$$

The term relative water content was originally defined as "relative turgidity" (WEATHERLEY 1950) or as "hydrability" (OCHI 1952). "Relative turgidity" has often been used (see a review by HEWLETT and KRAMER 1963) but it is an unsuitable term, as the measured value by no means expresses turgor.

The determination of water saturation deficit, or relative water content will be described in the following paragraphs.

2.5.2 Review and Discussion of the Methods which Have Been Used

The methods which have been used for the determination of water saturation deficit are summarized in Table 2.3. STOCKER (1928, 1929a, b) suggested a simple procedure for the determination of w.s.d. The plant or part of the plant is cut, rapidly weighed in darkness in a saturated atmosphere, and left with the stem or petiole in water to saturate to constant weight. After saturation, it is dried, weighed again and the dry weight determined.

This procedure of STOCKER has been modified and often simplified by many authors, with consequent redefinition of water saturation deficit (*e.g.* STEFANOFF 1931, VASSILJEV 1931, HALMA 1933, COMPTON 1936, EVENARI and RICHTER 1937, KILLIAN 1947, MOURAVIEFF 1959, 1964, STÅLFELT 1961, HARMS and McGREGOR 1962, CLAUSEN and KOZLOWSKI 1965, for details see HEWLETT and KRAMER 1963). We use the term water saturation deficit below only in the sense of Stocker's definition (Eq. 2.8).

Stocker's method has disadvantages which in some cases make its use impossible:

(1) More or less constant weight is reached during saturation of mature leaves only. Young leaves which are still undergoing expansion absorb water for a substantially longer period and attainment of a constant weight (usually after several days) corresponds to the beginning of death of the leaf (the so-called growth error).

Table 2.3 *Review of methods for determining water saturation deficit.*

Principle	Sample	Technique		References to the basic procedure	Remarks
Increase in weight after artificial saturation	whole leaf or plant	one weighing after saturation	saturation at barometric pressure	STOCKER (1928, 1929 a,b)	simple, but awkward; growth error considerable
			saturation after exposure at low pressure	HAMNER et al. (1945)	
		weighing several times during saturation		YEMM and WILLIS (1954)	usable for dissected leaves
	leaf discs or segments	immersed in water		SHREVE (1924)	growth error; edge infiltration error
		floating on water	—	WEATHERLEY (1950)	
			with correction for dry wt. decrease	WERNER (1954)	
			with correction for edge infiltration error	SPOMER (1972a)	edge infiltration error eliminated
			with inhibition of extension growth	BARRS and WEATHERLEY (1962), SLATYER and BARRS (1965), MILBURN and WEATHERLEY (1971)	growth error diminished or almost eliminated

Measurement of volume and weight of water taken up	leaf or plant	in agar sheet		ČATSKÝ (1959)	growth error, edge infiltration error
		in polyurethane foam sheet	one weighing after saturation	ČATSKÝ (1960)	growth error, edge infiltration diminished
			extrapolation	ČATSKÝ (1963, 1965)	growth error eliminated, edge infiltration diminished
				CZERSKI (1968)	rather complicated; usable for complex studies
Approximate methods	plant	visual comparison of plants with photographic standards		DOWNEY and MILLER (1971)	simple, rapid
	attached leaf *in situ*	measurement of leaf reflectance		THOMAS *et al.* (1971)	affected by many factors (*e.g.* leaf age)
	leaf or plant	actual water content expressed on the basis of early morning water content		VASSILJEV (1931) EVENARI and RICHTER (1937)	simple, reliable also for very high w.s.d.
	tissue sample	relation between w.s.d and proline amount in the tissue		PÁLFI and JUHÁSZ (1971)	rather laborious, indirect assessment; usable in complex studies

(2) Saturation of whole leaves or plants takes a long time, several hours or even days. Full saturation of the tissue is usually reached after 24 to 48 h. During this time the experimental material respires up to several per cent of its dry weight; this can result in a considerable error.

The other disadvantages are of rather a technical nature, and are less important at first sight: However, they add to the laboriousness of the whole procedure and thus limit the possible range of experimental work (*e.g.* large vessels are needed for saturation of a large number of samples, whole leaves or plants are destroyed, marking of individual samples during group saturation is difficult, drying the surface of moist leaves after saturation is not perfect, *etc.*).

These inadequacies and errors may be avoided in several ways:

(1) By reducing the time necessary for saturation of the tissue (*e.g.* by reducing the size of the sample),

(2) by chemical or physical inhibition of extension growth during the determination,

(3) by elimination of errors due to growth and respiration by calculation based on the course of the saturation curve.

Unsuccessful attempts have been made to accelerate saturation of whole leaves by using low pressure (HAMNER *et al.* 1945, ČATSKÝ 1959). The technique has been considerably improved by using small discs of leaf tissue which rapidly become saturated on immersion in water (*e.g.* SHREVE 1924), floating on water (*e.g.* MASON and PHILLIS 1942, WEATHERLEY 1950), or placing in a wet sheet of polyurethane foam (ČATSKÝ 1960, 1962).

Discs of leaf tissue are used for very varied kinds of studies both of water relations and of various physiological processes in plants. Many studies have shown that there are no objections to the use of discs from the physiological point of view, although the possibility of higher saturated water content of isolated pieces of tissue than of the intact leaf must be considered. It is often more suitable to measure the water saturation deficit of discs of leaf tissue alone than to determine the average value of the whole leaf or plant irrespective of its heterogeneity (*e.g.* when determining correlation between w.s.d. and physiological processes). WEATHERLEY (1950, 1951) reported that the rate and relative amount of water uptake was the same for discs and whole leaves of cotton. On the other hand, HEWLETT and KRAMER (1963) found in twelve species of hardwood trees a higher "full saturation" water content of discs as compared with whole leaves; this caused an error of + 3.7%, when the water saturation deficit varied between 0 and 27%.

The first convenient disc method was designed by WEATHERLEY (1950). The original, simplest technique consists of cutting a set of discs from the leaves, weighing the discs, and saturating them by floating on water. After 24 h they are surface-dried and weighed again, and then their dry weight is determined. More complex modifications (*e.g.* WERNER 1954) include cutting and weighing two parallel samples of discs from the leaves; one sample is saturated by floating on water, the other serves as the control and for the correction of dry weight losses by respiration during long exposures (24 h).

BARRS and WEATHERLEY (1962) reduced the time of saturation to 4 h and exposed the discs to an irradiance corresponding roughly to the compensation irradiance of photosynthesis of the discs. This avoided the error arising from loss of weight by respiration and limited the error resulting from growth of the discs. Nevertheless, accurate results cannot be obtained with young leaves, even with this modification. Furthermore, some technical difficulties reamin: *e.g.* the rather awkward manipulation of dishes containing water; the requirement for separate dishes for each sample, the possibility of infiltration into floating discs in some material; the possibility of slower saturation resulting from inadequate wetting of the cut edge, especially in discs from wrinkled leaves (the discs absorb water mainly through the cut edges, see *e.g.* WEATHERLEY 1947, OPPENHEIMER 1954), and transpiration which does not stop completely. A very detailed analysis of errors in the determination of w.s.d. by using floating discs (during determination of fresh, saturated and dry weights) was published by SLATYER and BARRS (1965). Their conclusions are generally valid and hence applicable to other methods.

Some inadequacies of the Weatherley method may be minimized by saturating the discs in holes in a wet sheet of foam polyurethane (ČATSKÝ 1960). In this system infiltration is usually negligible and full saturation of the tissue is usually reached after only 3 h, thus decreasing the so-called growth error. However, even this technique gives higher saturation water contents (Fig. 2.9) for the youngest, most rapidly growing leaves.

Inhibition of extension growth during saturation of the tissue, by chemical or physical means, is another possible way of avoiding the growth error. The effect od various concentrations of maleic hydrazide, KCN, indole-3-acetic acid or coumarine, and the effect of radiation of various wavelengths (ČATSKÝ, unpublished results), has been tested when saturating discs in polyurethane foam sheet. The effects of KCN, low temperatures and anaerobiosis have been studied with discs saturated by floating (BARRS and WEATHERLEY 1962). It was found that only low temperature is fairly specific, *i.e.* it inhibits the uptake of water required for the extension growth without also inhibiting that required for saturation of wilted tissue.

Chemical inhibitors are not therefore used in practice in the determination of water saturation deficit, although BARRS and WEATHERLEY (1962) considered this to be a promising idea. Saturation of discs at 1 °C has been followed recently in detail and evidence obtained that growth is satisfactorily inhibited at this temperature (MILBURN and WEATHERLEY 1971, PACARDO 1971). Errors caused by growth of the tissue, as well as errors caused by respiration, might be minimized considerably by this procedure.

These errors can be almost completely removed by the use of extrapolation methods. There are two phases of water uptake in the course of saturation of young tissue (Fig. 2.10). The initial, rapid uptake is related to satisfying the water deficit, whereas extension growth and other undefined factors cause the later, slower uptake (*e.g.* WEATHERLEY 1950, YEMM and WILLIS 1954, ČATSKÝ 1959, 1965, BARRS and WEATHERLEY 1962). YEMM and WILLIS (1954) obtained the wholesaturation curve by repeatedly weighing the leaves and determined the actual w.s.d. from the curve by graphical extrapolation. This procedure is very laborious but makes it

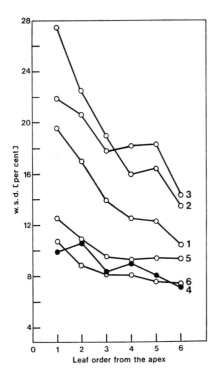

Fig. 2.9 Determination of the water saturation deficit in the same material (kale) by various methods. Methods: 1 — STOCKER, saturation of leaves for 24 h, 2 — STOCKER, saturation of leaves for 48 h, 3 — WEATHERLEY, 4 — YEMM and WILLIS, 5 — ČATSKÝ, saturation of discs for 3 h, 6 — ČATSKÝ, extrapolation method. Temperature: $22 \pm 1°C$, darkness. (ČATSKÝ 1965).

possible to use whole leaves, irrespective of their age, for the determination of the saturation deficit. This is important, as will be shown later, when working with material from which samples of the tissue (*e.g.* discs) cannot be cut.

The disc method with arithmetic extrapolation (ČATSKÝ 1962a, 1965) of the saturation curve can be used even with very young growing tissues and is sufficiently accurate. The method is based on the assumption that the second phase of the saturation curve (Fig. 2.10) very often has a quasi-linear character. The discs are weighed usually after 3 and after 6 h saturation in the foam polyurethane sheet. The weight of water taken up between the third and sixth hours of saturation is subtracted from the total weight after 3 h; this corrected weight is then taken as the saturated weight.

Difference in values found by different methods (Fig. 2.9) demonstrate clearly that methods with a long saturation time and no correction for growth are not suitable for young leaves and plants. Saturation of whole leaves, with graphical (YEMM and WILLIS 1954) and possibly even arithmetic extrapolation, provides relatively reliable estimates of the saturation water content.

An approximate method of determining water saturation deficit was used by VASSILJEV (1931) and EVENARI and RICHTER (1937), for technical reasons. They took the early morning content of water in the plant as a maximum and calculated

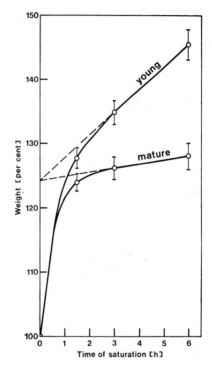

FIG. 2.10 The saturation curves of discs (8 mm), cut out from young and mature tissue (kale) with the same w.s.d. Temperature: $22 \pm$ $\pm\ 1°C$, irradiance (400—700 nm): *ca.* 15 W m^{-2}. (ČATSKÝ 1965).

instantaneous w.s.d. from the daily weight loss. They were od course aware that the early morning water content which they determined does not correspond to saturation according to STOCKER. Another approximate method of assessing water saturation deficit in maize has been described recently by DOWNEY and MILLER (1971). It is a modification of WEATHERLEY's (1950) method, and omits the oven dry weight from the calculation.

DOWNEY and MILLER (1971) also proposed another approximate assessment of water saturation deficit by using a series of photographic standards of plants at different water saturation deficits. Approximate estimates can then be made by visually comparing the plant measured and the standards.

PÁLFI and JUHÁSZ (1971) proposed an assessment of w.s.d. according to the changes in proline amount in the tissue.

A new principle for the determination of water saturation deficit was used by

CZERSKI (1968, *cf.* also 1964). The author proceeds from the assumption that the volume of water necessary for complete infiltration of the intercellular spaces must be equal to their volume in the fully saturated leaf. If the leaf absorbs more water during infiltration than that corresponding to the volume of intercellular spaces, some of the absorbed water must be used for the equilibration of the internal water stress in the cells. CZERSKI infiltrates the leaf with water under reduced pressure, collects the gas released from the intercellular spaces and measures its volume. The volume of infiltrated water is then determined from the weight difference of the leaf before and after injection. The volume [mm^3], and from it the weight [mg], of water required for full saturation of the leaf is found by subtracting the volume of gas from the volume of water. The

$$\Delta W_{(sat)} = \frac{\text{infiltration weight} - \text{volume of gas removed} - \text{initial weight}}{\text{infiltration weight} - \text{volume of gas removed} - \text{dry weight}} \times$$

$$\times\ 100\ [\%]\ . \tag{2.10}$$

CZERSKI (1964, 1968) designed equipment for this procedure to allow simultaneous determination of the w.s.d. of 10 leaves. Part of it is schematically shown in Fig. 2.11 (in the multiple device the T-piece (5) is replaced by a manifold connecting several vessels (3) to a common measuring capillary (6)).

The measurement is carried out in principle as follows:

(a) The device is filled with reboiled distilled water from vessel (1) (direction of water flow is changed by changing the position of the vessel) after opening taps (7) and (8); air is removed from the apparatus by manipulation with other taps and the level in vessel (3) is set a half the vessel height (13).

(b) All taps except (8) and (10) are closed and air dissolved in water removed by evacuating with an oil-filled pump to a pressure determined for the particular material in preliminary experiments (usually 40 mbar for 1 min).

(c) The weighed sample is placed in the cylinder (4) and the apparatus refilled with water by the procedure described in (a).

(d) Air from the intercellular spaces of the sample, which accumulates in the upper part of the cylinder (4), is released by switching on the pump while tap (8) is closed and (10) is open. The air is extracted for one minute and the pump is switched off. Any water deficit of the sample is satisfied during evacuation and during the following 1 to 15 min.

(e) Tap (10) is closed after equilibration with atmospheric pressure, taps (7) and (8) are opened and air is transferred to the measuring capillary (6) by manipulation with vessel (1); its volume is measured there and the air released through tap (12). The sample, infiltrated with water, is taken out, surface-dried, weighed and its dry weight determined.

The author found that it is not necessary to make corrections for temperature, air pressure and partial pressure of water vapour in the measured volume of air and gave evidence that the method yields accurate results. Nevertheless, this is a completely new method, the reliability of which has not yet been proved by general use. It is therefore necessary to make a careful check of its reliability and applicability to a particular material. On the other hand, the principle of the method makes it possible to check the water saturation deficit values obtained, which will

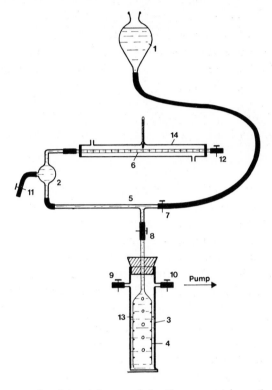

Fig. 2.11 The apparatus for determining w.s.d. by the gasometric method (CZERSKI 1968). 1 — water reservoir, 2 — glass vessel, 3 — glass pressure cylinder, 4 — glass cylinder for the sample, 5 — T-piece, 6 — measuring capillary, 7 to 12 — taps, 13 — water level, 14 — temperature controlled water jacket.

always, as already mentioned in the introduction, be influenced by the particular method used for their determination. It also has the great advantage of providing information on the tissue volume, cell volume and intercellular space volume at various water contents (cf. also KENNEDY and BOOTH 1958).

2.5.3 Practical Hints

Of the methods discussed in the previous Section, the disc extrapolation method is to be recommended as the most generally suitable. It can be modified for narrow

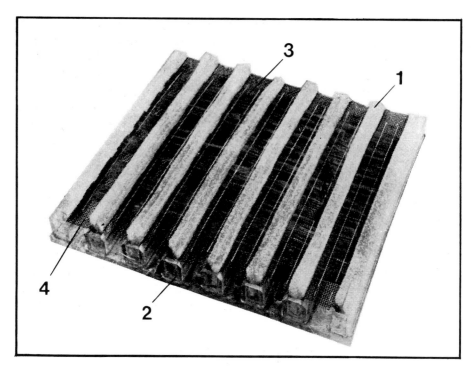

Fig. 2.12 A device for saturation of segments cut from leaves of cereals (NÁTR and ŠPIDLA 1961). 1 — polyurethane foam, 2 — plexiglass trough, 3 — tissue segments, 4 — metal net.

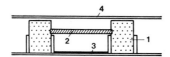

Fig. 2.13 Diagram of the location of segments during determination of w.s.d. in cereals. 1 — polyurethane foam in a plexiglass trough, 2 — tissue segment, 3 — filter paper, 4 — polyethylene foil.

leaves (*e.g.* those of grasses). For very dissected leaves of dicotyledons, from which neither discs nor segments can be cut, a modification utilizing small leaf parts and the extrapolation procedure may be used. Several practical recommendations will be described below.

Sampling

The most suitable size of discs should be determined for the particular material (*cf.* SPOMER 1972a).

Discs 8 mm in diameter are suitable for most plants. They are cut — with appropriate regard to sampling rules — with a sharp cork-borer or special sampler on a pad of softened PVC (*cf.* BARRS 1968, BROWN 1969, ŠETLÍK and ŠESTÁK 1971). Rubber bungs are less suitable as they fall to pieces too readily. The cork-borer should be rotated when cutting, in order not to damage the edges of the discs and thus to prevent later infiltration with water. For the saturation of leaves of grasses, conifers and similar plants, about one cm segments are cut from the leaves and the cut edges of the segments placed against strips of wet polyurethane foam (Figs. 2.12 and 2.13). A sample of 8 to 15 discs or segments is usually adequate for a determination; such a sample weighs 100 to 200 mg after saturation. Replicate samples (usually 3) are essential.

Saturation

Discs taken from dew-covered leaves are dried with filter paper (see later) immediately before measurement. After weighing the sample the discs are carefully transferred with forceps into holes (one disc per hole: for some species, *e.g.* maize, this is necessary to reach full saturation) in wet polyurethane foam sheet. The most suitable wetness of the foam sheet should be determined in preliminary experiments. For most plants completely saturated foam sheet is suitable. The foam sheet is covered with polyethylene film and pressed gently to express a small amount of

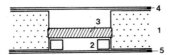

Fig. 2.14 A diagram showing the location of the disc in the polyurethane foam sheet. 1 — polyurethane foam, 2 — polyethylene ring, 3 — disc of leaf tissue, 4 — polyethylene foil, 5 — filter paper.

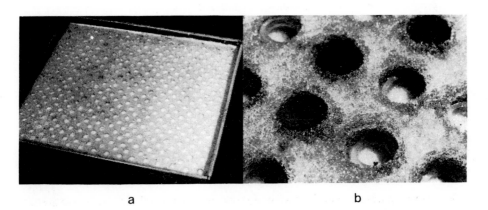

a   b

Fig. 2.15 Polyurethane sheet (a) with holes for saturation of the discs (b) in plastic dish. In three holes (dark ones) leaf discs have been inserted and laid on polyethylene rings, which may be seen in other holes.

water from the foam into the holes with the discs so as to provide liquid contact between the wall of the hole and the cut edges of the discs.

Sheets of a standard size of 200 × 250 × 7 mm with 15 series each of 12 holes 7.5 mm in diameter may be used (Figs. 2.14 and 2.15). Eight mm discs are 0.5 mm larger in diameter than the holes in the foam, so that there is perfect contact between the cut edge of the disc and the wall of the hole. It is possible to saturate

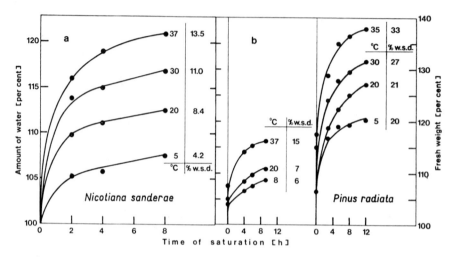

Fig. 2.16 The effect of temperature on saturation curves and on values of w.s.d. calculated in leaf discs of *Nicotiana sanderae* (a) and in mature needles of *Pinus radiata* (b). (MILLAR 1966 and ČATSKÝ 1969).

15 samples of 12 discs each in a sheet of the above dimensions. The sheets are placed in plastic dishes, at constant temperature (Fig. 2.16), at compensation irradiance or in darkness (see BARRS and WEATHERLEY 1962).

Drying and Weighing

The discs are weighed after two, usually 3 h periods (longer periods are usually necessary for segments of conifer needles). It is advisable to standardize the surface drying procedure, as uneven drying may cause errors. For example, discs are placed on a pile of four sheets of filter paper (*e.g.* Whatman No. 4) and covered with another four sheets. The "sandwich" is then inverted and a weight of 500 g applied for one minute to the top of the sandwich (SLATYER and BARRS 1965, Fig. 2.17). In many plants, however, the discs can also be dried satisfactorily with Kleenex or similar tissues without rigid standardization (*e.g.* PACARDO 1971). Finally the discs are oven-dried at 85 °C (3 h is usually sufficient).

Calculation

Arithmetic extrapolation of the saturation curve to the origin is used to calculate the water saturation deficit (ČATSKÝ 1965) as follows:

$$\Delta W_{(sat)} = \frac{2 \times W_3 - W_6 - W_0}{2 \times W_3 - W_6 - W_{(dry)}} \times 100 \, [\%] \, , \qquad (2.11)$$

where W_0, W_3, W_6, $W_{(dry)}$ are the weights of the sample after 0, 3 and 6 h, and the dry weight, respectively.

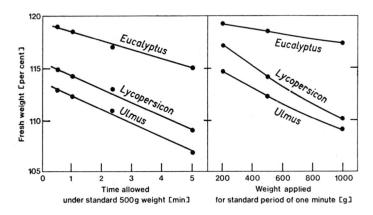

Fig. 2.17 The effect of different weights and times used for surface drying during determination of saturated weight of *Eucalyptus*, *Ulmus* and *Lycopersicon* tissue. Temperature: 20°C, irradiance: light compensation point (SLATYER and BARRS 1965).

Errors

Errors in the determination of w.s.d. by this method mainly result from errors in weighing. The error is usually *ca.* ± 10% of the estimated value for deficits between 0 and 10%, and *ca.* ± 3 to 5% for deficits above 15%. A small error (up to *ca.* ± 1.5%, Fig. 2.18) is caused if the uptake of water for extension growth between the 3rd and 6th hours is slower than in the period 0–3 h. On the other hand, extension growth is inhibited even by quite small w.s.d. (*e.g.* BOYER 1970, HSIAO *et al.* 1970): Hence the water uptake for extension growth may be considerably depressed at the beginning of saturation.

The edge injection error is more important. SPOMER (1972a) found a decrease in this error with increasing disc diameter from 4 to 18 mm (Fig. 2.19) and proposed a method to evaluate its magnitude, based on measurement of w.s.d. using discs of at least three sizes. The edge injection error may be, however, neglected in many species, if discs are saturated in polyurethane foam sheet.

Loss of dry weight during the experiment does not have to be taken into consideration: extrapolation excludes the effect of dry weight decrease provided that

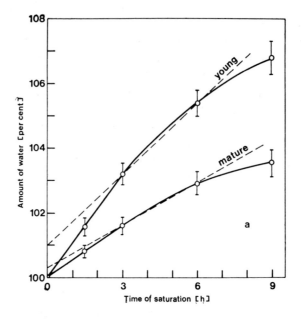

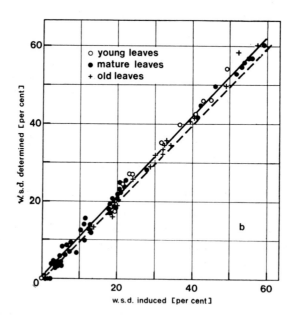

Fig. 2.18 a) Saturation curves of discs (8 mm) cut from artificially saturated tissue of young and mature kale leaves (ČATSKÝ 1965). The dashed line is used to calculate w.s.d. by the extrapolation procedure.

b) Determination of w.s.d. by the extrapolation method in kale tissues with w.s.d. artifically induced by wilting of the saturated tissue (ČATSKÝ 1965).

respiration proceeds at roughly the same rate during the 6 h. Dry weight loss in the six hours exposure at 25 °C is usually only 1 to 2% of the total dry weight (ČATSKÝ 1965). Any fluctuations of respiration rate in either part of the procedure therefore have practically no effect on the results.

The disc method is often used for determining the development of w.s.d. in the same leaves and plants. Equivalent samples can be taken during the experiment from the same leaves, if they are large enough. The total transpiring area of the leaves is of course decreased by repeated sampling so that a lower w.s.d. than in intact leaves is found, namely when the w.s.d. is between *ca.* 10 and 20%. This error is usually about 10% of the measured value. The decrease in deficit is of course not a general rule and should be verified experimentally for the particular material.

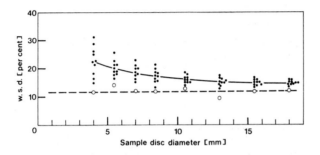

Fig. 2.19 Observed w.s.d. as a function of disc diameter (solid points). Open points and the dashed line represent the mean corrected w.s.d. calculated for each sample diameter (SPOMER 1972a) [floating discs, *Chrysanthemum morifolium* RAMAT. cv. Condor, 20°C, 2 h of saturation].

2.5.4 Determination of Harmful Water Saturation Deficits

2.5.4.1 Basic Terms

The same water saturation deficit can be innocuous or harmful, depending on the sensitivity of the particular plant or particular process to dehydration. Several "sublethal" limits have therefore been conventionally defined, differing by the extent of the damage caused (Table 2.4). The lethal water saturation deficit is then defined as the deficit causing the subsequent death of a tissue, organ or plant (*cf.* also references in OPPENHEIMER 1963).

Drying and discoloration of the tissue has usually been regarded as the criterion of its death. The limit between innocuous and lethal effects of dehydration is, however, by no means always marked by the discoloration of well-defined areas or spots. A loss of the capacity to attain full water content after resaturation has therefore often been used to detect the damage. This can be defined by "water resaturation deficit", w.r.d. (Table 2.4). 10-per-cent limit has usually been used to deter-

mine this value (RYCHNOVSKÁ - SOUDKOVÁ 1963, RYCHNOVSKÁ 1965, RYCHNOV-
SKÁ and KVĚT 1965), called also sublethal w.s.d. or critical w.s.d. by WEINBERGER
et al. (1972) (Fig. 2.20).

Tests of enzymatic activity, vital staining, autofluorescence, plasma rotation and
Brownian movement *etc.* have also been used as criteria of tissue death (*cf.* OPPEN-
HEIMER and JACOBY 1961, OPPENHEIMER and HALEVY 1962). However, leaves
damaged by drought usually die very slowly (*e.g.* OPPENHEIMER 1963, OPPEN-
HEIMER and SHOMER-ILAN 1963, OPPENHEIMER and LESHEM 1966). It is then futile
to try to define strictly reproducible, critical values of dehydration. However, it is

Table 2.4 *Terms and limits used for characterizing harmful water deficits.*

Term	State of the tissue	Reference
Initial damage point (i.d.p.)	W. s. d. at which some cells only are adversely affected.	PISEK and LARCHER (1954)
Critical range	Reparable damage with temporary protoplasmic and functional disturbances.	PISEK and LARCHER (1954)
Critical point	1 to 2% of the leaf area or of the cells are killed (= end of the critical range).	PISEK and LARCHER (1954)
Critical w. s. d.	Analogous to w.r.d.; saturated weight after wilting is 10% smaller than initial saturated weight	WEINBERGER *et al.* (1972)
Sublethal w.s.d.	5 to 10% of the leaf area killed (dry and discoloured).	TUMANOV (1930) OPPENHEIMER (1932)
	As critical w.s.d.	WEINBERGER *et al.* (1972)
Water resaturation deficit (w.r.d.)	Analogous to sublethal w.s.d. (determined visually) and to critical w.s.d.; saturated weight after wilting is 5 to 15% smaller than initial saturated weight	RYCHNOVSKÁ-SOUDKOVÁ (1963), RYCHNOVSKÁ (1965)
Permanent turgor loss point (p.t.l.p.)	A value beyond which the number of damaged cells increases steadily producing an increasing w.r.d. In this range, w.r.d. is more or less linearly related to water loss.	OPPENHEIMER (1963)

Table 2.4 *(continued)*

Term	State of the tissue	Reference
"One third kill" (Drittelschaden)	Death of one third of the leaf area	PISEK and LARCHER (1954)
—	Half the cells are irreparably damaged.	LEVITT (1956, 1963)
Lethal w.s.d. (Relative turgidity lethal points, foliage moisture content lethal points, soil moisture content lethal points)	Causes the subsequent death. (Expressed in relative turgidity, foliage moisture content or soil moisture content, respectively.)	PHARIS (1966), PHARIS and FERRELL (1966)
Sublethal water content	Water content at sublethal w.s.d., on dry weight basis [%].	PISEK and co-workers (1938, 1953, 1954)
Capacity for drought endurance (Aushartungsvermögen)	Sublethal w.s.d. divided by preceding average hourly water loss.	PISEK and co-workers (1938, 1953, 1954)
"Limiting value" ("Grenzwert")	= Sublethal w.s.d. = p.t.l.p.	ARVIDSSON (1951)
Non-recoverable w.s.d.	Loss in saturation capacity compared with the "natural" saturation (instead of artificial full saturation).	TSEL'NIKER (1955)
Recovery (Erholung)	Water content at resaturation divided by water content at presaturation. Recovery weight = r.w.c.	BORNKAMM (1958)
Drought stress (Trockenbeanspruchung)	Natural w.s.d. divided by critical (= sublethal) w.s.d.	HÖFLER et al. (1941)
Point of damage (Schädigungswert) Point of death (Lethalwert)		HUBER and ZIEGLER (1960)

possible to define points on a sliding scale, and the earlier idea that there is one definite point separating the living from the dead state must be abandoned. OPPENHEIMER improved the technique of viability tests and showed that lethal limits can be quite different in different tissues of the same organ. He also found some indication of differences between detached and attached leaves, but this was not proved (*e.g.* OPPENHEIMER and LESHEM 1966).

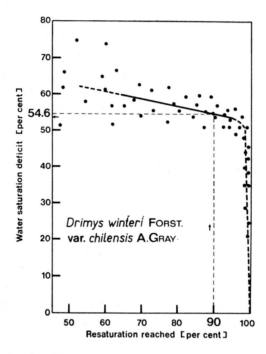

Fig. 2.20 Determination of water resaturation deficit (= critical or sublethal w.s.d.) according to WEINBERGER *et al.* (1972) [*Drimys winteri* FORST. var. *chilensis*].

2.5.4.2 Techniques

The "critical" water saturation deficit is measured, in principle, as follows: Saturated leaf or leaf samples are left to wilt to various w.s.d., and then resaturated. The samples damaged (according to the selected criteria) reach the critical value. Techniques, equipment and practical hints discussed earlier (Sections 2.5.2 and 2.5.3) are generally used also for determining "critical w.s.d.". The only essential feature is the assesment of tissue damage, which will be discussed in more detail.

In practice, whole leaves, shoots or plants were used originally (*e.g.* OPPENHEIMER 1932, BORNKAMM 1958). For preliminary experiments, ten leaves are sufficient, while for more exact work 50 to 150 are required, weighed individually. When whole leaves or plants are used, no bulk samples should be used because of the large individual variations in dehydration rate. The leaves are saturated in a moist chamber, weighed ($W_{(sat)}$) and exposed freely until they have developed deficits in the proposed critical range, and slight irreversible damage occurs. Then they are weighed ($W_{(fin)}$), resaturated and reweighed ($W_{(resat)}$), and oven-dreid ($W_{(dry)}$).

This techniques is, however, very unproductive. For this and other reasons (see discussion on STOCKER's method, Section 2.5.2), bulk samples of leaf discs (*e.g.* 50 to 200) have been preferred for routine measurements (*e.g.* JARVIS and JARVIS 1963). After saturation (see Section 2.5.3), the discs are weighed ($W_{(sat)}$) and subsequently exposed to dryish air for variable periods, or to variable humidity for a standard time (*e.g.* in microdesiccators, see Section 1.3.4.4)* to induce various water saturation deficits. Periods and humidities necessary are estimated for the material to be measured. The discs are then weighed again ($W_{(fin)}$), resaturated and reweighed ($W_{(resat)}$), and their dry weight is determined ($W_{(dry)}$). The final w.s.d. at highest dehydration, (w.s.d.$_{(fin)}$), and the water resaturation deficit (w.r.d.) are calculated:

$$\text{w.s.d.}_{(fin)} = \frac{W_{(sat)} - W_{(fin)}}{W_{(sat)} - W_{(dry)}} \quad [\%], \tag{2.12}$$

$$\text{w.r.d.} = \frac{W_{(sat)} - W_{(resat)}}{W_{(sat)} - W_{(dry)}} \quad [\%]. \tag{2.13}$$

Then w.s.d.$_{(fin)}$ equals the permanent turgor loss point if the water resaturation deficit has reached 10% (5% or 15% limits may be chosen conventionally).**

The results can be presented in graphs in which the relative size of the dead areas or the w.r.d. is plotted (on the ordinate) against the corresponding values of w.s.d.$_{(fin)}$ (on the abscissa). This often, but by no means always, results in a sigmoid curve with an inflection point near the sublethal deficit of the permanent turgor loss point. Lower values of w.s.d.$_{(fin)}$ usually correspond to water resaturation deficits of -5% to $+5\%$, while higher values of w.s.d.$_{(fin)}$ are on the quasi-linear ascending portion of the curve.

It should be stressed that the technique must be standardized very rigidly. Recovery depends not only on the w.s.d. reached, but also on the rate at which it is reached, the period for which it is held, preconditioning *etc.*

"Spot" and "resaturation" methods give very similar results for herbaceous plants. This is apparently not true of tree leaves, for which resaturation tests may indicate much smaller harmful deficits than do dry-spot tests (ARVIDSSON 1951,

---

* The latter technique gives not only w.s.d. at the point of nonrecovery, but also water potential.

** If whole leaves are used, visual assessment of 10% of the area damaged has often been taken as the criterion for having reached the critical limit ("spot method"); this limit has usually been called the sublethal w.s.d. It equals the permanent turgor loss point, determined by the "resaturation method".

BORNKAMM 1958). Microscopic check methods agreed much better with resaturation tests than with dry-spot tests in trees, so that the dry-spot test results are misleading (OPPENHEIMER, unpubl.).

The results obtained by the above techniques (visual "spot" or "resaturation" methods) must always be checked. Grave errors have often resulted from the use of macroscopic criteria alone, and cell physiological tests are considered to be essential. They should be extended to the internal tissues, at least in material with tough outer tissues. The examination of the epidermis alone is insufficient (GENKEL' and MARGOLINA 1952, MOURAVIEFF 1963). For example, the mesophyll of *Nerium oleander* dies when one third of the maximum water content has been lost; the epidermis dies only after two thirds have gone (OPPENHEIMER and LESHEM 1966). Pine needles behave similarly, *i.e.* the epi- and hypodermis are much more resistant than the endodermis, transfusion tissue and chlorenchyma (OPPENHEIMER and SHOMER-ILAN 1963).

The viability at all stages of dehydration may be examined microscopically, usually in molar sucrose solution to which 0.01% of neutral red in a weakly alkaline phosphate buffer has been added. Living cells stain red within 30 to 120 min and are often plasmolysed (for details see OPPENHEIMER and SHOMER-ILAN 1963).

Unfortunately, the examination of internal tissues using this so-called vital staining and plasmolysis rarely indicates general viability even in control sections of non-droughted material. The tender cells are partially damaged by knife or pressure, whereas in thicker sections high viability may be simulated even in damaged material. Some experience is needed to overcome these difficulties.

Incomplete vital staining of control sections may be the consequence of previous drought damage, especially in old leaves, which may also have been damaged by the onset of senescence. The viability of the control material is rarely checked as thoroughly as it should be. This may result in misinterpretation.

Another criterion of tissue viability is the 2,3,5-triphenyl tetrazolium chloride (TTC) test. It is generally carried out macroscopically (PARKER 1951). If no red formazan is formed, there has been total inactivation of the reducing enzymes involved. This will occur only at the stage of severe damage so this test is unsuitable for the diagnosis of incipient damage, unless it is complemented by microscopic observations showing the decrease of staining intensity at earlier stages.

Physiological or biochemical tests are a very promising means of investigation of drought damage (*e.g.* irreversible drop in photosynthetic activity, increase of respiration rate including enzymatic tests, *etc.*). These are, however, not yet generally used for testing for viability. See also MONTFORT and HAHN (1950), LARCHER (1953, 1957, 1960).

3

Water Exchange between Plant Roots and Soil

3.1 Availability of Soil Water

3.1.1 Introduction

3.1.1.1 Static Availability of Soil Water

Liquid water moves from soil into root cells along a gradient of decreasing water potential. The static availability of soil water to the plant roots is determined by the potential of soil water in the boundary layer closely surrounding the roots.

Total potential of soil water in the soil (Ψ_{total}) may have the following components:*

$$\Psi_{total} = \Psi_m + \Psi_p + \Psi_z + \Psi_s , \qquad (3.1)$$

where Ψ_m is the matric or capillary potential, Ψ_p is the pressure potential due to external pressure, Ψ_z is the gravitational potential, Ψ_s is the osmotic potential of the soil solution. For general definitions see Section 1.1.

Matric potential (Ψ_m) of soil expresses by its negative values a decrease of water potential compared with free pure water on the interface surfaces of the soil structure (including capillary forces — so that the matric potential in soil is sometimes called capillary potential). Matric potential is usually the most important of all the components of the total water potential of soil. The values of matric potential measured by some of the methods described in the Section 3.1.3.2 are therefore sometimes taken as values of the total water potential of a soil, the other components of water potential being neglected. Capillary potential (CP)** or the expression pF (SCHOFIELD 1935)

$$pF = \log_{10} h_w \qquad (3.2)$$

is often incorrectly used to express the total potential of water in soil.

As the matric potential is normally negative, it has been found convenient to refer to it in terms of a positive matric tension, designated τ. This results in confusion, as negative matric potential was designated τ by SLATYER and TAYLOR 1960.

---

* The terminology used here is based mainly on ASLYNG's report of the International Society of Soil Science (1963).

** Capillary potential is the negative pressure expressed in (positive) cm of water column (h_w).

Like other components of the total potential of soil water, the matric potential can be expressed in terms of pressure (bar) or in terms of free energy per unit volume or per unit mass (J kg^{-1}). The capillary potential (h_w or CP) (BUCKINGHAM 1907) can be considered as the first expression of matric potential in the literature. The term soil moisture tension (RICHARDS and WADLEIGH 1952) also corresponds to the term matric potential.

Pressure potential (Ψ_p) acts in soil mainly as a pneumatic potential caused by external gas pressure. An artificial pneumatic potential is applied in the pressure membrane apparatus (RICHARDS 1947, see Section 3.1.3.2) used to measure matric potential. Under natural conditions the air pressure in the soil pores is uniformly atmospheric pressure throughout so that it is not an important component of the total potential.

Gravitational potential (Ψ_z) is a component of total potential of soil water due to elevation.

Osmotic potential (Ψ_s) represents the depression of water potential due to osmotic pressure of the soil solution; it is negative.

If we do not consider the soil *in situ*, *i.e.* if we work with soil in the laboratory, for example, not all the components of the above-mentioned total potential of soil water are involved in affecting the soil water status.

Water potential in the soil (Ψ_w) is identical with the total potential of soil water (Ψ_{total}) in the absence of any difference in elevation, *i.e.* without the gravitational component (gravitational potential, Ψ_z, is zero). Then

$$\Psi_w = \Psi_m + \Psi_p + \Psi_s . \qquad (3.3)$$

The term total soil moisture stress (t.s.m.s.) [total soil suction] is positive, corresponding to the negative sum of matric and osmotic potentials. It is usually expressed in pressure units (bar) or in energy per unit volume or per unit mass (J kg^{-1}).

The sum of matric potential (Ψ_m) or submergence potential and the pneumatic potential is sometimes (according to the ISSS Committee of Soil Physics Terminology 1963) called pressure potential. This may cause confusion, because the term pressure potential is normally used only for the water potential components caused by real pressure, which is not true of the matric potential. The sum of gravitational potential (Ψ_z) and pressure potential (Ψ_p) is sometimes called hydraulic potential (Φ).

3.1.1.2 Dynamic Availability of Soil Water

The dynamic component of the availability of water to plant roots, *i.e.* not only the instantaneous water potential, but also the quantitative features of the uptake, is usually very important. This dynamic component of the availability of soil water is determined: (1) by the rate of water movement through the soil towards the absorbing roots and the absolute amount of water present in the soil rhizosphere which immediately surrounds them, (2) by the absolute absorption rate, (3) by the area of absorbing surface of the roots, which is partly determined by the rate of growth of the roots, *i.e.* their hydrotropic penetration into soil volumes with an as yet unexhausted pool of soil water and the rate of renewal of the active absorbing surface.

(1) Capillary conductivity* of soil decreases very considerably with decreasing soil moisture as the connecting sections in the system of soil capillaries are reduced; the pores of the largest diameter empty and air bubbles form. This means that a much larger potential gradient is necessary to supply soil water to the surface of absorbing roots at lower soil water contents.

(2) The absolute rate of water uptake is determined by the decreased water potential of the epidermal (rhizodermal) root cells and hence by the gradient of water potential between soil and roots. The water potential of roots does not decrease below the water potential of the above-ground parts under normal conditions, so that the gradient of water potential between soil and plant cannot reach very high values unless there has previously been irreversible damage to the plant by drying out. A rapid absolute rate of water uptake hence rapidly reduces the amount of statically available soil water and also its dynamically available amount, because of the limited movement of water in dry soil.

(3) The absolute rate of water absorption is also determined by the size of the absorbing surface of the roots. The size of the root system also determines the volume of the rhizosphere, *i.e.* the soil volume containing roots. The horizontal extent of the rhizosphere of one plant in a stand is determined primarily by competition with other plants. The vertical extent of the root system is determined by species-specific properties and also by two partially opposing factors; the depth of the soil water table and soil aeration. The depth of the root system is determined in many cases by a compromise between deep growth efficiently reaching the soil water and the dependence of root growth and also root permeability to water on a sufficient oxygen tension in the soil air.

The problem of the availability of soil water to plants and its relationship to the growth of plants has until recently been discussed from the following two different points of view. One group of authors claimed that all the water within the range between field capacity (absolute water capacity) and the permanent wilting percentage is equally available to plants (see Section 3.1.4.1). The other group suggested that the availability of water to plants decreases throughout practically the whole range, as soil moisture decreases.

The conclusions from discussion of this problem have been summarized by several authors (RICHARDS and WADLEIGH 1952, RODE 1952, VEIHMEYER 1956, STANHILL 1957, SLATYER 1960, VAADIA *et al.* 1961, GARDNER 1965, SLATYER and

---

* The term capillary conductivity is used for the water flow in the soil in unsaturated conditions, while the term hydraulic conductivity is preferred under saturated conditions.

GARDNER 1965). The availability of soil water decreases with decreasing water potential even at water contents higher than the permanent wilting percentage. This decrease of soil water availability is progressive, being greater at lower soil water contents. As the dynamic component plays a major part in the availability of soil water, as far as both soil (mobility of soil water, osmotic potential of soil solution), and, particularly, plants (rate of absorption, quantity and quality of roots) are concerned, the availability is not determined only by hydropedological constants, but also by the interaction between soil and plant.

Several models of soil water flux to the root and into the root have been suggested (PHILIP 1957, GARDNER 1960a, SLATYER and GARDNER 1965, COWAN 1965, reviews by GARDNER 1965, SLATYER 1967a, and others).

3.1.2 Brief Review of Methods Used to Determine the Soil Water Content

The soil water content is expressed as the mass of water per unit mass of dry soil (or as the volume of water per unit volume of moist soil), often as a percentage. The volumetric soil water content is designated by the symbol Θ, expressed as cm^3 water per cm^3 soil. A brief review of the methods for determining soil water content, including their advantages and inadequacies, is included, although the methods are not described in detail.

A. Sampling Methods

Sampling with a soil core sampler (tube sampler) is very simple for fairly heavy soils, near the soil surface. It is difficult only with rocky soils taken deeper below the ground. It is not possible to take a sample several times from exactly the same place. A sufficient number of replicate samples must be taken to reduce the error caused by the heterogeneity of soil water content and of soil structure (AITCHISON et al. 1951, TAYLOR 1955, STAPLE and LEHANE 1962).

(1) Gravimetric method (drying to a constant weight at 105 °C). This is the simplest and most reliable method, useful even for small samples. In soils with a high content of organic matter, drying at 60 °C in a vacuum oven is preferable. Sampling from shallow depths is very rapid, but drying takes rather a long time. The method is very reliable and serves to calibrate other methods.

(2) The methanol method (BOUYOUCOS 1931) consists in measuring the density of methanol containing a ground soil sample. Its accuracy is very limited and cannot be used with soil samples containing methanol-extractable compounds.

(3) The calcium carbide method (SIBIRSKY 1935) measures the pressure of acetylene formed from powdered calcium carbide after mixing with a ground soil sample. The method is very rapid, its accuracy is low (it yields results by 20% lower in some soils) and a special device (commercially available) is required.

(4) The electrical conductivity method (HANCOCK and HUDGINS 1954, HANCOCK and BURDICK 1957) measures the change of electrical conductivity of

a mixture of ethanol, acetone and sodium chloride mixed with a soil sample. The method is very rapid (7 min) and requires a conductometer. Its accuracy is limited and it is useful primarily for moist soils.

B. Methods Used to Determine the Water Content in Soil *in situ* Make it Possible to Measure Changes of Soil Water Content Continuously without Affecting the Soil Structure to any Considerable Extent

(5) Measurement of the thermal conductivity of soil, which is related to its water content (SHAW and BAVER 1939, HAISE and KELLEY 1946, MOMIN 1947, SKEIB 1950, DE VRIES 1952, VAN DUIN and DE VRIES 1954, BLOODWORTH and PAGE 1957, DE VRIES and PECK 1958, NEUWIRTH 1958, LINDNER 1964, PHENE et al. 1971). The method is very rapid; calibration is required for each soil; simultaneous measurement of soil water content is possible. Perfect contact of the measuring body with the soil is required. Results do not depend on the content of electrolytes in the soil.

(6) The electrical capacitance method is based on measuring the electrical capacitance of soil, which depends primarily on the soil water content (FLETCHER 1939, ANDERSON and EDLEFSEN 1942, CHILDS 1943, ANDERSON 1943, WALLIHAN 1946, PERSON 1952, SPAUSZUS 1955, VILKNER 1957, LE JEUNE and ARNOULD 1958). It is possible to take continuous measurements, which respond even to very rapid changes of soil water content, and measurement is rapid. The method requires perfect contact of the measuring device with the soil and the dielectric constant is very temperature-dependent.

(7) The method of neutron scattering is based on the fact that fast neutrons are slowed down considerably and scattered, much more by hydrogen atoms than by other atoms. The method is used to measure the concentration of hydrogen atoms in soil, which is many times higher in soil water than in living organic matter (roots) or non-living organic compounds (humus) (BELCHER et al. 1950, GARDNER and KIRKHAM 1952, UNDERWOOD et al. 1954, STONE et al. 1955, HOLMES 1956a, b, MORTIER and BOODT 1956, STOLZY and CAHOON 1956, VAN BAVEL et al. 1956, STEWARD and TAYLOR 1957, WEEKS and STOLZY 1957, VAN BAVEL 1958, HOLMES and TURNER 1958, HOLMES and JENKINSON 1959, HANKS and BOWERS 1960, MERRIAM 1960, VAN BAVEL et al. 1961, VAN BAVEL 1961, McGUINNESS et al. 1961, DAMAGNEZ 1962, HSIEH 1962, PERRIER and JOHNSTON 1962, VAN BAVEL et al. 1963, DE BOODT et al. 1963, HEWLETT et al. 1964, MAERTENS et al. 1965, COHEN and TADMOR 1966, PIERPOINT 1966, and others). A rather heavy, sensitive and costly device is needed for protection against radiation. Normally large volumes of soil are measured, so the method must be modified (e.g. by use of a covering plate) for measurements in surface soil layers or in the proximity of the water table. A radiation source and detector are inserted during each measurement into tubes permanently installed in the soil. The method is very accurate, in spite of the fact that unusually high contents of organic compounds, atoms of chlorine, boron or

Table 3.1. *Summary of the relationship between several soil hydrolimits and*

Soil water	Availability		capillary capacity %	available water %	Soil water potential ψ_w J g^{-1}	bar
Hygroscopic	Not available	Dry weight ~	0		$-\infty$	$-\infty$
					-92	-931
	For survival	Maximum hygroscopicity V_h ~				-223
					-9.2	-92.9
Capillary		Permanent wilting point p.w.p. ~		0	-1.5	-15
					-0.1	-1
	Available	Capillary (^{1}) capacity ~ Field capacity	100	100	-0.03	-0.33
						-0.1
Gravitational		Full saturation approx			0	0

(^{1}) Different values according to the various measuring procedures.
(^{2}) BUCKINGHAM 1907. Pressure of water column of the height h.
(^{3}) SCHOFIELD 1935. $pF = \log_{10} h$.

or iron in soil may cause errors. The results are not affected by a high content of salts, the method measures the volume water content directly (after calibration). Separate calibration is not necessary for individual soil samples in many cases.

(8) The gammascopic method (BELCHER *et al.* 1950, ASHTON 1956, ČABART 1958, ČABART and VÁLEK 1959, GURR 1962, FERGUSON and GARDNER 1962, HSIEH 1962, DAVIDSON *et al.* 1963, REGINATO and VAN BAVEL 1964, LIGON 1969, RYHINER and PANKOW 1969, RIJTEMA 1969) is based on measuring gamma ray absorption by soil water. It is not completely specific for water. Expensive and

3.1 AVAILABILITY OF SOIL WATER

values of static availability of soil water. Compiled from different sources.

Capillary potential [2] h_w	[3] pF	Matric potential ψ_m		Relative air humidity $\dfrac{p}{p_0} \cdot 100$
cm_{H_2O}		[4] mbar	[5] $\approx \Omega$	%
10 000 000	7.0	$(-\infty)$	—	0
1 000 000	6.0		—	50
220 000	5.4		—	85
	4.78		—	94.75 [6]
15 000	4.25	$(-15\ 000)$	*ca.* 75 kΩ	98.85
1 000	3.0	$-1\ 000$		99.93
500	2.7	$-\ 333$	*ca.* 600 Ω	
100	2.0			
0	0	0	~ 450 Ω	100

[4] Soil tensiometer.
[5] Resistance of a gypsum block of the A type (see section 3.1.3.2.).
[6] Relative air humidity above 10% H_2SO_4 solution or above a saturated Na_2SO_4 solution at 20 °C ($pF = 5.02$).

heavy equipment is required. The advantages and inadequacies are similar to those of the neutron scattering method.

(9) The measurement of the attenuation of β-radiation by water held in nylon pads in soil (NEWBOULD *et al.* 1968).

(10) The resistance block method (see Section 3.1.3.2) is often used to measure water content in soil *in situ*. Calibration is based on gravimetric determination of water content for each soil.

(11) The hygrophotographic method was suggested for semiquantitative determination of soil water content and its localization (SIVADJIAN 1957, 1960). It is

based on the detection of soil water by photographic plates treated in a special way.

(12) VAN DER WESTHUIZEN (1964) described a procedure using high frequency electro-magnetic waves.

It is not possible to deal here with the hydropedological methods. Specialized handbooks are recommended. Fig. 3.1 shows the relationship between soil water content and some hydropedological constants in sandy and clay soils. In addition, there is a review table of the relations between some hydropedological constants and the static availability of soil water (Table 3.1). Particular attention is devoted to parameters measured by methods described in the following sections

An exhaustive review of methods for measuring soil moisture has also been compiled by COPE and TRICKETT (1965).

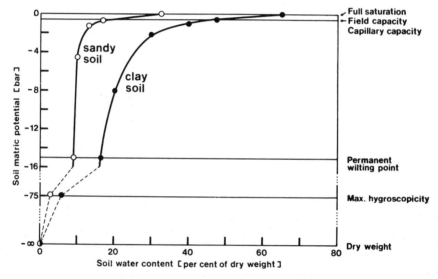

Fig. 3.1 Example of the relationship between soil matric potential (ordinate left) and certain hydropedological values (ordinate right) and soil water content (abscissa) for sandy soil (Panoche loam), clay soil (Chino silty clay loam). Adapted from KRAMER (1969) according to WADLEIGH *et al.*, (1946) and RICHARDS and WEAVER (1944).

3.1.3 Determination of Static Availability of Soil Water

3.1.3.1 Determination of Soil Water Potential

The laboratory psychrometric methods described in Section 1.3.4, used to measure the relative partial water vapour pressure above the soil sample after equilibrium has been reached, are the most useful methods for measuring the water potential of soil samples. The method (type B) of RICHARDS and OGATA (1958, 1961) modified by EHLIG (1962) may be used to measure water potential within a very wide range, while in the type A psychrometer (SPANNER 1951, MONTEITH and OWEN 1958, KORVEN and TAYLOR 1959) (for reference see again Section 1.3.4) the range is down to water potentials of about −75 bar because of the limiting Peltier cooling effect. The methods are technically rather challenging. However, they may yield an accurate picture of changes in water potential during changes of soil water content and may therefore serve as a basis for calibration of other methods, particularly field methods for the determination of soil content.

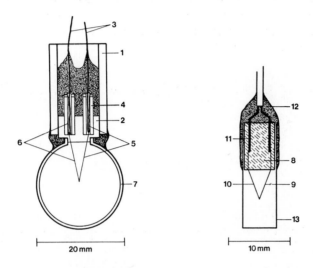

Fig. 3.2 Left: A cross section of the thermocouple psychrometer (type A) for measuring soil water potential (RAWLINS and DALTON 1967). 1 — acrylic tubing, 2 — Teflon insert, 3 — copper lead wires soldered with cadmium-tin "low thermal" solder using stainless steel solder flux, 4 — copper heat sinks, 5 — chromel-p wire 0.025 mm, 6 — constantan wire 0.025 mm, 7 — porous ceramic bulb 2 cm in diameter.
Right: Screen thermocouple psychrometer constructed by BROWN (1970). 8 — Teflon insert, 9 — chromel wire (25 μm diameter), 10 — constantan wire (25 μm diameter), 11 — copper lead wires, 12 — epoxy resin, 13 — screen cage made of fine-mesh stainless-steel wire.

The effect of temperature fluctuations has been a great obstacle to using psychrometric methods for measuring soil water potential *in situ* in the field. RAWLINS and DALTON (1967) discussed the reasons: (1) The relationship between wet bulb depression and the water vapour pressure in the chamber is temperature-dependent (KLUTE and RICHARDS 1962). (2) Changes in the sensitivity of the psychrometer with temperature are due to changes in thermal conductivity, in water vapour diffusivity of the air in the chamber, and in latent heat of vaporization of water (RAWLINS 1966). (3) Changes in ambient temperature cause temperature gradients between the chamber and the thermocouple reference junction. (See also CAMPBELL and GARDNER 1971.)

Most of these effects may be eliminated by using an A type psychrometer (*i.e.* utilising the Peltier effect) and the usual technique of subtraction of the e.m.f. measured immediately prior to cooling the wet junction from the e.m.f. after cooling. The remaining effect of a temperature increase is associated with the decrease in vapour pressure deficit with increasing air temperature, unless sufficient water vapour enters the chamber during the warming up. If water vapour is free to move to and from the chamber, this error can be minimized. RAWLINS and DALTON (1967) designed a thermocouple device in which the chamber is formed by a ceramic bulb shell with a sufficiently low resistance to water flux between the surrounding soil and the inner atmosphere.

Fig. 3.3 The thermocouple psychrometer for soil water potential measurement (RAWLINS and DALTON 1967). Right, without and left, with the porous ceramic bulb.

This type of soil psychrometer probe is shown in Figs. 3.2 left and 3.3. In a Teflon insert (1) with two copper tubing heat sinks (4), hollows for both thermocouple wires (chromel-p and constantan, 0.025 mm in diameter) (3) are made by means of a fine gauge hypodermic needle. The wires lead to heat sinks, to which they are soldered by means of a cadmium-tin solder using stainless solder flux, with two copper lead wires (4). The entire assembly is boiled in distilled water and allowed to dry. It is then inserted into a short piece of acrylic tubing (5) and filled with epoxy resin (6). After calibration above KCl solutions, the thermocouple assembly is inserted into the neck of a ceramic bulb (1 cm in diameter) and sealed with epoxy resin. Similar psychrometer was described by LOPUSHINSKY and KLOCK

(1970a, b). For calibration and measuring procedure see Sections 1.3.4.5 and 1.3.4.6. Fig. 3.4 shows the electrical circuit, similar to that shown in Fig. 1.23. A cooling current of 3 mA for 10 or 20 seconds and a microvoltmeter with a minimum range of 0.01 µV (Keithley Model 148) were used. Several commercially produced soil psychrometers are now available (*e.g.* soil psychrometer Wescor Inc., Logan, Utah, U.S.A.).

LANG (1968) and BROWN (1970) replaced the ceramic bulb with fine wire mesh screen (Fig. 3.2 right) to eliminate lag in water potential equlibration. But MERILL *et al.* (1968) and RAWLINS (1971) found sealed ceramic enclosures suitable for preventing possible microbial life on the wet junction. KAY and LOW (1970) designed a thermistor psychrometer for measuring soil water potential.

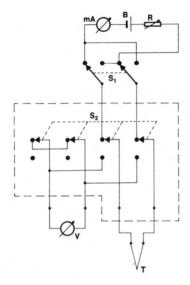

Fig. 3.4 Electrical circuit for soil thermocouple psychrometer (RAWLINS and DALTON 1967). µA — microampermeter (50 mA), B — 1.5 V dry cell, R — variable resistor, S_1 — 2-pole 2-throw switch, S_2 — 4-pole 2-throw switch connecting the thermocouple (T) either to the voltmeter (V) or to the cooling circuit, V — electronic voltmeter (minimum range 0.01 µV), T — thermocouple psychrometer (see Fig. 3.2). The circuit inside the dashed lines is mounted in a thermally insulated box.

RICHARDS (1965a, b) suggested a very interesting method for measuring soil water potential. He measured the temperature difference between two thermistors built into a sensor probe (Fig. 3.5). One of the thermistors is in a porous glass bead which is in water vapour equilibrium with the soil air and the second is in a standard dry atmosphere (*e.g.* dried with silica gel). The temperature difference is caused by the heat required to evaporate the moisture adsorbed by the thermistor surface from the ambient atmosphere. The difference can be calibrated against osmotic potential of NaCl solutions.

The capillary method of URSPRUNG and BLUM (1930) (see Section 1.4.2.3) cannot be recommended for standard determinations of soil water potential because of its poor reliability.

The method of paper hygrometers (HANSEN 1926, GRADMANN 1928, 1934, STOCKER 1930, GARDNER 1937, ECKARDT 1960, FAWCET and COLLIS 1967, MCQUEEN and MILLER 1968) also is not very reliable. Water absorption (measured gravimetrically) by a filter paper equilibrated above the soil sample may be calibrated in terms of soil water potential by equilibration of the same kind of paper

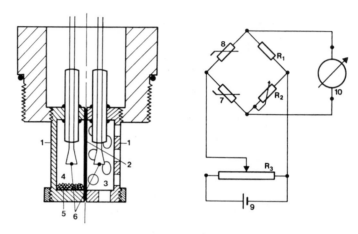

Fig. 3.5 Thermistor hygrometer for measuring soil water potential designed by RICHARDS (1965a, b).
Left: The probe. A split tube (1) was divided by a rubber membrane (2) into two compartments. Perforation in the wall of one compartment (3) permit establishing of air humidity corresponding to the water potential of the surrounding soil. The other compartment (4) is sealed at constant humidity near zero due to enclosed silica gel (5). A bead thermistor (6) (Stantel type F 23, Standard Telephones and Cables Pty, Ltd.) was located in the centre of each compartment.
Right: Circuit diagram used. 7 — thermistor situated in (3) 8 — thermistor in compartment (4), 9 — 1.35 V mercury cell, 10 — microvoltmeter ($\pm 5 \mu$V) with output of the recorder, $R_1 = 2$ kΩ, $R_2 = 2$ kΩ, $R_3 = 2$ kΩ.

above soil solution of known osmotic potential (for Ψ_w values lower [more negative] than -15 bar) or above soil samples brought to known Ψ_w on a pressure membrane extractor. ECKARDT (1960) calibrated discs of filter paper, diameter 30 mm, (weighed first after drying at 105 °C, before and after applying a standard drop of normal NaCl solution and repeated drying) by hanging ten or more pieces from the stoppers of wide-necked vessels, above graded NaCl solutions of known osmotic potentials. The vessels were left in a closed box at a constant temperature for about one month. A calibration is obtained between the osmotic potential of the samples and the weight changes of the discs. The samples are placed in similar sealed vessels containing soil up to the same height as the surface of the NaCl solutions used for calibration. Each vessel is then closed with a stopper from which hangs the calibrated disc, and sealed with molten paraffin. A set of vessels containing NaCl solutions is used as a parallel control. All the vessels are kept for about one month at a constant temperature, and then the weight changes of discs above the soil samples and above the solutions in the reference are determined. The water potential of the soil samples is determined from the calibration graphs. No complex equipment is necessary, so this method is worth testing, even though it is not very accurate.

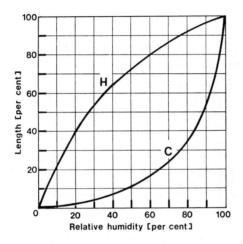

Fig. 3.6 Changes in length of hair (H) and cellophane film (C) as per cent total change (ordinate) related to relative air humidity (abscissa) (GEIGER 1952).

Simple cryoscopic methods (RICHARDS and CAMPBELL 1948, 1949, TAYLOR *et al.* 1961) have also been worked out for the determination of water potential of soil. However, the interpretation of the results obtained is not unambiguous and the methods will not be described here.

The total water potential of soil may also be measured by determining the relative humidity of soil air directly, in the soil, by means of a hygrometer. For these measurements HOFMANN (GEIGER 1952) developed a device utilizing the increase in length of a piece of cellophane film with increasing relative air humidity. Whereas the increase in length of the human hair used in standard hygrometers is less at high humidities, the length of a strip of the cellophane film (also of polyethylene film — HOFMANN 1962 — personal communication) increases progressively with increasing relative humidity within the range of 90—100% (see Fig. 3.6). This relationship does not itself depend on temperature, but the relationship between water potential and relative humidity of the air is influenced by temperature, as water potential is proportional to the partial water vapour pressure in the air. The reaction of the cellophane film is slightly slower than that of defatted human hair.

3.1.3.2 Determination of Matric Potential of Soil

Soil Tensiometers

Soil tensiometers, measuring directly the pressure under which the soil "draws" water from a porous vessel saturated and filled with water, have been constructed for direct determination of matric potential in soil *in situ* (ROGERS 1935, RICHARDS and GARDEN 1936, RICHARDS and LAMB 1937, RICHARDS *et al.* 1937, WALLIHAN 1939, STOECKELER and AAMODT 1940, RICHARDS 1942, VEIHMEYER *et al.* 1943, POST and SEELEY 1943, COLEMAN *et al.* 1946, RICHARDS 1949, FLEISCHHAUER-BINZ 1949,

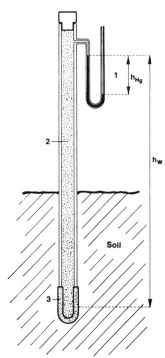

Fig. 3.7 Diagram of a soil tensiometer. 1 — mercury manometer, 2 — tensiometer tube, 3 — porous cup.

MILLER 1951, STOCKER and KAUSCH 1952, HAISE 1955, KAUSCH 1955, GARDNER 1960b, KLUTE and PETERS 1962, LEONARD and LOW 1962, and others).

A diagram of a soil tensiometer is shown in Fig. 3.7, The mercury manometer is connected by a water-filled tube (2) (no air bubbles should be present) to a water-filled porous pot (3), which is placed in the soil for long periods. Water diffuses along the gradient of water potential from the porous pot into the soil until an equilibrium is reached when the water potential of the soil equals the tension in the tensiometer. This tension is measured either with a mercury manometer (Fig. 3.7) or with a vacuum gauge (in all commercially produced tensiometers).

The measuring range of the tensiometer is limited, because when tension in the

pot reaches 1.2. to 1.3 bar, air enters through the walls, which were originally saturated with water, so that further measurement is not possible. KAUSCH (1955) claims
that this limit may be extended by using a porous ceramic pot of a suitable quality
(the pore size of the ROGERS' vessel is about $0.3 - 1$ μm). 15 bar ceramics are now
available commercially (Soil Moisture Equipment Company, Santa Barbara, California, U.S.A.). Matric potentials recorded by a soil tensiometer are slightly higher
(less negative) than the real matric potential of the soil, because, owing to the physical properties of the porous pot, a certain gradient of water potential between the
pot, the soil immediately surrounding the pot, and the soil some distance from the
pot remains in existence (KAUSCH 1955). Similar gradients are commonly found
between roots and the soil; but usually in the opposite direction.

Construction and operation of the tensiometer. Various types of soil tensiometer are commercially available. All joints must be completely air-tight. The tensiometer is filled with distilled water, which has been boiled to prevent bubbles forming from dissolved air. It is necessary to find the pressure at which each porous pot
(after saturation with water) is permeable to air. The pot is first saturated with
water. Either it is filled with water from a wash bottle, so that the outside is dry at
first and then wetted from the inside, or is placed in water, so that it is wetted from
the outside. The pot may also be filled by infiltration at reduced pressure. The
thoroughly wetted pot is connected to air at high pressure, with an attached manometer, and immersed in water; and the pressure at which air bubbles begin to leak
from the surface of the vessel is determined. This pressure is the upper limit of the
range of the tensiometer and should not be much less than 1 bar.

The ROGERS' (1935) type of tensiometer was filled with two liquids. The porous
pot in the soil was filled with boiled distilled water by means of two independent
inlets. An additional tube enclosed in a deflated thin-walled rubber bulb was inserted into the pot. The tube and the bulb were filled with boiled glycerol, which
served to transfer the tension to the manometer. It should be pointed out that the
more complex the tensiometer, the more difficult it is to remove air completely and
to ensure that air does not leak in after a time.

The porous pot is usually cylindrical (e.g. 2×10 cm) with walls about 2.5 mm
thick. Slightly conical vessels have also been used (ROGERS' vessels were 5×10 cm,
with walls 10 mm thick). Two conditions must be fulfilled when placing a filled pot in
the soil: (1) good contact between the vessel and soil is required, (2) the soil surrounding the vessel should not be compressed in any way. The equilibrium is
attained about an hour after the manometer falls to a level which remains constant.
The manometer can also be set zero initially by applying a positive pressure to the
refilling tube before closing it with, for example, an automatic rubber pipette bulb.
It is thus possible to accelerate the establishment of the equilibrium and to decrease the gradient between the pot and soil and within the soil. Equilibration takes
3 to 4 hours in moist soils and several days in drier soils. Tensiometers with a short
response time have been described by MILLER (1951), KLUTE and PETERS (1962) and

LEONARD and LOW (1962). Air intake by the tensiometer is shown during operation by considerable variation of the measured tension with temperature, and by the increasing length of the air-filled space in the filling tube.

As the tensiometer is sensitive to temperature changes (*e.g.* HAISE and KELLEY 1950), particularly to changes in pot temperature, it is better to use plastic (high density nylon impermeable to air) rather than metal tubing for connecting the manometer and the pot. It is advisable to read the manometer in the morning, when the temperature gradients between the porous pot and the soil are smallest. The presence of air must always be checked and air must be removed and replaced by distilled water. It is also advisable to change the position of the pot from time to time, as roots may grow around it preferentially.

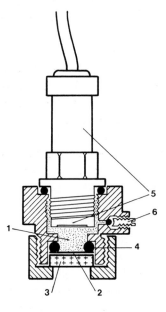

Fig. 3.8 Equipment for soil water potential measurement according to PECK and RABBIDGE (1966a, b). 1 — aqueous (initially hypertonic) solution of polyethylene glycol (MW 20 000), 2 — semipermeable membrane, 3 — 15-bar bubbling porous ceramic plate, 4 — O-rings, 5 — pressure transducer, 6 — fitting for calibration and pressure checking.

A multiple fast-response tensiometer system for field use connected to a number of tensiometers through a hydraulic scanning valve was described by RICE (1969).

The accuracy of the tensiometer is about 1 mbar, provided that it contains no air bubbles and that the soil water content does not change too rapidly. This accuracy is much greater than that of other methods, and is about the same as for gravimetric determinations of the soil water content. Tensiometers are extremely useful for continuous measurement of matric potential *in situ*.

Osmotic Tensiometer

A new tensiometric technique was suggested by PECK and RABBIDGE (1966a, b) for measuring soil matric potential *in situ* (Fig. 3.8). A commercial pressure transducer (5) is used to measure the changes of internal pressure in a small chamber (1) filled with a solution of polyethylene glycol, molecular weight 20 000. The probe cell is placed in the soil; the solution chamber is separated from the soil by a dialysis membrane (2) impermeable for the PEG, permeable for the solutes in soil solution. The dialysis membrane is supported by a 15-bar bubbling pressure ceramic membrane (3), which is readily permeable to liquids. Changes of solution volume are brought about by equilibration between the internal osmotic potential of the solution and the external matric potential of the surrounding soil. These changes are then measured as pressure changes by a pressure transducer. The reading decreases linearly with decreasing matric potential.

Resistance Block Method

The block method, which also measures the matric potential of soil, is very widely used in the field *in situ*. It is based on measuring the electrical conductivity (resistance) of a porous block between built-in electrodes. The block is inserted in the soil for long periods (BOUYOUCOS and MICK 1940).

The blocks are relatively easily prepared, particularly those made of plaster. Permanent emplacement of the blocks in soil is advisable for continuous measurements. A light, transferable conductimeter is used for the measurement. The resistance measured may be converted to matric potential or water content by calibration with suction plate or pressure membrane equipment.

Contact between the block and the soil should be as close as possible. The poorer the contact, the slower the response. Equilibration takes several hours. This may be a disadvantage, particularly during a period of rapid increase in soil water content (after rain or irrigation). This delay, however, does not affect measurements made during drying. Hysteresis is a more serious disadvantage, as the measured resistance does not correspond exactly to the same soil water content or matric potential during both drying and wetting (*e.g.* BOUGET *et al.* 1958). Hence it is advisable to calibrate the blocks during drying-out as changes in soil water content during drying-out of the soil are more important to the physiology of the plant. Differences caused by hysteresis do not exceed a difference of 0.5 to 1 bar in matric potential (TANNER and HANKS 1952, TAYLOR 1955).

The electrical conductivity of the medium between and around the two electrodes is measured. The medium is made of porous material (*e.g.* plaster of Paris, fibreglass or nylon). The measured resistance depends on (1) the content of water in the porous material and (2) the content of salts in the solution wetting the block. Varying concentration of salts in the soil solution which infiltrates the blocks may influence the measurement. To prevent (at least partly) the blocks from becoming saturated with the soil solution, particularly at the beginning, they are inserted into the soil after previous wetting with distilled water. One of the advantages of plaster of Paris blocks as compared with others is that the solubility of the calcium sulphate is sufficient to maintain a more or less stable concentration of electrolytes in the block. Concentrations of soil solution in normal soils are negligible in comparison with the calcium sulphate concentration and hence influence the resistance of the blocks only very slightly. The error of course may become larger in soils with more concentrated soil solutions (WEAVER and JAMISON 1951, COPE and TRICKETT 1965). The osmotic potential of the soil solution thus normally has only a negligible effect on the resistance of the blocks: the conductivity of the block increases, so that the measured water potential increases, when the concentration of the soil solution is higher [*i.e.* during an actual decrease in water (osmotic) potential]. It

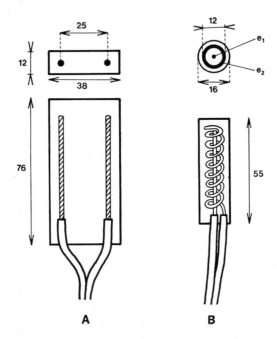

Fig. 3.9 Two types of gypsum blocks. The original dimensions of the block of the A type (BOUYOUCOS and MICK 1940) were 64 × 32 × 13 mm, the electrodes were 57 mm long, the distance between them being 19.5 mm. Type B (CLOSS and JONES 1955), see the description in the text and Figs. 3.10 and 3.11.

should be pointed out in this connection that the block method does not measure water potential of soil but its predominant component − matric potential. (3) In flat blocks the conductivity of the surrounding soil may interfere (see for instance ZADRAŽIL and DAREBNÍK 1958). In this respect, blocks with concentrically arranged electrodes are better. (4) Finally, the resistance of the block depends on its temperature, which should be measured. A correction for temperature is then made on the basis of a previously established relationship.

The effect of temperature may be eliminated by recalculation according to the following equation:

$$R_{(15)} = \frac{R_T}{(1 + \alpha)\Delta T},$$ (3.4)

Fig. 3.10 A brass device for the preparation of gypsum blocks of the B type (CLOSS and JONES 1955). See the description in the text. Inner and outer electrodes are first put in position (with the inserted parts shown on the bottom right) and gypsum is poured in. Subsequently the device is split up into individual components, the inserted parts taken out, the leads soldered on (Fig. 3.11), the device reassembled and the outer gypsum layer poured in to surround the outer electrode.

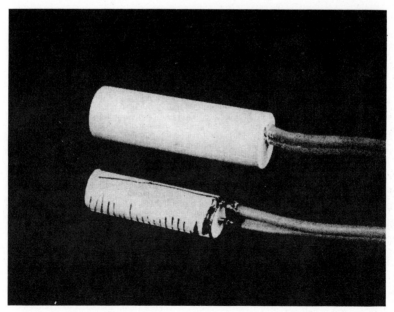

Fig. 3.11 Completed block of the B type and the same block without the outer gypsum layer and with soldered leads.

where $R_{(15)}$ is the resistance at a reference temperature of 15 °C, R_T is the resistance measured at a temperature T, $\Delta T = T - 15$ [°C], α is the temperature coefficient, which has a mean value of 0.029 within the range of -20 to $+40$ °C. The temperature T is usually measured with a thermistor sealed in plastic directly in the block.

The blocks. Plaster blocks are readily available. They can be prepared from dental alabaster plaster, and may be of even better quality than those available commercially (Fig. 3.9). The electrodes are formed by tinned brass litz wires.

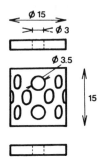

Fig. 3.12 The copper jacket forming the outer electrode and PVC lids with a hole used to insert the central electrode of gypsum blocks (NEUWIRTH 1958). Dimensions in mm. See the description in the text.

Smaller cylindrical blocks (type B) with concentrically arranged electrodes (CLOSS and JONES 1955) were found to be better. The inner electrode is a straight wire along the axis of the block, the outer is a spiral made of the same wire, 12 mm in diameter. The blocks of the B type are prepared in a device shown and described in Figs. 3.10 and 3.11.

NEUWIRTH (1958) described the shape and preparation of another type of plaster block with concentrically arranged electrodes. As shown in Fig. 3.12, a cylinder 15 mm high and 15 mm in diameter with circular holes 3.5 mm in diameter, forming the outer electrode, is made of copper foil 0.12 mm thick. Upper and lower covers are made of polyvinyl plastic; they have a central hole 3 mm in diameter for the insertion of the inner electrode, which is a tinned wire 3 mm in diameter. The holes of the envelope are temporarily sealed with paper, the whole structure is filled with dry plaster which is then wetted with distilled water and allowed to set.

It is vital to use the same mixture (weight per weight) of plaster and water for each block. The mixture should be very runny when made. Precautions must be taken to avoid trapping air bubbles in the block: the gauze electrodes should be degreased carefully with detergent beforehand and then be moved up and down several times in the mixture.

Plaster blocks are subject to corrosion in soil, so that their smooth surface is destroyed after several months of operation. Bouyoucos (1935) described a method of impregnating plaster blocks by rapid immersion in a solution of nylon in ethanol, which substantially increases durability without changing their good properties. A whole series of blocks made of other porous materials is described in the literature; nylon (Bouyoucos 1949), fibreglass (Bouyoucos and Mick 1948,

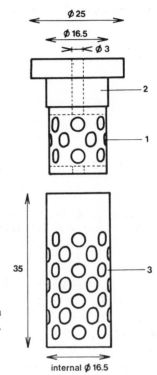

Fig. 3.13 The device for the calibration of gypsum blocks (Neuwirth 1968). See the description in the text.

Coleman and Hendrix 1948, England 1965), combined in such a way that the electrodes are situated between two fibreglass layers (Youker and Dreibelbis 1951), or between two nylon layers, the whole thing being sealed in plaster, *etc.* Some types of blocks are placed in a perforated aluminium cover. The fibreglass laminate blocks are perhaps slightly more sensitive than plaster blocks and, like the nylon ones, are also more stable. However, both types are more sensitive to the content of salts in the soil solution than are plaster blocks. It is sometimes difficult to rewet plastic blocks after drying, so their contact with soil is less effective than that of plaster blocks.

Different types of blocks were also described and discussed by Anderson and Edlefsen (1942), Perrier and Marsh (1958), Pisek and Heizmann (1962), Bouyoucos (1964), Cannel and Ashbell (1964) and others.

Calibration of the blocks. It is advisable to select the blocks carefully prior to calibration by putting them in distilled water and excluding those differing from the average by more than 50 ohm. The A type and the B type blocks usually have a resistance of 450 to 550 ohm and 170 to 250 ohm, respectively, in distilled water. For accurate determination of soil water content each block should be calibrated for a particular type of soil. It is possible to use the mean calibration from a calibration of several blocks for a whole series of simultaneously selected and evaluated blocks only if the blocks have been extremely carefully prepared and selected.

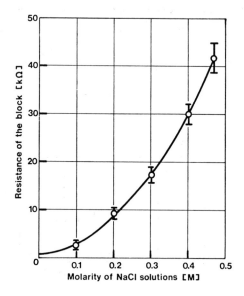

FIG. 3.14 Example of the calibration graph showing the relationship between the resistance of the block (NEUWIRTH 1958, Fig. 3.12) (ordinate) and molarity of test NaCl solutions (abscissa).

Calibration in kaolin on a pressure membrane (RICHARDS 1941) in terms of matric potential is the best technique. However, calibration by vapour pressure equilibration (NEUWIRTH 1958) is also satisfactory. The block (1) in Fig. 3.13 is inserted by means of a polyvinyl attachment (2) into an outer cover (3) made of perforated copper foil. This is wrapped with filter paper and the whole device is lowered gradually into NaCl solutions of different concentrations, so that the filter paper is just in contact with the solution, and maintained in a closed space at constant (± 0.01 °C) temperature. Equilibration between the water vapour pressure above solutions of various osmotic potentials and the matric potentials of the plaster blocks is complete after two to three weeks. An example of the calibration graph is shown in Fig. 3.14. Calibration based on conductivity in distilled water was described by TANNER et al. (1969).

When calibrating the blocks in terms of water content each block (or each series of blocks) should be calibrated for a particular type of soil. It is to calibrate the blocks in terms of water content in the following way: The blocks, previously saturated with distilled water, are placed in soil in a wire basket which may be lined with linen. The thinner the layer of soil surrounding the blocks, the more rapid the calibration, since the required equilibration between the water content of the soil and that of the blocks is achieved more rapidly. A layer 1.0 to 1.5 cm thick is sufficient. The soil is wetted to about its capillary capacity and the whole basket, including the blocks, is weighed. The resistance of the blocks is measured at the same time. The soil is allowed to dry for some time (several hours, depending on the soil volume), and is then placed for several hours in a space almost saturated

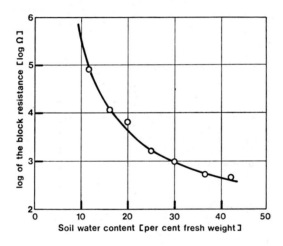

Fig. 3.15 Example of the calibration graph of the gypsum blocks of the A type. The relationship between log of resistance of the block (ordinate) and soil water content (abscissa).

with water vapour (*e.g.* in a desiccator) in order to prevent further evaporation and to allow equilibration. The whole range of soil water content is covered in this way. Soil water contents at each weighing are calculated after subtracting the weight of the basket and blocks (with a correction for water content of the blocks). This gives the relationship between the resistance of each block and the water content of the soil. On the basis of the relationship between soil water potential and water content, it is also possible to draw a calibration curve of the relationship between the resistance of the block and water potential. An example of such a calibration curve is given in Fig. 3.15.

Installation of the blocks in soil. The importance of close contact of the blocks with soil has already been pointed out. It is easier to place cylindrical blocks than rectangular ones, as probe tubes of the same external diameter as the blocks can be used. When burying blocks in soil it is necessary to ensure that the soil

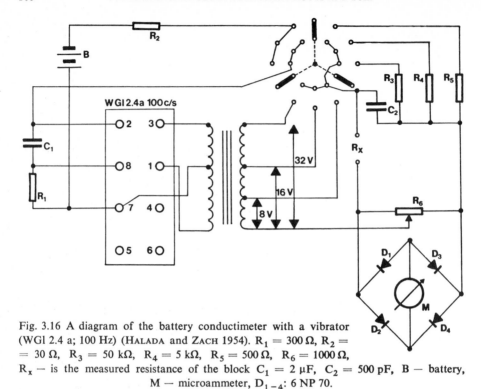

Fig. 3.16 A diagram of the battery conductimeter with a vibrator (WGl 2.4 a; 100 Hz) (HALADA and ZACH 1954). $R_1 = 300\,\Omega$, $R_2 = 30\,\Omega$, $R_3 = 50\,k\Omega$, $R_4 = 5\,k\Omega$, $R_5 = 500\,\Omega$, $R_6 = 1000\,\Omega$, R_x — is the measured resistance of the block $C_1 = 2\,\mu F$, $C_2 = 500\,pF$, B — battery, M — microammeter, D_{1-4}: 6 NP 70.

occupies the same volume as it did before disturbance. It is advisable to insert the blocks into the vertical walls of the hole, in a sufficient distance from the disturbed soil volume. It is difficult to place blocks in stony soil and in this case it is best to surround them with fine soil.

Measuring devices. Conductimetric bridges for field use are battery driven. For detection of the null point, meters, magic eyes or earphones can be used.

A simple measuring device for field use was described by HALADA and ZACH (1954, Fig. 3.16). It uses a 3 V battery, a vibrator WCl 2.4 V 100 Hz as a source of alternating current, and a compensation bridge for measurement. The resistance at which the bridge is balanced is found with a rotating potentiometer. (A null point is often not reached completely, because of capacitance in the block circuit.) The apparatus is very reliable and the 3 V battery lasts for a surprisingly long time if it is switched on only for measurement. Descriptions of other devices have been published by CRONEY et al. (1951), PEREIRA (1951) and HINSON and KITCHING (1964) and a diagram of a sophisticated meter by KITCHING (1965) is shown in Fig. 3.17. In addition, a number of factory-made instruments are available.

Accuracy of measurement. The block method makes it possible to make reliable measurements of the matric component of soil water potential within the range of 0 to 15 bar and with an accuracy of about 5 to 10% of the measured value.

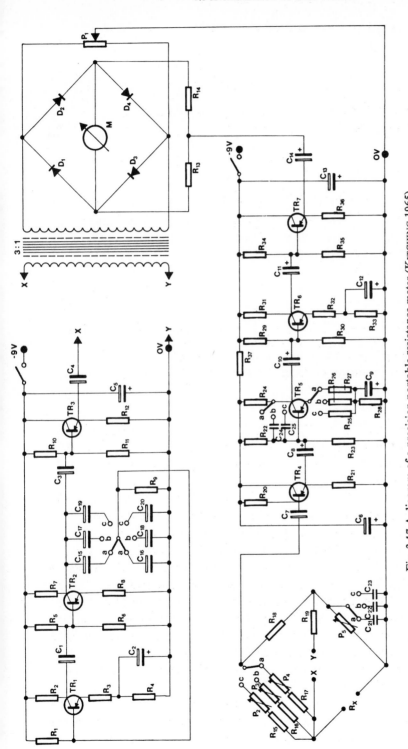

Fig. 3.17 A diagram of a precision portable resistance meter (KITCHING 1965).

$R_1 = 27$ kΩ, $R_2 = 4.7$ kΩ, $R_3 - 1$ kΩ, $R_4 = 1$ kΩ, $R_5 = 39$ kΩ, $R_6 = 10$ kΩ, $R_7 = 8.2$ kΩ, $R_8 = 3.3$ kΩ, $R_9 = 4.7$ kΩ, $R_{10} = 33$ kΩ, $R_{11} = 47$ kΩ, $R_{12} = 1$ kΩ, $R_{13} = 470$Ω, $R_{14} = 470$Ω, $R_{15} = 120$ kΩ, $R_{16} = 10$ kΩ, $R_{17} = 1$ kΩ, $R_{18} = 4.7$ kΩ, $R_{19} = 1.5$ kΩ, $R_{20} = 2.2$MΩ, $R_{21} = 3.9$ kΩ, $R_{22} = 82$ kΩ, $R_{23} = 22$ kΩ, $R_{24} = 319$ kΩ, $R_{25} = 560$ Ω, $R_{26} = 4.3$ kΩ, $R_{27} = 2.7$ kΩ, $R_{28} = 1.2$ kΩ, $R_{29} = 100$ kΩ, $R_{30} = 27$ kΩ, $R_{31} = 3.3$ kΩ, $R_{32} = 330$Ω, $R_{33} = 1.2$ kΩ, $R_{34} = 33$ kΩ, $R_{35} = 68$ kΩ, $R_{36} = 1$ kΩ, $R_{37} = 150$ kΩ, $P_1 = 1$ kΩ, $P_2 = 30$ kΩ, $P_3 = 5$Ω, $P_4 = 500$ Ω, $P_5 = 10$ kΩ, $C_1 = 0.5$ µF, $C_2 = 50$ µF, $C_3 = 0.25$ µF, $C_4 = 1.5$ µF, $C_5 = 100$ µF, $C_6 = 1000$ µF, $C_7 = 0.5$ µF, $C_8 = 16$ µF, $C_9 = 500$ µF, $C_{10} = 5$ µF, $C_{11} = 5$µF, $C_{12} = 500$ µF, $C_{13} = 200$ µF, $C_{14} = 250$ µF, C_{15} and C_{16}: 0.005 µF, matched pair, C_{17} and C_{18} : 0.05 µF, matched pair, C_{19} and C_{20}: 0.44 µF, matched pair, C_{21}, C_{22} and C_{23} — small triming capacitors, $C_{24} = 0.001$ µF, $C_{25} = 0.002$ µF, TR_1- OC 75, TR_2- OC 75, TR_3- GET 103, TR_4- 2 G 309, TR_5- 2 G 309, TR_6- 2 G 309, TR_7- 2 G 309, D_1 — GEX 54, D_2 — GEX 54, D_3 — GEX 54, D_4 — GEX 54, M: $0.5/0/0.5$ mA, R_x — measured unknown resistance.

Mean standard deviation of the block method in TAYLOR's comparative experiments (1955) was about 12% higher than that of the gravimetric method and the neutron-scattering method, which was considered to be the most accurate method for the determination of soil water content within the required range.

Other Methods for Matric Potential Estimation

INGVALSON et al. (1970) embedded a salinity sensor (see p. 185) in the wall of the ceramic shell of the soil psychrometer by RAWLINS and DALTON (1967) described earlier. The psychrometer measures the sum of osmotic potential and matric potentials and the salinity sensor estimates osmotic potential alone. Matric potential of the soil may be calculated from both data using a calibration between electrical conductivity of the salinity sensor and osmotic potential of the soil solution.

PHENE et al. (1971) designed, similarly to the methods mentioned in (5) of the Section 3.1.2, an equipment which measures the heat dissipation rate from a porous block heated by a fixed heat pulse, this dissipation rate being dependent on the water content of the block inserted into the soil. The authors found a single calibration curve between the matric potential of the soil and the rate of cooling measured as the difference of the temperature read before and after the application of the heat pulse.

OSTER et al. (1969) measured independently matric and osmotic potentials of soil samples with a thermocouple psychrometer similar to that of RAWLINS and DALTON (1967) using a ceramic shell with a bubbling pressure of 15 bar. The shell has a vent to the atmosphere and is inserted into a pressure chamber filled up with the soil sample to be measured. After equilibration the psychrometer measures sum of matric and osmotic potentials. Then the air pressure in the pressure chamber with the soil is increased until free soil solution (at zero matric potential) appears on the inside surface of the ceramic bulb shell: then the osmotic potential of this solution is measured by the psychrometer. The standard error of the method was found to be as low as about ±0.04 bar.

Suction Plate and Pressure Membrane

The matric potential of a soil sample can be controlled, not measured, down to −2 bar by means of the so-called suction plate apparatus (RICHARDS 1948), in which the soil solution is sucked from the moistened soil sample on a porous ceramic plate by the application of underpressure below the plate (Fig. 3.18). Recently, even 15 bar ceramic plate devices "extractors" have become available commercially.

The pressure membrane apparatus, now available commercially in various types (often referred to as pressure membrane extractors) may be used (RICHARDS 1947 — Fig. 3.19) for matric potentials of less than −2 bar. The desired matric potentials are reached by applying pressure to the soil samples, so that the soil solution is removed.

Both methods make it possible to calibrate indirect methods (water content determinations, block resistances, etc.) in terms of matric potential (*e.g.* in bars). Both types of instruments are produced commercially. More detailed descriptions are to be found in soil science manuals.

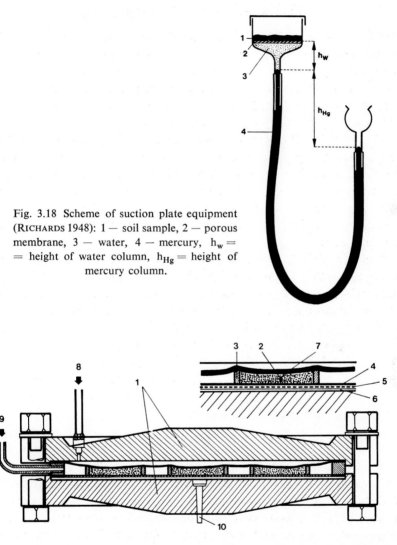

Fig. 3.18 Scheme of suction plate equipment (RICHARDS 1948): 1 — soil sample, 2 — porous membrane, 3 — water, 4 — mercury, h_w = = height of water column, h_{Hg} = height of mercury column.

Fig. 3.19 Cross section of the RICHARDS' pressure membrane apparatus (RICHARDS 1947). Two brass halves (1) of the pressure chamber containing (see also detail on right upper part of the figure) soil samples (2) which are placed in metal rings (3) on cellophane visking film (4) lying on metal mesh (5) and a brass plate (6). The samples are covered with a sheet of rubber (7) which is pressurised through an inlet (8). The pressure is controlled through the inlet (9). Excess water is expressed from the samples through the cellophane visking film, brass mesh and water outlet (10).

3.1.3.3 Determination of the Osmotic Potential of Soil Solution

Preparation of a soil solution was first studied by SAUSSURE (1804), who saturated soil with water and then obtained a solution by applying pressure. SCHLOESING
(1866) later worked out a method of preparation of the soil solution by the removal
of water.

Methods of determination of the soil osmotic potential may be divided into
four groups:

(1) Measurement of the osmotic potential of the soil solution directly in soil,
e.g. by measurement of the depression of freezing point (BOUYOUCOS and MCCOLL
1915, 1916, 1918). Soil cryoscopy is subject to many errors: the principal one is
caused by the difficulty in obtaining a perfect temperature equilibration.

(2) Preparation of a soil extract, measurement of its osmotic potential and calculation of the osmotic potential of the original soil solution. The so-called "saturated" extract is usually prepared, *i.e.* a small volume of distilled water is added to
the soil sample to form a soil paste. The solution is then extracted with a Büchner
funnel, and its osmotic potential determined by a conventional method such as
cryoscopy. The dilution is calculated from the water contents of parallel soil
samples and of the paste, and the osmotic potential of the original solution is then
calculated from the measured osmotic potential of the dilute solution from the
paste. However, this procedure involves an error in principle, in that, for example,
a solution of inorganic salts which is ten times diluted does not have an osmotic
pressure one tenth of the original, because there is a different degree of dissociation
in solutions of different concentrations. It is also possible that during preparation
of the "saturated" soil extract more compounds are dissolved than in the original
soil sample. The drier the original soil sample and the greater the difference between the water contents of the sample and of the soil paste, the greater will be the
error.

To prepare the "saturated" soil extract a small volume of water is added to
a soil sample (about 250 g) in a dish, continuously stirred with a spatula. The
mixture is "saturated" with water when its surface is glossy, when it flows down
slowly when the dish is tilted and (except with clay soils) smoothly and completely
flows down from the spatula. This mixture is left to stand for about an hour, its
saturation is checked again according to the above criteria, a sample (25 to 50 g)
is taken for the determination of the water content and the residue transferred to
a Büchner funnel and filtered under reduced pressure.

As the concentration of the soil solution is usually very low, it is advisable to
use a very accurate procedure for the determination of the osmotic potential. The
choice of the method depends on the available quantity of the solution. The cryoscopic method is the most frequently used. As there is a linear relationship between osmotic potential and conductivity of the soil solution (different for different solutes), it is possible, and advantageous in some cases, particularly for large

numbers of samples, to determine the conductivity of the solution using a conductivity meter. Measured resistances are converted directly to osmotic potentials by
using an empirical calibration curve made on the basis of several simultaneous
cryoscopic and conductimetric determinations.

(3) Preparation of the solution by direct separation of the liquid and solid
phases. Methods of this type avoid the shortcomings of the methods already described. They yield a soil solution unchanged either qualitatively or quantitatively.

The methods mainly involve applying direct pressure by means of a press. The
drawbacks of methods based on this principle are that large, complex devices and
high soil water contents are required. SHMUK (1921 – 1923) attempted to avoid the
requirement for high soil water contents by mixing the soil with a mineral oil,
pressing the soil and separating the solution obtained from the oil by centrifugation.
However, the use of oil is no improvement on simple pressing, using low pressure, and it makes the operation more complex. KRYUCHKOV (1947) constructed
a simple device for obtaining soil solutions by applying high pressures from
1 kg cm^{-2} to 20 000 kg cm^{-2}. This is very important for investigating the energy
of binding of the liquid phase by the solid one.

Attempts to obtain soil solution by centrifugation or by filtration through
a Pasteur-Chamberlain filter or ceramic vessel have been made (*e.g.* WAGNER 1962).
In the method designed by RICHARDS (1949), the soil is placed in a pressure membrane apparatus and the solution which passes through the cellophane membrane
as a result of applying an air pressure of 16 bar is collected. The solution contains
no phosphate, which is bound by the cellophane (REITMEIER and RICHARDS 1944).

Methods involving the use of a liquid for the removal of the soil solution belong
to this third group, which includes SCHLOESING's (1866) method. Different authors
have used different liquids. The method described by KOMAROVA (1956, successfully used by MOSS (1963), based on the replacement of the soil solution by
ethanol, is the most useful so far used. Soil is put in a tube 1.5 to 3 cm in diameter
(the soil column should be about 15 cm high) and 96% ethanol is added at the top.
The first part of the solution which is displaced contains no ethanol and is separated
from the remaining, mixed portion. Ethanol replaces the solution more thoroughly
than liquids used in other methods and the equipment required is very simple.

Methods using presses, or based on the replacement of the solution by ethanol,
are the most practical of those mentioned.

(4) For the estimation of osmotic potential of the soil *in situ* indirect methods
were suggested using the measurement of electrical conductance of a thin ceramic
wafer with pores fine enough to remain saturated with soil solution through the
normal range of the matric potential. The ceramic wafer is in closed contact with
the soil. Then saturated with the soil water (soil solution) its electrical conductance
is a unique function of the osmotic potential of the soil (so-called salinity sensors,
KEMPER 1953, RICHARDS 1966, INGVALSON *et al.* 1970).

3.1.4 Estimation of Dynamic Availability of Soil Water

The availability of soil water depends not only on its static component, determined primarily by the water potential of soil, but also on the dynamic components, which (for soil) are determined primarily by the low unsaturated conductivity of soil water. As the mobility of soil water decreases progressively with decreasing soil water content, it is of most importance at low water contents at about the limit of availability of soil water.

Many authors have attempted to charaterize the so-called "water supplying power of soil" by weighing various models of "roots", usually porous bodies inserted for a short time (*e.g.* 1 hour) into the soil (LIVINGSTON and KOKETSU 1920, and many subsequent authors, see KRAMER 1949 and 1969). TEPE and LEIDENFROST (1963) characterized this property of soil in terms of the weight increase of previously dried, treated filter paper, placed in the soil for 15 min, for example, in special metal holders, so that only the torn edge of the paper is in contact with the soil. The area of the paper was 0.206 cm^2, corresponding, according to the authors, to an absorption surface of 200 000 root hairs and initial water potential of -40 to -50 bar.

3.1.4.1 Permanent Wilting Percentage

Permanent wilting percentage (p.w.p.) is defined as the percentage soil moisture content at which the plant wilts permanently, *i.e.* it does not recover its turgor because of its inability to absorb the necessary water from the soil even if transpiration is prevented. P.w.p. does not mean the absolute limit of availability of soil moisture, as the plant may survive (not grow) in even drier soil. The principle of the determination of the permanent wilting percentage is as follows. The plants are grown (usually in special containers filled with the soil) to a certain growth or developmental phase in conditions of optimum soil moisture. Watering is then stopped and evaporation from the soil surface is prevented. At the first sign of wilting (or another agreed symptom) the containers are transferred into darkness in an atmosphere saturated with water vapour to prevent further transpiration. If the plants do not recover turgor in 12 to 24 hours, the soil moisture corresponding to the p.w.p. has been reached (or exceeded).

The principles of the method used to determine p.w.p. were worked out in 1912 and 1913 by BRIGGS and SHANTZ, who considered p.w.p. as a hydropedological constant independent of the properties of the plants studied. This idea was based on the fact that the values of water or matric potential which determine the static availability of soil water increase unusually steeply with decreasing soil water content, at water contents around the p.w.p. Thus large differences in water potentials of roots of various plant species (*e.g.* between -10 and -30 bar), and the corresponding water potentials of soil, correspond to relatively small differences in soil water content, often small enough to be within the range of measurement error. This does not affect the fact that the p.w.p. is influenced by the properties of plants.

Wilting itself reflects zero turgor, *i.e.* zero hydrostatic pressure, and the water potential of cells of wilted tissues then equals the osmotic potential of which the maximum values in various plants are well known. The water potential of roots is higher than that of above-ground parts during normal transpiration. It can decrease to the above-ground value of the water potential, *i.e.* − at zero turgor − to the osmotic potential of the above-ground parts, during rapid transpiration or insufficient water absorption from soil (SLATYER 1957). As soon as transpiration is prevented, gradual equilibration of water potential throughout the whole soil-plant system takes place: the water potential of roots and above-ground parts increases to that of the bulk soil. During the determination of p.w.p., this occurs when the wilted plant is transferred in its container into a saturated atmosphere in the dark. If the soil water potential is still higher than the root water potential and if a sufficient amount of soil water in the rhizosphere is available to increase the water potential of the above-ground parts and so to restore their turgor, the p.w.p. has not yet been reached.

It follows from this discussion that the soil moisture content determined as the p.w.p. depends on: (1) the osmotic potential of the above-ground parts during wilting and when wilting symptoms appear, (2) the water saturation deficit of above-ground parts when the plants are transferred to a humid atmosphere, since the corresponding amount of water has to be absorbed from the soil in order to restore their turgor. This water deficit depends on the transpiration rate before wilting, and thus on the environmental conditions. All this confirms that the p.w.p. also depends on (1) properties of the plant, (2) experimental conditions in which the plants grow before and after wilting and (3) on the dynamic as well as the static availability of soil water. The dynamic component of soil water is affected by (4) the root density in the soil, which depends not only on properties of the plant but also on cultivation conditions (size of vessel, soil fertility, number of plants and their rate of growth *etc.*). It should be pointed out here that the properties of plants are not determined only by their taxonomic specifity but also by their ontogenetic development. These have only a minor effect on p.w.p. when expressed in terms of soil water content, but the opposite is true when the soil moisture is expressed as water potential.

The techniques of determining p.w.p. must therefore be accurately and carefully standardized.

Technique Used to Determine P.W.P.

Many procedures used for the determination of p.w.p. have been described. Theoretically, it is also possible to determine p.w.p. under field conditions. However, the density of roots in soil may be insufficient, and it is very difficult to prevent transpiration during the first symptoms of wilting. Reproducible results can be obtained only by using containers. The dimensions of the container must encourage a high root density in the whole soil volume during the growth phase. The presence of volumes of soil in which there are no roots erroneously increases the measured value of p.w.p. It is necessary to prevent evaporation from the soil surface at the beginning of measurement, *i.e.* when watering has stopped. Sealing this surface with paraffin wax, or a mixture of paraffin wax, vaseline and bees wax (1 : 1 : 1) prevents any evaporation, but it may result in a shortage of soil oxygen. Covering the soil with a layer of cotton wool or a sheet of PVC is also recommended. It is necessary to ensure that soil is in darkness if containers made of plastic or paraffin paper are used. Porous vessels (flower-pots *etc.*) permeable to water cannot, of course, be used.

It is important to define the symptoms of wilting and the part of the plant to be examined (since not all leaves wilt at the same time). The intervals between observations should be shorter, *i.e.* the material inspected several times a day, at the time when the first symptoms of wilting are expected. It is also useful to choose two or more phases of wilting, the middle one being taken as the selected symptom. During the period when watering has stopped it is necessary to keep the plants under standard conditions of low transpiration, *i.e.* constant vapour pressure deficit and dimmed radiation. High transpiration rates speed up the determination but the results obtained are less accurate.

The space in which the plants are placed after the selected wilting symptoms have appeared should be dark and saturated with water vapour to prevent evaporation. A dark cover lined with wet filter paper is sufficient. This cover must not be exposed to radiation. It is advisable to place a vessel containing warm water inside the cover. The plants are left to stand for 24 hours, to allow equilibration of water potential and the opportunity to restore their turgor, if possible. The plants are then transferred to the experimental conditions in the light again and examined to see if they are still wilting. If they are, the soil is taken out from the container, the surface root-free layer and the roots are rapidly discarded and the soil water content (gravimetric determination) or the water potential of the remaining soil sample is determined.

Examples of Standardized Procedures Used by Different Authors

The following procedures are presented as examples, which may form the basis of a method suitable for a particular experimental investigation.

BRIGGS and SHANTZ (1912a): Seeds (of barley, for example) were placed in glass dishes (250 cm^3) containing wet soil and the soil surface was covered with a thin layer of paraffin wax. The germinating plants push through this layer. The dishes were placed in larger containers of water, ensuring a moist atmosphere. The determination was carried out when the experimental plants had 2 to 3 leaves.

FEDOROVSKIĬ (1948) used 1 kg soil samples with 10 flax plants (or 3 cucumber plants *etc.*). The soil was fertilized, and watered twice a day to 60% of full capacity. Watering was stopped before flowering and the soil surface sealed with a melted mixture of paraffin wax and vaseline.

FURR and REEVE (1945): Sunflower (*Helianthus annuus* L.) seedlings were taken out from a sand culture at the time when the pericarp becomes detached and planted one by one in soil in 500 ml cans, each with a 20 mm hole in its lid. Air-dried fine soil, continuously spread in the can, was used; a small amount of KH_2PO_4 was added to clay soils and the soils were watered with water containing KNO_3, sandy soils were watered with a complete nutrient solution. The plants were protected from the sun at first, and were later exposed to solar radiation up to the time when the third leaf was fully developed. Water was added up to field or capillary capacity during this phase (on the basis of a preliminary gravimetric determination). The hole in the cover was plugged with cotton wool around the stem, watering was stopped and the plants were placed under a canvas cover. Wilting of the bottom leaf pair was then observed.

KOCHERINA (1948): Air-dried soil was placed on a drainage layer of glass chips in aluminium vessels 50 mm high and 38 mm wide. Three germinated wheat seeds were planted in each vessel and a nutrient solution was added through a glass tube up to capillary capacity. When the coleoptiles were about 1 cm long, the soil was covered with parchment with holes for the coleoptiles and a layer of cotton wool on the surface. Watering was stopped during the appearance of the second leaf, and wilting was determined as the third leaf grew out. The top 25 mm of soil were re-

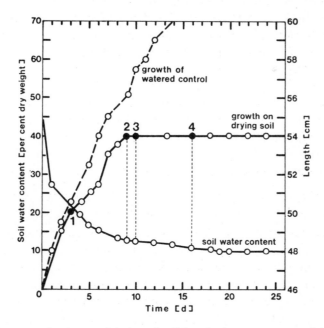

Fig. 3.20 Diagrammatic picture of drying of soil (at glasshouse temperature) by a stand of spring wheat (cv. Podbořanka) in the developmental stage of initial shooting. 1 — initial wilting, 2 — slowing down of growth, 3 — cessation of growth, 4 — permanent wilting, (PENKA 1956).

moved when the p.w.p. was reached, before determination of the soil moisture content.

Váša (1959): 100 ml beakers were filled with soil to 1 cm from the top and a glass tube of 5 mm diameter was inserted down the centre to the bottom. Five to six germinated barley grains were planted about 2 cm deep, and 25 ml of water added through the glass tube. The soil surface was kept loose and the vessels were placed in a glasshouse or other suitable cultivation conditions. When the second leaves had become longer than the first, 4 plants were left in the container, the soil surface was covered with cotton wool and watering was stopped. Three phases of wilting were defined: (1) the ends of the first leaves wilted, (2) all leaves wilted, *i.e.* bent in the middle of their blades, (3) all leaves bent completely. Phase (2) was taken as the selected symptom of wilting.

Penka (1956) suggested to follow the rate of longitudinal growth of the plant with respect to decreasing soil moisture in the pot. The soil moisture (determined gravimetrically by weighing the pot) at which the growth rate (1) slowed down and then (2) ceased represent limit values of soil moisture for extension growth of the plant used (Fig. 3.20).

Several methods of calculating p.w.p. are based on the assumption that p.w.p. is a hydropedological constant dependent only on soil properties (*e.g.* multiplication of the value for maximum hygroscopicity by a coefficient of 1.5 (2−3) *etc.*). However, such calculations cannot be regarded as determinations of p.w.p. but are only an approximation of the so-called critical moisture content.

3.1.5 Experimental Control of Soil Moisture

One of the most difficult technical problems is maintaining soil at a constant soil water content or soil water potential. If the desired water content is more or less around capillary capacity, the problem is less difficult. Various types of so-called autoirrigators have been designed: Livingston (1908, 1918) used a device similar to a porous tensiometer cone buried in the centre of pot containing soil and distributing water drawn up from a reservoir. The height difference between the cone in the pot above and the water reservoir below enabled control of the water potential of the soil to a limited extent. Instead of the cone, double-walled pots with a porous inner wall have been used to improve the unsatisfactory distribution of water in the pot (Wilson 1929, Richards and Blood 1934, Richards and Loomis 1942, Read *et al.* 1962 Hack 1971).

Various methods of controlled irrigation have been designed. Earlier techniques are described by Post and Seeley (1943), the most recent by Kramer (1969). They range from simple wicks drawing up water through the holes in the bottom of plastic pots containing soil, to systems of timing irrigation by plastic tubes (Stice and Booher 1965, Arnold 1970) or by sprinkling (Aljibury *et al.* 1965), using tensiometers as sensors.

The problem of maintaining a constant soil water stress is much more difficult. Small water stresses were achieved by MOINAT (1943) by placing pots on top of sand columns of various heights (up to 100 or 200 cm) standing in pans with water. EMMERT and BALL (1933), HUNTER and KELLEY (1946) and others tried to add water through vertical tubes reaching separately to several horizontal soil layers separated from each other by layers of paraffin wax or mixtures of paraffin wax, vaseline and beeswax which are impermeable to water but permeable to growing roots. The results were not particularly good. VÁCLAVÍK (1966) was able to get a better distribution of soil moisture by random injection of the water to be added with a large syringe into different parts of the soil volume. WHITEHEAD and HOOD (1966) suggested a method suitable for sandy soils. In temperature-controlled conditions, air saturated with water vapour is passed through the soil at a constant rate found by trial and error.

The osmotic potential of nutrient solutions may be decreased by adding more or less inert solutes, but it may not be possible to find really inert solutes (see Section 1.3.2.1). Even the promising polyethylene glycol has been found both to have some toxic effects and to be absorbed.

Very promising devices which maintain constant matric potential in thin vertical layers of soil separated by dialysis membranes from solutions (mostly of polyethylene glycol) have been described by GARDNER (1964), PAINTER (1966), ZUR (1967) and LAWLOR (1969). Rapid deterioration of the membranes and the thin soil layers are the weak points.

3.2 Determination of the Water Potential of Roots

As soil solution is a natural environment of roots, the most natural methods of determination of the water potential of roots are the compensation methods in liquid phase (see Section 1.3.2). Difficulties caused by properties of the material arise when measuring the water potential of roots in this way. The method requires knowledge of the accurate initial value of the concentration (refractive index) of each test solution. This condition is hard to fulfil for roots. Soil particles (organic or mineral) stick to the roots, especially the root hairs, so that it is not possible to clean them perfectly by mechanical means, nor even by washing the roots in the solution in which they are to be immersed. The transferred soil particles then cause a change in the initial concentration, if they contain soluble compounds. This change is small in absolute terms, *i.e.* it is shown by a change in the fourth decimal place of the refractive index, but it could be significant when using the compensation method. This difficulty may be avoided by taking the first drop of the test solution after inserting root samples, mixing and then determining the initial value of the refractive index (SLAVÍKOVÁ 1963a). The refractive index after exposure is then compared with the initial value.

The principle as well as the procedure of the refractometric method is described in Section 1.3.2.3 and it is only necessary to include some comments about the determination of water potential of roots here. Test solution gradations of 0.1M, corresponding to almost 3 bar of water potential, are usually used. However, smaller gradations, *i.e.* 0.025 to 0.1M (equivalent to about 0.6 to 0.25 bar) are required for more detailed work. Such a series may be used if the range of the expected values of the water potential is known at least approximately, so that a series of about 8 test solutions is sufficient. Smaller gradations require very accurate reading of the refractive index, as the changes of the refractive index around the isotonic solution are very small. It is very hard to make an advance estimate of the required range of a series of test solutions for use with roots (SLAVÍKOVÁ 1963c). The estimate of water potential is obtained from a whole series of simultaneous measurements, as in all compensation methods. In order to exclude heterogeneity of the material, caused, for example, by a gradient of water potential along a single root (branch), roots should be sampled at the same distance from the axis and at the same depth for each measurement (SLAVÍKOVÁ 1963b).

The procedure is as follows. If possible, a completely intact root branch with a large number of absorbing roots (roots with primary anatomical structure) is exposed and cut off. This small sample branch is rapidly cleaned with a fine brush and immediately transferred to a tube containing the test solution and sealed with a rubber stopper. A whole series of sucrose concentrations is gradually filled in this way. When using a large series of concentrations, it is better to place the cut root branches in the air space above the test solution and only to immerse them immediately before the initial measurement. (Equilibration in the vapour phase is negligible, as the time is relatively short compared with subsequent equilibration in the liquid phase.) Insertion of the roots into the tube containing the test solution, pushing them together with the glass tube (the solution can be thoroughly mixed around the roots by blowing through the tube), drawing off the excess solution and measuring the initial refractive index should not take more than one minute. The whole series of solutions should be finished after about 5 to 6 min, as concentration changes caused by the equilibration between roots and the test solutions would otherwise already have taken place.

The method described above enables roots with intact rhizoderm (except for a single cut area) and mostly with primary anatomical structure to be in contact with the test solution, so that there is continuous and rapid equilibration of water potentials between the tissue and the test solution. Difficulties in using this method for measurements with leaf tissue (decrease of values during the incubation and choice of a suitable time for the incubation), in which equilibration takes place only at the cut edges of leaf discs and in which there is also swelling of the epidermal tissues (see Section 1.3.2.3) are thus avoided.

A suitable ratio between the amount of test solution and the surface area of the roots is one of the requirements for accurate measurement. The test solution should form only a thin layer around the root surface, maintained by surface tension of the solution. The solution adheres evenly over the surface, in contrast to leaves, which are not easily wettable. The roots in the tube should be closely but lightly pressed together, so that there is only a thin layer of the test solution between them. This layer of roots should be about 5 to 8 mm deep, so that there is enough solution for two refractometric measurements (initial and final). Readings can be taken after 3 to 4 hours. Other details of measurement by the refractometric method are described in Section 1.3.2.3.

FISCUS (1972) designed a modification of an A-type thermocouple psychrometer for *in situ* measurement of root-water potential. For psychrometric methods see detailed description in Section 1.3.4. The measuring chamber for roots consists of two Teflon parts: a cylindrical "sleeve" (0.95 cm inner diameter) and a bottom cap with calibration "well"; both have two opposite slits each, allowing circular holes for the intact root to be inserted through. When calibrating, the bottom cap is rotated by 90° so that the holes are closed.

Root water potential can also be determined by extrapolating the relationship between the rate of water uptake by roots from solutions and the osmotic potential of the solutions to the value equivalent to zero uptake (BROUWER 1953).

3.3 Determination of Water Absorption by Roots

There are two types of equipment used to determine the rate of water absorption by roots from the loss of water or nutrient solution from an enclosed volume in which the roots are immersed. One type consists of instruments suitable for measuring the rate of absorption by the whole root system of one or several plants. The second type includes instruments used to measure gradients in the rate of absorption in various parts of a single root.

In both types, absorption of water or a nutrient medium is determined from the movement of a water column in a horizontal graduated capillary connected to a vessel containing the root system, single root or part of a root.

3.3.1 Potometric Determination of Rate of Water Absorption by a Complete Root System

A diagram of a simple potometer used to measure absorption by the complete root system of a plant is shown in Fig. 3.21. The point at which the stem or part of the root of the plant is sealed into the potometric vessel must be air-tight. The enclosed volume of water (or nutrient solution) must not contain air bubbles, which would cause distortion of the results by any pressure or temperature changes.

It is usually very difficult to seal the stem or part of the root, which is generally wrapped with cotton wool impregnated with Leick potometric sealing material (see

Section 5.6.4.3) or a similar mixture at the point of fixing. KÜSTER (1956) recommends a mixture of bees wax and dehydrated lanolin (1 : 2). A special sealing material, Terostat, has excellent properties. Packing with a strip of polyurethane foam saturated with vaseline and covered over with plasticine was also successful (KOZINKA 1966). A sealing mixture of bees wax, petroleum jelly and tacki wax also works well (O'LEARY 1965). O'LEARY did not give the proportions of the components in the mixture but we have successfully used the proportions 12 : 12 : 1.

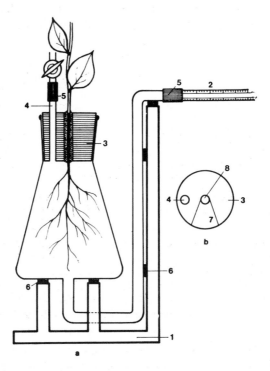

Fig. 3.21 A diagram of the potometer for determining the rate of water absorption by the whole root system of the plant (a) and a rubber stopper used for air-tight closure of the potometer (b). 1 — stand, 2 — measuring capillary, 3 — rubber stopper, 4 — double deaeration tube, 5 — rubber tubing, 6 — rubber pads, 7 — triangular notch, 8 — hole used to fix the plant.

O'LEARY used it to seal apical root segments in the measuring capillary when studying rates of root bleeding.

PETTERSSON (1966) used a commercially available water-proofing grease (Bostik 292, Bostik A. B., Hälsingsborg, Sweden) to seal the roots for induced water uptake. A silicone grease produced by Silicone Lubricant (Dow Corning Corporation, Midland, Michigan, USA) has become very popular recently (HU JU and KRAMER 1967). It is useful particularly when working with no pressure differences or with slight positive or negative pressure.

EVANS and VAUGHAN (1966) analyze two procedures for air-tight sealing of apical root segments in glass chambers. They describe simple heat sealing and pressure sealing. A mixture of 5 per cent paraffin wax (m.p. 50 to 52 °C) and 95 per cent dehydrated pharmaceutical lanolin was used. The melting point of the mixture is 38 to 42 °C. A hypodermic syringe with electrically heated subcutaneous needle (No. 17) is filled with the mixture, which can be then applied directly onto the required place. Apical root segments were sealed in root chambers made from 8 mm glass tubing with 3 mm long capillary tubing at the end. The diameter of the capillary was only slightly larger than that of the root and the first one mm of the capil-

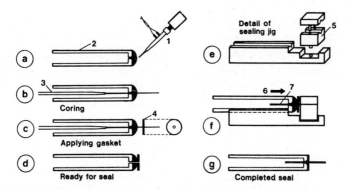

Fig. 3.22 Preparation of pressure seal: (a) heated hypodermic needle (1) is used to apply a dome of sealing compound to root chamber (2); (b) dome is cored using 0.98 mm coring rod (3); (c) applying Parafilm gasket (4); (d) root chamber ready to reactive root; (e) detail of pressure sealing jig using a sealing block (5); (f) cross section through root chamber (being forced against seal block, (6) and sealing jig showing inserted root (7) just prior to sealing; (g) completed pressure seal (EVANS and VAUGHAN 1966).

lary was extended at a 45° angle. This makes it possible to form a space for the pressure sealing. However, simple heat sealing may also be useful: the sealing material is applied from the syringe around the root into the capillary. Pressure sealing requires a more elaborate procedure (Fig. 3.22). The details may be found in the paper cited. If the stalk is perfectly smooth, it is useful to cement a thin rubber sheet to the upper or lower surface of a longitudinally cut rubber stopper, the sheet having a round hole cut out after it has been stretched. The root system of the plant is drawn through this hole by further stretching the rubber sheet. The tension is then released and the rubber hermetically seals round the stalk, which is inserted through the longitudinally cut stopper, to which the rubber sheet is then glued. This method of sealing can be used only when there is a small excess of water (or nutrient solution) in the potometric vessel. Otherwise, the method must be combined with the use of one of the sealing mixtures described above.

Air is removed from the whole system by a deaeration tube with a tap. The top of the tube is shaped like a small funnel and is used to add water to the system when the meniscus in the measuring capillary reaches the end of the scale. A hypodermic syringe with a needle stuck into rubber tubing connected to a double deaeration tube has also been successfully used to top up the potometer (KOZINKA 1966). This modification is also suitable for adding various agents (osmotic, respiratory inhibitors, growth factors *etc.*) to the root environment, without taking the plant out of the potometer. It is possible to continue the readings immediately after refilling the measuring capillary only if (1) the closed space does not change in volume because of slight pressure differences and (2) it contains no air bubbles.

The measuring capillary. Its size is determined by the sensitivity required and the range of measurement. Accurate calibration is carried out by drawing a small volume of mercury into it, to form a mercury column about 10 mm long. This small column is gradually moved step by step along the capillary, which is graduated in millimeters, and the length between the two menisci is measured in each position as accurately as possible. The mercury column is then weighed, the weight (in g) is divided by 0.013546 (the weight of 1 mm^3 of mercury at 20 °C), and the volume of the column is calculated. This volume is divided by the length of the column, to give the relation between length in mm and volume in mm^3 at individual positions along the capillary.

The meniscus of the water column should be well defined and there should be no water losses from spreading of water on the walls. It is therefore useful to coat the measuring capillary with silicone oil. Evaporation of water is prevented by closing the capillary with a large hypodermic syringe, or by observing the movement of an air bubble in the uninterrupted water column of the capillary, which is bent and the end immersed in a vessel containing water. There are irregularities in the values measured if the capilllary is not sufficiently clean. These are caused by variations in adhesion of water to the walls of the capillary.

Temperature correction of the volume is required, and for accurate measurements the potometric vessel should be placed in a water bath. The variation of the temperature of the vessel should be as small as possible. In order to prevent errors caused by a temperature difference between water in the potometric vessel and the measuring capillary, the capillary can be covered with glass tubing, through which water of the same temperature as the water bath is pumped by an ultrathermostat.

The accuracy of results obtained by so-called "open potometers", which detect a decrease in the level of a liquid root medium in an open vessel, and not the volume changes in an enclosed root space, is very poor, even if the accuracy of reading is increased by connecting the vessel below the level of the liquid surface with a long measuring capillary the far end of which is slightly higher. As the surface level goes down the meniscus in the capillary goes down, too.

KUPERMAN *et al.* (1968) attempted to increase the accuracy of measurement using an open potometer, shown diagrammatically in Fig. 3.23. Vessels 1 and 2 are connected by a one-valve tap (5) and a rubber tube (6). A syringe (3) is connected to the system by rubber tubing. The piston is connected to a micrometer screw (4).

The device is filled with water while the tap (5) is open. When the levels have become steady, the reading (n_1) on the micrometer scale is taken. A root system of the experimental plant is then inserted into vessel 1. The levels are left to equilibrate and the tap is closed. The time when the tap is closed is taken as the beginning of the measurement. The level in vessel 2 is adjusted up to a mark in a capillary by rotating the micrometer screw. The second reading (n_2) on the micrometer scale is then taken. The level in the vessel is returned to the n_1 position and the tap opened.

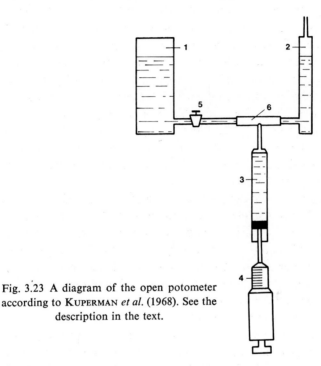

Fig. 3.23 A diagram of the open potometer according to KUPERMAN *et al.* (1968). See the description in the text.

The tap is closed after the exposure and the level in vessel 2 raised up to the mark on the capillary. A third reading (n_3) on the micrometer scale is taken. The difference $n_2 - n_3$ corresponds to the volume of water taken up by the plant in time t. This may be calculated according to the equation

$$V = \frac{A_P(A_1 + A_2 + A_x)}{A_2}(n_2 - n_3),\qquad(3.5)$$

where A_P is the area of cross section of the pump (3), A_1 the area of cross section of the vessel (1), A_2 the area of cross section of the vessel (2) with the capillary, A_x is the area of the cross section of the plant stem at the surface of the nutrient solution.

$$\frac{V}{n_2 - n_3} = \frac{A_P(A_1 + A_2)}{A_2} - \frac{A_P}{A_2}A_x,\qquad(3.6)$$

where $V/(n_2 - n_3)$ is determined by the ratio between the micrometer unit and the unit volume. It follows from equation 3.6 that $V/(n_2 - n_3)$ depends directly on A_x. It is hence possible to draw a curve of the relationship between the micrometer scale units and the cross section of the plant stem after measuring the A_P, A_1 and A_2 values. The device is particularly useful for comparative measurements of the rate of water uptake, and in cases when frequent readings are not required.

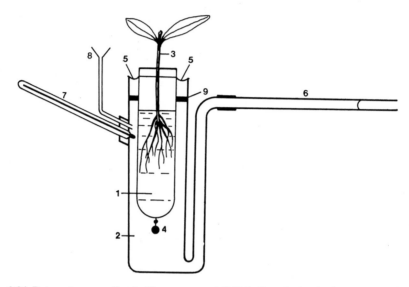

Fig. 3.24 Potometer according to KLESHNIN *et al.* (1954). Description in the text. 7 — thermometer, 8 — funnel for filling up the vessel (2) and hence controlling the position of the meniscus in (6), 9 — oil layer.

A special and unusual modification of the "open potometer" was described by KLESHNIN *et al.* (1954) (Fig. 3.24). A vessel (1) with a plant (3) in water or nutrient solution is floating in a cylindrical vessel (2) filled with 5 per cent solution of the emulgator OP-10. Its vertical stability is secured by a weight (4) and its adherence to the wall is prevented by threads (5). The movement of the meniscus in the measuring capillary (6) represents the volume changes in (2) due to weight changes of (1).

A further type of open potometer was described by PETTERSSON (1971). It was designed for measurement of water and ion uptake. Its construction and the measuring procedure are described in Fig. 3.25. In the open potometer by EHWALD (1971) the absorbed water was monitored as small changes in concentration of dextran in the potometric vessel, measured by a flow-through polarimeter. Nutrient solution aerated by water vapour-saturated air was circulating through the potometric vessel and the Perkin Elmer Polarimeter 141M (Fig. 3.26). The author describes the possibility of measuring changes in water uptake at 1 to 2 minute intervals.

Time limitation is a common deficiency of all potometric measurements. Water (or nutrient solution) cannot be aerated in a closed space, so that sufficiently accurate measurements cannot be made for more than 6 to 8 hours even in water which was well-aerated before being put in the potometer. The preparation period, from enclosing the root system to the beginning of measurement, should be included

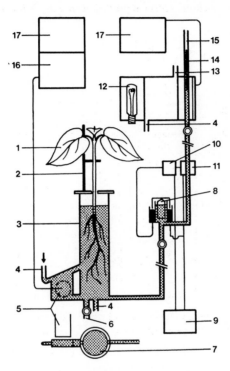

Fig. 3.25 The equipment used for continuous recording of the phosphate and water uptake in a plant root (PETTERSSON 1971). 1 — plant, 2 — plant holder, 3 — plexiglass vessel containing labelled nutrient solution, 4 — air inlet, 5 — end-window GM-tube, 6 — solution outlet, 7 — plexiglass vessel. vertical view, 8 — float switch, 9 — DC-set, 10 — relay, 11 — magnet valve, 12 — lamp, 13 — air outlet, 14 — float, 15 — glass burette, 16 — ratemeter, 17 — recorder.

in this total possible period of measurement. The rate of water uptake is usually lower after enclosing the root system and then increases slightly. Measurements may then begin.

FALK (1966b) used small (20 ml) potometric vessels, with frequent changes of the nutrient solution, made relatively rapidly (less than one minute) without taking the plants out. The potometric vessel and the reservoir containing the nutrient solution were placed in the same water bath, so that there were no temperature differences caused by the exchange. A well-aerated nutrient solution can be used for the exchange. LARKUM (1969) designed a potometer enabling determination of water absorption in aerobic and anoxic conditions in a flowing medium (Fig. 3.27).

The results of potometric measurements are distorted by an error originating in visual evaluation of the movement of the meniscus in the measuring capillary. It seems necessary to replace visual evaluation by an objective method, preferably including recording. Experiments along these lines have recently been described by KÜSTER (1956). A method of photoeletric detection of volume changes in the

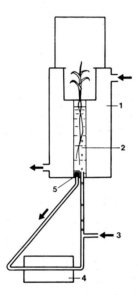

Fig. 3.26 A method for continual measurement of water uptake by roots using a recording polarimeter (EHWALD 1971). 1 — thermostatic water jacket, 2 — root chamber with nutrient solution and 5% dextran, 3 — moistened gas inlet, 4 — polarimeter cell, 5 — filter.

measuring capillary of the potometer designed by FALK and SANGREN appears to be promising for recording potometric measurements. The method has not been published, nevertheless it has already been used by FALK (1966b) for studying changes in water uptake during rapid changes of osmotic potential of a nutrient solution. It appears that the method is useful for determining changes in the rate of water uptake which are as rapid as changes in transpiration rate, determined by gasometric methods (FALK 1966a, b). See also recent papers by JOHNSSON (1973) and BROGÅRDH and JOHNSSON (1973b).

An ingenious method of photoeletric recording of the volume changes in the measuring capillary of the potometer has been developed by RUFELT and HELL-KVIST (unpublished) in the Institute of Plant Physiology at the University of Uppsala. The pen of a recorder is coupled photoelectrically to the movement of the meniscus of the water column in the capillary. A new type of recording potometer was described by CZERSKI (1972). BROGÅRDH and JOHNSSON (1973b) detected the changes in measuring capillary minimum water volume difference of about 0.02 mm^2 by means of photodiodes. The electrical signal controls a stepper motor which drives a piston into a cylinder, thereby replacing the water taken up by the plant. Every step of the motor means that a small discrete volume of water is pumped into the potometer vessel. The number of steps is recorded.

Potometry may also be used with the technique of separated root systems. Combining two methods allows the possibility of measuring water and ion uptake by parts of a root system (MICHEL and ELSHARKAWI 1970, KALRA 1971), studying lateral water transport in roots (JACKSON 1956, BAKER and MILBURN 1965, SLAVÍKOVÁ 1967) and studying root resistance to water flux (KOZINKA 1970 — see Fig. 3.28 —, STOKER and WEATHERLEY 1970).

3.3.2 Micropotometric Determination of the Differences in the Rate of Water Absorption in the Various Parts of a Single Root

Figures 3.29 to 3.34 show several modifications of potometers used to measure the gradient of water absorption in the various parts of a single root. Waterproof separation of individual parts of the root frequently seems almost impossible if

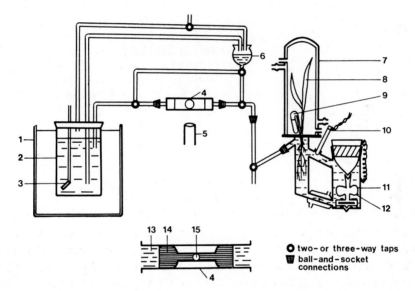

O two- or three-way taps
⊟ ball-and-socket connections

Fig. 3.27 Diagram of the apparatus used to measure the rate of water absorption and changes in the level of oxygen around the root of a young barley plant (LARKUM 1969). 1 — water bath, 2 — large reservoir, 3 — sintered glass bubbler, 4 — capillary system (enlarged in the inset), 5 — travelling microscope, 6 — small reservoir for refilling the root chamber and manoeuvring the position of the air bubble in the capillary system, 7 — shoot chamber with water jacket, 8 — barley seedling, 9 — glass phial containing potassium hydroxide, opening into the air space of the root chamber, 10 — oxygen electrode, 11 — root chamber, 12 — stirring device with rotor and glass-encased magnet, 13 — nutrient solution, 14 — xylol, 15 — air bubble. The system of by-passes and other connections between (2) and (11) were for the purposes of exchanging the nutrient solution, gassing the nutrient solution, levelling the solutions in these two chambers and manoeuvring the air bubble in the capillary system.

sufficiently natural conditions are to be preserved. SIERP and BREWIG (1936) separate the individual root parts in the capillary chambers with lanolin-saturated cotton wool.

The system of micropotometers designed by GREGORY and WOODFORD (1939) consists of several cylindrical glass chambers, about 1 cm high, with two glass

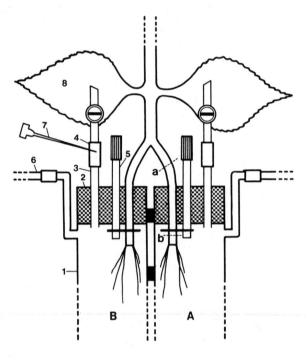

Fig. 3.28 Scheme of equipment for the measurement of the water uptake by halves of split roots *Coleus blumei* BENTH. (KOZINKA 1970). 1 — potometric vessel, 2 — rubber stopper with three orifices, 3 — trifurcated air tube with one-way rubber pipe (4), 5 — brass stick with razer for cutting off the roots without removing the plant from the potometer, 6 — measuring capillary, 7 — needle of the syringe for application of synthetic auxin to the potometric vessel, 8 — the plant with split root, (a) the point of cutting off half of the split root in the air, (b) the point of cutting off half of the split root inside the potometer.

tubes attached at the sides (Fig. 3.29). One tube connects the chamber to the measuring capillary, the other is used for adding water or nutrient solution to the chamber. Each chamber has a ground glass ring by which it is mounted on to the chamber below. The chambers are separated from each other by a membrane made of soft rubber, with a hole for insertion of the root in its centre. A collar of soft rubber tubing is glued inside the hole. Tubing which makes it possible to seal the root in the collar efficiently without damaging it mechanically should be used (GREGORY 1938). The rubber is pushed on to a slightly thicker glass tube prior to inserting the root and is then carefully slipped onto the part of the root where the membrane is to be fixed. This procedure makes it possible to prevent any damage to the root during its insertion into the membrane.

A similar division of the root into several zones was made by BROUWER (1953) using more elaborate apparatus for the measurement of the rate of absorption of water and nutrients in the various zones (Fig. 3.30). Many original ideas were used by LEISTER et al. (1970) and BURLEY et al. (1970) in designing an isolation unit for supplying radioisotopes to specific segments of an intact root. The unit will surely be modified in future for measurement of water uptake as well.

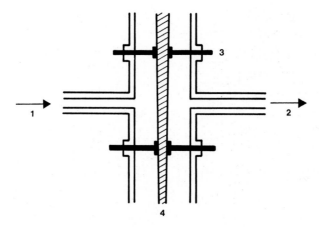

Fig. 3.29 The micropotometer according to GREGORY and WOODFORD (1939). 1 and 2 — sites of inflow and outflow of liquid from the potometric chamber, 3 — membrane separating the potometric chambers from the root parts, 4 — root.

A micropotometer described by HAYWARD et al. (1942) consists of brass housing (1) with a micropotometric chamber formed by a piece of thick-walled rubber tubing (2), calibrated capillary tube (3), glass stopper (4) and adjusting screw (5) (Fig. 3.31). A part of the rubber tube is firmly glued in the rings of the brass housing with DE KHOTINSKY's cement. A half notch is cut in the rubber potometric vessel on the side where the screw is situated. The centre of the notch is connected with the hole where the part of the root to be studied is inserted. This hole is made with a red-hot needle. The root is inserted into the hole through the notch, with the housing completely open. A set of chambers should be prepared with holes of various diameters corresponding to the various parts of the root. Seals are checked by applying air pressure at the open end of the capillary. If the root is not sufficiently sealed, air bubbles leak through around it and the adjusting screw should be gently tightened. The chamber is then filled with water (or nutrient solution) by moving the glass stopper. A set of several potometers is attached to the various zones of a single root and placed, with the whole root system, in water at constant temperature. Only the open ends of the measuring capillaries

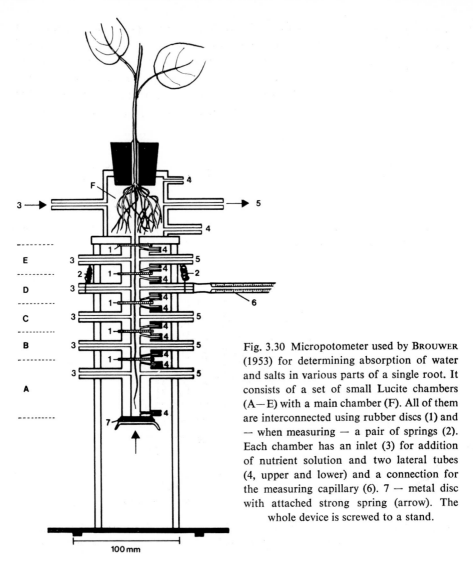

Fig. 3.30 Micropotometer used by BROUWER (1953) for determining absorption of water and salts in various parts of a single root. It consists of a set of small Lucite chambers (A–E) with a main chamber (F). All of them are interconnected using rubber discs (1) and — when measuring — a pair of springs (2). Each chamber has an inlet (3) for addition of nutrient solution and two lateral tubes (4, upper and lower) and a connection for the measuring capillary (6). 7 — metal disc with attached strong spring (arrow). The whole device is screwed to a stand.

project from the side wall of the water bath. VERETENNIKOV (1957) made some improvements in the original device and combined mechanical sealing with cacao butter.

A micropotometer designed by ROSENE (1937, 1941) is a special modification of the equipment used to measure the rate of water absorption in various parts of a single root. It consists of several capillaries of the same diameter. One end is sealed and pushed back, to make a "bed". A hole of the same diameter as that of the part of the root to be studied is blown out or drilled out in the "bed". This hole is connected with the hole in the measuring capillary. Varying numbers of these micropotometers are attached to a single root (Fig. 3.32). Construction of the apparatus is simple and it is useful for laboratory measurements on terminal, non-branched parts of roots of about the same size. As it functions in air, relative humidity must be close to saturation point. A similar principle was used by KAWATA and LAI (1968). Instead of a glass capillary, a polyethylene capillary tube was used with the end orifice adapted for attachment to the root surface. CAILLOUX (1944, 1966) used a similar device to ROSENE, designing micropotometers for measuring

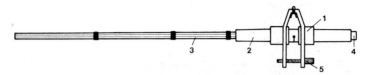

Fig. 3.31 Micropotometer according to HAYWARD *et al.* (1942). 1 — brass box, 2 — potometric chamber, 3 — measuring capillary, 4 — glass stopper, 5 — adjusting screw.

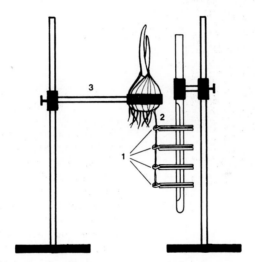

Fig. 3.32 The device used by ROSENE (1937, 1941) for measuring water uptake in various parts of a single root. 1 — set of capillary tubes, 2 — root, 3 — stand.

water uptake by a single root hair or even a part of a root hair. The whole root hair or its apical part only is inserted into a tiny measuring capillary (Fig. 3.33, a, b, c, d). Water uptake of part of a root hair is measured by attaching the tip of the measuring capillary to the part of the root hair surface, or the part is positioned in a ring at the tip of the measuring capillary (Fig. 3.33, f). General view of the equipment used by CAILLOUX is in Fig. 3.34.

A potometer for application of labelled elements to a particular part of a root (Fig. 3.35), constructed by WIEBE and KRAMER (1954), is also of this kind. The root is inserted through a hole in a soft rubber (1) and sealed with lanolin at the place of insertion. A nutrient solution containing the isotope washes a limited root zone.

The solution is agitated by bubbling air. The rest of the root and the whole root system are in the control nutrient solution.

Identical procedure was used by RUSSELL and SANDERSON (1967) and CLARKSON and SANDERSON (1970) for designing an absorption cell for synchronous measurement of different parts of several roots. The roots pass through the openings in the wall of polyethylene tubes and are sealed by anhydrous lanolin warmed to 45 °C followed by a mixture of 50 per cent resin colophony and 50 per cent paraffin wax (melting point 45 °C).

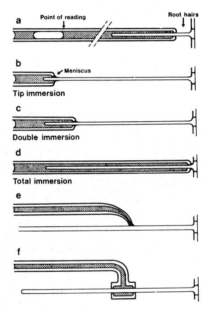

Fig. 3.33 The straight micropotometers (a, b, c, d), bent-(e) and ring-(f) micropotometers (CAILLOUX 1971).

3.3.3 Potometric Measurement of Experimentally Induced Water Uptake

The measurement of the rate of induced water uptake at an increased hydrostatic pressure, or substitution of the suction force of above-ground organs by low pressure, involves many technical problems which also arise in potometry. MEES and WEATHERLEY (1957a, b) designed a device (Fig. 3.36) in which the root system, with no above-ground organs, is placed in a Perspex container (volume about 1 200 ml) and aerated by forcing air under pressure through the tube (7). The remainder of the plant stem is fixed in a polyethylene cone which provides air-tight enclosure of the whole system. The detailed arrangement is shown in Fig. 3.37.

The rate of sap flow is measured from movement of an air bubble introduced into the measuring capillary (3) through the side arm (12). The end of the capillary is below the surface of the water in the container (4). A device for changing the root medium is part of the pressure container. The old solution is ejected by opening the tap (8). Fresh nutrient solution is then added through the hole (9) under pressure.

In the apparatus by PETTERSSON (1966) (Fig. 3.38), the detopped root (6) fixed in a plastic holder is placed in a nutrient solution (12) which is continuously aerated through a sintered glass device (5). The plant stem is inserted in a glass tube (3) and sealed in by waterproof grease (Bostik 292). A scale for measuring the volume of the exuded sap is attached to the tube. The hypodermic syringe (1) with a needle (canulla) (2) is used to take samples of the liquid. Low pressure in the glass tube is attained by removing air through a tube (8) connected in series with a pressure gauge, a pressure levelling vessel, and a vacuum pump.

A simpler device was used be STUART and HADDOCK (1968) for measuring induced water uptake and loss simultaneously with thermoelectric measurement of sap flow rate.

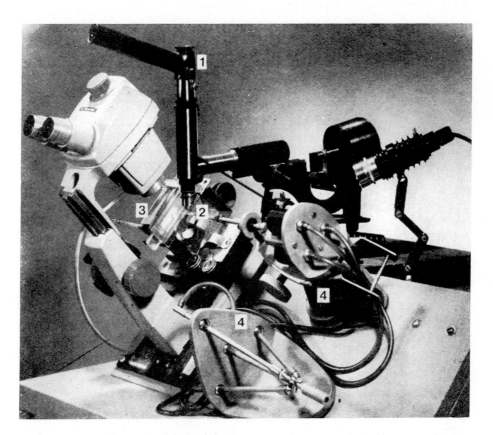

Fig. 3.34 Complete equipment for measuring water uptake by individual root hair and its parts by CAILLOUX (1971). 1 — measuring microscope, 2 — micropotometer, 3 — humid microchamber for the plant under observation, 4 — CAILLOUX micromanipulator.

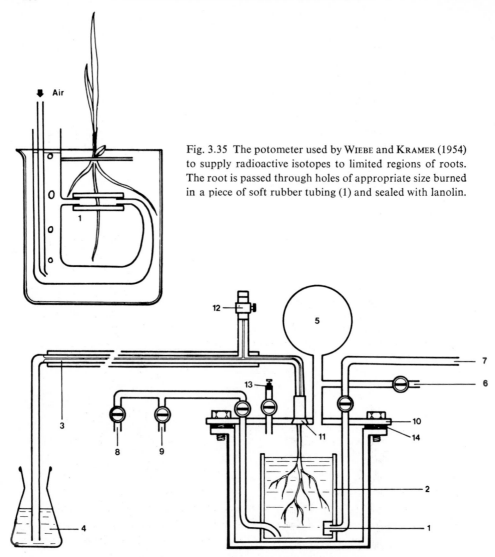

Fig. 3.35 The potometer used by WIEBE and KRAMER (1954) to supply radioactive isotopes to limited regions of roots. The root is passed through holes of appropriate size burned in a piece of soft rubber tubing (1) and sealed with lanolin.

Fig. 3.36 The pressure apparatus. 1 — "stone" aerating block, 2 — Perspex beaker containing the root system, 3 — measuring capillary, 4 — flask containing water, 5 — Bourdon tube pressure gauge, 6,7,8 and 9 — inlets of the capillary, 10 — lid of the container, 11 — polystyrene cone, 12 — side arm of the capillary, 13 — pressure release valve, 14 — circular rubber washer (MEES nad WEATHERLEY 1957a,b).

3.3.4 Potometric Determination of Water Balance

Determination of water balance using potometry involves simultaneous measurement of water uptake and water loss of a plant. In the past such devices were called "weighing potometers" (Wägungspotometer, MONTFORT 1922, LACHEN-

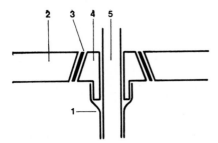

Fig. 3.37 Method of sealing the stem in the lid of the container. 1 — rubber sleeve, 2 — lid of the container, 3 — conical rubber washer, 4 — polystyrene cone, 5 — stem of the plant (MEES and WEATHERLEY 1957a, b).

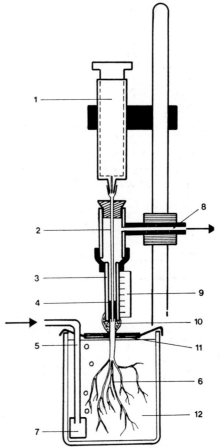

Fig. 3.38 Suction apparatus. 1 — all-glass syringe, 2 — syringe needle canulla, 3 — measuring glass tube, 4 — root exudate, 5 — air inlet, 6 — root system, 7 — aerating block, 8 — air evacuation tube, 9 — calibrated scale, 10 — waterproof grease, 11 — plastic holder supporting the root, 12 — tested solution. The function of the apparatus is described in the text. (PETTERSSON 1966).

MAYER 1932). The procedure is well described by KNIGHT (1965) (Fig. 3.39). A more advanced potometer was designed by KÜSTER (1958) (Fig. 3.40). The amount of water absorbed is determined as weight difference of the device.

In measuring water uptake IVANOV (1962) used the so-called "drop principle": a water droplet is added automatically to the potometric vessel each time the weight of a water drop is lost. So the weight of a water drop (about 20 mg) determines the sensitivity of the method (Fig. 3.41). The loss component is measured as weight loss of the whole device.

Another gravimetric method of recording simultaneous water absorption and transpiration rates was designed and gradually improved by TARŁOWSKI (1955, 1956, 1961). The principle is as follows (Fig. 3.42): The weight decrease of a vessel (1) containing the nutrient solution and the plant, as a result of transpiration, is determined on one balance (A) and the decrease in weight of water in a beaker (2) from which water is drawn into a sealed flask (3) placed outside the balance is determined on the other balance (B). Flask (3) is connected by air-tight rubber

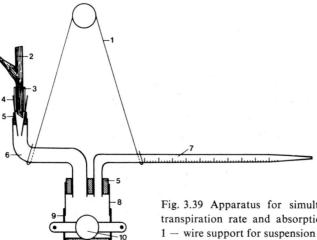

Fig. 3.39 Apparatus for simultaneous determination of transpiration rate and absorption rates (KNIGHT 1965). 1 — wire support for suspension on balance arm, 2 — base of the shoot, 3 — rubber tube, 4 — tapered glass tube, 5 — rubber bung, 6 — glass tube, 7 — bent 1 ml pipette, 8 — wide glass tube, 9 — wide rubber (reservoir), 10 — screw clip (for adjusting the position of meniscus in pipette).

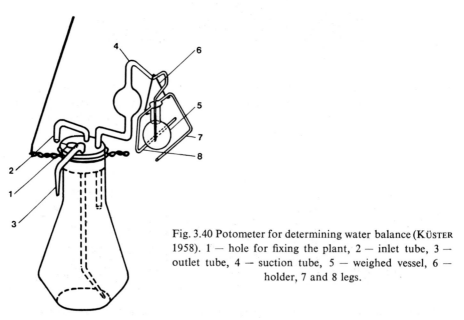

Fig. 3.40 Potometer for determining water balance (KÜSTER 1958). 1 — hole for fixing the plant, 2 — inlet tube, 3 — outlet tube, 4 — suction tube, 5 — weighed vessel, 6 — holder, 7 and 8 legs.

tubing to the enclosed air space above the nutrient solution (1). Water is drawn from beaker (2) through a water-filled tube and rubber tubing to vessel (3) when the pressure in the air space in the air tight flask (1) decreases as a result of absorption of water by the plant roots. This transfer of water from beaker (2) to flask (3) is proportional to the absorption of nutrient solution by the plant roots in the

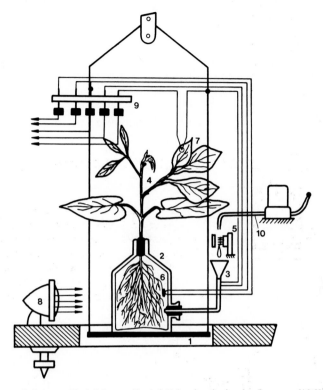

Fig. 3.41 Scheme of the so-called "drop principle" in the device by IVANOV (1962). 1 — balance pan, 2 — chamber with nutrient solution, 3 — inlet for drops being added, 4 — plant, 5 — drop counter, 6 — sensor of the thermoregulator, 7 — air humidity sensor, 8 — infrared heater, 9 — contacts, 10 — dropping device.

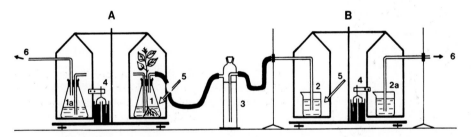

Fig. 3.42 Scheme of the TARŁOWSKI apparatus for simultaneous and continuous gravimetric measurement of transpiration and absorption rates. See the description in the text.

flask (1). Tubes (6) are mutually interconnected (not shown in the Figure). Whenever the equilibrium of the balances is impaired by transpiration (balance A) or absorption (balance B), the mercury contacts (4) are connected. An electronic relay device causes a small steel ball of a known precise weight to fall onto the appropriate pan of the balance, each fall of a ball being recorded by a counter. The error of the measurement does not exceed 5%.

REDSHAW and MEIDNER (1970) combined potometric measurement of water uptake and loss with continuous determination of sap flow by heat pulse method in intact plants.

3.3.5 Soil Potometers

Potometric principles have also been used in determining water absorption from soil. The basic construction has been described by LEBEDEV and SOLOVEV (1964) (Fig. 3.43). The equilibrating vessel (1) is filled with water or nutrient solution.

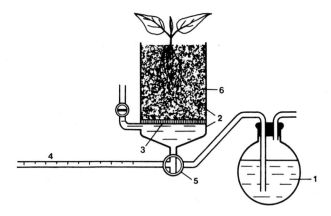

Fig. 3.43 Soil potometer (LEBEDEV and SOLOVEV 1964). See the description in the text.

After the tap (5) is opened, water flows to the bottom part of the potometric vessel (2) up to above the sintered glass filter (No 2) (3). Air bubbles are expelled. The equilibrating vessel is then connected with the measuring capillary (4) by a three-way tap (5) and then again the potometric vessel is connected with the equilibrating vessel. The upper part of the potometric vessel is filled with soil (6) with plants, the surface of the soil is covered to prevent evaporation. When measuring water absorption the measuring capillary is always first connected with the equilibrating vessel and the volume is read, then the measuring capillary is connected with the potometric vessel and subsequent readings are taken. Further modifications of this soil potometer were described by LEBEDEV et al. (1964a). Two types of recording soil potometers (terrapotometers) were designed by IVANOV (1963, 1965). A simple device for measuring water uptake from soil at capillary capacity was described by KLOZ (1953). BRAUN (1970, 1971, 1972) estimated the amount of water taken up by the plants from the changes in water table level in a soil potometer.

3.4 Xylem Sap Exudation from Excised Roots and Root Systems

3.4.1 Introduction

If the above-ground parts of a healthy growing plant, which is not suffering from water shortage, are cut away, it can be observed that the stump of the stem exudes sap from the cut vessels. This xylem sap exudation (root pressure exudation, bleeding) is influenced by a number of external and internal factors.

Root bleeding may be understood as an osmotic process, as follows from the equation of SABININ (1925)

$$J_w = k_{exud}(\Psi_{s(sap)} - \Psi_{s(e)}) , \tag{3.7}$$

where J_w is the rate of exudation, $\Psi_{s(sap)}$ osmotic pressure of exuded liquid, $\Psi_{s(med)}$ osmotic pressure of the solution surrounding the roots and k_{exud} a constant reflecting the conductance of the roots for water. It follows from the equation that the difference between the water potentials ($\Delta\Psi$) is the driving force of exudation. This difference is maintained by active transfer of ions from the external medium into the roots (CRAFTS and BROYER 1948, LUNDEGÅRDH 1950, VAN ANDEL 1953, JARVIS and HOUSE 1970). There is ample evidence of a dual mechanism operating in ion uptake (ANDERSON and REILLY 1968, MINCHIN and BAKER 1970).

3.4.2 Measurement of Sap Volume

Methods Using Measuring Tubes or Capillaries

A simple arrangement was described by LUNDEGÅRDH (1960). A rubber tube was attached to a short stump of coleoptile and leaf parts of young decapitated wheat plants. A glass capillary of inner diameter 0.4 to 0.5 mm was inserted into the rubber tube. The movement of the meniscus in the measuring capillary was followed with a horizontal microscope. The root system was immersed in a wide test tube.

HOUSE and FINDLAY (1966a) used primary maize roots with their basal ends inserted directly into glass capillaries of uniform bore. Additional sealing was achieved with a lanolin-paraffin mixture (95 per cent lanolin) or with a paraffin-resin mixture (50 per cent paraffin) applied at a temperature just above its melting point (about 40 °C) (ANDERSON and HOUSE 1967). The capillaries were mounted on vertical, graduated Perspex plates so that the roots were entirely surrounded by the aerated medium. The height of the exudation column can be read directly from the graduations (ANDERSON and COLLINS 1969).

At Aberdeen (WEATHERLEY, BAKER, BOWLING, JACKSON), 4 week old *Ricinus* plants were detopped below the first node in the hypocotyl region. Tissues external to the cambium were removed and a piece of rubber tubing was fitted over the stump. To this was joined a capillary U tube, directing the exuded fluid into a tared specimen bottle (BAKER and WEATHERLEY 1969).

To introduce the meniscus into the measuring capillary more easily, VAN ANDEL (1953) used a three-way tap, two arms of which were formed by a horizontal capillary and by a vertical tube connected by a rubber tube to the stump of the plant. The third outlet served to empty the system or replace the solution. A similar apparatus was described by STUART and HADDOCK (1968).

FILIPPOV (1956) studied exudation of cotton plants in field conditions in a similar way. A calibrated tube was attached to the stump of a decapitated plant by a rubber tube and was fixed in a horizontal position. It was necessary to half-fill the tube with water as there was absorption of water from the tube into the stump (so-called negative exudation) if soil moisture decreased below a critical value.

Measuring the Dimensions of Exudate Drops

GOEDEWAGEN (1967) estimated the ability of individual roots to absorb and transfer water by maintaining the roots in their original soil block and freezing their cut, proximal ends. The soil block was saturated to field capacity and kept in a moist atmosphere. The occurrence of a drop of exudate on the cut surfaces of the roots was inspected at regular intervals. Each time the drop was measured and removed. The relative activity of the individual roots was thus found after several days. The roots were then washed and examined morphologically.

Measuring the Volume of Exuded Fluid

GROSSENBACHER (1939) put short rubber tubes onto the stumps of the experimental plants. The liquid accumulating in the tubes was drawn from a number of plants at regular intervals and the volume obtained in a given time was determined.

Weighing a Pad of Absorbing Material

SPEIDEL (1939) determined the amount of exudate by measuring the increase in weight of a glass cap filled with cotton wool and placed for a period on the detopped stump. MARKOV (1960) used the same method with parchment cups, under field conditions.

Measuring the Size of a Wet Spot

Small amounts of radioactive exudate produced by short root segments projecting from the experimental solution into an atmosphere saturated with water vapour (SMITH 1970) were collected by touching the top of each with fresh discs of filter paper at intervals. The spots were measured and the volumes of exudate were estimated from the dimensions of the spots. The discs were later assayed for radioactivity.

Using Measuring Strips

The method has been used under field conditions (ÚLEHLA J. 1963a, b) and in the laboratory (ÚLEHLA J. 1961). The sap exuded by the stem stumps is drawn up into a strip of filter paper (*e.g.* 10 by 100 mm) inserted at its narrow end into a vertical slit made in the stump, the surface of which may be cut in the shape of a trough. The measuring strip is protected from evaporation by a tightly fitting polyethylene bag and is kept shaded. The position of the advancing boundary of the wetted zone is read at intervals against marks made on the strip. The strip may be impregnated with green ink diluted 1 to 50 or with a suitable dye, which is washed out and moves with the advancing front. The course of wetting is plotted graphically. The resulting curve makes it possible to estimate the duration of exudation lag, if any, and the rate of exudation, from the slope of the second ascending shoulder of the curve. The exudation rate, in mg per hour, is estimated as the weight increase of the strip and its cover bag divided by the net time of exudation, *i.e.* the total time minus exudation lag. Paired data on the exudation rate and the rate of wetting may be used for constructing a calibration graph, valid for the particular brand of paper used for the strips. The calibration graph may be used for the estimation of exudation rate without weighing the strips.

Using a Fraction Collector

Sap exudation may be recorded by means of a fraction collector, as was demonstrated by BARANETZKY as early as in 1873, according to SPEIDEL (1939).

Recording Drops of Exuded Sap

WENT (1944) described a device in which dropping tubes from six stumps of experimental plants lead in parallel to a single strip of filter paper on a rotating drum. The exudate always passed through a chamber containing a dye so that the traces of the evaporated drops were easily visible. The rate of exudation can be determined from the rate of paper movement, the number of drops per unit of length, and the average weight of the drops.

3.4.3 Other Experimental Techniques

Osmotic Permeability of Individual Roots

The transient phenomenon may be used for the estimation of osmotic permeability, L_p, of individual roots, as described by HOUSE and FINDLAY (1966b). After an equilibration period of at least 1 hour in a medium, (C_1) the exudation rate, J_{w_1}, was measured over six 5 min intervals (to check constant exudation rate). The external medium was then replaced by one of lower osmolarity, C_2. Within about 150 s the exudation had reached a minimum rate, J_{w_2}. Assuming that salt transport remained constant during the experiment, L_p was calculated using the expression

$$L_p = \frac{J_{w_1} - J_{w_2}}{RT(C_2 - C_1)}.$$ (3.8)

Application of External Pressure and Suction

External pressure of suction may be used to induce a change of exudation rate independent of the ionic composition of the medium (KRAMER 1932, MEES and WEATHERLEY 1957a, b, JACKSON and WEATHERLEY 1962a, b, JENSEN 1962, LOPU-SHINSKY 1964, O'LEARY 1965, KLEPPER and GREENWAY 1968). In contrast to suction applied to the stem stump, application of external pressure may induce pressure differences across the cortex of more than 1 bar (MEES and WEATHERLEY 1957a). The pressure apparatus described by JACKSON and WEATHERLEY (1962a) consisted essentially of a heavy nickel-plated steel canister of about 7 l capacity with a tube at the bottom to enable changing the root medium. There were separate inlets for aeration and pressure maintenance. The lid, holding a Bourdon tube pressure gauge, could be clamped down onto a rubber washer to make an air-tight seal. A Perspex holder for the the stem fitted a conical rubber washer forced by the pressure in the canister into a tapered opening in the centre of the lid. A seal between the stem and the holder was achieved by means of a rubber sleeve. The rate of sap exudation was measured by collecting and weighing the exudate.

To measure any leakage through breaks in the root system, mannitol was added to the medium (MEES and WEATHERLEY 1957a) and its concentration in the sap was estimated using periodic assay (REEVES 1941).

Roots in a Water Vapour-Saturated Atmosphere

JARVIS and HOUSE (1970) observed prolonged exudation from excised maize roots placed in an atmosphere saturated with water vapour. The excised roots were washed for 3 min in either distilled water or a solution of potassium chloride. Each root was sealed into a glass capillary. The capillaries were attached to a Perspex plate so that the roots passed through a narrow slit in the cover of a glass chamber. The chamber contained a layer of distilled water, aerated so that space occupied by the roots was saturated with water vapour.

The prolonged steady flux of potassium and calcium was thought to provide evidence of the symplasm-linked part of the dual ionic pathway. The initial high rate of exudation was considered to be a consequence of the absorption of ions from the prewashing medium through the extracellular space.

Plasmolysed Roots

According to JARVIS and HOUSE (1970), COLLINS and LINSTEAD (unpubl.) recommended pretreatment of roots with a hypertonic mannitol solution to disrupt the plasmodesmata in the cortex by plasmolysis and thus to block the symplasmic route of ion transport. In the experiments of JARVIS and HOUSE (1970), the roots were pretreated with 0.8M mannitol solution for 2 min and then rinsed five times, for a total of 10 min, in the solution used for subsequent exudation measurements.

Root Segments

SMITH (1970) studied the sap exudation from short root segments of maize. "Apical" segments comprised the terminal 35 mm of the root, "basal" segments were taken between 15 and 50 mm from the root tip of primary roots averaging 120 mm in length. The distal ends of the basal segments were sealed with wax. A stirrup made of polyethylene tubing, hooked over the rim of a beaker containing the medium (a solution of 0.1M RbCl containing ^{86}Rb), allowed the segments to project 5 to 8 mm above the solution. A collar of melted vaseline prevented the experimental solution from mixing with the exudate. Comparison of exudation from apical and basal segments demonstrated considerable variation in water and solute fluxes along the root axis.

Inverted Segments

SMITH (1970) compared the exudation from "basal" (see above) and "basal-inverted" segments. Basal-inverted segments were cut in the same way as the basal segments, but they were sealed with wax at the proximal ends (furthest from the tip) and placed in the ^{86}Rb solution upside-down, *i.e.* with the distal ends pointing upwards. The differences in water and solute fluxes between basal and basal-inverted segments were not significant, and so, according to SMITH, do not support the hypothesis of a direct relation between the exudation rates from each longitudinal portion of a root and the physiological activities in the vessels of the same portions.

Roots Deprived of Epidermis

Several authors have studied the exudation of roots stripped of epidermis and the outer cortex (SANDSTRÖM 1950, O'LEARY 1965, ANDERSON and REILLY 1968, STUART and HADDOCK 1968). In the procedure of ANDERSON and REILLY, an intact maize seedling with a root of 8 to 15 cm in length was placed on a wet glass plate. A sharp razor blade, slanted in the direction of the strokes, was gently drawn along the length of the root. This was repeated as the root was rotated slightly along its axis between the strokes, until most of the root surface was removed or destroyed. This treatment considerably decreased the salt fluxes at low external salt concentrations.

STUART and HADDOCK (1968), following the procedure of SANDSTRÖM (1950), removed the epidermis of sugar beet plants by treating them in 20μM n-diamyl acetic acid. In contrast to the controls, the exudation of treated roots was not inhibited by 0.25 mM ammonium carbonate at pH 8.4.

Isolated Steles and Cortical "Sleeves"

Physical separation of anatomically different root structures appears to offer a direct approach towards the separation of the contributions of different root tissues to the exudation process. It is possible to remove the complete cortex from excised roots, so obtaining isolated steles and hollow cortical cylinders — the "sleeves". The water and ion transport of the isolated steles (BRANTON and JACOBSON 1962, LATIES and BUDD 1964, LÜTTGE and WEIGL 1964, LÜTTGE and LATIES 1966, COLLINS and HOUSE 1969, JARVIS and HOUSE 1969) and the "sleeves" (GINSBURG and GINZBURG 1970a, b) have been studied in detail. The roots of maize 10 to 12 cm in length are suitable for the preparation of steles and "sleeves". Each root is bent at a distance of about 9 cm from the root tip until the cortex becomes disrupted and the stele can be gently pulled out.

Experiments on isolated steles and "sleeves" require special techniques; tritiated water and radioactive salts are often used.

Precultivation of Maize

There have been a large number of laboratory experiments on sap exudation from primary roots of maize plants grown according to HOUSE and FINDLAY (1966a, b). The seeds are surface-sterilized by a short rinse in alcohol followed by a rinse in 0.5 per cent $HgCl_2$. After thorough washing in several volumes of distilled water, the grains are left in aerated distilled water in the dark. The soaked grains are placed on nylon gauze supported by Styrofoam frames floating on continuously aerated 0.1 mM $CaCl_2$. The roots are harvested after about four days in the dark at 25 °C.

3.4.4 Determination of Root Pressure

It is possible to prevent root bleeding by increasing the pressure of air or of a mercury column on the stump of the decapitated plant. It is assumed that the required pressure, measured by a manometer, corresponds to the difference between the water potential of the root medium and that of the connecting elements in the roots (STOCKING 1956, BROUWER 1965).

3.4.5 Mathematical Models of Sap Exudation

Since the introduction of SABININ equation (see 3.4.1), mathematical models were often used to describe the phenomena participating in sap exudation (ARISZ et al. 1951, MEES and WEATHERLEY 1957a, b, HOUSE and FINDLAY 1966b, ANDERSON and HOUSE 1967, ANDERSON and REILLY 1968, MINCHIN and BAKER 1970, GINSBURG and GINZBURG 1970). Therefore, it is quite natural that the use of computers found its way also into the study of root exudation (ANDERSON et al. 1970) as a logical extension of experimental techniques. The dialogue between model and experiment (TYREE 1973) will undoubtedly hasten the process of understanding the complex mechanism of root exudation.

4

Liquid Water Movement in Plants

4.1 Determination of Hydraulic Conductivity (Water Permeability) of Cytoplasm, Cell Membrane and Parenchymatous Tissues

In the liquid phase the driving force for the flux of water (q_w) is the difference ($\Delta \Psi_w$) or gradient $d\Psi_w/dz$) of water potential:

$$q_w = - K_w \frac{d\Psi_w}{dz}, \quad [\text{g cm}^{-2}\text{s}^{-1}] = [s] \frac{[\text{g cm}^{-1}\text{s}^{-2}]}{[\text{cm}]}, \tag{4.1}$$

where the hydraulic conductivity K_w (often designed L_p) has the dimension [s] and the hydraulic resistance r_w [cm s^{-1}].

The hydraulic conductivity (water permeability) of a natural membrane (*e.g.* plasmalemma, tonoplast, cell wall, tissue segments) can generally be determined from the results of three types of experiments, differing in the driving forces causing the water flux across the membrane (STADELMANN 1956, 1963):

(1) "Osmotic" measurements, *i.e.* measurements in which the driving force is the difference in osmotic potential on opposite sides of the membrane.

(2) "Pressure" measurements, in which the driving force is an induced hydrostatic pressure (pressure potential) difference.

(3) "Diffusion" measurements, in which concentration differences of water molecules are used (HTO, HDO, H$_2$^{18}O).

The conventional equation valid for all types of permeability measurements is as follows:

$$K_w = \frac{dW}{dt} \frac{1}{AE}, \tag{4.2}$$

where W is the amount of water transported across the membrane in the time t, A is the area of the membrane, and E is the driving force. The dimension of the permeability coefficient (hydraulic conductivity) K_w depends on the driving force and the way it is measured and used in the calculations. The diffusive conductivity K_{w1} is expressed in [cm s^{-1}], see Diffusion Measurements, page 222.

4.1.1 "Osmotic" Measurements

If hydraulic conductivity of cytoplasmic membranes is to be measured, the cell (s) is (are) immersed in a solution which is hypotonic, for example, to the vacuolar solution previously measured. In the case of tissue segments, the tissue is situated so that it is separating solutions of different osmotic potentials or pure solvent and a solution, all of known osmotic potential.

Generally, results obtained in "osmotic" measurements may be calculated according to

$$\frac{dW}{dt} = K_{w2} \cdot A[\Psi_{s(e)} - \Psi_{s(i)}], \qquad (4.3)$$

where for cell plasmolysis (see Section 1.1), $\Psi_{s(e)}$ is the osmotic potential of the external osmotic solution, $\Psi_{s(i)}$ is the osmotic potential of the internal vacuolar sap (both expressed in bars). In this case K_{w2} has units of [cm s^{-1} bar^{-1}]. The factor F_2 is used for converting K_{w2} to K_{w1}. For water, 1 mol $\triangleq$ 18 cm^3 and $\Psi_s = cRT$ ($R = 83.143$ bar cm^3 K^{-1} mol^{-1}, $T = 293\,°C$), $F_2 = 1.352 \cdot 10^3$ bar (STADELMANN 1963).

This applies to experiments where the driving force is measured as the difference in concentration of a solute for which the particular membrane is highly selective, i.e. it has a reflection coefficient σ close to 1 (see Section 1.3.2.1).

$$\frac{dW}{dt} = K_{w3} = A(C_{s1} - C_{s2}), \qquad (4.4)$$

where W is expressed in [mol] of water, C_{s1} and C_{s2} are the concentrations of the substance used [mol cm^{-3}] and K_{w3} is then expressed in [cm s^{-1} mol^{-1}]. The factor F_3 used for converting K_{w3} to K_{w1} in $K_{w1} = F_3 K_{w3}$ then equals $F_3 = 5.556 \cdot 10^{-2}$ mol cm^{-3} (STADELMANN 1963).

The measurement of the rate of plasmolysis or deplasmolysis (see Section 1.1 and 1.4.1.1) is usually used for the measurement of the permeability of the cytoplasmic membranes of individual cells (e.g. HÖFLER 1917, 1918a, b, DAVSON and DANIELI 1943, HEINRICH 1962 and other authors). Deplasmolysis may be complete if the cell is surrounded with pure water, or partial if the cell is placed in a hypotonic solution. During measurement of hydraulic conductivity of cytoplasmic membranes as above, there is gradual plasmolysis during which plasmometric measurements may be made i.e. the changes of the volume of the vacuole and/or the cell can be determined (see Section 1.4.1.2).

Plasmometric measurement of the course of plasmolysis or deplasmolysis as described in Section 1.4.1.2 (HÖFLER 1917, 1918a, b, 1930, HUBER and HÖFLER 1930) is very suitable for determining the permeability of the cytoplasm, the plasmalemma or the tonoplast to water: the water permeability constant K_w may be calculated from the equation

$$\frac{dG_1}{dt} = -K_w \left[\Psi_{s(e)} - \frac{\Psi_{s(i)}}{G_1} \right], \qquad (4.5)$$

where G_1 is the degree of plasmolysis at time t_1, $G_1 = V_1/V_0$, V_1 and V_0 are cell volumes, t is time. From equation (4.5), after integration and rearrangement,

$$K_w = \frac{1}{\Psi_{s(e)}(t_2 - t_1)}\left(G_1 - G_2 + G\ln\frac{G_1 - G}{G_2 - G}\right),\qquad(4.6)$$

where G is the final degree of plasmolysis after reaching equilibrium, $G = V_1/V_0 = \Psi_{s(i)}/\Psi_{s(e)}$.

The following equation may be used for the calculation of K_{w1} in plasmometric experiments (partial deplasmolysis, total and partial plasmolysis) (STADELMANN 1963):

$$K_{w1} = 32.0\frac{a}{C\left(G_0 l + \dfrac{a}{3}\right)(t_2 - t_1)}\left[G_0 l \lg\frac{G_1 - G_0}{G_2 - G_0} + \frac{a}{3}\lg\frac{G_1 l + \dfrac{a}{3}}{G_2 l + \dfrac{a}{3}}\right],\qquad(4.7)$$

where K_{w1} is expressed in [cm s^{-1}], numerical coefficient 32.0 [mol l^{-1}], a is the internal width of the cell [cm], l is the internal length of the measured cell [cm], C is the concentration of the external solution [mol l^{-1}] and G_0, G_1 and G_2 are the degrees of plasmolysis at times t_0, t_1 and t_2 [s].

LEVITT et al. (1936) calculated the permeability constant (K_w) for water in isolated protoplasts from

$$K_w = \frac{\Delta V}{\bar{A}(\Psi_{s(i)} - \Psi_{s(e)})t},\qquad(4,8)$$

where $\bar{A}$ is the average area [cm^2], ΔV the volume change [cm^3]. A similar equation was used by MYERS (1951) for calculating the permeability of thin non-plasmolysed sections of sugar beet roots.

KAMIYA and TAZAWA (1956) designed an elaborate glass apparatus for the measurement of the permeability of a single cell of *Nitella* to water by transcellular osmosis, *i.e.* a transcellular water flux was induced by differences in osmotic potential at the two ends. An internodal cell of *Nitella* was placed between two chambers in which the test solutions could be changed. The chambers were connected to each other by a side tube which served to indicate the direction and rate of movement of water according to the induced osmotic gradient. Similar devices for giant algal cells were used by DAINTY and HOPE (1959) and DAINTY and GINZBURG (1964).

In osmotic measurements stirring of the external solutions applied is a necessary condition ensuring uniform concentration of the test solutions. Even in such cases an unstirred boundary layer of the thickness 20 to 500 µm usually remains, the concentration of which may differ from the bulk solution. This layer can influence the apparent permeability measured, the error also depending on the arrangement of the experiment (DAINTY and HOPE 1959, DAINTY 1963).

4.1.2 "Pressure" Measurements

Here the driving force is an induced difference of hydrostatic pressure. KA-
MIYA et al. (1962) and TAZAWA and KIYOSAWA (1970) measured the permeability of
the cell wall of Nitella in a simple device by artificially applying a difference of
hydrostatic pressure (potential) Ψ_p across the cell.

4.1.3 "Diffusion" Measurements

When the driving force is represented by a water concentration difference,
K_{w1} is expressed in [cm s^{-1}] according to

$$\frac{dW}{dt} = K_{w1}A(C_{w1} - C_{w2}),\qquad(4.9)$$

where dW/dt is the water flux [mol s^{-1}], A is area of membrane [cm^2], and C_{w1}
and C_{w2} are the two water concentrations in [mol cm^{-3}].

Several kinds of labelled water, HDO, $H_2{}^{18}O$ and HTO, are suitable for direct
measurement of permeability to water by "diffussion" experiments (selfdiffusion).
As the scope of this book is limited, it is not possible to describe the techniques
which have been suggested (see for example WARTIOVAARA (1944), BUFFEL (1952),
COLLANDER (1954), ORDIN et al. (1956), DAINTY and HOPE (1959), LEBEDEV (1959),
HÜBNER (1960), ORDIN and GAIRON (1961), WOOLEY (1965), JARVIS and HOUSE
(1967), HOUSE and JARVIS (1968).

In diffusion experiments using labelled water,

$$\frac{dW}{dt} = K_{w4}A(C_{w1} - C_{w2}),\qquad(4.10)$$

where W is expressed in [mol], C_{w1} and C_{w2} in [mol cm^{-3}], and K_{w4} in [cm s^{-1}].
The corresponding factor (for $K_w = F_4K_{w3}$) $F_4 = 2$, because of the coupled back
diffusion, since for every molecule of HDO, for example, passing in one direction
across the membrane, another molecule of water passes in the opposite direction.

In selfdiffusion as well as "osmotic" techniques, thorough stirring of the solutions is necessary to eliminate any effect of a diffusional boundary layer (an unstirred layer) at the membrane surface (DAINTY 1963, DAINTY and HOPE 1959,
SLATYER 1967a).

Autoradiography using labelled compounds is a technique more suitable for use
at the tissue level than with individual cells, as the available photographic emulsions
are not yet good enough.

Most measurements of the water permeability constant (K_w) in plant material
(cytoplasmic membrane and parenchymatous tissues) are within the range of 1 to
15×10^{-1} cm s^{-1} (STADELMANN 1963). The highest values measured by trancellu-
lar osmosis in internodal cells of Characeae (e.g. Nitella) are 100 to 400×10^{-4}
cm s^{-1}.

4.2 Measurements of the Velocity of Water Movement in Xylem

The linear velocity of the transpiration stream can be measured by labelling it with a soluble compound or heating it, and then examining the rate of transfer of the compound or of heat by the transpiration stream.

For transparent plant stems, it is possible to use dyes (*e.g.* fluorochromes) added at a known time to the root medium or injected into the stem. The first occurrence of the dye at a selected distance up the stem is then detected (by UV fluorescence in the case of fluorochromes). LUNDEGÅRDH (1954) measured the time which elapsed between administration of NaCl solution and sudden increase in conductivity between electrodes introduced into the stem at a certain distance above the place of administration. The use of radioactive compounds to label the transpiration stream has some advantages. The movement of radioactivity after application can be measured with a GM tube placed at a certain distance above the site of administration of the labelled compound. The technique is now commonly used, but it has the disadvantage that it is not easy to determine accurately the time of the first appearance of radioactivity at a certain selected point along the plant stem, because of the "background" radiation. All the techniques described so far have the common disadvantage that foreign compounds are injected into the plant and usually injure the stem at the point of injection or even at the point of detection by certain techniques (*e.g.* conductivity methods).

The water stream in xylem may also be labelled by heat. The principle of the method involves measuring the upward velocity of a heat pulse, which is related to the velocity of water moving in the xylem vessels.

Average maximum velocities of the transpiration stream in xylem are 1.2 to $2.2 \, \mathrm{m \, h^{-1}}$ in ferns, $1.2 \, \mathrm{m \, h^{-1}}$ in evergreen conifers, 1 to $6 \, \mathrm{m \, h^{-1}}$ in leafy trees with scattered porous wood (*e.g.* birch, beech, lime), 14 to $44 \, \mathrm{m \, h^{-1}}$ in ring porous wood (oak, ash trees), 10 to $60 \, \mathrm{m \, h^{-1}}$ in herbs and $150 \, \mathrm{m \, h^{-1}}$ in lianes (HUBER 1956).

4.2.1 Heat Pulse Methods

The heat pulse method was originally described by HUBER (1932) and HUBER and SCHMIDT (1937), then empirically improved by many others and theoretically developed by MARSHALL (1958, 1962) and CLOSS (1958). It remains the simplest, relatively most accurate and most elegant method for determining the velocity of the transpiration stream.

The stem, and hence the water of the transpiration stream moving in it, is heated at a particular place. The time taken for a temperature increase to be detected at a distance from the point of heating, by a thermocouple placed in the stem, is measured. In addition to a convectional mass flow of water causing movement of heat in the transpiration stream, there is conduction of heat by the stem tissue. MARSHALL (1958) writes, for trees:

$$v = \frac{Au\varrho_{sap}c_{sap}}{\varrho c} ,$$ (4.11)

where v is velocity of the heat pulse (by convection and conduction), A is fraction of the unit cross-sectional area of wood occupied by moving sap streams, u is sap velocity, ϱ_{sap} and c_{sap} are density and specific heat of sap, respectively (both may be assumed to be unity), ϱ and c are density and specific heat of wet wood, respectively. (For *Pinus radiata* MARSHALL gives $\varrho = 1.07$ and $c = 0.77$). This means that although v is not identical with u, Au (the flux of sap per unit cross-sectional area of sap wood) can be calculated from the measured v.

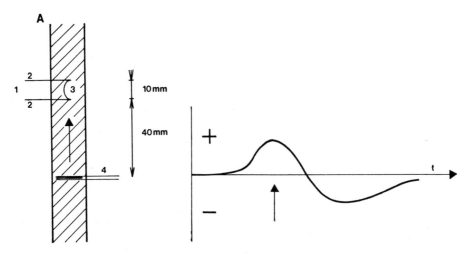

Fig. 4.1 A diagram of the arrangement (A) used for thermoelectric measurement of the velocity of water movement in xylem vessels. The arrangement is suitable for rates higher than $60\,\mathrm{cm\,h^{-1}}$ (HUBER 1932). 1 — thermocouple, 2 — copper wire, 3 — constantan wire, 4 — heater.

Fig. 4.2 The time course of deflection of the galvanometer used for arrangement A in Fig. 4.1. Abscissa: time after heating, ordinate: galvanometer deflection. See the description in the text. The arrow indicates the time at which the transpiration stream passed the heated point between 4 and the lower thermojunction (2) of the thermocouple (1) (40 mm) (see Fig. 4.1) (HUBER 1956).

Measuring Procedures

It is possible to use the arrangement shown in Fig. 4.1 (HUBER 1932, BAUMGARTNER 1934/35) when the velocity of water movement in the xylem is higher than 60 cm s^{-1} and conduction of the heat pulse can be ignored. Deflection of the galvanometer then corresponds to the scheme shown in Fig. 4.2. The lower junction, which is closer to the point of heating, heats up earlier, and there is a deflection of the galvanometer in one (+) direction. (The deflection is proportional to the temperature difference between the two junctions of the thermocouple). Further movement of the warmed sap in the direction of the more distant thermojunction results in a decreased deflection. Finally, the more distant junction is heated, by which time the lower junction has cooled down, resulting in the opposite galvanometer deflection (−).

It is necessary to use compensation arrangement for both junctions of the thermocouple (arrangement B) shown in Fig. 4.3 (HUBER and SCHMIDT 1937), at low velocities of water movement of less than 60 cm per hour. The measuring and compensating junctions are located above (a fairly long distance away) and below the

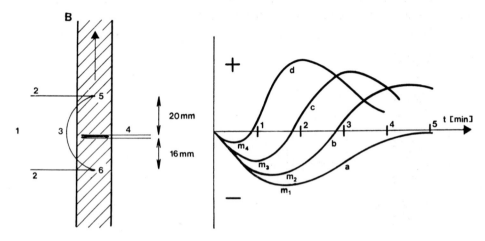

Fig. 4.3 A diagram of the compensation arrangement (B) used for thermoelectric measurement of the velocity of water movement in xylem vessels (HUBER and SCHMIDT 1937). The arrangement is suitable for rates lower than 60 cm h^{-1}. Symbols as in Fig. 4.1. 5 — measuring junction, 6 — reference junction of the thermocouple.

Fig. 4.4 The time course of the galvanometer deflection when using arrangement B shown in Fig. 4.3 Abscissa: time after heating in minutes, ordinate: galvanometer deflection. See the description in the text. Curve a corresponds to zero velocity of water movement in xylem vessels, curves b, c and d correspond to gradually increased velocities (HUBER 1956).

point of heating (a short distance away), respectively. Heat conduction through the stem heats lower junction (6) first, resulting in a negative deflection of the galvanometer (Fig. 4.4). The deflection then changes to positive as the upper junction (5) heats up as a result of the movement of the heated transpiration stream. The lower minimum (m) may be readily detected from the relationship between the galvanometer deflection and the time after heating (Fig. 4.4). However, the lower minimum does not correspond to the time when the main convection heat wave reached the upper junction but occurs much earlier. The upper maximum corresponds to the

velocity of water movement. However, it is only with difficulty that it can be iden-
tified with reasonable accuracy, so that the time when the lower deflection is minimal (t_{min}) is used. The velocity of the transpiration stream, v [cm h^{-1}] is then:

$$v = \frac{d}{t_{min}} - \frac{d}{t_0},$$ (4.12)

where d is the distance of the upper junction of the thermocouple above the point
of heating, t_{min} is the time taken for the lower minimum of the galvanometer deflection (m) to be reached and t_0 is the time taken to reach the minimum deflection

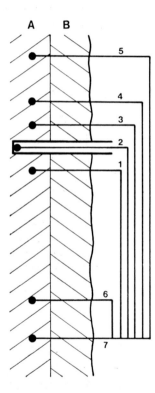

Fig. 4.5 Diagram of the arrangement
of measuring sites on a cross section
of a tree trunk for the registration of
water flux in tree xylem (VIEWEG and
ZIEGLER — (1960). A — xylem, B —
bark, 1 and 3 to 7 are leads from the
thermojunctions, 7 — reference junction, 2 — heater (see Fig. 4.6).

when the velocity of water movement is zero. Results obtained when there is no
transpiration may be used to find t_0 (e.g. in the morning before sunrise). However,
this calculation is necessary only if absolute values are sought.

LEYTON (1970) obtained most consistent results in trees with the thermojunctions
equidistant from the heater and by reading the time t of occurrence of a maximum
temperature difference. It was shown that

$$v^2 = \frac{x^2}{t^2} - \frac{2k}{t},$$ (4.13)

where x is distance between heater and thermocouples (usually one or two cm) and k is the heat diffusivity of wet wood, which is determined from t when $u = 0$.

BAUMGARTNER (1934/35) described a procedure for calibrating the thermo-electric method. A known volume of water is passed through a plant by means of a pump while the rate of this "transpiration stream" is measured by the thermo-electric method.

VIEWEG and ZIEGLER (1960) described an arrangement for recording the velocity of xylem water movement (Fig. 4.5). Using a continuously acting heater (2) inserted into the active xylem of a tree trunk (for details of the heater see Fig. 4.6) the steady state heating was measured by recording the temperature differences between the reference thermojunction (7) and a row of junctions (3 to 5) situated above the heater and junctions (1) and (6) situated below it. The method was modified by SADDLER and PITMAN (1970).

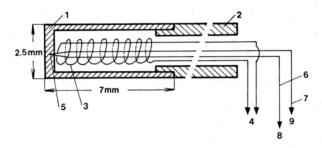

Fig. 4.6 Heater used for measuring the water flux in the xylem vessels of a tree trunk (VIEWEG and ZIEGLER 1960). 1 — brass capsule, 2 — plexite, 3 — heating wire, 4 — wires to the battery, 5 — constantan/copper thermojunction, 6 — copper wire, 7 — constantan wire, 8 — connection to galvanometer, 9 — connection to the second (reference) thermojunction of the thermocouple.

Heating Devices

Theoretically, heating the thinner stems of herbaceous plants with the fingers is sufficient, but such heating is not accurately localized. An insulated wire, 0.1 mm diameter, twisted round the stem and heated with a pocket battery can be used. BAUMGARTNER (1934/35) used an insulated copper wire (0.1 mm) twisted twelve times around a pencil and flattened cross-wise to a rod-like shape. A resistance wire (nickelin, diameter 0.2 mm, 30 mm long, heated with a 4 V accumulator) attached to the stem may also be used (KLEMM 1962). The period of heating is usually 1 to 8 seconds. The heating wire may also be made of 2 cm of constantan wire (0.3 mm diameter) soldered between two copper wires (0.3 mm diameter) about 50 cm long. SWANSON (1962) mounted a nicrom wire coated in epoxy cement in a hypodermic needle. The source is suitable for tree trunks. In Fig. 4.6 a heater used in a device for measuring the velocity of water movement in tree xylem is shown (VIEWEG and

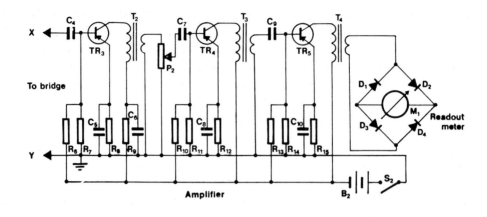

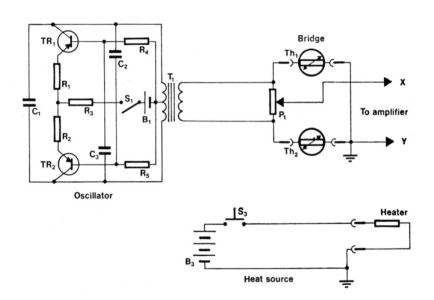

Figs. 4.7 and 4.8 A diagram of the oscillator, bridge and heat source, amplifier and read-out meter used by Swanson (1962) for measuring water movement in tree trunk xylem. $R_1 = 33\,\Omega$, $R_2 = 33\,\Omega$, $R_3 = 240\,\Omega$, $R_4 = 100\,k\Omega$, $R_5 = 100\,k\Omega$, $R_6 = 2.2\,k\Omega$, $R_7 = 10\,k\Omega$, $R_8 = 100\,\Omega$, $R_9 = 1\,k\Omega$, $R_{10} = 10\,k\Omega$, $R_{11} = 2.2\,k\Omega$, $R_{12} = 1\,k\Omega$, $R_{13} = 10\,k\Omega$, $R_{14} = 2.2\,k\Omega$, $R_{15} = 1\,k\Omega$, Th — thermistor (2 kΩ at 25 °C matched ±2% of each other), $P_1 = 1\,k\Omega$, $P_2 = 100\,\Omega$, (all resistors and potentiometers 0.5 W ±5%), $C_1 = 0.2\,\mu F$, $C_2 = 0.02\,\mu F$, $C_3 = 0.02\,\mu F$, $C_4 = 10\,\mu F$, $C_5 = 25\,\mu F$, $C_6 = 25\,\mu F$, $C_7 = 10\,\mu F$, $C_8 = 25\,\mu F$, $C_9 = 10\,\mu F$, $C_{10} = 25\,\mu F$, $D_{1,2,3,4}$: IN 270 or equivalent, $M_1 = 0-50\,\mu A$ meter, T_1 primary = 48 Ω, secondary = 8 and 16 Ω, $T_{2,3,4}$: primary 20 kΩ, secondary 1.2 kΩ, $B_1 = 1.5$ V, $B_2 = 3$ V, $B_3 = 6$ V, $S_{1,2,3}$ — switches; $TR_{1,2,3,4,5}$: 2 N 104, 2 N 109 or 2 N 217.

ZIEGLER 1960). LADEFOGED (1960) used short-wave radiation from a diathermia apparatus (200 to 300 W for a period of 20 to 40 seconds) to heat the transpiration stream in intact trees. The radiation was passed through the trunk from rubber-backed electrodes fastened to the trunk. The trunk is also heat-insulated at this point, if necessary. In any case, the stem should be protected against direct insolation in the neighbourhood of the measurement.

Temperature Sensing Elements

LADEFOGED (1960) used a thermoelement consisting of two rows of about 40 copper/constantan junctions located 2.5 cm apart around the tree trunk: the length of the element was about 15 cm. The thermoelement was insulated by means of a thick felt pad. The distance between the heating elements and lower junctions of the thermoelement was 5 to 20 cm.

Instead of thermocouples, BLOODWORTH and coworkers (1955, 1956) used a pair of thermistors 1.9 cm apart, attached by a special clamp to the plant stem. Another arrangement using thermistor probes was described by SWANSON (1962). Bead thermistor probes were assembled in hypodermic needles (20 gauge and thicker). Diagrams of the oscillator, bridge, heat source and amplifier used in his measurements in trees are shown in Figs 4.7 and 4.8.

A simple device for application on thin stems was described by MICHLER and STEUBING (1968).

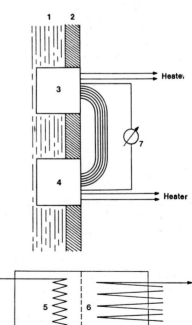

Fig. 4.9 A diagram of the arrangement for recording the water flux in a stem (LEYTON 1970). Vertical section through a tree stem. 1 — active wood, 2 — bark, 3,4 — permanently heated units, each consisting of resistance wire heater (5) and five thermojunctions (6), forming a thermopile which measures the temperature difference between the two units, 7 — potentiometer.

LEYTON 1967, 1970) developed another method for recording sap flow velocities in tree trunks. The method is based on the measurement of heat exchange between the moving sap and two permanently heated units which are inserted vertically one above the other (Fig. 4.9). A heated unit is made of a thin brass plate folded over a simple resistance wire heater (2 Ω, heated continuously by a current of about 200 mA) and five chromel-constantan thermojunctions and cemented with epoxy resin. The difference in temperature $(T_2 - T_1)$ is recorded. The sap reaching the upper unit is warmer than when it reached the lower unit as it has already been warmed up a little by the lower unit heater. So the upper unit is less cooled by heat loss than the lower one. An additional thermocouple is incorporated in the lower unit and connected to a similar thermocouple in the stem well below the unit, so that the temperature difference between the sap and the lower unit $(T_1 - T_{sap})$ can be recorded. It was shown thermodynamically that the ratio $(T - T)/(T_1 - T_{sap})$ is a sole function of the sap flow past the units.

TYREE and ZIMMERMANN (1971) induced changes in ions and water transport by pulses of electric potential gradients. The resulting and measured pulses of electrical current (depending mainly on cation transport) were decreased by water transport if this had opposite direction to the induced ion transport and *vice versa*. From both data the velocity of water movement in the xylem could be calculated.

4.2.2 Radioactive Tracer Method

KLEMM and KLEMM (1964) made a successful attempt to use radioactive ^{32}P isotopes for continuous measurement of the velocity of the transpiration stream in trees. Through a T-cut in the bark, 1 ml of a solution of 0.3 mCi $H_3{}^{32}PO_4$ was administered under the loosened bark in the morning before sunrise. The radioactivity was measured every 10 minutes throughout the day with a GM tube inserted 50 cm above the point of injection of the solution. There was a significant correlation between the velocity of the transpiration stream $[\text{cm h}^{-1}]$ measured simultaneously by the thermoelectric method (HUBER and SCHMIDT 1937) and the ratio $\Delta A/t$ (A is radioactivity increase per hour, t is time after injection of $^{32}PO_4{}^{3-}$). The procedure yields only relative values of the velocity of the transpiration stream in the stem.

4.2.3 Magnetohydrodynamic Method

This method already used in technical flowmeters and blood flowmeters was applied to the measurement of the transpiration stream in shoots by SHERIFF (1972). If a uniform magnetic field is applied at right angles to the direction of the fluid stream in a pipe, the fluid flow cutting the lines of magnetic flux causes a voltage to be induced in the same way as induction in solid conductors. A pair of electrodes in a Perspex assembly is placed on either side of the measured shoot at right angles to both the lines of magnetic flux and to the transpiration stream direction.

The voltage measured by means of an electronic detection system is directly proportional to magnetic flux density, to internal radius of the "pipe(s)" and to the instantaneous velocity of the stream. The magnetic field is generated by an a.c. electromagnet. For detailed description see the original paper. The author found a good agreement between readings and sap flux measured potometrically. The method causes minimum disturbance to the plant and may be used for instantaneous as well as continuous measurements.

4.3 Measurements of the Volume Flux of Water in Xylem Vessels

Whereas there are several suitable methods for the determination of the linear velocity of water, there are only indirect procedures for the determination of the rate of the transpiration stream. The linear velocity (v_w) (*e.g.* expressed in m per hour) is the translocation velocity of the water columns in the conducting pathways. The flux (dW/dt) is the amount of water transported through a length of conducting tissues per unit of time (*e.g.* in cm^3 per hour).

The volume flux of water through vascular bundles depends primarily (1) on the resistance of the conducting elements of the bundles and (2) on the linear velocity (v_w) of the transpiration stream in them. The area of cross section of the vessels (in mm^2) per g fresh weight of the green parts which they supply with water is designated as the relative conductive cross section of the vascular bundles, which is usually about 0.5 mm^2 g^{-1} in woody plants, about 0.2 to 0.3 mm^2 g^{-1} in mesophytes and 1 to 7 mm^2 g^{-1} in desert plants.

By making the simplifying assumption that in conducting tissues there is only negligible adsorption of compounds dissolved in the transpiration stream, the rate of water movement may be estimated from the quantities of certain compounds accumulated in transpiring organs a certain time after their application at the base of the length of stem in which the flux of water is to be measured. The compound chosen is one which is easy to analyse quantitatively and which is either absent from the tissues or present in negligible concentrations. Suitable compounds include labelled phosphate ($^{32}PO_4^{3-}$) and other labelled compounds, or lithium chloride (LiCl), which is easily detected with a flame spectrophotometer, (or an excess of NaCl which may also be determined conductimetrically as a conductivity difference *etc.*).

Generally, the water flux through the plant is roughly proportional to the simultaneous loss of water by transpiration from organs supplied with water by corresponding conducting pathways, over a fairly long period of time, provided that the water content of the tissue is the same at the beginning and end of the period. The total amount of water transpired may thus be measured during the period between two consecutive sunrises, since it can be assumed that the water content of transpiring organs is the same, *i.e.* roughly maximum, at sunrise.

4.3.1 Heat Balance Methods

An interesting method for the determination of the volume flux of water transported through the stem of a tree was described by DAUM (1967). The method is
based on the heat balance of the stem: during the day, the heat (Q_H) entering the
part of the cylinder of active xylem between points A and B (Fig. 4.10) must be
balanced by the heat transferred to inactive xylem (Q_w) (towards the centre of the
tree trunk), the heat transferred away by water transported upwards by the transpiration stream in the living xylem (Q_T), and the heat stored or released by the
active xylem (Q_s).

$$Q_H = Q_w + Q_T + Q_s . \qquad (4.14)$$

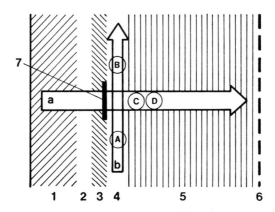

Fig. 4.10 Diagrammatic representation of the measurement of the water flux in the xylem
of a tree stem (DAUM 1967). A vertical radial section of half a tree trunk. 1 — bark, 2 — phloem
3 — cambium, 4 — active xylem, 5 — inactive xylem, 6 — geometrical axis of the trunk,
7 — heat flux plate, a — heat flux during daytime, b — upwards heat flux resulting from
water flux in the active xylem, A,B,C, and D — points where thermocouples are inserted.

Heat entering the active xylem (Q_H in [J cm^{-2} min^{-1}]) is measured directly by
heat flow plates, inserted under flaps of bark and sealed with a caulking compound.
The heat lost by transfer to the inactive xylem (Q_w) is calculated from the temperature differences between points C and D

$$Q_w = k_t(T_C - T_D)Ad^{-1} , \qquad (4.15)$$

where k_t is the thermal conductivity of the intact xylem, T_C and T_D are temperatures at points C and D, A is log. of the mean area of the cylinder between A and B,
and d is the radial distance between points C and D.
     The flux of heat into or out of the active xylem is

$$Q_s = C_x g \frac{dT}{dt}, \qquad (4.16)$$

where $C_x g$ is the thermal capacity of the active xylem per °C, (C_x is the specific heat, g is the total weight of the active xylem segment). $C_x g = (Q_H - Q_w) (dT/dt)^{-1}$ when $T_B - T_A = 0$, with changing T_A and T_B, dT/dt is the slope of the temperature curve (T_A and T_B). The heat flux is thus determined by the slope dT/dt.

The heat transferred upwards by the transpiration stream (Q_T) is

$$Q_T = C_w \varrho E (T_B - T_A), \qquad (4.17)$$

where C_w is the specific heat of the active xylem, ϱ is the density of the sap ($\varrho = 1$), E is the rate of transpiration [cm^3 min^{-1}], $T_B - T_A$ being the temperature difference between points A and B.

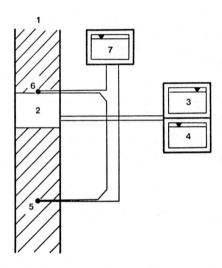

Fig. 4.11 A diagram of the arrangement used for measuring water flux rate in tree trunk from the heat balance using constant and measured heat input (ČERMÁK et al. 1973). 1 — — radial section through active xylem, 2 — parallel electrodes made of stainless steel inserted vertically into the trunk, fed by alternating current of measured and recorded intensity (3) and voltage (4), 5 and 6 — thermojunctions of a copper/constantan thermocouple, 7 — recorder of the thermocouple output.

The water flux dw/dt [cm^3 min^{-1}] in the segment of the active xylem then equals

$$\frac{dw}{dt} = \frac{Q_H - Q_w - C_x g \dfrac{dT}{dt}}{C_w \varrho (T_B - T_A)}. \qquad (4.18)$$

All heat fluxes into the tree trunk (see arrow 6 in Fig. 4.10) are taken as positive.

The temperatures at points A, B. C and D are measured by copper/constantan thermocouples (wires 0.2 mm in diameter) inserted through holes drilled into the stem and then filled with caulking compound. This part of the stem is protected against insolation. The temperatures and data obtained by the heat flow plate (a), are recorded.

ČERMÁK *et al.* (1973) used uniform and measured heat input into active xylem of the trunk through two or four parallel stainless-steal plate electrodes (alternating current) inserted vertically into the xylem. They calculated steady state heat balance from the difference of the temperature in the active xylem measured by means of copper/constantan thermocouple between the point 6 (Fig. 4.11) just above the heated segment and below it (point 5). Neglecting radial heat loss into the active xylem, minimizing the tangential loss into the neighbourhood by additional parallel electrodes and minimizing the outward radial loss by thermal insulation the authors write:

$$q_\mathrm{w} = \frac{Q}{\Delta T c A},\qquad(4.19)$$

where q_w is vertical water flux in the active xylem per unit area of the segment, Q is the measured heat input, ΔT is the temperature difference between the two points, c is specific heat of water and A is the area of the segment. This method is very promising and capable of refinements.

A thorough mathematical analysis of steady-state heat step methods was offered by PICKARD and PUCCIA (1972) and PICKARD (1973) who doubt about the reliable practical use of these methods for quantitative measurement of total sap flux. They conclude that the best experimental realization would be that which involves the measurement of the decay of temperature with distance on the apical side of the heater which seems never to have been tried.

4.3.2 Radioactive Tracer Methods

Rate of longitudinal water transport in tree trunks were measured with tritiated water by KLINE *et al.* (1970) with the aim of assessing the transpiration of individual trees. As a basis, they used the theoretical analysis of tracer dynamics in steady-state systems given by ZIERLER (1964), BERGNER (1966) and ORR and GILLESPIE (1968). Like LJUNGGREN (1969), the authors used the equation:

$$M = F \int_0^\infty f'(t)\mathrm{d}t\qquad(4.20)$$

which states that for a completely mixed steady-state system the product of the flow rate (F in ml day^{-1} per tree) and the total integral of the specific activity in leaf samples $f'(t)$/time curve is equal to the total amount of tracer originally applied to the trunk (M). This equation is valid for instantaneous as well as prolonged administration of the tracer. It applies to complete mixing of the tracer, a condition which is not fulfilled in the tree but can be approximated through replicated sub-sampling of many leaves from the entire crown canopy. Furthermore, it holds for steady flow, which is of course not a property of the daily course of transpiration, so that the resulting values correspond to an average rate over the whole period.

The technique was as follows: 1 ml tritiated water (HTO) was introduced into each of several holes (0.6 cm in diameter and 2.5 to 9 cm deep — depending on the trunk diameter, which was 6 to 55 cm) about 1 m above ground level. The holes were sealed. Triplicate leaf samples were collected three times per day every 3 hours during the first week, then once daily and later once weekly through the 70 days. Leaf samples were wrapped in polyethylene and frozen. Leaf water was extracted with a vacuum freeze-drier and counted for tritium using a standard liquid scin-tillation counter. The data were corrected for decay and expressed in c.p.m. ml^{-1} water and plotted against time. The value M in equation 4.20 is replaced by $(M - R)$ where R is the residual amount of tritium at the end of the experiment, estimated from the biomass and the measured average relative water content. Error analysis gave values $\pm$ 6 to $\pm$ 12%, evidently caused by (1) incomplete mixing of the tracer, (2) leaching of tritium from leaves by rain, (3) enrichment of tritium in leaves by isotopic mass effect, (4) sampling errors.

5

Water Exchange between Plant and Atmosphere

5.1 Introduction

Plants lose water from the surface of their aerial parts to the atmosphere under most ecological conditions. Water is lost most frequently as water vapour in transpiration: the active outflow in the liquid phase — guttation — is less frequent. The substantial decrease in water potential at the plant-atmosphere boundary and the supply of sufficient heat energy (for latent heat of vaporization of water) is usually responsible for transpiration. Pressure in the xylem brought about by absorption activity of roots causes guttation.

Water exchange between plant and atmosphere may also be bidirectional: Water transfer from the plant as evaporation of water (transpiration) and guttation; transfer from the surrounding atmosphere to the plant surface occurs through the condensation of atmospheric water on the plant surface and its possible uptake by plant cells.

In the liquid phase in the plant the flux of water is directly proportional to the gradient of water potential (*i.e.* to the difference of water potential per unit of path length) and inversely proportional to the resistance of the medium through which the water movement takes place. In the vapour phase similarly, the water vapour flux is directly proportional to the gradient of water vapour concentration and inversely proportional to the resistance. As there is a change of phase during water loss by evaporation, and during the opposite process of condensation of water in dew formation on the plant surface, water exchange between plant and atmosphere also depends on the heat balance of the plant surface.

Thus water exchange depends basically on three groups of factors:

(1) Firstly it depends on the energy balance of the plant surface because sensible heat changes to latent heat of vaporization, and *vice versa*, during vaporization and condensation, respectively. The heat balance of a plant leaf may be expressed by the following equation:

$$R_n + H + lE + G + aP_N = 0, \tag{5.1}$$

where R_n is net radiant heat flux, H is convective transfer of sensible heat, lE the major component of heat loss as latent heat of vaporization, G heat flux into the leaf by conduction, aP_N chemical energy conversion (photosynthesis and respiration), a the chemical energy storage coefficient [15.1 kJ g^{-1}], l the latent heat of vaporization of water [about 2.5 kJ g^{-1}]. Fluxes towards the plant are taken as positive. At the same time,

$$R_n = R_s(1 - r) + R_l, \tag{5.2}$$

where R_s $(1 - r)$ is the net short-wave flux, r is the short-wave reflectance and R_l is the net long-wave flux.

The heat balance of the leaf determines the leaf temperature, which in turn influences the reradiation component of convective heat transfer (H), heat conduction (G), and latent heat transfer, since it also influences the partial vapour pressure of water in the leaf (see below).

(2) The second group of factors determining water exchange between plant and atmosphere includes the direction and magnitude of the gradient of water potential (Ψ_w). Water potential of water vapour (*i.e.* water in the gaseous phase) is determined by the expression

$$\Psi_w = \frac{RT}{\overline{V}_w} \ln \frac{p}{p^0}, \tag{5.3}$$

It thus depends directly on absolute air temperature (T is in $^\circ$K) and relative air humidity p/p^0. R is the universal gas constant and $\overline{V}_w$ is partial molal volume of water. The water potential of the liquid/air interface in the plant depends on its matric (Ψ_m) and osmotic (Ψ_s) component. Matric potential (Ψ_m) depends primarily on the structure and ultrastructure of the surface of the plant body. Concave structures decrease the partial water vapour pressure, whereas convex structures increase it. Thus, for example, the water vapour pressure at the surface of a water drop of radius 10^{-5} cm is by 12.5% higher than that above a flat water surface. Concave cavities of ultracapillaries (micellar structures *etc.*) present at plant surface could theoretically decrease the water vapour pressure below that at a free water surface. Osmotic potential depends on the concentration and type of the dissolved compounds in the aqueous phase saturating the transpiring surface (see Sections 1.1 and 1.4). In most cases the water potential of the transpiring surface of above-ground plant parts approaches zero.

(3) The third group of factors influencing water exchange between the plant and the surrounding atmosphere includes resistances to the water vapour flux between the active surface and the bulk air. Water movement along this path is of two types, depending on its physical basis: molecular diffusion or convection of water molecules in air. Inside the plant the water vapour diffuses between the cell walls surrounding the intercellular spaces and the outer margin of the guard cells of the stomata. A further resistance to diffusion lies in the boundary air layer adhering to the outer leaf surface. A third component is represented by the water vapour flux by convection in the bulk air.

5.2 Air Humidity

The origin of land plants from originally submersed water plants was without doubt one of the most complex changes involved in phylogenetic development. The relatively low atmospheric humidity surrounding the above-ground organs of land plants is the main reason why this change was associated with fundamental rearrangement of the plant body.

5.2.1 Expressions for Air Humidity and a Review of Methods Used for its Determination

Air humidity may be expressed absolutely, either as partial water vapour pressure (p) in pressure units such as millibars [mbar] or the height of a mercury column (mm Hg $\approx$ torr*); or as the mass of water vapour per unit volume of moist air, $e.g.$ $kg_{H_2O} \, m_{air}^{-3}$. Direct determinations of absolute humidity as partial water vapour pressure are difficult. The measurement of absolute air humidity in terms of the weight of water vapour in a volume of air is rather laborious, although simple in principle. A known amount of air is passed through U-tubes containing, for example, $CaCl_2$, phosphorus pentoxide, magnesium perchlorate, and molecular sieves, arranged in series. The amount of adsorbed water vapour is calculated from the increase in weight of the U-tubes. Partial water vapour pressure may be calculated from the temperature and absolute air humidity. In addition, indirect gasometric methods (infrared analysers, corona hygrometers, lithium chloride hygrometers, microwave dew point hygrometers $etc.$) described in Section 5.5.3.1 may be used after calibration to measure absolute humidity. At a particular temperature and atmospheric pressure, air contains water in the form of vapour only up to a certain partial pressure, the saturation water vapour pressure, the temperature-dependence of which is shown in Fig. 5.1. The saturation water vapour pressure is reached when air is saturated with water vapour: such air is in dynamic equilibrium with a water surface at the same temperature.

Relative humidity (r.h.) is the absolute humidity or vapour pressure of the air (p) expressed as a fraction or percentage of the absolute humidity or vapour pressure of air saturated with water vapour at the same temperature (p^0). It is dimensionless. Thus relative humidity is the relative saturation of air with water vapour. The relationship between relative humidity and water potential of a solution (Ψ_s) is shown in Fig. 1.2. As there can be different water vapour pressures at the same relative humidity but at different temperatures, relative humidity without temperature data is a useless expression of air humidity and is not one of the factors determining evaporation.

---

* 1.333 mbar $=$ 1 torr $\approx$ 1 mm Hg.

Evaporation depends on the saturation deficit (s.d.), *i.e.* the difference between water vapour pressure at saturation (p^0) and actual water vapour pressure of the air (p). This is usually expressed in mbar (or torr, or mm Hg):

$$\text{s.d.} = p^0 - p. \qquad (5.4)$$

Dew point temperature is another expression of humidity. It is the temperature at which the air becomes saturated with water vapour (water vapour pressure p^0), *i.e.* relative humidity 100% and saturation deficit 0. Condensation takes place from air cooled below its dew point. The dew point temperature is therefore important in drying air by freezing.

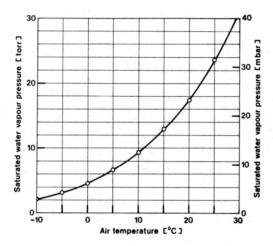

Fig. 5.1 Maximum (saturated) water vapour pressure in air as related to air temperature.

Psychrometric measurement is the simplest, reasonably accurate method of determination of air humidity. The temperature difference between the "dry" and "wet" bulb of a pair of thermometers, is measured. The dry thermometer is a normal accurate thermometer (*e.g.* with 0.1 °C gradations): the mercury bulb of the "wet" thermometer is covered with wet muslin. The temperature of the wet thermometer is lower than the dry one, and equilibrates at the point at which the intake of heat equals the loss of heat in evaporation.

The two heat fluxes are dependent (in different ways) on ventilation (wind speed), resulting in the psychrometric constant γ [mbar °C^{-1}] so that

$$p = p^0 - \gamma(T - T'), \qquad (5.5)$$

where p^0 is the saturated water vapour pressure, p the ambient water vapour pressure, both [mbar], T and T' are surrounding air temperature (dry temperature) and wet thermometer temperature, respectively, γ for a sheltered and non-ventilated

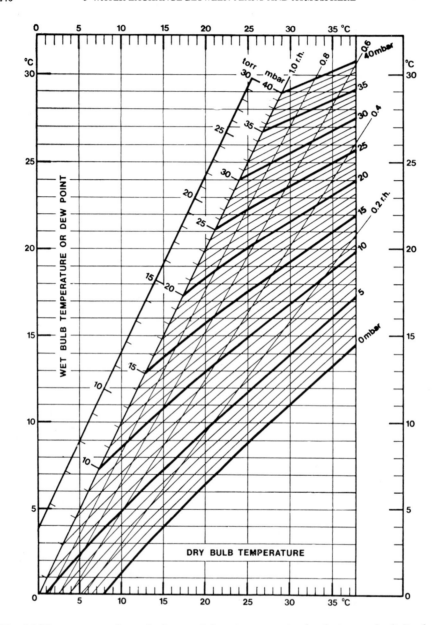

Fig. 5.2 Nomogram used to calculate partial water vapour tension in torr and mb (1 mb ≈ ≈ 0.75 torr), dew point and estimation of the relative air humidity from the data of the dry (abscissa) and wet (ordinate) thermometer of the aspirated psychrometer. The lines designated r.h. (0.2; 0.4; 0.6; 0.8; 1.00) correspond to relative air humidity, lines 0 to 40 correspond to water vapour pressure in mbar. The bevelled diagonal scale serves to convert mbar to torr. The dew point is calculated in such a way that the temperature of the wet thermometer is subtracted horizontally from the respective point representing the resulting water vapour pressure in mbar or torr on the line 1.0 r.h. Adapted from SLATYER and McILROY (1961).

August psychrometer equals 0.001047 mbar °C⁻¹, at a mean wind speed of 0.8 m s⁻¹. As the psychrometric constant depends on the ventilation rate (wind), it is much more accurate to measure air humidity with an aspirated (ventilated) Assmann psychrometer, in which both thermometers are ventilated at about 4 m s⁻¹. Then

$$p = p^0 - 0.667(T - T') \frac{b}{1013} \qquad (5.6)$$

(b is atmospheric pressure in mbar).

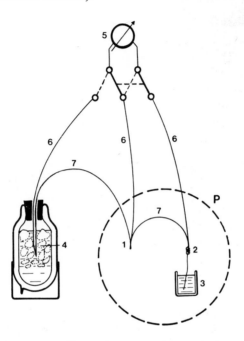

Fig. 5.4 Scheme of the thermocouple psychrometer (P). 1 — dry junction, 2 — wet junction wrapped with a thread drawing water from the vessel containing distilled water (3), 4 — junction placed in a vacuum flask containing distilled water with ice, 5 — meter, 6 — copper wires, 7 — constantan wires (about 0.05 to 0.1 mm in diameter). Constantan wires should be as short as possible to avoid a decrease of the measured voltage due to their resistance.

Psychrometric tables (for instance Aspiration-Psychrometer-Tafeln 1955, Smithsonian Meteorological Tables 1963) may be used to calculate humidity from the wet and dry bulb temperatures of an aspirated psychrometer. A nomogram for the calculation of partial water vapour pressure in torr and mbar (1 mbar = = 0.75 mm Hg), dew point and estimation of relative humidity is shown in Fig. 5.2 (adapted and redrawn from SLATYER and McILROY 1961). Fig. 5.3 (after the page 240) is a graph used to determine relative air humidity from the temperatures of the dry and wet bulb thermometers of an aspirated psychrometer.

Aspiration of very small psychrometers, *e.g.* fine wire thermocouples, is not necessary (DIEM 1953), and calculations for an aspirated psychrometer can still be used. The wet junction of the thermocouple may be made, for example, by tying a fine thread or tissue closely around the junction to act as a wick. This is wetted before measurement, or one end is led into a vessel containing distilled water (Fig. 5.4). For micrometeorological purposes, very many more or less miniature psychrometric sensors have been constructed using thermistors (*e.g.* DIEM 1953, UNGER 1953) or platinum resistance thermometers (BERGER-LANDEFELDT 1954, MCILROY 1955).

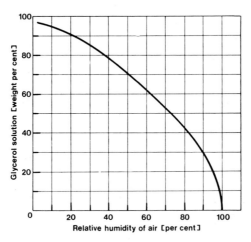

Fig. 5.5 The relationship between the concentration of water solution of glycerol (in weight per cent) and relative air humidity above them (per cent) at 25°C. (International Critical Tables III: 291, 1928).

There is a large number of other indirect methods used to measure air humidity, based on various principles. Some involve measuring the volume or length changes of hygroscopic materials which are in equilibrium with the air (*e.g.* defatted human hair, cellophane, polyethylene *etc.*). A detailed description and theory of hygrometric methods has been given by SPENCER-GREGORY and ROURKE (1957) and WEXLER (1965).

5.2.2 Control of Air Humidity

It is very often necessary to expose plants to controlled atmospheric humidity.

Saturated solutions of some inorganic salts, with a constant water vapour pressure at a particular temperature, may be used for an enclosed volume of air in a constant humidity chamber, when it is possible to work with unchanged air (Table 5.1). The same water vapour pressure is maintained above such salt solutions after vaporization of some of the water from the solution (excess salt precipitates from the supersaturated solution) or after any uptake of water from the

Table 5.1 *Saturated water solutions of salts exhibiting at 20 °C defined constant tension of water vapour [mbar] corresponding to a certain relative humidity [% at 20 °C]. (According to Physico-Chemical Tables, Prague 1953b).*

A substance in the saturated solution	Water vapour tension [mbar]	Relative humidity [%]
$Pb(NO_3)_2$	22.9	98
$CaSO_4.5 H_2O$	22.9	98
$Na_2SO_3.7 H_2O$	22.3	95
$Na_2HPO_4.12 H_2O$	22.3	95
$(NH_4)H_2PO_4$	21.7	93.1
$Na_2SO_4.10 H_2O$	21.7	93
$NaBrO_3$	21.5	92
K_2HPO_4	21.5	92
$ZnSO_4.7 H_2O$	21.1	90
K_2CrO_4	20.5	88
KCl	20.1	86.4
$KHSO_4$	20.1	86
KBr	19.6	84
$(NH_4)_2SO_4$	18.9	81
NH_4Cl	18.5	79.2
$Na_2S_2O_3.5 H_2O$	18.3	78
$CH_3COONa.3 H_2O$	17.7	76
$(HCOO)_2.2 H_2O$	17.7	76
$NaClO_3$	17.6	75
$NH_4Cl + KNO_3$	16.9	72.6
$NaNO_2$	15.5	66
$(CH_3COO)_2Mg.4 H_2O$	15.2	65
$NaBr.2 H_2O$	13.6	58
$Ca(NO_3)_2.4 H_2O$	13.1	55.9
$NaHSO_4.H_2O$	12.2	52
$Na_2Cr_2O_7.2 H_2O$	12.2	52
$KCNS$	10.98	47
KNO_2	9.81	45
$Zn(NO_3)_2.6 H_2O$	9.81	42
CrO_3	8.18	35
$CaCl_2.6 H_2O$	7.54	32.3
CH_3COOK	4.68	20
$LiCl.H_2O$	3.51	15
$ZnCl_2.1^1/_2 H_2O$	2.32	10

plant material (more of the undissolved substance dissolves until the solution is saturated again). This applies particularly to well-stirred solutions. The required relative air humidity in the constant humidity chamber may be maintained very accurately by connecting the chamber to a closed circuit with a wash bottle con-

Table 5.2 *Relative air humidity above water solutions of sulphuric acid. Column A — relative air humidity (r. h.) in per cent at 20 °C, column B — g of H_2SO_4 per 100 g of H_2O, column C — g of H_2SO_4 in 100 g of the solution, column D — specific weight of the solution at 15 °C (according to* WALTER *1931b).*

A r.h. [%]	B	C [%]	D	A r.h. [%]	A	C [%]	D
99.5	0.97	0.96	1.006	89.0	20.47	17.0	1.119
99.0	1.95	1.91	1.012	88.0	22.15	18.2	1.128
98.5	2.92	2.84	1.019	87.0	23.90	19.3	1.137
98.0	3.90	3.75	1.025	86.0	25.60	20.4	1.145
97.5	4.87	4.65	1.031	85.0	27.20	21.4	1.153
97.0	5.85	5.53	1.037	84.0	28.80	22.4	1.161
96.5	6.82	6.38	1.043	83.0	30.40	23.3	1.168
96.0	7.80	7.24	1.049	82.0	32.00	24.2	1.175
95.5	8.78	8.08	1.054	81.0	33.55	25.1	1.182
95.0	9.75	8.88	1.060	80.0	35.00	25.9	1.189
94.5	10.70	9.67	1.066	79.0	36.45	26.7	1.195
94.0	11.60	10.40	1.071	78.0	37.90	27.5	1.201
93.5	12.50	11.10	1.076	77.0	39.30	28.2	1.207
93.0	13.40	11.80	1.081	76.0	40.68	28.9	1.213
92.5	14.30	12.50	1.086	75.0	42.02	29.6	1.219
92.0	15.18	13.17	1.091	74.0	43.38	30.2	1.224
91.5	16.08	13.84	1.096	73.0	44.70	30.9	1.229
91.0	16.95	14.50	1.101	72.0	46.00	31.5	1.234
90.5	17.85	15.15	1.106	71.0	47.30	32.1	1.239
90.0	18.72	15.75	1.110	70.0	48.60	32.7	1.244

taining a solution at constant temperature with air passing over it (see *e.g.* OSBORNE and BACON 1961).

Solutions may also be used to control the humidity of a slow stream of air. Horizontal tubes (1.5 m long for flow rates up to about 2 l min^{-1} are filled to one half their diameter with the solution and air is passed through them above the solution (ANDERSSON *et al.* 1954, BJÖRKMAN and HOLMGREN 1966, JARVIS and SLATYER 1966b). Such solutions may also be used in sintered glass wash bottles. The temperature of the solutions must be controlled. Solutions of sulphuric acid are also commonly used for humidity control. Concentrations, expressed in various ways, and the corresponding values of relative humidity above the solutions are shown in Table 5.2. Solutions of glycerol may also be used (see Fig. 5.5).

The best and most accurate procedure for obtaining constant air humidity in an open system (see Section 5.5.3.1) is by cooling air previously saturated with water vapour (at a higher temperature) to the dew point corresponding to the required absolute humidity. This may be done by bubbling the air stream through distilled water in sintered glass tubes or brass tubes with a perforated septum at room temperature (*i.e.* at a temperature higher than the required dew point temperature) in order to saturate it and then cooling it down again either in the same kind of tube or in cooling coils in a well-stirred water bath with accurate temperature control ($\pm$ 0.05 °C), at the dew point temperature. For medium and lower humidities cooled water baths or special dehumidifiers should be used. The accuracy of the result depends on the constancy of the bath temperature.

For humidity control of streams of air, two air streams with known different humidities may be mixed. The simplest procedure is to mix dry air (dried in a condenser or over a desiccant) and saturated air, using rotameters (flow meters) and

Table 5.3 *The effectivity of some drying agents expressed in* mg *of water vapour remaining in 1 l of dried air. (Compiled from different sources).*

Drying agent	Formula	Water vapour residue [mg 1^{-1}]
Magnesium perchlorate, anhydrous	$Mg(ClO_4)_2.0.12\ H_2O$	0.000 001** 0.000 2 *
Anhydrone	$Mg(ClO_4)_2.1.48\ H_2O$	0.001 5 *
Phosphorus pentoxide	P_2O_5	0.000 02 ** 0.003 6 *
Potassium hydroxide, melted	KOH	0.002 **
γ-Aluminium-oxide	$\gamma-Al_2O_3$	0.003
Sulphuric acid	H_2SO_4 100%	0.003
Magnesium perchlorate, anhydrous indicating	88% Mg ClO_{42}. .0.86% $KMnO_4$	0.004 4 *
Drierite at 25 °C		0.005 0 ***
Silica gel at 30 °C		0.006 **
Calcium chloride, anhydrous	$CaCl_2.0.18\ H_2O$	0.067 *
Calcium chloride, granulated		0.14 to 0.25
Calcium oxide	CaO	0.2 ***
Sulphuric acid	H_2SO_4 95.1%	0.3 **
Calcium chloride melted		0.36 **
Cuprum sulphate, anhydrous		1.4 **
Cooling at -30 °C		0.330
-20 °C		0.880
-10 °C		2.154
0 °C		4.846

 * Tested by Trusell and Diehl (1963) using air flow rate 1.35 l h^{-1} through U-tube 14 mm i.d. and 150 mm deep, at 25 °C.
 ** Equilibrium data of air enclosed above the desiccant.
*** Luck (1964).

needle valves, in the required proportions. CROSS (1968) has described a servo-operated system.

An air stream with a fairly stable but not precisely controlled humidity can be obtained by heating outside air, the water vapour content of which does not vary over a short time (*e.g.* 1 hour), to a constant temperature. Very dry air may be obtained in winter by heating outside air to a constant laboratory temperature. To control humidity of a large volume of air ($50 \, l \, min^{-1}$ and more), water vapour can be supplied from a flask with a wire or electrode heater which is switched on by a contact hair hygrometer through a reliable mercury relay *(e.g.* SLAVÍK and ČATSKÝ 1963) or heated continuously using controlled voltage.

Drying by condensation caused by cooling may be used to lower the humidity of the air stream. Air is cooled to the appropriate dew point temperature at which the absolute humidity corresponds to the required water vapour pressure at the temperature to which the air is subsequently heated (see Section 5.2.1). This procedure is used for calibrating gasometric devices used to measure air humidity. Condensers based on this principle (Sirigor Wasserdampfabscheider) using the Peltier effect for accurate setting of the surface temperature of the cooling system are produced commercially *e.g.* by Siemens Co. in Erlangen (KOCH and WALZ 1967, KOCH *et al.* 1968).

An enclosed volume of air can often be dried more or less perfectly with solid drying agents, reviewed in Table 5.3.

5.3 Water Absorption by Aerial Parts of Plants

5.3.1 Introduction

Our knowledge of water absorption by above-ground parts of plants is still insufficient, although considerable attention has been devoted to this problem.

Drops of water on the plant surface, or water vapour at high relative humidity, provided that the temperature gradient is suitable (leaf temperature $\leq$ air temperature), are possible sources for water absorption by above-ground parts. There is almost always a favourable gradient of water potential in the first case as the water potential of pure condensed water is higher than that of the epidermal cells. In the second case, however, there is a decrease of water potential in the direction of the epidermis only at very high relative air humidities ($> 98\%$) or at low leaf surface temperatures compared with air temperature. For example, Table 1.3 (page 11) shows that at a relative air humidity as high as 98% and isothermic conditions ($20 \, °C$), the water potential at the plant surface would have to be less than -27.2 bar for absorption of water vapour to occur. For these reasons, it is also difficult to demonstrate water absorption by roots from saturated air experimentally. It is very difficult to maintain a favourable temperature gradient as any increase of root temperature (*e.g.* from respiration) may annul or even reverse the necessary gradient of water potential.

External sources of liquid water for absorption are: dew, fog (so-called horizontal precipitation), rain or wetting of leaves by spraying. The wetting of the cuticle of the leaf, which is essential for absorption of liquid, water, is a function of the contact angle: the smaller the contact angle, the greater the wettability. The contact angle depends on the surface tension of the water and the type of surface and ist properties, *e.g.* epidermal hairs, hydration of the cuticle, presence of wax films *etc.*

It appears that water absorption by above-ground parts, whether by dew, rain, horizontal precipitation or artificial water spray, is not quantitatively important in water uptake. However, the indirect effect of dew and wetting in general is unquestionable. This effect may act in two ways: first, the wetting of the plant surface decreases the transpiring surface; and, second, the relative humidity near the leaf surface increases as a result of evaporation and thus decreases the evaporative power of the surrounding atmosphere. In both cases, transpiration decreases and the water balance of the plants is improved.

The importance of dew has been reviewed *e.g.* by KREČMER (1951), STEUBING (1955), STONE (1957a, b, 1965), WAISEL (1958), MONTEITH (1963) and WALLIN (1967) and others.

5.3.2 Review of Methods Used to Measure Dew

Under normal conditions dew is only a small part of the total annual precipitation. It hase been shown that most of the dew deposited on vegetation usually originates from condensation of water evaporated from the soil or from transpiration by the lower leaves elsewhere in the canopy. This is caused by the usually negative temperature gradient at night between the soil (which is warmer) and the surfaces where most dew is deposited. The use of accurate lysimeters has confirmed this. The rate of dew formation is relatively low: the maximum possible rate is about 0.1 mm h^{-1} ($0.07-0.15$ mm h^{-1} depending on temperature). The maximum possible amount of dew per night in temperate zone is less than 0.5 mm.

Micrometeorological conditions at the surface determine the localization and amount of dew fall. The ecological and physiological importance of dew to plants is mainly determined by the amount of dew which condenses directly on plant surfaces. This may differ considerably from amounts on other surfaces *e.g.* the soil and the various devices used to measure dew (drosometers). Only measurements on the plants themselves are of interest ecologically. It should be pointed out that dew is sometimes mistaken for guttation water secreted from plant leaves by specialized tissues. Its amount may be substantial in certain plants. Absorption of water by above-ground parts has been discussed by GESSNER (1956) and SLATYER (1960).

Many methods for measuring dew have been used but so far there are no methods which are generally applicable for all types of work. The methods will not be described here in detail. Detailed reviews have been published by Uhlíř (1948) and Mäde (1956). Modern methods include lysimetric measurements (Jennings and Monteith 1954, Collins 1961), determination of weight differences of samples from the top soil layer taken in the afternoon and before sunrise the following morning (Rahman and Hahidy 1958), weight registration of dew condensation on shoots (Hirst 1954), instruments measuring time of wetting of leaves (Uhlíř and Stružka 1948, Uhlíř 1957, Uhlíř and Uhlíř 1959, Lomas and Shashoua 1970), the thermoelectric drosometer (Hofmann 1955, 1958, Roth 1960), direct measurements of dew on plants (Arvidsson 1958), the interesting volumetric method of Scott (1962) and the thermodynamic approach (Hofmann 1955, 1958).

5.3.3 Determination of Water Absorption by Aerial Parts of Plants

No general methods of measuring the quality and the quantity of absorption by above-ground parts of plants have yet been developed. Most procedures are based on the determination of the weight of absorbed water (Steubing 1949, Arvidsson 1951, Haines 1953, Breazeale et al. 1950, 1951, Meidner 1954, Slatyer 1956, Stone and Fowells 1955, Stone 1957b) or the use of radioactive water, HTO (Vaadia and Waisel 1963, Plaut and Reinhold 1967, Waisel 1969, Ketel et al. 1972). Jensen et al. (1961) measured water absorption volumetrically. Stone et al. (1950) estimated water absorption from the decrease in humidity of the air surrounding a plant shoot enclosed in a sealed space.

5.4 Ecological Factors in Transpiration

Transpiration is the biologically controlled evaporation of water from the plant surface into the ambient environment. It would be very advantageous to be able to characterize by a particular measurement the entire set of factors that influence transpiration, i.e. to measure the "tendency" of an environment to remove water from the above-ground plant parts in a given situation. However, it has already been pointed out that transpiration depends not only on a number of environmental factors but also on the properties of the transpiring surface. Consequently, it is not theoretically possible to construct a physical model of transpiration that will characterize the interaction between the environment and an artificial water surface in a way analogous to transpiration under natural conditions.

5.4.1 Evaporimeters. Equivalent Evaporation Power

A large number of physical models have been constructed to measure evaporation for standard methods for meteorological network purposes. The temperature and energy exchange characteristics of evaporimeters are the main problems in construction. The water surface of the pan of the Wild evaporimeter used by the

Meteorological service in many countries, and the water surface of other evaporimeters with a free surface, have energy exchange characteristics which are different from those of porous surfaces of the LIVINGSTON and BELLANI atmometers and even more different from those of a flat transpiring leaf or a conifer needle. The PICHE evaporimeter is often thought to be more suitable for biological measurements. It consists of a disc of wet filter paper. However, the energy exchange properties of the disc are quite different from those of a flat transpiring leaf: *e.g.* it is

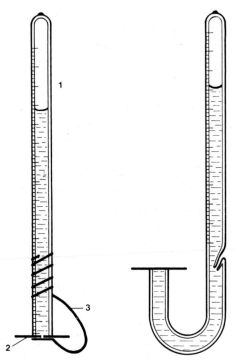

Fig. 5.6 The PICHE evaporimeter. Left: 1 — glass tube with a sealed upper end and ground to a plain surface at its lower end, 2 — perforated evaporation disc, 3 — positional spring. Right: The PICHE evaporimeter of the French Meteorological Service (WEHRLE 1942). See the description in the text.

substantially cooler than the air during insolation whereas the opposite is true of the majority of leaves, even if they are transpiring rapidly. The "humidity" surface properties of most evaporimeters are less important. Differences in structure between wet porous ceramic material, wet filter paper and a free water surface cause relatively small differences in evaporation rate.

Evaporimeters do not measure the actual evaporation from a natural surface, but allow estimates of the relative rates of evaporation from a certain standard surface to be made. As we are mainly interested in plant surfaces, a suitable evaporimeter should approximate evaporation from the plant leaf as far as possible.

The PICHE evaporimeter (Fig. 5.6) has been considered, with numerous reservations, to be useful for ecological and physiological studies. It has a standard evaporating surface of more or less defined physical properties. There are some plants and ecological conditions in which the rate of evaporation and the energy exchange properties of the evaporating disc of the PICHE evaporimeter are similar to those of the leaves. The PICHE evaporimeter has the advantages that it can be assembled from simple, easily available parts. The short life of the evaporating disc, which is affected by rain and grazed by wasps and hornets, is associated with one of its

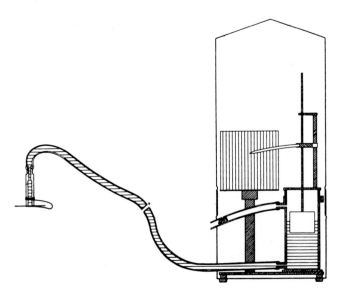

Fig. 5.7 Schematic section of the PICHE atmograph (KLAUSING 1957) made by J. A. Bosch Nachf. S. Bosch, Freiburg i. Br.

advantages: it must be replaced fairly frequently with the result that its evaporating surface does not become changed by permanent exposure to pollution, growth of algae etc. which may be the case with other types of evaporimeters. The PICHE evaporimeter must be freely exposed at the same height as the plant leaves under consideration and not shaded. Paper of the same type should be used for all the measurements to be compared. A fairly thick dark green blotting paper is usually used, although this is in no way identical to leaves in optical properties. The French Meteorological Service ONM (Paris) designed a PICHE evaporimeter with a U-tube and a disc with no openings. The air replacing the evaporated water enters the tube through a side opening located at the level of the evaporating disc (Fig. 5.6, right). The PICHE evaporimeter is not sensitive to horizontal shocks. Methods of recording have been developed by STRUŽKA (1956) and KLAUSING (1957). The KLAUSING atmograph has a float device which records the loss of water in a cylinder

from which the evaporating disc is supplied with water through a soft PVC tube
(Fig. 5.7). STRUŽKA's evaporimeter counts the air bubbles coming up through the
central hole in the disc. A simple modification of the PICHE evaporimeter with
plastic tube, only one side of the filter paper disc acting as an evaporating surface,
and gravimetric measurement of water loss, was designed by WARING and HER-
MANN (1966).

The theory of errors, construction and thermodynamic properties of the PICHE
evaporimeter were described by ROTH (1961) who also summarized the various

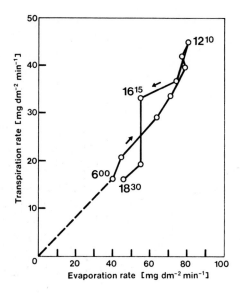

Fig. 5.8 An example of the so-called PISEK diagram: dependence of the transpiration rate
on evaporation rate, measured by a PICHE evaporimeter during the day (time course according
to the arrow along the line).

comparative measurements and the experience gained by Boss (1952), BAUMBACH
(1952), WÄCHTERSHÄUSER (1954), UHLIG (1955), HEIGEL (1957) and RICHTER (1958).
For comparison with other types of evaporimeters see *e.g.* CARDER (1960), PELTON
(1964), and many others.

5.4.2 Relative Transpiration

Attempt to isolate the physical and biological components of transpiration involve measurements of transpiration rate in comparison with physically measured
evaporation rates, reflecting the relative evaporative demand of the atmosphere.
Relative transpiration is taken as the transpiration rate divided by the simultaneously measured evaporation rate (both rates expressed per unit of surface and time).
However, the method used to measure evaporation (type of evaporimeter) deter-

mines its absolute magnitude, *i.e.* different evaporimeters give different evaporation rates under the same conditions. Relative transpiration therefore depends not only on the plant, and thus on the biological component of transpiration, but also on the type of evaporimeter used. Values of relative transpiration are only comparable if evaporation has been measured in the same way. Relative transpiration as a standard value expresses to a certain degree the biological component of transpiration.

The PISEK diagrams expressing the dependence of the transpiration rate on the simultaneous rate of evaporation (equivalent evaporation) during one day are very objective. (See *e.g.* PISEK 1956.) An example of such a graph is shown in Fig. 5.8.

5.5 Transpiration

5.5.1 Introduction

Transpiration is loss of water as vapour from the above-ground surface of the plant to the surrounding atmosphere. It is caused by the usually negative gradient (decrease) of water potential between the transpiring surface and the adjacent air layer. As the evaporation is regulated by biological processes and structures, it is both subject to the laws which determine the rate of evaporation (the physical component of transpiration) and it is also influenced by the biological state of the transpiring tissues (the biological component of transpiration).

Transpiration may be determined either in terms of its instantaneous rate, *i.e.* the amount of water transpired during a short period of time, or (integrally) as the amount of water transpired during a longer period of time regardless of the variation of the transpiration rate during that time.

The transpiration rate of a particular plant organ or its part varies considerably with time. This is determined not only by the taxonomic classification of the plant, but also by variation between individual plants, by the ontogenetic age of the plant and the organ during the vegetative season (seasonal pattern) and, particularly, by physiological processes and diurnal changes (diurnal pattern). The atmospheric environment influences the transpiration rate through the factors determining the physical component (see also Section 5.1) and through the external factors affecting the biological component (*e.g.* illumination influencing the state of the stomata *etc.*). These factors change considerably with time. The soil environment also influences the transpiration rate: the availability of soil moisture is the main factor determining the uptake component of plant water balance. The soil environment is also subject to considerable changes with time.

5.5.2 Measurement of Instantaneous Transpiration Rate

Procedures useful for the determination of instantaneous transpiration rates more or less *in situ* will be described first. There are only two methods of practical use for field work: short-term weighing of parts of plants which have been cut off, and the gasometric method using an electric humidity detector for working with intact plants. The thermometric method described below is only of a theoretical nature.

A single measurement of the instantaneous transpiration rate in the field may be considerably influenced by short-term changes in meteorological conditions (cover of sun by clouds, gust of wind *etc.*) which are often not detected by instruments (*e.g.* they are not detected by an evaporimeter) but may still distort the results obtained. This error may be avoided by repeated measurements and, if possible, by more detailed investigation of the daily pattern of transpiration rate.

5.5.2.1 *Method of Short-term Weighing of Detached Parts of Plants*

In the detached leaf method (IVANOV 1918, 1928, HUBER 1927, STOCKER 1929b) the weight loss of the detached part of a plant (leaf, shoot, branch or possibly the whole above-ground part) is measured for several minutes after cutting. The method is based on the assumption that during the first few minutes after cutting the previous transpiration rate is maintained with no significant change. This assumption has been tested many times since the method was introduced. It is now usually checked by direct gasometric methods or detailed analysis of transpiration curves. IVANOV, one of the pioneers of this method, showed (1928) that a temporary sudden increase of the transpiration rate is sometimes observed several minutes after cutting (after 2 to 5 minutes). In some cases this increase takes place only after a longer period (after 8 to 15 minutes). Theoretically, there are several possible explanations, but the shock itself and the mechanical effect of cutting may be neglected. This "IVANOV effect" is often attributed to the release of tension in the cohesive water columns in the vessels after cutting; after a short delay this results in increased water uptake by the leaf. In conditions of rapid transpiration, this may be reflected by the increase in water potential at the transpiring surface and thus by a temporary increase in transpiring rate. It was found using gasometric measurements (ANDERSSON *et al.* 1954), that the closer the cut was made to the leaf, the higher the observed increase. This would seem to support the suggested explanation of the IVANOV effect.

However, immersion of the cut surface in melted paraffin (see the description of the method below), which seals the conducting elements, does not always prevent the IVANOV effect. Furthermore, the increase occurs most frequently when there is a low water deficit (in the morning or in the evening) thus suggesting an alternative explanation. Immediately after cutting, water is lost more readily from the epidermal cells than from the guard cells, thus leading to an increase in guard cell turgor relative to the surrounding cells, and hence to a larger stomatal aperture, so that the transpiration rate increases. A supraoptimal deficit (terminology according to STÅLFELT 1955) develops during the subsequent minutes, resulting in hydroactive closure of the stomata and decreased transpiration rate.

As the transpiration rate generally decreases after cutting in proportion to the water loss (see also Section 5.5.5) and in some plant species hydroactive stomatal closure takes place relatively soon (after about 10 minutes or exceptionally even less depending on the conditions), the exposure interval (*i.e.* the time between both weighings and the time between cutting and the first weighing) should be as short as possible, so that the difference in weights is small and a sensitive balance is required. As total transpiration rates may vary within the range of 0 to 40 mg g^{-1} min^{-1}, it is necessary to weigh with a sensitivity of 1 mg when using 0.5 g of detached plant material and an exposure of four minutes, in order to determine a mean transpiration rate of 10 mg g^{-1} min^{-1} with an accuracy of 5%, whereas a sensitivity of 0.1 g is sufficient if 50 g of plant material is used. The selection of a suitable balance thus depends on the required accuracy of the determination of the transpiration rate, on the rate itself, on the weight of the sample and on the exposure time.

To compare physiological changes, it is advantageous to measure leaves or leaf blades only, making it possible to calculate the transpiration rate per unit leaf area (see Section 5.5.8). Whole shoots or branches are preferable if the transpiration rate of a whole plant, shrub or tree is to be determined. The transpiration rate may then be recalculated either per unit weight (fresh weight, dry weight, or weight of water) or the area of the transpiring parts may be measured subsequently. It is useful to use large pieces of plant, because the temporary increase in transpiration is less as the distance between the cut surface and the transpiring leaves increases.

Rapid weighing is necessary: weighing should not take longer than about 5% of the exposure time, *e.g.* a maximum of 12 seconds for 4 minutes exposure. There are two reasons for this: (1) The first weighing should be as rapid as possible to give the maximum possible exposure time. (2) The time the plant is on the balance, particularly during the second weighing, may affect the measured transpiration rate, as the actual exposure time is the time between the end of the first and the end of the second weighing.

Having selected the size of the sample and the balance, we may then make preliminary measurements of the detached sample every minute under conditions resembling the future experimental measurements. If weighing is sufficiently rapid, between weighings the sample can be exposed in the natural position and under natural conditions, away from the balance. If this is not possible, the balance is positioned so that the plant is exposed under relatively natural conditions. Weighing is continued for 10 to 20 minutes, depending on the time of the longest exposure. The values of weight loss obtained are best expressed in absolute units and plotted against time. The pattern and the magnitude of the decrease in transpiration rate, or the possible occurrence of the IVANOV effect, can thus be found. This series of measurements should be repeated several times, so that the variability of the material can be estimated, as well as the possible size of methodological errors. The absolute accuracy of weighing and the maximum possible exposure time are then chosen according to the accuracy required. In the example shown in Fig. 5.9, the initial transpiration rate, which was 7.5 mg min^{-1} for an initial fresh weight of 0.463 g (corresponding to a transpiration rate of 16.22 mg g^{-1}min^{-1}),

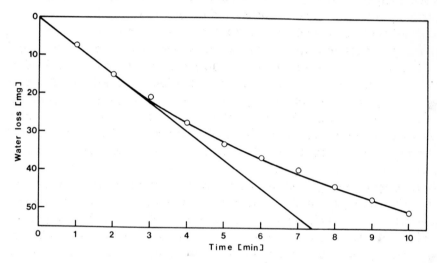

Fig. 5.9 An example of the course of water loss from a cut leaf during the first ten minutes after cutting. The upper curve represents the real course, the lower curve the theoretical course with a non-decreasing transpiration rate.

was calculated from the mean transpiration rate during the first two minutes. The weight loss after 5 minutes was only 33 mg, although the loss would be 37 mg if the calculated transpiration rate had been maintained. The transpiration rate calculated from the water loss during the first 5 minutes is 10.8% lower than the initial transpiration rate calculated from the first two minutes. An exposure of 4 minutes results in an 8.4% decrease, whereas the error increased to 18.0% for an exposure of 6 minutes.

In the example in Fig. 5.9, if the transpiration rate is to be determined with an accuracy of 5%, samples of about 0.5 g fresh weight must be weighed with an accuracy of 1.5 mg (29 : 20), 1.8 mg (37 : 20) and 2.2 mg (44 : 20), for exposure times of 4, 5 and 6 minutes, respectively, so that a torsion balance with a maximum of 500 mg and accuracy of 1 mg is suitable for detached plant samples of about 400 to 500 mg exposed for 5 minutes. If ± 5% accuracy is to be achieved at half the transpiration rate, it would be necessary either to double the exposure time (to 10 minutes) or to double the accuracy of weighing (to 0.5 mg). The number of necessary replicate measurements can be calculated using a normal statistical procedure. More accurate estimates of the transpiration rate can be obtained by considerable reduction in the exposure time (*e.g.* to one minute or even less) but such a reduction introduces new errors: (1) the plant remains on the balance for a considerable part of the exposure time, (2) there are errors in determining the short time interval and (3) the estimate of change in weight is relatively less precise.

Measuring procedure. Suitable healthy and outwardly comparable experimental material (leaves, shoots, branches with several leaves, whole above-ground parts of plants) is chosen for the whole experiment, particularly when measuring at different times of day. The leaf or other plant parts are cut smoothly with a razor or possibly stripped off the plant. If preliminary experiments showed that there is an IVANOV effect after cutting, it is useful to find out whether cutting under melted paraffin (IVANOV *et al.* 1950) would prevent it. For field measurements, paraffin (m.p. 50 to 58 °C) can be melted in a flat copper dish above an alcohol burner and the selected branch or shoot bent so that the part which is to be cut is immersed under the surface of the melted paraffin before cutting. Sealing the cut surface with vaseline, cocoa butter *etc.* is not recommended as any wiping between measurements would cause considerable errors which could not be checked.

Leaves that are oriented more or less horizontally can be exposed on very thin nylon threads or brass gauze fixed horizontally at the original position of the samples. To minimize water loss, the plants should be transferred in aluminium foils or polyethylene bags if the balance is far from the site of exposure. However, this is undesirable and the balance should be as close as possible. When exposing the plants it must be borne in mind that the equivalent evaporative power both outside and inside the stand changes considerably with the height above the ground (SLAVÍK 1960). The exposure should therefore be at exactly the original height. It is necessary to transfer and expose the leaves very carefully when working with dew-covered plants.

HUBER (1927) suggested a compensation technique for weighing. The pointer of the torsion balance is set from 10 to 20 mg or less after the first weighing and the time necessary to reach zero is measured by a stop watch. This procedure reduces the exposure time considerably, but the plant part remains on the balance all the time, so its is not under natural conditions. In addition, since the exposure times vary, variable errors are caused by the decrease in the transpiration rate. A similar compensation procedure may also be used with medicinal balances, suitable for heavier samples used to decrease the biological variability (*e.g.* simultaneous weighing of several leaves) or when working with whole shoots. This compensation procedure is also useful when weighing by means of taring (dry sand, ballotini *etc.*).

The sample is transferred immediately after the first weighing (time zero) to the exposure position, when measuring by means of a non-compensation procedure.

The sample is put back on the balance just in time for the next weighing, which can be made more quickly by previously setting the balance to an estimated value. If time is not critical, a third weighing can be made after a further exposure, making it possible to estimate the error caused by the decrease in the transpiration rate with time, or possibly any additional irregularities during any individual measurement. The area and dry weight of the sample are measured.

The balance should be placed in a protective box, which is closed during weighing, and an additional screen can be used to protect the balance from wind and sun.

The method of short-term weighing may be also used for pieces cut out from leaf blades (SLAVÍK 1963b). This makes it possible to localize the transpiration rate within a single leaf blade or to work with small samples when other parameters or processes are measured (*e.g.* water saturation deficit, water potential, photosynthetic rate etc.). Discs cut from the leaf tissues with a cork-borer are suitable material. The determination of the weight loss of the leaf discs is carried out on torsion balances, single discs or series being weighed, depending on the purpose of the experiment. Single discs are exposed on a thin net. For ease of calculation, it is usually assumed that the thickness of the leaf blade is the same in all the leaf samples to be compared. The transpiration rate of two kinds of leaf discs with a different ratio between the intact surface area of leaf tissues and the cut surface area is determined. For example, whole discs of 15.2 mm diameter were compared with annuli made from the same-sized discs by cutting out a central disc of 8.2 mm diameter. Discs and annuli cut from the blade were weighed on a torsion balance with an accuracy of 0.1 mg before and after free exposure for 2 to 5 minutes. A disc 15.2 mm in diameter had an intact transpiration area (upper and lower) of $2 \times$ $\times$ 1.815 cm^2 and the length of the cut edges (perimeter) is 4.77 cm. An annulus with diameters 15.2 mm and 8.2 mm had an intact transpiration area (upper and lower) of $2 (1.815 - 0.528)$ cm^2, *i.e.* 2.574 cm^2 and cut edges of $4.77 + 2.57 =$ $= 7.34$ cm. If a is the transpiration rate of the intact leaf surface in mg cm^{-2} and b the water loss in mg per cm of cut edge, then the total gravimetrically determined water loss from the whole disc $W_d = 3.63\ a + 4.77\ b$ and total water loss from the annulus $W_a = 2.57\ a + 7.34\ b$. From these measurements a and b could be calculated as follows:

$$a = \frac{W_d - 4.77\ b}{3.63} \quad \text{and} \quad b = \frac{W_a - 0.71\ W_d}{3.96}. \qquad (5.7)$$

5.5.2.2 Short-term Gasometric Measurements

An interesting gasometric method suitable for measuring the instantaneous transpiration rate in field conditions was designed by GRIEVE and WENT (1965). An electric humidity sensor is used to measure a short-term change in relative humidity of the air in a transpiration chamber (Fig. 5.10) which contains a part of an intact plant *in situ* for a short time. The chamber is fixed on a holder

(1) and has double Perspex walls. By means of an iris diaphragm with a large diameter (2) it is possible to enclose a whole plant, branch or leaf in the chamber (by constricting the plant stem or the leaf petiole). The motor (3) drives the fan (4) which provides rapid and effective mixing of the air in the chamber. Any suitably sensitive sensor with short response time, *e.g.* a lithium chloride hygrometer

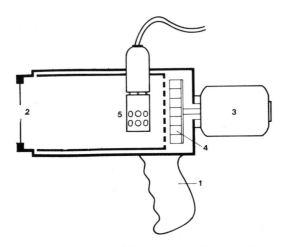

Fig. 5.10 The transpiration chamber for the short-term gasometric measurement of the transpiration rate (GRIEVE and WENT 1960). See the description in the text.

(5) may be used – the authors used the DUNMORE (1938a, b) sensor Aminco. Humidity in the chamber is measured when the plant is enclosed, and every 10 s for the first 30 s. The transpiration rate is obtained from increase in absolute humidity, calculated from the rate of change in relative humidity and the simultaneously determined air temperature. As the sensitivity of the measurement of relative humidity must be as high as possible, it is necessary to choose a sensor with a suitable range. Very short exposure of the plant inside the transpiration chamber reduces overheating of the plant and the air if insolation is high. The measured values correspond to transpiration rates in rapidly moving air as the chamber air must be thoroughly mixed during exposure.

5.5.3 Continuous Measurement of the Transpiration Rate of Intact Plants

5.5.3.1 Gasometric Methods

The principle of most gasometric methods used for continuous determination of transpiration rate is the same: the increase in absolute humidity of the air stream passing over the transpiring plant is measured.

Most of the principle and many of the techniques used in gasometric transpira-

tion measurements are similar or even identical with those in the gasometric method for measuring CO_2 exchange, and the reader will find a very good theoretical
and practical treatment of the methodological problems concerned in JARVIS,
ČATSKÝ et al. 1971. General aspects and principles are reviewed here, and special
techniques for transpiration measurements are described.

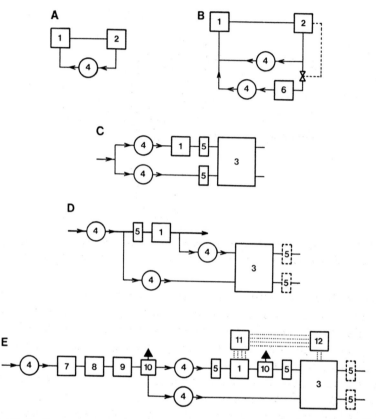

Fig. 5.11 Block diagrams of different types of gasometric systems measuring transpiration
rate and using transpiration chambers. 1 — transpiration chamber, 2 — absolutely measuring
humidity analyser, 3 — differentially measuring humidity analyser, 4 — air pump, 5 — flow
meter, 6 — dehumidifier, 7 — humidifier, 8 — cooler, 9 — thermostat, 10 — blow-off chamber, 11 — light, ventilation, leaf and air temperature control, 12 — data recorder. Principal
components in heavy lines, additional components in broken lines.
A — Example of a closed system. B — Semi-closed system. C to E: Open systems.
C and D: No air conditioning is shown.
C — The difference in humidity of two parallel air streams is measured, one of them going
through the transpiration chamber. D — Difference in air humidity before and after the transpiration chamber in two sample air streams is measured.
E — Example of a more sophisticated system with control of the incoming air humidity, and
temperature, illumination and ventilation of the transpiration chamber. (Adapted from
JARVIS, ČATSKÝ et al. 1971.)

Gasometric Systems

It is possible to use either a closed air circuit with a circulating air stream or an open system in which a unidirectional air stream is analysed and released after passing through the transpiration chamber containing the plant.

In a fully closed system a perfectly airtight transpiration chamber is inserted in a completely airtight circuit with circulating air (Fig. 5.11, A). The rate of increase in absolute humidity, *i.e.* the amount of water vapour in the total volume of circulating air, is measured. Even with a large volume of air enclosed, the air humidity usually increases rapidly, with an effect which is difficult to evaluate. This type of closed system is hence not suitable for measuring transpiration rate and has seldom been used for this purpose.

There are various semi-closed systems which are particularly suitable for CO_2 exchange measurement but less suitable for transpiration. In some of these systems (Fig. 5.11 C) measured volumes of dry air replace partly the air humidified by transpiration. The humidity sensor or water vapour analyser is used as a null point instrument controlling the doses of dry air or the volume of air to be dehumidified (*e.g.* KOLLER and SAMISH 1964).

An alternative procedure uses a bypass where the circulating air is desiccated using a desiccant or a cooled dehumidifier (*e.g.* a Peltier water-vapour trap produced commercially by Siemens, Erlangen, FRG, see also Section 5.2.2). Here the humidity sensor controls the dew temperature to which the bypass air is to be cooled in order to maintain constant air humidity in the main air stream *i.e.* in the chamber (*e.g.* Sirigor gas exchange equipment by Siemens, Erlangen, see KOCH *et al.* 1968, 1971). The transpiration rate is then calculated from the amount of water removed in the bypass, *e.g.* from the difference in absolute humidity of the air in the bypass before and after the water-vapour trap and from the flow rate in the bypass (usually high enough). A great advantage of the above compensation system is that the effective air humidity (see page 263) is maintained at the level of that of the incoming air.

Semi-closed systems are not suitable for detecting and measuring short-term changes or fluctuations in transpiration rate. They require a perfectly airtight system, a condition which is not easily attained.

Open systems are more widely used (Fig. 5.11, C to E). The difference between the water vapour content in the air entering the transpiration chamber and leaving it, or the difference between two parallel air streams, one of which passes through the transpiration chamber while the other serves as reference, is measured. The air is drawn through the transpiration chamber at a constant measured flow rate. The accuracy of the air flow rate measurement is critical, whereas the requirement for airtightness is considerably less important. It is sufficient to make sure that the resistance in the air flow line of the "transpiration" air stream is substantially less than the resistance of possible pathways of leakage from the chamber and tubing, so that the amount of air released or taken up through leakage is negligible compared with the flow rate of the "transpiration" air stream. The transpiration chamber does not have to be completely airtight if the air is pushed rather than pulled through the chamber, or, especially, if the air is pulled through the chamber and the chamber is surrounded by the controlled (reference) air.

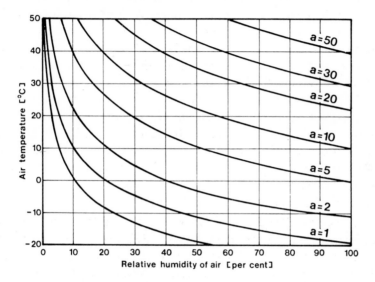

Fig. 5.12 Relationship between relative air humidity and air temperature at various absolute air humidities (a) in mg of water vapour per litre of humid air.

The transpiration rate q_w in an open system is calculated

$$q_w = F \frac{\Delta C_w}{A} \quad [\text{g cm}^{-2} \text{ s}^{-1}], \tag{5.8}$$

where F is air flow rate $[\text{cm}^3 \text{ s}^{-1}]$ corrected for temperature and pressure, ΔC_w increase in absolute humidity of the air due to transpiration $[\text{g cm}^{-3}]$, and A leaf area $[\text{cm}^2]$.

For the relationship between absolute and relative air humidity and temperature of the air see Fig. 5.12, and Tables 6.6 and 6.7 in the Table Appendix.

In open systems — in comparison with closed and some semi-closed systems — there is less trouble arising through differences of atmospheric pressure in the chamber or at the measuring device, which may influence the results. However, there may still be variations in pressure depending on the location of the pumps and the major resistance in the air line. The air in both the transpiration chamber and at the sensor should be as close to atmospheric pressure as possible, since a decrease in atmosphere pressure results in a decrease in partial water vapour pressure and this can cause appreciable error in the estimate of transpiration rate.

Transpiration Chambers

Gasometric procedures require that the plant or part of a plant is enclosed in a chamber, which is more or less isolated from the outside atmosphere. As chambers are usually designed for an attached part of the plant, some difficulties arise — especially in type D air circuits (Fig. 5.11, D) — in sealing the petiole of the leaf or the stem of a shoot of an intact plant. WOLF et al. (1969) suggested that the air inlet through which the air is pumped into the chamber should be at the slot for the petiole so that if there are any leaks there it is only the incoming air that escapes.

The chamber must allow transmittance of the desired irradiation from the natural or artificial source used. If it is made of a material with low transmittance to infrared radiation (e.g. glass or acrylic plastic) the chamber may act as a trap for secondary infrared reradiation from the leaf and the other walls of the chamber, becoming an additional source of infrared radiation and of sensible heat, so increasing the leaf temperature and the temperature of inside air. (This is the so-called "glasshouse effect", or "cuvette effect"). This undesired heat load on the plant may be eliminated to a large extent by using thin film coated polypropylene (Propathane C) instead of acrylic plastic. When no simultaneous CO_2 exchange measurements are made, polyethylene may be used, which has a similarly favourable high infrared transmittance combined with low absorptance and reflectance, but high permeability to CO_2.

Two other measures are in current use for eliminating these undesired effects in the chamber. The window and walls of the chamber (usually made of Perspex) can be artificially cooled either by cool air, by inner Peltier cooling or by a water jacket containing flowing cooled and thermostatistically controlled water (or other coolant). The incident radiation can be filtered either by the water layers alone or by glass filters absorbing infrared radiation (e.g. BG 17 and BG 19 by Schott and Gen., Mainz or VEB Schott, Jena). These filters must be cooled, for example by flowing water or air.

The chamber must be so designed that its size, shape, and the position and diameters of the air inlet(s) and outlet(s) ensure a continuous and regular flow of air past all the transpiring plant surface. If there were no supplementary mixing in the chamber there would be only limited turbulent mixing, so that the boundary layer of still air would be thick and its resistance (r_a — see Section 5.5.6.1) very high. This would not only reduce the transpiration rate but also very substantially reduce the sensible heat transfer from the leaf to the passing air, bringing about an increase in the leaf temperature. Furthermore, there would be large gradients of humidity in the chamber, causing large variations in transpiration rate.

It is much more satisfactory to stir the air inside the chamber, e.g. by a propeller or a small fan. It has been shown (JARVIS 1971) and experimentally proved (TRANQUILLINI 1964) that as the boundary layer resistance decreases, the difference between the temperature of a freely exposed leaf and a leaf enclosed in the chamber decreases, so that appropriate ventilation can considerably alleviate any increase in leaf temperature.

Various types of chambers have been designed: for parts of flat leaf blades, for complete leaves, for parts of both surfaces of a leaf separately, for narrow monocotyledonous leaves, for tree branches, for whole plants or pieces of the plant cover, and for large shrubs or even tree crowns. The bigger the chamber the more difficult it is to ensure the thorough mixing which is essential. Even with adequate stirring, it is necessary to ensure rapid mixing in of the incoming air.

Internal stirring also largely solves another difficult problem, that of the lack of homogeneity of the air humidity in the chamber and the definition of the so-called "effective humidity" or the "mean" humidity surrounding the surface of the transpiring plant. For an unstirred chamber, GAASTRA (1959) and AVERY (1966) concluded that in the estimation of CO_2 assimilation, which is analogous to transpiration in this respect, the mean CO_2 concentration is

$$\bar{C}_a \approx \frac{C_r - C_s}{\ln C_r - \ln C_s} \text{ rather than } \bar{C}_a = \frac{C_r + C_s}{2}, \qquad (5.9)$$
$$(5.10)$$

where $\bar{C}_a$ is the mean (and "effective") ambient CO_2 concentration, C_r the reference CO_2 concentration and C_s the concentration of the outcoming air (sample air). This mean concentration in an unstirred chamber depends on the air flow rate through the chamber in relation to the leaf surface, on the geometry of the chamber and on the position of the leaf itself. Suitable internal stirring of the chamber air independently from the air flow through the chamber ensures a uniform humidity through the chamber at the transpiring surfaces. This air humidity is that of the outgoing air, which is measured and compared with the humidity of the reference incoming air. This "effective" humidity is of course dependent on the incoming air humidity, the amount of transpired water and the flow rate. Thus, if a particular, definite air humidity is required (e.g. a definite water vapour pressure at a definite air and/or leaf temperature), the humidity of the incoming air and/or the flow rate are changed until the desired humidity of the outgoing air is achieved.

On the other hand, the artificial ventilation of the chamber, together with the air flow rate through the chamber to a lesser degree, results in ventilation of the leaves which is artificial in comparison to the natural conditions. Even a high air flow rate alone generally results in insufficient ventilation and a high boundary layer resistance. In an unstirred chamber, the boundary layer resistance changes in relation to the air flow rate. In a well-stirred chamber, the boundary layer resistance is reduced to a small, more or less constant value (e.g. 0.1 or 0.2 s cm^{-1}) which is not affected by any change in air flow rate through the chamber. This may be advantageous in many respects. In a well-stirred chamber, the transpiration rate is independent of the air flow rate.

A very instructive and largely exhaustive description of examples of the main types of assimilation chambers was compiled by JARVIS, ČATSKÝ et al. in ŠESTÁK *et al.* 1971. As the design and construction requirements are similar for transpiration chambers, no special review and description of transpiration chambers is included here; instead the reader is referred to the manual edited by ŠESTÁK *et al.* (1971).

Other Components of Gasometric Systems

Very accurate measurement of the air flow rate through the transpiration chamber is necessary: the accuracy of gasometric measurements of transpiration rate is generally limited by the accuracy of the air flow rate measurements. With leaf chambers, flow rates of 50 to $100\,l\,h^{-1}$ are usual for leaf surfaces up to $20\,cm^2$.

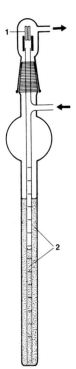

Fig. 5.13 The capillary flow meter with the exchangeable capillary (1) (after removing the ground joint cap). The flow is measured according to the decrease of the level of ethylene glycol (2) in the central tube (SLAVÍK and ČATSKÝ 1965). Eudiometry (Fig. 5.14) serves for calibration.

For larger plant surface areas, air flow rates of about 100 to $300\,l\,h^{-1}$ or more are necessary. Rotameters are the most commonly used type of flow meters. Rotameters with various ranges can be obtained commercially. Reliable types of capillary flow meters (see Figs. 5.13 and 5.14) have been found to be a useful alternative to rotameters. All flow meters used should be carefully recalibrated in position in the system (see ŠESTÁK *et al.* 1971).

Tubing. Glass, plastic (polyethylene, tygon, PVC, nylon) or copper tubing is most commonly used. Polypropylene and butyl rubber tubing have low water adsorption properties, and polyethylene to a lesser extent. Polyethylene, PVC and tygon are permeable to CO_2, so that they are of limited use for simultaneous CO_2

determinations. Copper tubing may be made permanently flexible by previous heating to a dull red. A very serious difficulty, especially in transpiration rate measurements, is water vapour condensation. Condensation in the transpiration chamber and in the tubing is a very serious source of error and must be prevented. The condensation tends to occur frequently in conditions of high relative humidity or high transpiration rate. Cooling down to dew point temperature must be avoided. A very reliable way of preventing condensation is to heat the line by means of a resistance wire. If PVC or polyethylene tubing is used, a chromenickel conductor about 0.5 to 0.75 mm^2 in diameter (resistance about 5 ohm per metre length) is drawn through the tubing (most easily by means of a strong magnet, nail and

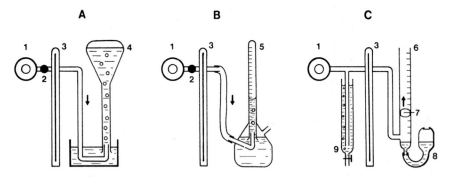

Fig. 5.14 Three various modes used to calibrate capillary flow meters (A,B,C). 1 — (membrane) pump, 2 — needle valve, 3 — capillary flow meter to be calibrated, 4 — calibrated flask used as eudiometer, 5 — all-glass eudiometer (STREBEYKO 1965a), C — soap-bubble flow meter: 6 — measuring burette, 7 — soap bubble, 8 — rubber bulb with soap solution, 9 — air pressure regulator. See also page 315.

a thread). The wire is then heated with 220 V directly from the mains supply. For a short line, which would have a total resistance of less than 200 ohm (the resistance wires of all lengths of tubing are arranged in parallel), a resistance is added to the circuit to give a total resistance of 200 ohm, in order to obtain a final input of 5 W per metre of tubing; this ensures heating to several degrees above the surrounding air temperature, so preventing condensation (KOCH 1957). The outlet of the heating wire from the polyethylene tube is shown in Fig. 5.15 (according to KOCH 1957). Care should be taken to avoid short circuits during rainfall in field conditions.

The tubes may also be heated by means of heating tapes or by using a jacket of water from an ultrathermostat circulating through wider tubes adjacent to the tubes through which the air flows. It must be emphasized that any temperature decrease below the dew point in the air line between the chamber and the sensor measuring the air humidity will cause condensation and very unreliable results.

The length of the pathway from the chambers to the instrument determines the response time: the shorter the distance, the faster the response.

Pumps play important roles in gasometric systems: (1) They push or suck the air through the air conditioning devices — humidifiers, thermostatic baths, dehumidifiers and driers — in which there are often significant resistances to air flow, in which case pushing is to be preferred. (2) The same or additional pumps are used to push (this is preferable to sucking — see Section on transpiration chambers) the air-conditioned air through the transpiration chamber and through the parallel tubing for the reference air sample. (3) A third pump for each stream of air is used after the blow-off on the transpiration chamber outlet and before a rotameter and

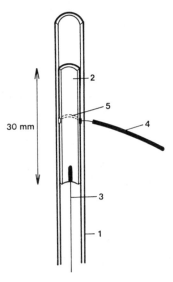

Fig. 5.15 Section through the end of heating in the tube (1) made from softened PVC. 2 — copper tube 30 mm long tightly inserted into the tubing. The end of the heating wire (3) is soldered to it. The inlet wire (4) is soldered to a groove (5) turned into the tube from the outside and pulled through the wall of the tubing (KOCH 1957).

the humidity analyser. This ensures a constant flow rate and a small excess pressure through the analyser. When more than one pump is used for each stream of air, every section is separated by blow-off chambers to atmosphere pressure at the end of the previous section. Pumps in positions other than at the very beginning (i.e. before air conditioning) or the very end (after the humidity analyser, and placing pumps in this position should be avoided) must be leak-proof and (with the exception of those at the end) must not contaminate the air with oil, as often happens with piston pumps. In most gasometric systems, membrane pumps and, more recently, jet pumps are used, usually with a flow rate of 30 to $250 \, l \, h^{-1}$ or more. In closed systems, peristaltic pumps can be useful.

Leaf and air temperature in the chamber must not only be controlled but must also be measured and recorded. For measuring leaf temperature, thermistor beads or copper-constantan thermocouple junctions are usually pressed against the leaf surface, or small needles with the thermistor or thermojunction, or the thermojunctions themselves, are cautiously inserted into the leaf blade or its vein. Unfortunately, we can never make the ideal measurement of the temperature of the transpiring surface itself. This is not only because of the volume, size and heat conductivity of the temperature sensor and of the leaf but also because it is the temperature of the liquid/air interface within the leaf which is relevant from the point of view of transpiration. The care necessary in measurement and evaluation of leaf temperature depends on the particular problem under investigation by gasometric measurement of the transpiration rate. There is a very instructive recent review of theory and techniques involved by PERRIER (1971).

Infrared Analysers

Determination of humidity by infrared analyser (IRGA) is a very precise, modern gasometric method of measurement of transpiration: earlier papers include SCARTH et al. 1948, HUBER and MILLER 1954, DECKER and WETZEL 1957, DECKER and WIEN 1960, POLSTER and FUCHS 1960, POLSTER et al. 1961, DECKER et al. 1962. The principle, construction and systems used have been described in detail by JANÁČ et al. 1971, with special reference to IRGAs used for CO_2 analysis. Only the most important features of IRGAs for determining water vapour will therefore be mentioned here.

In comparison with other methods of water vapour analysis, IRGA has much greater resolution of very small differences. (Some makes of water vapour IRGAs are, however, very sensitive to fluctuations in background concentration of water vapour.) The main disadvantages are the high price and the rather difficult, but necessary frequent calibration.

Other parameters (response time, stability, repeatability etc.) are more or less comparable with other methods of water vapour analysis, and depend on the construction of the IRGA, maintenance, system used, etc.

Infrared analysers with positive filtration (Figs. 5.16 and 5.17) are commonly used for the analysis of water vapour in the air as well as CO_2, CO, SO_2 etc. The infrared analyser measures the difference between the density of the gas, i.e. the difference in absolute humidity of two air streams. This makes it possible to make measurements of two types: (1) Measurement of the absolute concentration when the air in the measuring tube is compared with an inert gas or with a gas containing a known concentration of water vapour in the sealed reference tube. (2) Measurements in which the difference between the concentration of the gases in the two tubes is measured (i.e. differential measurement).

Infrared analysers produced at present by a number of manufacturers (see list in JANÁČ et al. 1971), usually have a minimum range for H_2O of $0\ldots0.5$ g m^{-3} and a maximum of $0\ldots$ saturation. It is also possible, by suppressing the zero, to obtain greater sensitivity over a small part of the range, e.g. $10\ldots20$ g m^{-3} is useful for measuring transpiration rate.

In some instances the air analysed may contain a gas having some absorption bands in common with those of water vapour (a so-called interfering gas; Fig. 5.18). CO_2 is an example arising in plant physiological studies. However, its effect may be neglected at the usual concentration of up to $1\ 000\ \mu l\ l^{-1}$. At higher concentrations, the so-called filter tube (Fig. 5.16 left) is filled with the interfering gas. The filter tube is common to both radiation sources and gas tubes. However, this technique of eliminating interference decreases the sensitivity of the analyser so that it is preferable to determine and allow for the effect of the interfering gas experimentally.

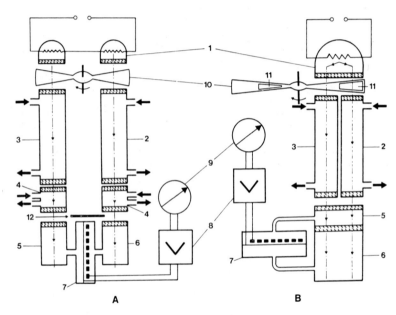

Fig. 5.16 Construction of infra-red gas analyser with detector absorption chambers in parallel (A) and in series (B).

1 — source of radiation, 2 — reference tube, 3 — sample tube, 4 — filter tube, 5,6 — absorption chamber of the detector (7), 8 — amplifier, 9 — galvanometer or recorder, 10 — rotational shutter, 11 — openings in the shutter, 12 — optical shutter for setting zero.

At present, type A is most frequently used for water vapour determination (see JANÁČ et al. 1971 for details): A non-selective source of infrared radiation (1) is a basic component of the analyser; the radiation passes through a reference tube (2) usually containing dry nitrogen and through a sample tube (3) containing the air to be analysed. Water vapour contained in the analysed air absorbs infrared radiation of certain wavelengths (namely around 6 and 3 μm, Fig. 5.18) to an extent corresponding to its concentration. The other radiation components pass through the tube (3) and enter one chamber of a detector (6). The radiation passing through the reference tube (2) at a non-reduced flux density within the whole range of the spectrum then enters the other, control, chamber (5) of the detector. The radiation of wavelengths absorbed by water vapour is selectively absorbed in both chambers of the detector. The gas contained in the detector absorbs radiation in certain specific absorption bands so

Fig. 5.17 The lower part of the infrared analyser URAS 2 (Hartmann & Braun, Frank-furt/Main) with horizontally arranged radiation sources (1), measuring tubes covered with a metal jacket (2) and detector (3) with a preamplifier (4).

that the detector does not respond to radiation in other wavelengths, absorbed by other gases. The detector of a water vapour analyser should therefore be filled with water vapour. However, this is impractical as changes in the volume concentration of water vapour, and condensation of water on the walls of the detector chamber cannot be prevented if the temperature changes. Another gas, *e.g.* ammonia (NH_3), with similar absorption bands to water (6.1 to 6.3 μm) (Fig. 5.18), is used instead. A pressure difference develops between the two chambers of the detector because of the different amounts of absorbed energy. The difference corresponding to the concentration of water vapour is converted to an electrical signal by means of a membrane condenser (7) between the two chambers of the detector and after amplification (8) is measured or recorded with a galvanometer or recorder (9).

The analyser of type B has been developed only recently (LUFT *et al.* 1967, produced *e.g.* by Maihak A. G., Hamburg). It has greater stability and better selectivity of measurement than the type A. The main feature of this analyser is that the radiation is absorbed by two successive layers of gas in the detector. The measurement is based on the fact that the gas absorbs radiation more intensively in the centre of the spectral band than at its tails (see JANÁČ *et al.* 1971 for details).

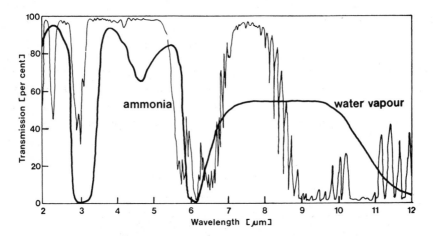

Fig. 5.18 Infrared transmission spectra of water vapour and ammonia.

Psychrometric Methods

The principles of psychrometric measurement of air humidity are described in Section 5.2.

The psychrometric device shown in Fig. 5.19 for gasometric measurement of transpiration was described by VAN OORSCHOT and BELKSMA (1961). SLATYER and BIERHUIZEN (1964) and BIERHUIZEN and SLATYER (1964) constructed a differential psychrometer for continuous measurement of transpiration rate, on the basis of a paper by WYLIE (1963). The device consists of two wet thermocouples. Air passes over one of them before entering the transpiration chamber and air leaving the transpiration chamber passes over the other. Fig. 5.20 shows a cross section of one of the two units drilled in a block of acrylic plastic. As the temperature (T) of both air streams is equilibrated in a copper coil in the water bath prior to passing over the wet thermocouples, the rates of ventilation of the two thermocouples are identical, and the pressure differences negligible, the psychrometric constant A depends on the barometric pressure and is identical for the two psychrometers. The following equation, using the same symbols as equation 5.5 then applies to both psychrometers:

$$p_1^0 - p_1 = \gamma b(T - T_1') \text{ and } p_2^0 - p_2 = \gamma b(T - T_2') \tag{5.11}$$

(subscript 1 refers to air before it enters the transpiration chamber, subscript 2 to air leaving the transpiration chamber b is expressed here as fraction of the normal atmospheric pressure for which γ is calculated). Subtraction of the two equations and substitution of Δp for $p_1 - p_2$ then yields

$$\Delta p = \gamma b(T_1' - T_2') + (p_1^0 - p_2^0) \tag{5.12}$$

and after rearrangement

$$\Delta p = (\gamma b T'_1 + p^0_1) - (\gamma b T'_2 + p^0_2).$$ (5.13)

These two equations demonstrate two ways of calculating Δp. Using equation 5.12 the difference in wet bulb temperature $(T'_1 - T'_2)$ is measured and values of p^0_1 and p^0_2 corresponding to these measured temperatures are found in meteorological tables. When using equation 5.13 it is possible to calculate Δp directly for different

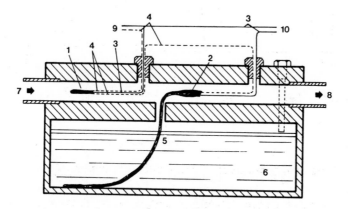

Fig. 5.19 Psychrometer sensor for measuring the transpiration rate according to VAN OORSCHOT and BELKSMA (1961). 1 — "dry" thermocouple junction, 2 — "wet" thermocouple junction, 3 — copper wires, 4 — constantan wires, 5 — thread for wetting the wet junction, 6 — distilled water reservoir, 7 — air inlet, 8 — air outlet, 9 — outlet wires for air temperature measurement, 10 — outlet wires for measuring the psychrometric difference.

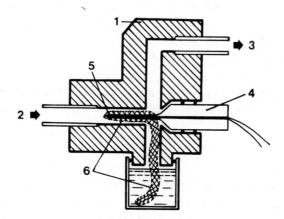

Fig. 5.20 A scheme of the differential psychrometer (SLATYER and BIERHUIZEN (1964). 1 — Perspex block, 2 — air inlet, 3 — air outlet, 4 — insertion plug sealed with O-rings and holding the thermocouple (5) wetted with a wetting wick. The whole described device should be visualized as doubled, drilled out and arranged in parallel in the same Perspex block.

values of T' by means of previously prepared tables for $(\gamma b T' + p^0)$. The difference in absolute humidities $(a_1 - a_2)$ (in mg H_2O per litre air) may then be calculated from the following equation:

$$a_1 - a_2 = \varrho \, \frac{\Delta p}{p_T^0}, \qquad (5.14)$$

where ϱ is the density of water vapour in saturated air at temperature T, and p_T^0 is the saturation water vapour pressure at temperature T, i.e. at the bath temperature.

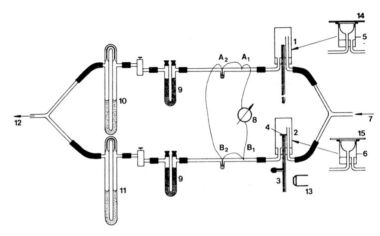

Fig. 5.21 Scheme of GLOVER's method for measuring the transpiration rate psychrometrically (GLOVER 1941). 1 and 2 — calibration chambers, a calibration capillary (3) is localized in the lower one, 4 — filter paper disc, 5 and 6 — two leaf chambers; one (6) can be attached to a leaf (15), the other (5) is sealed to a plastic plate (14), 7 — common air inlet, 8 — galvanometer, 9 — drying tubes, 10 and 11 — capillary flow meters, 12 — exhausting outlet, 13 — horizontal microscope, 14 — plastic plate, 15 — leaf, A and B — thermocouples, (A_1 and B_1 — dry junctions, A_2 and B_2 — wet junctions).
The movement of the meniscus of the water column is controlled with a horizontal microscope when calibrating the device. Two leaf chambers (5 and 6) are connected instead of the chambers 1 and 2 during the measurement proper. One chamber (6) is attached to a leaf (15), the other (5) is sealed to a plastic plate (14).

The thermocouple shown in Fig. 5.20 is a copper/constantan element with a tip protected by dissolved Perspex or Araldite and projecting 2 to 3 mm into the air path, which changes its direction at this point, so that good ventilation of the wet junction is ensured. The constantan wires of both junctions are connected together, so that they measure $(T_1' - T_2')$, one being connected by means of another constantan wire to a thermojunction placed in the constant temperature water bath or the air path just before the wet bulb. This arrangement makes it possible to measure $(T - T_1)$ and calculate $(T - T_2)$ or vice versa. Such psychro-

meters are sufficiently sensitive even at flow rates of less than 0.25 l per minute and
have a response time of less than 1 minute.

Simpler pychrometric methods for determination of transpiration rate have been
described by KATO et al. (1960) and KATO (1967). They enclosed whole plants in
transpiration chambers and used a high flow rate. THOMAS and HILL (1937) measured evapo-transpiration of stands in big chambers in a similar way.

Another type of psychrometric device was suggested by GLOVER (1941) for accurate measurement of the transpiration rate of a single leaf. In this method (Fig.
5.21) air coming from a common source passes as a single stream controlled by two

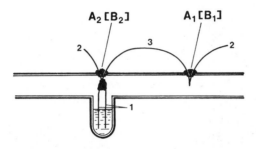

Fig. 5.22 A detailed scheme of the psychrometric thermocouple of the GLOVER's apparatus.
A_1, B_1 and A_2, B_2 are the "dry" and "wet" junctions, respectively. The wet junctions are
wetted with a thread (1), 2 — copper wires, 3 — constantan wires.

capillary flow meters through two narrow parallel tubes. There is a "dry" constan-
tan/copper thermocouple junction and a "wet" junction (wetted from a short side
tube with a thread wound around the connection) (Fig. 5.22) in each tube. All four
thermocouples (A_1, A_2, B_1, B_2) are connected in such a way that the e.m.f. between the pair in one tube is compensated for by the e.m.f. of the second pair in
the other tube, resulting in zero current. If the air stream in the two tubes is identical, the zero setting is to a considerable extent independent of the variations in
temperature and humidity of the air passing. However, as it is practically impossible to measure the actual rate of flow of air round the wet junctions and to control
their wetting, the flow rate in one tube is adjusted with a needle valve in such
a way that zero galvanometer deflection is reached in the second tube at a given
flow rate. Any difference in humidity between the air passing through one tube
and that passing through the other then causes a galvanometer deflection proportional to the difference.

The system is calibrated using an interesting procedure. A calibration chamber
(1 and 2, Fig. 5.21) closed with a rubber stopper is attached to both branches, instead of transpiration chambers (5 and 6). A T-shaped capillary (3) filled with
water leads to one of them (2). Water wets a small piece of filter paper (4) placed
on the upper open end of the capillary entering the chamber. The bottom meniscus

of the water column in the calibrated capillary rises during evaporation of water from the filter paper. This movement is observed by means of a horizontal microscope, so that the volume of water evaporated during a given time in one of the parallel air streams passing through the two tubes is known. The galvanometer deflection corresponding to this amount of water is determined; this is almost completely independent of the relative humidity and thus also of the temperature of air entering the common inlet of the system.

ABOU RAYA (1955) described a modification of GLOVER's method for measuring the transpiration separately from each surface of a single leaf.

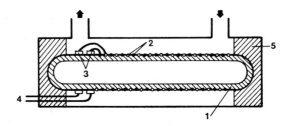

Fig. 5.23 Chamber with lithium chloride sensor by SPRUIT (GAASTRA 1959). 1 — glass tube sealed at both ends, 2 — platinum wires, 3 — brass rings, 4 — inlet wires of the measuring device, 5 — rubber stoppers.

Lithium Chloride and Other Resistance Hygrometers

Most are based on the fact that the electric resistance of a film of lithium chloride (LiCl) is a function of the water vapour pressure of the air in equilibrium with the film. Numerous instruments of this type for the measurement and recording of air humidity have been designed (DUNMORE 1938a, b, STEIGER 1951, BRASTAD and BORCHARDT 1953, LIENEWEG 1955, TANNER and SUOMI 1956, SPRUIT in GAASTRA 1959). Generally they consists of a plastic base fitted with electrodes and coated with a lithium chloride solution in a binding mixture. Some are commercially available, e.g. humidity sensor Aminco by DUNMORE 1938 is now produced by Hygrodynamics Inc., Silver Spring, Maryland, U.S.A. For a list see page 279.

The sensor designed by SPRUIT (GAASTRA 1959) (Fig. 5.23) consists of two platinum wires, 120 mm long and 0.1 mm thick wound in parallel around a glass tube 10 mm in diameter, sealed at one end. The tube is heated to soften the glass and slightly widened by blowing until the wires are partially sealed in the outer surface of the tube. The other end of the tube is then also sealed and a small amount of 0.1N LiCl solution (4.238 g in 100 ml solution) is applied to its surface and left to dry. The ends of the conductors are then soldered to a single brass ring connected to the inlet of the measuring apparatus.

The humidity sensor is enclosed in a glass chamber with air inlet and outlet (Fig. 5.23) which is placed in a constant temperature bath, at a temperature several degrees higher than the air in the transpiration chamber during measurement. The incoming air is heated to this higher temperature by passing through a copper tube in the measuring chamber prior to entering the chamber containing the sensor. This heating is necessary because the accuracy of the sensor decreases at high relative humidities; the relative humidity decreases with increasing temperature, if the same absolute humidity is maintained. The instrument is calibrated with air of known relative humidity and temperature. The error of the humidity measurement

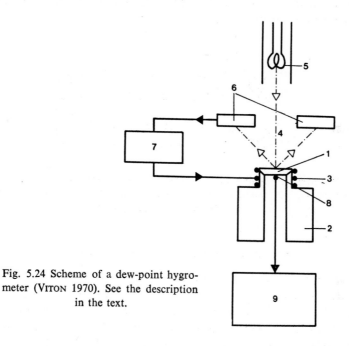

Fig. 5.24 Scheme of a dew-point hygrometer (VITON 1970). See the description
in the text.

is ± 1%, which corresponds to about ± 2% error when measuring the transpiration rate as the difference between the humidity of the control and transpiration air streams. The error increases up to ± 7% at low transpiration rates.

For example of a conductimeter to be used for lithium chloride sensors of this type see Fig. 5.34 (Page 304).

JONES and WEXLER (1960) described an electronic hygrometer element using a thin film of barium fluoride over metallic film electrodes on a glass substrate. The electric impedance of the element varies by about three orders of magnitude with the relative humidity within the range of 20 to 100% r.h., with a very short response time of a few seconds.

A sulphonated polystyrene hygrometer with a rapid response, and other types of electric hygrometers, were described by WEXLER (1957).

A very small hygrometer fitting inside the tip of a hypodermic needle was constructed by BRASTAD and BORCHARDT (1953).

Another type of lithium chloride humidity sensor consists of a plantinum wire thermosensor which is enclosed in a fibre glass envelope soaked with lithium chloride solution. Two wires are wound in parallel around it. The electric current between these two wires is controlled automatically by changes of lithium chloride concentration resulting from water vapour exchange between bypassing air flow and solution. At a point where no net exchange of water vapour takes place (at about 10% relative humidity at the surface of the sensor) the heating by the electric current ceases due to low conductivity of dry LiCl and a sensor temperature is reached corresponding to the absolute humidity of the bypassing air (the higher the humidity the higher the temperature). This temperature [in the range between about 36 °C and 65 °C for the dew point range 0 to 20 °C, respectively] is measured with the Pt-thermosensor and recorded.

Dew-point Hygrometers

Modern automatic versions of the old type dew-point hygrometer have been designed recently and are now commercially produced. The principle is the following (Fig. 5.24): The stream of air to be measured is passed through a chamber containing a fine metal mirror (1) which can be cooled by the Peltier effect or, alternatively, heated by a resistance wire (3) [block (2)] or the air. A light beam (4) from an incandescent lamp (5) falls on the mirror and is diffused so that light falls on a photocell photoresistor (6) if dew forms on the mirror. The signal from the photocell is amplified (7) and used to control the cooling or heating of the mirror, so that the temperature of the mirror is maintained at the dew-point temperature. This

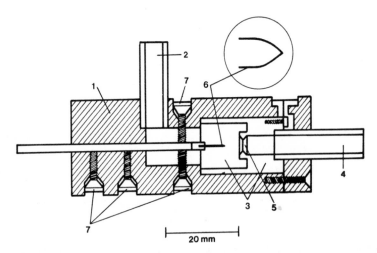

Fig. 5.25 Diagram of the corona chamber (ANDERSSON and HERTZ 1955). 1 — Perspex block, 2 — air inlet, 3 — brass chamber, the inside of which is connected with a glass tube air outlet (4) by means of a hole (5), 6 — platinum needle, the enlarged tip of which is shown on the upper right, 7 — screws for adjusting the position of the platinum needle.

temperature is measured by a thermocouple or thermistor (8) and recorded (9).

A great advantage of this instrument is that it needs no calibration. The dew-point temperature is measured directly and the corresponding air humidity or vapour pressure is found from tables. It covers a very wide range of humidities (dew-points of -40 to $+30\,°C$). The accuracy of the result depends on the precision of the measurement of the surface temperature of the mirror, and on the extent to which the control system causes this to fluctuate, $i.e.$ whether it is one, two or three term controller.

Commercial dew-point hygrometers are produced by Salford Electrical Instruments Ltd., Manchester, United Kingdom, by Cambridge Instruments Inc., Newton, Massachusetts, U.S.A., by Aqmel, Pau, France (VITON 1970).

Corona Hygrometer

The use of the corona hygrometer is a very sensitive method for determining transpiration rate by measuring relative humidity (ANDERSSON et al. 1954, ANDERSSON and HERTZ 1955). The corona discharge from a positively charged tip (direct voltage about 3 000 V) towards a negatively charged area depends on the air humidity. The frequency of discharges between specially adjusted poles in a chamber through which air containing the transpired water passes is measured.

The corona chamber is shown in Fig. 5.25. Air is supplied to a cylindrical Perspex chamber with a centrally located platinum needle (6) 0.3 mm in diameter sharpened as shown in the Figure. The shape and properties of the tip are critical and, along with the geometry of the chamber, influence the discharges. The

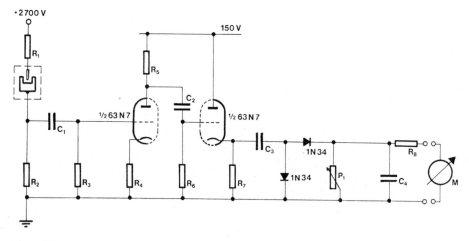

Fig. 5.26 Diagram of the electronics of the corona hygrometer (ANDERSSON et al. 1954). Value of the protective resistance R_1 in the series with the corona is not essential as long as it remains higher than 50 MΩ. $R_2 = 50\,k\Omega$, $R_3 = 300\,k\Omega$, $R_4 = 10\,k\Omega$, $R_5 = 500\,k\Omega$, $R_6 = 30\,k\Omega$, $R_7 = 1\,k\Omega$, $R_8 = 15\,k\Omega$, $C_1 = 100\,pF$, $C_2 = 220\,pF$, $C_3 = 0.1\,\mu F$, $C_4 = 0.02\,\mu F$, M — recorder, $P_1 = 50\,k\Omega$.

authors point out that some chambers do not function well because of the unsuitable shape or surface of the platinum tip. A brass ring (3) and (5) forming the negative pole is located opposite the needle. The corona depends on the voltage and the range of measurement may be changed to a certain extent by changing the voltage. Discharge of a current of 1 to 5 μA originates at 3 000 V when the distance between the tip and the brass plate is 3 mm. The discharge is directed to the hole (5), so preventing any damage to the plate. It is not so important to be able to change the distance between the tip and the plate.

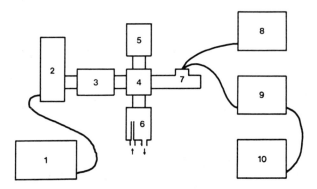

Fig. 5.27 Block diagram of the microwave hygrometer (FALK 1966a). 1 — regulated power supply (Oltronix LS 14 B), 2 — microwave generator: reflex klystron (Philips 2 K 25), 3 — ferrite isolator (Philips PP 4022 X) connecting the klystron with a hybrid tee (4 — Philips PP 4050 X) through a ferrite isolator, 5 and 6 — frequency meters (Sivers SL 5205), one (5) used as a reference cell with the cavity closed, the second (6) used as measuring cell with two holes drilled in the lid: for the measured air stream which is flushing this cell, 7 — adjustable crystal holder with microwave diode (type 1 N 23) giving an output of a low frequency alternating current, 8 — oscilloscope, 9 — differential operational amplifier (G. A. Philbrick Model P 65), rectifier and zero point depressor (for wiring diagrams see the original paper), 10 — pen recorder.

The corona hygrometer is extremely sensitive to particles of dust. It is also sensitive to carbon dioxide concentration. In our experience, this sensitivity is sometimes greater than that to air humidity. All these factors depend primarily on the type of tip and geometry of the discharge. When setting up the corona hygrometer it is therefore necessary to make several chambers and select several well-adjusted tips. The frequency of discharges is measured by an electronic device, shown in Fig. 5.26.

The corona hygrometer responds instantaneously, so there is only a short delay in recording a change in transpiration rate, determined by the air flow rate and the distance between the transpiration chamber and corona chamber. It is very useful for the accurate determination of short-term changes of transpiration rate of intact plants in controlled conditions following application of an experimental treatment. The air may be either pushed or pulled through the system, the rate of air flow is not limited by the procedure itself within the range normally used. However, the behaviour of the corona discharge at very high flow rates is not known.

Microwave Hygrometer

A microwave hygrometer was designed by FALK (1966a) based on the measurement of the dielectric properties of the air by means of microwaves (wavelength about 3 cm). Unlike the corona hygrometer, it is insensitive to dust particles and carbon dioxide. It covers the whole range of relative humidities, its sensitivity is better than 0.05 mm Hg water vapour pressure, and the calibration in terms of water vapour pressure is almost linear. A diagram and description are shown in Fig. 5.27.

List of Manufacturers of Different Commercial Types of Hygrometers

Many different types of hygrometers are now produced commercially by a number of manufactures. Here is a list compiled by P. G. and M. S. JARVIS. It is not intended to be a comprehensive directory of all firms but only to yield some information and to give the addresses:

Aqmel, Pau, France
  Mirror hygrometer cooled by the Peltier effect.
Auriema Ltd. Impectron House, 23/31 King Street, London W. 3, England.
  Dew-point Hygrometer Model 880.
Cambridge Systems Inc., 50 Hunt Street, Newton, Mass. 02158, U.S.A.
  Model 992 Dew-point hygrometer, using a stainless steel mirror cooled by the Peltier effect.
DuPont: Instrument Products Division, E. I. DuPont de Nemours & Co. (Inc.), Wilmington, Delaware 19898, U.S.A.
  DuPont 510 Moisture Analyser.
Elliot Process Automation Ltd., Blackwall Lane, Greenwich, London SE 10, England.
  26—303 Portable moisture monitor. Cell with phosphorus pentoxide.
Foxboro Company, Neponset Avenue, Foxboro, Massachusetts, U.S.A.
  Dew-point recorders.
Hartmann & Braun AG, 6 Frankfurt/Main-West 13, Postfach 1361, BRD.
  Lithium chloride psychrometer, hair hygrometer.
Hygrodynamics Inc., 949 Selim Road, Silver Spring, Maryland 20910, U.S.A.
  Dew-point meters; electronic hygrometer systems based on the hygrosensor, the operation of which is based on the ability of a hygroscopic film to change its electrical resistance instantly with microchanges in moisture condition. Many types of hygrosensor available.
M.E.L. Equipment Company Ltd., Manor Royal, Crawley, Sussex, England.
  Philips Humidity Cell Type PR 6010 H. Continuous dew-point measurement. Lithium chloride.
Minneapolis-Honeywell Regulator Company, Minneapolis 8, Minnesota, U.S.A.
  SSP 129 A—D Dew probe sensors (lithium chloride).
Moisture Control and Measurement Ltd., Thorp Arch Trading Estate, Boston Spa, Yorkshire LS 23 7 BJ, England.
  Hygrometers with sensor consisting of aluminium wires anodised by a special process and formed into a moisture sensing capacitor. All traces of electrolytes are removed.
Molecular Controls Ltd., Scottish Life House, Leeds I, England.
  MCL hygrometer GP-10. Sensor with porous aluminium oxide coating.

Negretti and Zambra Ltd., Stocklake, Aylesbury, Buskinghamshire, England.
   Gregory Hygrometer Element — consisting of a length of very fine fibres wound over platinum clad electrodes and impregnated with a hygroscopic salt.
Panametrics, 221 Crescent St, Waltham, Mass. 02154, U.S.A.
   Aluminium oxide hygrometer Model 1000.
Rotronic AG, Kernstrasse 8, 8004 Zurich, Switzerland.
   Electronic hygrometer, including digital. Sensing elements based on change of resistance of hygroscopic substances.
Salford Electrical Instruments Ltd., Times Mill, Heywood, Lancashire, England.
   Thermoelectric Dew-point Hygrometer.
Shaw Moisture Meters, Rawson Road, Bradford, England.
   Dew-point meters, including digital meters.
Wallac oy, Turku, Finland.
   Ersec air humidity measuring system with lithium chloride dew-point detector cells.
Yellow Springs Instrument Co., Yellow Springs, Ohio 45387, U.S.A.
   Dew-point hygrometer with miniature lithium chloride moisture sensors.

Thermoflux

Thermoflux is an instrument measuring the heat of absorption of the water vapour in air in concentrated sulphuric acid. It was constructed by Badische Anilin und Sodafabrik Company in cooperation with HUBER (HUBER and MILLER 1954). The instrument is based on the following principle. Air is drawn at a known, controlled and measured rate from the transpiration chamber into the instrument through a coil in a water bath. Its temperature is equilibrated exactly with that of a stream of concentrated sulphuric acid in another glass coil line. Both streams (air with water vapour, and concentrated sulphuric acid) are mixed, water vapour is absorbed by the acid and the temperature of the acid increases by the heat of absorption. The rise in temperature is measured by a thermopile composed of 200 constantan/iron thermojunctions. The "cold" junctions measure the temperature of the sulphuric acid before it is mixed with the air; the "hot" junctions measure the temperature of the acid after the mixing. The temperature difference measured with a galvanometer is then proportional to the heat of absorption corresponding to the quantity of water vapour absorbed. Control and transpiration air streams are measured alternately. A closed circuit of sulphuric acid is used as it was found that the heat of absorption is practically independent of the concentration of sulphuric acid (4.068 J per g of water vapour in 100 g of 100% sulphuric acid and 4.021 J in 95% sulphuric acid). The instrument measures the amount of heat absorbed, and therefore the absolute amount of water vapour supplied, per unit time. Sensitivity may be decreased by a "shunt" introduced before the recording galvanometer, increasing the range of measurement. However, it is better not to change the sensitivity but to change the range by shifting the zero point electrically. It is also possible to transfer a known and measured part instead of the whole air stream to the Thermoflux. This is often necessary with the transpiration air stream if transpiration is rapid.

The instrument is relatively costly (it is produced by Hartman and Braun, Frankfurt am Main, at present) but according to data in the literature it is reliable and very accurate.

Gravimetric Absorption Methods

VOTCHAL (1940) described a simple method using gravimetric determination of differences in absolute air humidity between control and transpiration air streams for determination of the average transpiration rate of individual parts of intact plants. Both streams must have exactly the same flow rate. Water vapour is absorbed by a desiccant placed (see Fig. 5.28) as a thin layer on two pieces of net (1)

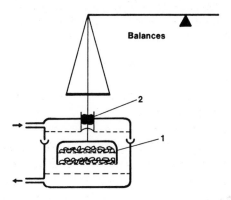

Fig. 5.28 Box with a two-floor net (1) carrying granulated $CaCl_2$ and hanging on one balance arm during continuous determination of transpiration rate by the method according to VOTCHAL (1940). 2 — mercury seal.

hung on the two balance beams of an analytical balance by thin wires. The wires pass through mercury seals in the lids of the boxes enclosing an air-tight space around the absorbers. The boxes (one for the control stream, the other for the "transpiration" stream) have double lids serving for the uptake of the air and double bases through which the air flows out. Decelerations of the flow rate of air in the wide space around the absorbers ensures complete absorption of the water vapour. The instrument operates with a delay determined primarily by the air flow rate, which is limited by the desiccant used. VOTCHAL worked with velocities of 100 or 200 litres per hour.

5.5.3.2 Gravimetric Methods

Gravimetric methods designed for continuous measurement of transpiration rate must include recording. Such devices have been called transpirographs. A detailed summary and description of the older types was given by LEICK (1939).

Recently, a large number of devices have been designed, using load cells, displacement transducers, strain gauges etc. Most of these are more or less similar to lysimeters and so are beyond the scope of our manual.

5.5.4 Measurement of Integrated Transpiration

The recording methods mentioned in Section 5.5.3.2 may of course also be used to measure total amount of water transpired over a fairly long period of time (hour, day or longer). In addition, gravimetric methods, in which large or small plant containers are subjected for a given time to experimental conditions and weighed periodically, may also be used.

Relatively small plant containers may be called small lysimeters. Small lysimeters described by HESSE (1952, 1954a, 1954b etc.) are glass, cylindrical vessels 100 ×
× 280 mm with an outlet tube with a tap at the bottom and a wide ground edge at the top (Fig. 5.29 left). A perforated porcelain plate is placed in the lower third; the upper two thirds are filled with soil the surface of which is covered with a double

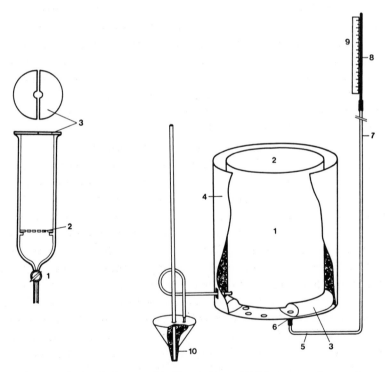

Fig. 5.29 Left: A small glass lysimeter according to HESSE (1952). 1 — tap, 2 — perforated porcelain plate, 3 — double lid.
Right: A simple hydrostatic lysimeter COURTIN and BLISS (1971).
See the description in the text.

lid with a hole for the plant stem. These lysimeters are placed in soil in such a way that the top of the lysimeter is level with the surface of the surrounding soil. If the lysimeters are to be exposed under normal stand conditions, they should be suitably placed in an undisturbed stand of the same plants as those cultivated in the lysimeter. Such small lysimeters were weighed by HESSE at intervals (*e.g.* 30, 90 minutes, 12 or 24 hours), with a stated accuracy of $\pm$ 50 mh. However, this accuracy is somewhat doubtful, because of the likelihood of contamination of the container from the outside.

EIDMANN (1943) used rather larger glass containers, 160 $\times$ 300 mm, with crude crushed quartz for drainage, sealed at the soil surface by a double layer of a sealing mixture (guttapercha : colophony : bees wax : vaseline $-$ 30 : 25 : 10 : 35).

Similar small lysimeters or weighed plant containers may be prepared in many ways. Sealing the soil surface may be achieved more or less perfectly either with a melted mixture of paraffin, bees wax and vaseline, 1 : 1 : 1, or by covering with a sheet of PVC or a thick layer of cotton wool. Ideally, the seal should not change the aerobic conditions of the soil environment significantly. Soil moisture, maintained by periodic additions of water to the container, should therefore not exceed the capillary capacity of the soil. This is ensured in HESSE's small lysimeters by the free space at the bottom and the outlet taps, through which gravitational water can run out after watering. Evaporation from the soil surface may also be roughly compensated for by subtracting the weight loss of control containers without plants.

Automatic shop balances may be used for limited accuracy weighing of five to seven kg containers. The accuracy of weighing, and hence of the estimate of transpiration, decreases with increasing weight of the container unless larger amounts of water are transpired between weighings.

A very interesting simple and inexpensive hand-made type of a small hydrostatic lysimeter for determining integrated evapotranspiration (transpiration) was designed by HANKS and SHAWCROFT (1965) and modified by COURTIN and BLISS (1971). It consisted of double cans (Fig. 5.29 right): the inner can (1) (25 cm in diameter) containing the plant and soil system (2) was resting on a sealed ring made of a part segment of bicycle inner tube (3) (by resplicing it with neoprene cement) made to fit exactly in the bottom of the outer can (4) (29 cm diameter). A copper tubing (5) about 5 mm diameter and 2 m long was soldered onto the valve stem (6) of the bicycle tube and formed a standpipe (7) ending with a glass capillary (8). The tube and tubing was evacuated and then filled with coloured antifreeze to an appropriate level according to the weight of the plant and soil system. Movements of the upper meniscus of the liquid in the capillary were measured by a scale (9). The pressure applied to the bicycle tube by the inner can with soil and plants was equal to the product of the weight of the liquid column and of the density of the liquid. This pressure can be calibrated in terms of water weight. The reported accuracy was 0.25 mm of water. A percolation vessel (10) situated below the lysimeter enabled the drainage of the lysimeter of excess water.

Increase in the dry weight of plants must be taken into consideration in measurements of total transpiration over long periods*. This correction is made either by determination of dry weight of parallel, periodic harvested samples (of aboveground parts and roots) or by interpolation on the basis of the final dry weight at the end of the experiment.

5.5.5 Water Loss by Detached Leaves

The analysis of water loss by detached leaves consists in the determination of the long-term course of water loss by transpiration, by the weight changes of detached plant parts under standard conditions during "artificial wilting". Quantitative, stomatal and cuticular transpiration rates and diffusion resistances may be calculated from the rate of water loss during drying out of the leaves as the stomata close.

The course of water loss is followed under standard conditions which if necessary can be adjusted to be similar to ecological conditions prevailing in the appropriate habitat. By beginning at full water saturation, the method eliminates any non-specific effect of water balance (temporary daily or permanent water deficits, hydroactive closing of stomata), so that different plants or their organs (taxons, eco-types, experimental treatments *etc.*) are compared under the same conditions.

Under the empirically derived assumption that the transpiration rate from detached leaves in constant conditions is exponentially related to the water potential at the transpiring surfaces and that this water potential is proportional to the relative water content, FUKUDA (1935) and HYGEN (1951, 1953) worked out an analysis of water loss curves. When at the beginning of the determination stomata are hydroactively fully open (the weighing starts with previously fully water-saturated leaves) and their aperture has not yet hydroactively changed, the logarithm of the relative water content plotted against time is linear (the so-called stomatal phase). The same is true of the third phase when the stomata are already hydroactively closed (the so-called cuticular phase). The second, transient phase reflects the gradual hydroactive closing of the stomata (the closing phase of the water loss curve) (Fig. 5.30).

The above analysis of water loss curves, described in more details by FUKUDA (1935), HYGEN (1951, 1953) and SLAVÍK (1958a, b) is rather empirical. Nevertheless in controlled conditions (*i.e.* at constant illumination high enough for photoactive opening of stomata, controlled water vapour pressure and temperature of the air, constant ventilation [small wind tunnel] and measured leaf temperature) the determination of water loss of detached leaves may give some data on comparative transpiration rates, mainly on the relationship between transpiration rate and the

---

* Strictly, it is the difference between dry weight increase and the nutrient content taken up from soil which should be added.

relative water content of the plant part. It may serve for estimates of transpiration (stomatal and cuticular) resistance (see Section 5.5.6) and its dependence on relative water content, *i.e.* estimates of the hydrosensitivity of the stomata.

The term water holding capacity may be based on the data derived from the rate of loss of water by detached leaves. In the cuticular phase with hydroactively closed stomata, the transpiration rate is *ceteris paribus* inversely proportional to the cuticular resistance, which may serve as a measure of water holding capacity. CETL (1953, 1957) used the cuticular loss of water from detached leaves after hydroactive closure of the stomata to test the drought resistance of plants.

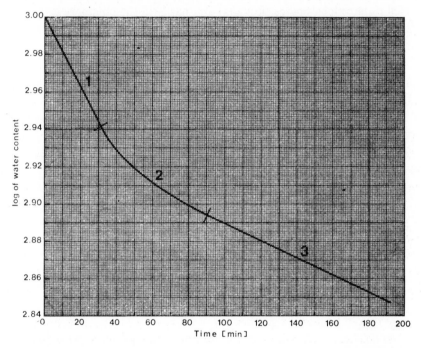

Fig. 5.30 The same example of the water loss curve plotted as the relationship of the log of water content [log $(W - W_{dry})$] with time (t). Ordinate: log of water content where the initial water content at zero time equals 1000. 1 — stomatal phase, 2 — closing phase, 3 — cuticular phase.

ARLAND (1929, later also 1952) and his coworkers measured the total transpiration of detached plant parts during the first thirty minutes after cutting (ARLAND's "wilting method" ["Anwelkmethode"]). These values were used to test agricultural procedures (fertilizing *etc.*). They do not correspond to actual transpiration rates such as the rates measured by the cut-leaf method, and they cover a wide range of stomatal apertures. POLSTER (1951), CETL and PENKA (1958) and others have criticized this method.

5.5.6 Determination of Transpiration Resistance

5.5.6.1 Introduction

SEYBOLD (1929, 1930) was the first to introduce the reciprocal of transpiration rate as a transpiration resistance, measured under standard conditions of evaporation (still air with a standard water vapour gradient). WELTEN (1933) expressed it numerically by an equation similar to equation 5.17.

The diffusional portion of the water vapour transfer in transpiration, *i.e.* the transfer of water vapour by molecular diffusion in still air, is localised partly in the plant itself (intercellular spaces, stomata) and partly in the boundary layer of air at the surface of the plant.

$$r_{total} = r_1 + r_a , \tag{5.15}$$

where r_{total} is the sum of the transpiration resistances, r_1 is the resistance situated in the leaf and r_a is the resistance of the boundary layer.

The transpiration rate or net flux of water vapour transfer from the leaf (q_w in [g cm^{-2} s^{-1}]) may be expressed by the following equation:

$$q_w = - D_{vap} \frac{dC}{dz} , \tag{5.16}$$

where D_{vap} is the molecular diffusivity (diffusion coefficient) of water vapour in still air (0.26 cm^2 s^{-1} at 20 °C), C is water vapour concentration in the air [g cm^{-3}] and z is the length of the segment of the diffusion pathway [cm]. Also:

$$q_w = \frac{C_1 - C_2}{r_w} , \tag{5.17}$$

where C_1 and C_2 are water vapour concentration at each end of a segment [g cm^{-3}] and r_w is the diffusive resistance of the segment [s cm^{-1}].

The total (r_{total}) and leaf (r_1) transpiration resistances and also the resistance of the boundary layer (r_a) may be measured and calculated or estimated by several procedures:

(1) by measurement of steady state water vapour exchange, *e.g.* by steady state gasometric measurements of transpiration rate in controlled conditions (Section 5.5.6.2),

(2) by non-steady state measurements of water vapour exchange, *e.g.* with transpiration porometers (Section 5.6.4.1), or by weighing detached leaves (Section 5.5.5),

(3) by calculation from the leaf energy balance (Section 5.5.6.3),

(4) by crude relative estimations using cobalt chloride paper (Section 5.5.6.4).

The diffusive resistance to water vapour transfer during transpiration are the three serial diffusive resistances involved in stomatal transpiration: the intercellular resistance (r_i) in still intercellular air, the stomatal resistance (r_s) to diffusion through the stomatal pore and the resistance of the boundary layer (r_a) at the surface of the transpiring leaf.

The parallel resistance involved in the cuticular component of transpiration comprises two resistances in series: the cuticular resistance (r_c) and boundary layer resistance (r_a).

If transpiration from both sides (adaxial = "upper" and abaxial = "lower") of a flat leaf lamina is considered separately the two resistances of the two surfaces are in parallel, so that

$$\frac{1}{r_{\text{total}}} = \frac{1}{r^{\text{ad}}} + \frac{1}{r^{\text{ab}}}, \tag{5.18}$$

where r^{ad} and r^{ab} are the total resistances of the adaxial and abaxial sides of the leaf respectively. The resistances of the adaxial and abaxial sides of one and the same leaf, may be different, because of differences in structure or in environmental conditions.

The resistance of the intercellular spaces (r_i) is a function of their shape, geometry and communication with adjacent intercellular spaces. In amphistomatous leaves, r_i may be calculated, for example, from measurements made by gas diffusion porometers (see Section 5.6.4.1), from the equation

$$r_{\text{total}} = r_a^{\text{ad}} + r_s^{\text{ad}} + r_i + r_s^{\text{ab}} + r_a^{\text{ab}}, \tag{5.19}$$

where the superscripts ad and ab denote adaxial and abaxial sides, respectively. [This r_i is double the intercellular resistance involved in transpiration from one side of the leaf only.] It is also possible to make calculations from the dimensions of the spaces in the leaf (JARVIS et al. 1967, MILTHORPE and PENMAN 1967). Estimates by JARVIS et al. for a cotton leaf 200 μm thick give values of r_i of 1.8 s cm^{-1}.

The diffusive resistance of the stomatal pores (r_s) is measured by suitable porometric methods (see Section 5.6.4), making allowance for the resistance of the boundary layer (r_a). The stomatal resistance so measured usually includes the intercellular resistance (r_i). Alternatively, r_s may be calculated from the dimensions and shape of the stomatal pores.

Since r_s and r_i are difficult to determine separately and the cuticular resistance is often much larger, all these are sometimes combined into a single resistance of the leaf (r_l).

Cuticular resistance (r_c) is probably mostly a liquid phase resistance, i.e. a hydraulic resistance in the cuticle and not a resistance to water vapour diffusion. Nevertheless, either assuming a diffusive component in the vapour phase or for convenience, the cuticular resistance is often expressed in terms of a water vapour diffusive resistance, and estimates of 10 to 10^3 s cm^{-1} have been made.

Treating the stomatal (r_s) and cuticular (r_c) resistances as in parallel, and neglecting the intercellular resistance,

$$\frac{1}{r_1} = \frac{1}{r_s} + \frac{1}{r_c}, \tag{5.20}$$

so that

$$r_s = \frac{r_1 r_c}{r_c - r_1}. \tag{5.21}$$

The cuticular resistance may be measured as the total resistance in leaves with artificially closed stomata (by darkness, high carbon dioxide concentration of the ambient air, or application of suitable antitranspirants).

Boundary layer resistance (r_a) is measured as total resistance (r_{total}) of a wet leaf or, more usually, a wet leaf replica (e.g. wet blotting paper), in place of a transpiring leaf (plant surface).

Further data may be found in Section 5.6.4; a detailed description of the theory of resistances to the transport of carbon dioxide and water vapour is given by JARVIS (1971).

The boundary layer thickness, and hence resistance, depends on ventilation (i.e. wind speed) and leaf surface relief and structure (the presence and size of hairs, veins, waxes etc.). For calculations of leaf or stomatal resistances (r_1 or r_s) from measurements of the total resistance (r_{total}), it is highly desirable to keep r_a low by using artificial ventilation. Diffusive resistances of broad leaves in stirred leaf chambers are usually about 0.1 to 1.0 s cm^{-1}, with lower estimates of 0.01 to 0.1 s cm^{-1} for needles and dissected leaves.

5.5.6.2 Calculation of Transpiration Resistances from Measurements of Steady State Transpiration Rate

Steady state measurements of transpiration rate from a leaf may be used for calculating transfer resistances from

$$r_{total} = \frac{C_{sat} - C_a}{q_w} \tag{5.22}$$

and

$$r_1 = \frac{C_{sat} - C_a}{q_w} - r_a, \tag{5.23}$$

where C_{sat} is the water vapour concentration equivalent to the saturated vapour pressure at the temperature of the leaf (see later), and C_a is the water vapour concentration of the ambient air (RASCHKE 1958, GAASTRA 1959).

In using these equations the following assumptions are made:

(1) Water vapour concentration at the liquid-air interfaces in the leaf (C_{sat}) does correspond to the saturation water vapour pressure at the temperature of the interface. Theoretically, this cannot be true in most cases, as the water potential of the evaporating surface is seldom zero. With the normally small depressions of water potential (-10 bar), this error is small and decreases as the relative humidity of the air decreases (MILTHORPE 1961).

(2) The temperature of the liquid-air interfaces in the leaf can be measured as the leaf (or leaf surface) temperature, because of the high thermal conductivity and low thermal capacity of the leaf. Appropriate sensors should be used (see Section 5.5.3.1 and PERRIER 1971). For ease of calculation, Tables 6.6 and 6.7 may be used. The following equation is used to calculate C from the partial water vapour pressure:

$$C_{sat} = \frac{273e}{bT} \varrho_w ,\qquad (5.24)$$

where e is the partial water vapour pressure, b is atmospheric pressure in the same units (e.g. mbar, mm Hg), T is Kelvin temperature [K], ϱ_w is the specific density of water vapour $18/22\,414$ g cm^{-3}).

Critical points of the procedure — in addition to those connected with the accurate estimation of water vapour flux — are:

(a) Parallel measurements of q_w with a wet leaf replica of known surface temperature for calculating r_a.

(b) Precise determination of the ambient air humidity and temperature, allowing correct calculation of C_a.

(c) Reliable measurement of leaf temperature (making the above assumption (2) for correct calculation of C_{sat}).

A great emphasis must be put on the measurement of leaf and porometer air temperature. SLATYER (1971) showed how big the error in estimated value of r_{total} may be due to an error in the measurement of leaf temperature. The error in r_{total} increases with increasing T_{leaf} and with increasing humidity of the air. Error in the measurement of C_a results in error of r_{total} which is increasing with the increasing difference between humidity of incoming and outgoing air.

5.5.6.3 Calculation of Transpiration Resistance from Leaf Energy Balance

IMPENS (1966), MILLER and GATES (1967), IMPENS et al. (1967), UNGER (1968), and JARVIS (1971) describe procedures using calculation from the leaf energy balance to determine r_{total}. For constant radiation each induced change in transpiration, i.e. in latent heat exchange, results in a change in leaf temperature. The reader is referred to these papers.

5.5.6.4 Cobalt Chloride Paper Method

Cobalt chloride paper was originally used for very crude estimations of relative transpiration rates. This method is based on the fact that blue, anhydrous cobalt salts $[CoCl_2, Co(NO_3)_2]$ are very hygroscopic and change to pink (red) hydrates $[CoCl_2 . 6H_2O, Co(NO_3)_2 . 6H_2O]$ by absorption of the hydration water. This colour change was utilized in the method originally proposed by STAHL (1894) and developed by LIVINGSTON (LIVINGSTON and BAKKE 1914, LIVINGSTON and SHREVE 1916). Filter paper saturated with an aqueous solution of the cobalt salt, well dried and kept in a desiccator was held against a living intact leaf on the plant by a Perspex clid and the rate of transpiration from that part of the leaf surface was estimated from the rate at which the blue paper turned pink. The method was originally used to estimate the transpiration rate in the field. However, the cobalt chloride paper interferes with normal transpiration in several ways: The difference in water vapour pressure is substantially increased in comparison with the original natural conditions because of the hygroscopicity of the anhydrous cobalt salt. The boundary layer is disturbed by the paper on the leaf surface and the diffusion path length and gradient of the water vapour pressure is decreased. The exclusion of light may result in closure of the stomata. In addition, the water vapour pressure of the paper increases appreciably with the gradual hydration of the cobalt salt during the exposure so that evaporation decreases very strongly; thus the instantaneous. actual transpiration rate is by no means accurately measured by the cobalt chloride paper method. This method has been criticized by BAILEY et al. (1952) and MILTHORPE (1955).

In spite of these criticisms cobalt chloride papers have been used to get a crude estimation of relative transpiration resistance. The techniques of preparing cobalt chloride papers and the colour standards are described by HENDERSON (1936) and more recently, by MEIDNER and MANSFIELD (1968).

5.5.7 Transpiration Ratio and Productivity of Transpiration

The total amount of water transpired in a vegetation period, divided by the total dry weight produced (in the same weight units), is called the transpiration coefficient.

The transpiration coefficient is often considered to reflect the production requirements of a particular plant species or variety for water but it also depends very considerably on the environmental conditions during the vegetation period. Examples of transpiration coefficients of some plant species given here depend to a great extent on soil and climatic conditions: sorghum about 270, maize 370, sugar beet 400, wheat 500, barley 520, cotton 560, vegetables 550 to 700, oat 600, rye 630, potato 640, lucerne 860, weeds 270 to 1 000 etc.

The reciprocal of the transpiration coefficient is sometimes called the productivity of transpiration. However, it would be more useful to reserve this term for the ratio between the photosynthetic assimilation rate of carbon dioxide and simultaneous transpiration rate (*e.g.* KOCH 1957, RÜSCH 1959 *etc.*).

5.5.8 Calculations of the Transpiration Rate and Examples

The transpiration rate may be expressed in several ways. It may be expressed in terms of the surface, since evaporation takes place from the plant surface. Plant surface is understood here, rather than area, as not all plants have flat transpiring organs and the total transpiration surface should be considered. The surface is therefore double the measured area of flat amphistomatous leaf laminas. The reader is referred to KVĚT and MARSHALL (1971) for methods of measuring leaf area. In terms of transpiring surface the rate of transpiration is therefore expressed in $g\,cm^{-2}\,s^{-1}$ or, more often, in $mg\,dm^{-2}\,min^{-1}$ or in $g\,dm^{-2}\,day^{-1}$. It is also

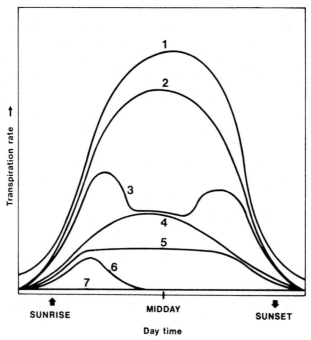

Fig. 5.31 Scheme of theoretically possible types of daily courses of transpiration rate (STOCKER 1956b). Curve 1: stomata permanently open, transpiration rate proportional to the evaporation power, 2: midday depression of transpiration rate due to a temporaty closure of stomata, stomata are closed at night, 3: stomata are completely closed around noon, 4: stomata remain closed, cuticular transpiration is linearly proportional to the evaporation power, 5: midday depression of the cuticular transpiration, stomata remain closed, 6: permanent decrease of cuticular transpiration caused by water deficit originated during the morning hours, 7: transpiration rate almost zero at a permanent and considerable water deficit. The curves are asymmetric.

possible to express the transpiration rate in terms of unit mass. Fresh weight is most readily measured. Expression in terms of water content [mg $g_{H_2O}^{-1}$ min^{-1}] is the most logical procedure. When calculating transpiration per unit of weight, the following units are used: mg per minute [mg g^{-1} min^{-1}] or g per g per 24 hours [g g^{-1} d^{-1}].

Fig. 5.31 shows the diurnal course of transpiration under natural conditions (STOCKER 1956b). This is a theoretical scheme, not an actual example. No actual example is presented, because of the considerable number of internal and external factors which influence transpiration and which would change the course in various ways.

5.6 Stomata

5.6.1 Introduction

The exchange of gases between leaf tissues and the ambient environment depends on the permeability of the stomata and the cuticle to the gases.

The diffusive resistance to water vapour (in completely still air) of a single stoma in the multiperforated septum as it is formed by the leaf epidermis with stomata, is expressed by

$$r = \frac{\Delta C}{q_w}, \tag{5.25}$$

where ΔC is the difference in water vapour concentration [g cm^{-3}] between the liquid-air interface and the ambient air and q_w is the net diffusive flux of water vapour through the stoma [g s^{-1}] so that the resistance of the stoma r has the dimension [s cm^{-3}].

According to FICK's first law of diffusion

$$q_w = \frac{D A \Delta C}{l}, \tag{5.26}$$

where D is the diffusion coefficient of water vapour in still air (0.26 cm^2 s^{-1} at 20 °C and 1 013 mbar*), A is the cross-sectional area of the diffusion pathway [cm^2] and l is the length of the diffusive path [cm].

---

* In narrow channels, with dimensions approaching the mean free path of the diffusing molecules (about 0.06 μm) the effective D_{eff} is smaller than the free diffusion coefficient D:

$$\frac{1}{D_{eff}} = \frac{1}{D} + \frac{1}{D_1} \tag{5.27}$$

and

$$D_1 = \frac{8}{3} h \frac{\delta}{k_1} \left(\frac{2RT}{M}\right)^{1/2} \tag{5.28}$$

where D_1 is diffusion coefficient for narrow channels, h is the mean hydraulic radius, δ and k_1 are shape factors, and M is molecular weight of water (COWAN and MILTHORPE 1968).

Hence

$$r = \frac{l\Delta C}{DA\Delta C} = \frac{l}{DA}.$$ (5.29)

The total diffusive resistance of a single stoma comprises two components. The first is the resistance of the stomatal tube itself (r_{tube})

$$r_{tube} = \frac{l}{DA}$$ (5.30)

secondly, because of the fan-like geometry of the diffusive flow lines at the ends of the stomatal tubes the diffusion from and to a tube in perfectly still air is not proportional to the cross section, as is diffusion through the tube, but to the diameter (or circumference).

According to STEPHAN's law, the resistance outside the end of the tube is:

$$r_{end} = \frac{\Delta C}{2D\Delta Cd} = \frac{1}{2Dd},$$ (5.31)

where d is the diameter of the tube [cm].

As (in equation 5.30) $A = \pi d^2/4$ we get

$$r_{tube} + r_{end} = \frac{4l}{\pi Dd^2} + \frac{1}{2Dd} = \frac{4}{\pi Dd^2}\left(l + \frac{\pi d}{8}\right)$$

and hence

$$r_{tube} + r_{end} = \frac{1}{DA}\left(l + \frac{\pi d}{8}\right).$$ (5.32)

The term $(l + \pi d/8)$ represents the effective length of the pathway [cm], l being the actual measured length [cm] and $\pi d/8$ the end correction. This "end correction" must be applied to both ends of a tube, so that

$$r_{stoma} = \frac{1}{DA}\left(l + \frac{\pi d}{4}\right),$$ (5.33)

where r_{stoma} is the resistance of a stoma, including the end correction of the pore (under the simplifying assumption that the shape of the end of the pore is the same inside and outside) in perfectly still air. (MEIDNER and MANSFIELD 1968).

For elongated stomata the part of r played by stomatal geometry was found to be

$$r = \left(\frac{d}{\pi ab} + \frac{\ln (4a/b)}{\pi a} \right) \Big/ (Dn) , \qquad (5.34)$$

where a, b, d and n are the semilength, semiwidth, depth and density of the stomata (PARLANGE and WAGGONER 1970).

5.6.2 Frequency, Size and Shape of Stomata

Not only the stomatal aperture but also the frequency and size of the stomata are of importance. The ratio between the total stomatal area and the corresponding leaf area is referred to as the relative stomatal area usually expressed as a percentage. Values from 0.1 to 3.6%, most frequently 0.6 to 1.0%, are common. The stomatal frequency is usually expressed as the number per mm^2 of leaf surface (generally between 40 and 300). About one third of plant species have amphistomatous leaves, with stomata on both surfaces. There are usually more stomata on the lower surface. However, the reverse is true of some cereals. About two-thirds of more than 3 000 species examined, (including most broad-leaved trees), have stomata only on the lower surface of the leaves (hypostomatous leaves). Leaves floating on the surface of water are epistomatous, with stomata only on the upper surface. There are usually fewer stomata in the apical part of the leaf blade of dicotyledonous, but these stomata are usually larger than those in the basal part.

The stomatal index is the ratio of the number of stomata to total number of epidermal cells, expressed as a percentage.

The length of the stomatal openings can be measured reliably. The average length is 15 to 40 μm, but even larger stomata are found in some species, *e.g.* up to 80 μm in wheat.

The stomatal frequency can be determined either by direct microscopic observation of the leaves or of epidermal strips, or by examination of replicas, using a ×40 objective. The advantage of replicas is that they can be examined at any convenient time. It helps in examining the material to accentuate the contours of the guard cells with a solution of chlor-zinciodine (20.0 g zinc chloride, 6.5 g potassium iodide, 1.3 g iodine in 10.5 ml H_2O in a brown dropper bottle), which stains the cell walls brown.

The number and size of the stomata depend on the type of plant, the conditions in their habitat and the location of the leaf. Plants from dry localities, leaves exposed to the sun and leaves higher up the stem have more, smaller stomata than plants from wet localities, shade leaves and leaves located lower down the stem.

5.6.3 Measurement of Stomatal Aperture

The stomatal aperture can be measured directly as the width of the slit between the two guard cells of a stoma and can be expressed as the relative aperture of the stoma, in comparison to the width of the fully open slit. The absolute dimensions of the stomatal slit are required for calculations concerned with the effect of the width of the slit on the rate of diffusion of water vapour, CO_2 etc.

The width of the aperture may be measured: (1) by direct microscopic observation of the leaf, (2) microscopic observation of epidermal strips, (3) microscopic observation of a replica of the epidermal surface.

5.6.3.1 Direct Observation

Microscopic observation of intact leaves or parts of leaves has the following advantages: (a) by focusing at different depths, optical sections of the stomata can be found, and so dimensions in stomata of complex structure can be measured, (b) leaf hairs, emergences etc. do not have a very disturbing effect.

As living material is under observation, work must be rapid; and cutting off part of the leaf blade, or transfer of the material to different conditions, of light for example, primarily the intense illumination under the microscope (no matter whether incident or transmitted light is used for observation), may sometimes cause very rapid changes in stomatal aperture. However, intact living material can be used in rapid work with a known material and with such an experimental arrangement that frequent transfers of the plants are not necessary. It is possible to repeat a measurement during the course of the experiment at the same place on the same leaf of an intact plant under the microscope (assuming that the necessary manipulations do not cause rapid changes in the stomatal aperture that would be detected during measurement). This makes reliable comparisons possible.

Transmitted light can be used for observations with thin and pale leaves. Green light may be the least active in causing photoactive changes in stomatal aperture. The intact leaf can be held on the microscope stage by means of a weak spring (*e.g.* from a wrist watch), rubber devices or plastic foam. In choosing the part of the lamina to be examined differences in the frequency and size of the stomata and in their reactivity must be taken into account. STÅLFELT, who was an outstanding investigator of the physiology and ecology of stomata, recommended working with leaf segments (about 0.5 cm^2) cut as quickly as possible from the intact leaf blade. The segments were placed in a drop of liquid paraffin on a miscroscope slide and observed with an oil immersion objective. Further enlargement with a high-power compensating eyepiece may also be an advantage. The widths of the slits of as many stomata as possible are rapidly measured in one or more microscope fields. The method is limited only by the sensitivity of stomata to the treatment: cutting, transfer and intense illumination under the microscope. The sensitivity and reaction rate of the stomata must be tested in preliminary experiments.

So-called opaque illumination, in which the leaf surface is illuminated through an observation objective, may be used for thick, dark leaves which are not too hairy. Even illumination of the whole microscope field, allowing sufficiently rapid and accurate focusing is essential.

For direct observation, and still and time lapse photography, ELKINS and WILLIAMS (1962) suggested the use of a combination of a single lens reflex camera (or possibly a reflex cine camera) with microscope optics, using a low-power objective (with a working distance as long as possible) and a 20 × eyepiece. A leaf holder is mounted onto the camera body by means of a focus rod or a micrometer, which allows the leaf surface to be brought into focus. Exposure is by an electronic flash through the leaf. It should be remembered that such intense illumination with flash may influence the subsequent movements of the stomata.

Until recently, direct microscopic observation of epidermal strips, rapidly stripped off mechanically and immediately fixed in absolute ethanol (LLOYD 1908) was a method in fairly frequent use. After fixation, pieces of epidermis were transferred into a HEATH solution, prepared as follows. Enough water is added to 2.5 g iodine and 100 ml "melted" phenol just to dissolve the mixture, and an excess of potassium iodide is then added. However, changes in pressure are necessarily caused by stripping off the epidermis. This frequently results in change in stomatal aperture, even before fixation. The fixation only occasionally influences the size of the stripped epidermis, so that the LLOYD method may be used to determine the number or even the size of the stomata. However, it should not be used for measurement of stomatal apertures (OPPENHEIMER 1935, NADEL 1940).

Transverse sections (hand-made, or cut with a freezing microtome) are used for evaluation of the dimension and shape of the guard cells and stomatal pore geometry. These data are necessary for calculation of the diffusivities of stomatal pores, in connection with gas exchange of plant leaves (PENMAN and SCHOFIELD 1951, BANGE 1953, LEE and GATES 1964, WAGGONER 1965, JARVIS et al. 1967, MILTHORPE and PENMAN 1967, PARLANGE and WAGGONER 1970).

The scanning electron microscope, which has a greater depth of field than either the high magnification of the light microscope or the transmission electron microscope, is very suitable for showing minute structures of the surface of stomata and cuticle (e.g. MOZINGO et al. 1970 and ROOK et al. 1971, TROUGHTON and DONALDSON 1972, SANCHEZ-DIAZ et al. 1972).

5.6.3.2 Microrelief (Replica) Methods

A drop of a solution of a transparent solid compound in a volatile solvent or a polymerising mixture is applied as a thin layer to the leaf surface, where a transparent film with a replica print of the epidermis with stomata is formed after evaporation of the solvent or after polymerisation. The replica is then examined under the microscope, or if it is opaque, a further transparent replica of the replica is

made. Depending on the viscosity of the fluid it penetrates more or less the external parts of the stomatal slits, so that some of its dimensions may be measured.
There is evidence that in many cases the solution does not enter the pores themselves and so the replica corresponds to the antechamber, for example. The method is
not suitable for hairy leaves, leaves with sunken stomata or with stomata of small
apertures.

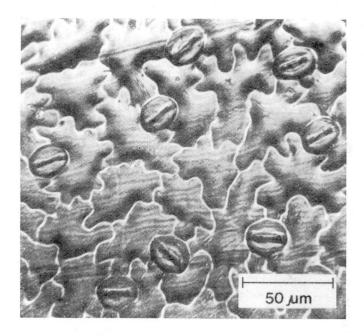

Fig. 5.32 Microphoto of a cellulose acetate replica of a lower epidermis of *Chrysanthemum maximum* leaf with stomata reprints (oblique illumination-photograph by J. Pazourek).

It is an advantage that the results can be evaluated later at a convenient time, as
a properly prepared replica retains its transparency for a long time. In addition,
many replicas can be taken at the same time; so that a large number of samples
can be collected. It is necessary, however, to verify for each species whether the
replicated structures really represent the stomatal pores, as often only the outer
parts of the guard cells are replicated, and these may be open even when the pores
are fully closed. Comparison of the results with direct observation of the stomata and
with other measurements, *e.g.* porometry (Solárová 1965, 1968), is recommended.
    The replica method was originally suggested by Buscalioni and Pollacci
(1901a and 1901b), who used a solution of collodion in ethylether. A commercial
(4%) solution of collodion in chloroform may also be used. However, it should first
be concentrated twice by evaporation, so that a solid replica is formed rapidly
after a drop is spread on the leaf surface. The solutions used should have the fol-

lowing properties. (1) The leaf must not be locally cooled and thus injured by eva-poration and (2) the aperture of the stomata must not be changed before solidi-fication by "narcotization" by the solvent. Ether is the least suitable solvent in view of this requirement.

A solution of methacrylic acid can be prepared by dissolving 5 to 10 g of colour-less methacrylate polymer in 100 ml of pure anhydrous chloroform. WENZL (1939) used a 5% celloidin solution in ethanol-ether mixture (1 : 2 to 1 : 3). It is also pos-sible to use commercially produced solutions of celloidin, with the trade names Rellon, Rubbit and Rudol 33. VÁCLAVÍK (1955) and NORTH (1956) used concen-trated solutions of cellulose acetate in ethylacetate. Colourless nail-varnish is also useful (e.g. PAZOUREK 1962). MOLOTKOVSKIÏ (1935) used celluloid solution.

Shrinkage of the replicas may occur during drying, so that the size of the repli-cated structures may be slightly decreased. This can be avoided by using the ad-hesive print method (WOLF 1939). An acetone solution of celloidin is used (made for studying microreliefs of animal tissue preparations), the foil is removed by transparent adhesive tape.

A drop of the solution is applied to the leaf surface (with a rod or a small brush) and spread, preferably with a finger pad, so that the drop is distributed over an area of about 1 cm^2. When this has solidified it lifts from the leaf surface in some places so that air penetrates underneath. A piece of adhesive tape is attached, and then, together with the adhering replica, is stripped off the leaf and stuck down on a marked slide. It is necessary to make preliminary tests of the quality of the ad-hesive tape, in order to avoid any decrease in transparency after long storage. The preparations are best observed in a drop of water under a cover slip under opaque illumination or with an almost closed iris diaphragm.

SAMPSON (1961) and ZELITCH (1963) used silicone rubber impressions of the leaf surface as "negatives". "Positives" were made with cellulose acetate solution. A viscous fluid silicone rubber (the kind used by dentists for making replicas of teeth is most suitable), well stirred and then mixed with a sufficient amount of catalyst, is gently spread over a small part of the leaf surface and left to polymerize (2 or 3 minutes). This is the "negative", which is peeled from the surface, dried in a desiccator and painted with a thin film of a commercial colourless nail varnish diluted with one-third the volume of acetone. The "positive" is dried and stripped in the same way as collodion replicas. The use of silicone rubber plastic for leaf surface replicas was also described by GROOT (1969). Scanning electron micro-scopy of replicas was also used for measurement of stomatal aperture (IDLE 1969a).

5.6.3.3 Infiltration Methods

Infiltration methods are particularly useful for field work. When applied to a leaf, liquids with low surface tension penetrate the intercellular spaces; this pe-netration shows as a dark spot in incident light. The ability of the liquid to infiltrate through stomata into the intercellular spaces depends on its surface tension, viscosity and dissolving capacity. The first and third properties influence the extent to which the surface is wetted: this increases with decreasing surface tension (*e.g.*, water has an exceptionally high surface tension: 76 g s^{-2}), and with the capacity of the liquid to dissolve the fatty or waxy layers of the cuticle (URSPRUNG 1925). SCHÖNHERR and BUKOVAC (1972) found that only liquids with surface tension lower than about 25 to 30 dyne cm^{-1} give zero contact angle and wet the leaf surface. The viscosity of the liquid determines the rate of movement through the stomatal pores into the intercellular spaces. In addition, the rate of infiltration depends on the density, size and, particularly, the aperture of the stomata and also on the size of the intercellular spaces, from which the infiltrating liquid displaces air. Infiltration does not therefore depend only on the ability of the liquid to enter the intercellular spaces but also on the leakage out of the intercellular air, so it is not possible to immerse leaves in the infiltration liquid. STÅLFELT (1939) in his work with needles of conifers, using ether, allowed the needles to become wetted on one side only so that air could leak freely through the intercellular spaces and the other side. The method was originally designed by HABERLANDT (1905) and later extended by MOLISCH (1912), DIETRICH (1926), SCHORN (1929), ALVIM and HAVIS (1954), HLUCHOVSKÝ and SRB (1958), RUTTER and SANDS (1958), WORMER and OCHS (1959), OPPENHEIMER and ENGELBERG (1965), MICHAEL (1969) and others.

When working with one infiltration liquid, the time for significant infiltration, *i.e.* formation of spots of a previously selected type, is measured (SHMUELI 1953 used liquid paraffin for banana leaves).

However, it is better to use a series of infiltration liquids graded according to decreasing "infiltration ability" and to find the liquid which just causes infiltration and the next in the series, which just does not. The liquids should be stored in stoppered bottles with glass rods used to apply a drop of each liquid in a marked order to several leaves. It is necessary to work quickly and in order through the series to best utilize the advantages of the method.

MOLISCH (1912) used a series of xylene, benzene and ethanol in order of decreasing infiltration ability. The first member enters even partially closed and small stomata, the last only into completely open stomata: STEIN (1912) used a series of petroleum ether, kerosene, liquid paraffin in order of decreasing viscosity. DIETRICH (1926) used petroleum ether, xylene, benzene, turpentine, paraffin, mixture of castor oil and turpentine 1 : 3, the same mixture in a ratio of 1 : 2, ethanol, castor oil and turpentine 2 : 1, liquid paraffin. SCHORN (1929) and DALE (1961) worked with a series of solutions with a continuous gradation of viscosity, capillarity and wettability made up of a mixture of readily infiltrating isobutanol and non-infiltrating glycolate in proportions 1 : 9, 2 : 8, 3 : 7 *etc*. WORMER and OCHS (1959) suggested a series of mixtures of isopropanol and water in proportions from 100% isopropanol to a mixture of 35% isopropanol and 65% water, with 5% gradations. ALVIM and HAVIS (1954), for very small stomata, used a series of 11 mixtures of xylene, or n-dodecane or n-tetradecane, and liquid paraffin from 100% by volume of xylene to 100% liquid paraffin, with gradations of 10% by volume in between. HLUCHOVSKÝ and SRB (1958) stained the infiltration liquids with Sudan III, to increase the clarity of the spots. RUTTER and SANDS (1958) used ethanol stained with crystal violet. OPPENHEIMER and ENGELBERG (1965) recommended the following procedure for conifer needles: A solution is prepared of 0.9 g crystal violet dissolved in 2 ml ethanol, to which 9 ml ethylether and 9 ml chloroform are added. The needles are immersed in this solution for 2 to 5 minutes (rubber gloves should be used) and then washed with secondary butanol. The type, degree and size of infiltration spots are evaluated by the authors using a scale based on the staining of the mesophyll cells under the stomata.

BOYSEN-JENSEN (1928) observed water infiltration after evacuation of intercellular spaces. Measurements are made of the negative pressure necessary to remove gases from the intercellular spaces sufficiently to bring about infiltration of water into an immersed leaf when normal atmospheric pressure applies again. The negative pressure is applied by using a mercury pump. This method has not been used by other authors for standard measurements.

Infiltration may be also caused by excess pressure. FROESCHEL and CHAPMAN (1951) and FROESCHEL (1953) designed a so-called pressure "stomata meter" based on measuring the pressure under which infiltration of water into an immersed leaf occurs, resulting in release of air bubbles from the stomata. The method avoids the disturbing effect of the different wetting properties of infiltration liquids and to same extent avoids errors caused by differences in properties of the intercellular spaces. It gives relative results. The equipment required can be used in the field.

A pressure infiltration device suitable for measuring the relative stomatal apertures in conifer needles was described by FRY and WALKER (1967) and modified by LOPUSHINSKY (1969) for field measurements. Pressure is applied from a small cylinder of nitrogen and the leaf is immersed in ethanol (50 to 60 volume per cent) − water mixtures (50 to 60 : 50 to 40). The pressure necessary for a predetermined infiltration picture is measured.

Results obtained by infiltration methods are relative, and can hardly be compared in different species. The methods are not reliable for hairy plants. They give a picture of the relative openness of the stomata and also reflect the size and density of stomata, and the extent and communication of the intercellular spaces. The procedure based on determination of the time necessary for infiltration is further complicated by a substantial error, as it depends on the rate of air leakage, mostly through the opposite side of the leaf. In procedures based on finding the particular liquid in a series which just enters the leaf, the observed degree of infiltration is mainly related to the stomata on the side of the leaf to which the drop is applied. Infiltration methods are simple to do and they yield results immediately; but the results obtained are only relative and to a certain extent are subjective.

SCHÖNHERR and BUKOVAC (1972) proved that spontaneous infiltration of liquids (with low surface tension) takes place when the contact angle is smaller than the wall angle of the stomatal aperture wall, the degree of stomatal opening being of little importance. This would mean that the stomatal aperture cannot be assessed by spontaneous infiltration.

5.6.4 Porometry

Stomatal pores represent a major resistance to the diffusive flux of water vapour in transpiring leaves. Measurements of the diffusive resistance of stomatal pores (plus that of the substomatal intercellular spaces and of the stomatal antechambers) are therefore essential to assessing the influence of stomata in relation to transpiration (and/or exchange of other gases between the leaf tissues and the surrounding atmosphere).

Because of technical difficulties, only a few techniques have been developed for direct measurement of diffusive resistance of stomata by diffusion porometry. Diffusion porometers measure the diffusive flux of a gas (water vapour) in an amphistomatous leaf. This flux can be caused by an artificially established differences in concentration (and partial pressure) of a diffusing gas between the atmosphere surrounding the upper and the lower surfaces of the leaf blade (see Section 5.6.4.1). Water vapour diffusion porometers have also been constructed for measuring and calculating the total diffusive resistance of the leaf from the transpiration rate under known conditions of ambient humidity and leaf temperature (see also Section 5.6.4.2).

Gases, the concentration of which can easily be measured (usually by physical methods), are most frequently used in diffusion porometers. For example, radioactive argon (MORESHET et al. 1967, 1968), helium (GALE et al. 1967), hydrogen (GREGORY and ARMSTRONG 1936, SPANNER 1953, LOUGUET 1965, MILTHROPE and PENMAN 1967), nitrous oxide (SLATYER and JARVIS 1966, JARVIS and SLATYER 1966a, 1970) have been used. It is important to avoid any difference in total gas pressure between the two surfaces of the leaf, as this would result in mass flow.

The difficulties encountered when trying to make direct measurements of diffusion rates and hence diffusive resistances have led to the development of a large number of techniques for measuring the mass (viscous) flow of a gas (usually air) through amphistomatous leaves. Such mass flow porometers can hardly be used on leaves with stomata on only one surface.

The air flow rate at a known pressure difference is most frequently measured in mass flow porometers. The pressure difference causing the flow is critical: it should be low so that it does cause any changes in stomata aperture or resistance of the intercellular spaces (RASCHKE 1965a). Too large a quantity of air forced through the leaf mesophyll may also cause occasional dehydration and consequent change in stomatal aperture.

5.6.4.1 Water Vapour Diffusion Porometers

Assuming that the transpiration rate is controlled exclusively or mainly by changes in the aperture of the stomata, the transpiration resistance is often used as a measure of stomatal resistance. These methods use so-called water vapour diffusion porometers comprising a leaf chamber containing an electric hygrometric sensor, a device for predrying the chamber air, and thermistor(s) or thermocouple(s) for measuring leaf and air temperature. The transit time or time lapse, Δt[s] necessary for the humidity of the enclosed air to increase from the known and measured starting value to a previously chosen higher humidity — usually 1 to 5 per cent relative humidity higher — resulting from transpiration from the enclosed part of the leaf surface is measured.

$$q_w = \frac{(C_2 - C_1) \, V}{A \, \Delta t}, \tag{5.35}$$

where C_1 and C_2 are the original and final water vapour concentrations of the air in the cup, V is the volume of air enclosed [cm^3], and A is the leaf area enclosed [cm^2]. Hence

$$r_{total} = \frac{(C - \bar{C}) \, \Delta t A}{(C_2 - C_1) \, V}, \tag{5.36}$$

where $\bar{C}$ is the average air humidity in the cup during Δt:

$$\ln \bar{C} = 0.5(\ln C_1 + \ln C_2). \tag{5.37}$$

The assumption is made that the slight decrease in transpiration rate during Δt, resulting from the increase in humidity of the air enclosed in the cup, does not change the leaf temperature.

The general measuring procedure is as follows: The leaf cup is closed (e.g. inserting a plastic plate in place of the leaf in the clamp, see Figures 5.33 to 37) and the inside air is dried by pumping in dry air from an external drying device (in unventilated porometers) or by introducing a device containing a desiccant into the air stream in a ventilated porometer. The cup is then attached to the leaf surface and the time for an arbitrarily chosen increase in air humidity from C_1 to C_2 is measured*, using the humidity sensor resistance bridge (e.g. see Fig. 5.34). Knowing the volume of the chamber (V) and the area of the exposed leaf surface (A), r_{total} (for a leaf) or r_a (for wet filter paper) can be calculated using equation 5.36.

Most diffusion porometers measure non-steady state transpiration rate, i.e. the rate of increase in humidity in a closed cup clamped to the transpiring leaf surface. They may be either unventilated, with still air in the porometer cup so

---

* A circuit for a self-timing diffusion porometer was described by KENNY and McGRUDDY (1972).

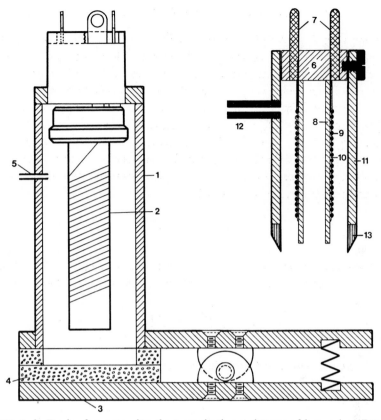

Fig. 5.33 Left: Device for measuring the transpiration resistance of leaves (van BAVEL *et al.* 1965). 1 — chamber containing a hygrometric sensor (2) which can be attached to the leaf by means of a clamp (3). The leaf is sealed by means of a sponge rubber gasket (4). 5 — air inlet for drying the air prior to measuring.
Right: Section through a humidity sensor chamber used for estimating relative stomatal apertures (WALLIHAN 1964). 6 — PVC basal plate, 7 — connection plugs of the humidity sensor, 8 — plastic cylinder, 9 — wire helix, 10 — humidity sensitive coating, 11 — plexiglas tube, 12 — copper tube for connection to drier, 13 — soft rubber gasket.

that water vapour flux in the cup air from the leaf/air boundary layer to the humidity sensor takes place by diffusion, or (see later) ventilated, where diffusive vapour flux is limited to the leaf and boundary layer only and the bulk air in the cup is thoroughly mixed by a propeller ensuring that the boundary layer resistance is reproducible and low.

For review see also STIGTER (1972).

Unventilated Porometers

In the unventilated diffusion porometers the diffusion of water vapour from the transpiring leaf surface to the humidity sensor takes place in more or less still air assuming isothermal conditions in the cup. Thus the geometry of the cup between the leaf surface and the sensor is critical (see later).

WALLIHAN (1964) used the cup illustrated and described in Fig. 5.33 right, again with an Aminco-DUNMORE sensor covering a specified range, *e. g.* to 30% or 39 to 43%. The cup was first pressed onto a plastic ruler instead of the leaf,

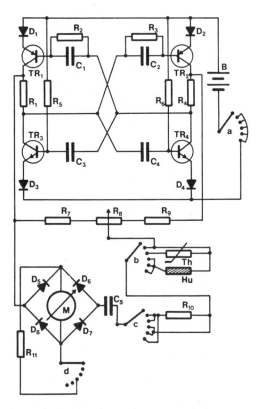

Fig. 5.34 Diagram of an a.c. resistance meter for humidity sensor in the water vapour diffusion porometer (KANEMASU *et al.* 1969). $R_1 = 470\,\Omega$, $R_2 = 32\,K\Omega$, $R_3 = 32\,k\Omega$, $R_4 = 470\,\Omega$, $R_5 = 32\,k\Omega$, $R_6 = 32\,k\Omega$, $R_7 = 68\,k\Omega$, $R_8 = 1\,k\Omega$, $R_9 = 1\,k\Omega$, $R_{10} = 900\,k\Omega$ (1%), $R_{11} = 1\,k\Omega$, $C_1 = 0.5\,\mu F$, $C_2 = 0.5\,\mu F$, $C_3 = 0.22\,\mu F$, $C_4 = 0.22\,\mu F$, $C_5 = 0.1\,\mu F$, D_1,D_2,D_3,D_4 : IN 4002, D_5,D_6,D_7,D_8 : 4 U 191, TR_1 and TR_2 : 2 N 3251, TR_3 and TR_4 : : 2 NL 893, Th — thermistor, Hu — humidity sensor, a,b,c,d — switches, B — battery 13.5 V, M = 10 µA meter. Frequency: about 90 Hz.

to close the cavity. The sensor was connected to a portable battery-operated A. C. resistance meter. If the first reading exceeded 3.0 on the 20 µA meter, the air in the cup was dried by passing a series of short puffs of dry air through the inlet until a steady reading of about 3.0 was reached. The cup was then placed on the leaf surface and the transit time necessary for the reading to change from

0.4 to the reading corresponding to an increase of about 1% relative humidity was determined.

Van Bavel *et al.* (1965) used a leaf cup (Fig. 5.33 left) which can be directly clamped on a leaf, with an opening and free volume (*i.e.* excluding the sensor volume) of 2.84 cm^2 and 14.4 cm^3, respectively. The sensor was an Aminco-Dunmore element, range 14 to 27% relative humidity (Hygrodynamics, Inc., Maryland). The authors measured the transit time for a change in relative humidity of about 4%. Similar type was used by Djavanchir (1970).

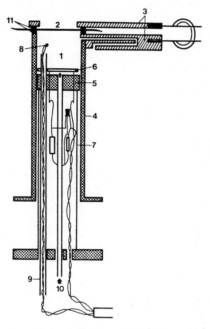

Fig. 5.35 Unventilated porometer with adjustable diffusive path length (Meidner 1970). 1 — porometer cup, 2 — leaf, 3 — Perspex parts of the leaf clamp with buldog clip, 4 — syringe barrel, 5 — syringe plunger supporting the sensor element (6), 7 — syringe stem with three electronic components, 8 — thermistor supported by sliding tube (9) enabling to measure either leaf or air temperature, 10 — drying air inlet, 11 — silicone rubber washers.

Kanemasu *et al.* (1969) improved the geometry of the unventilated type of porometer by placing a commercial, cylindrical Aminco-Dunmore humidity sensor parallel to the leaf in a hemicylindrical receptacle, so shortening the diffusion pathway from the leaf surface to the sensor. A phenolic plastic insert holding a 0.076 mm thick stainless steel plate with uniformly spaced holes 1 mm in diameter is placed in the porometer opening. This reduced convection in the chamber. The improved electronic circuit of the resistance meter is shown in Fig. 5.34.

The sophisticated types of unventilated diffusion porometer in use at present are those described by Meidner (1970) and Stiles (1970). Both have flat home-made humidity sensors (Pope, or sulphonate polystyrene hygrometers, WEXLER 1957) with a short response time, situated parallel to the leaf surface. They can be constructed relatively easily following the detailed description in the papers, although they are now commercially available (*e.g.* Phys. Chem. Res. Corp, New York).

The type designed by MEIDNER (Fig. 5.35) has a cup with a diffusive path length which can be adjusted with the clamp made of a plastic syringe, both for calibration and for adjusting to different ranges of resistances, and a thermistor, which is also adjustable for measuring leaf surface and air temperature alternately. One jaw of the leaf clamp has a swivel lid for closing the orifice. STILES (1970) described a porometer with non-adjustable diffusive path length but with two different sizes.

Recently a condensation type of porometer was described by MORESHET and YOCUM (1972). The time lapse is measured until the humidity of the predried air in the porometer cup reaches saturation at 0 °C dew-point temperature. This is visualized by condensation on the surface of a small disc of cellulose acetate film with an aluminium coat on its internal surface placed laterally. This surface is visible through a transparent opposite window in the cup. The external surface of the disc is cemented to a copper rod protruding into an ice chamber made of two polyethylene jars, one inside the other, insulated with polystyrene and filled with a mixture of crushed ice and cold water. A double cup porometer held together with a surgeon's clamp was designed for amphistomatous leaves. Each cup has inlet and outlet enabling the predrying of the air by several flushes of air through a tygoon hose with silica gel and magnesium perchlorate before the measurement, and a leaf thermistor. Calibrating procedure similar to that of KANEMASU *et al.* (1969) gave linear relationships between Δt and dummy perforated resistances for different temperatures of evaporating surface. Only leaf temperatures need to be measured and respected in calibration since the air temperature at condensation point is constant (0 °C).

The theory of unventilated porometers has recently been exhaustively treated by MONTEITH and BULL (1970). They conclude that the sensor should be plane and parallel to the leaf surface. The square root of the time lapse $[\Delta t]$ is then linearly proportional to the diffusive path length. For different stomatal apertures and/or numbers the distance between the leaf and sensor (diffusive path length) should be adjustable. The authors suggest a fairly high initial relative humidity of the air (50 to 70%) so that transpiration rates are not too high. If the temperature of the whole leaf cup is known to vary within about 1.5 °C, the possible error in the effective length of the diffusive pathway is $\pm 2\%$.

Calibration of Unventilated Porometers

On the assumption that there is no thermal convection in the cup, VAN BAVEL *et al.* (1965) calculated the total transpiration resistance

$$r_1 = \frac{k \, \Delta t - l}{D},$$

(5.38)

where k is an experimentally determined sensitivity constant for the instrument expressed in [cm s^{-1}], l is the diffusive length [cm] determined experimentally for the cup used. They calibrated the porometer by inserting different diffusive path lengths simulated by tubes of different lengths between the leaf and the sensor.

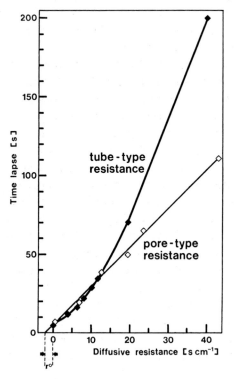

Fig. 5.36 Comparison of calibration curves for an unventilated porometer, obtained by using tubes of different length (a) and by using perforated plates as dummy resistances (KANEMASU et al. 1969). r_0 — diffusive resistance of the porometer cup.

KANEMASU et al. (1969) calibrated using artificial resistances (acrylic plates with small drilled holes) placed between a wet filter paper and the porometer opening. Low resistance plates (4 s cm^{-1}) were made by cutting holes of the same size as the porometer opening; higher resistances were plates of various thickness with varying numbers of 1 mm holes. The resistance of the plates was calculated using the following equation:

$$r = \frac{l_{\text{eff}}}{D} = \frac{4A\left(l_0 + \dfrac{\pi d}{8}\right)}{Dn\,\pi d^2},$$ (5.39)

where l_{eff} is the effective diffusive pathway in the acrylic plates, D the diffusivity of water vapour in still air, A the area of leaf sampled, l_0 the length of each hole, d the diameter of the holes, n the number of holes, $\pi d/8$ a correction for one "end effect" in each hole (for explanation see Section 5.6.1). The total cross-sectional area of the holes $(n\pi d^2/4)$ must be greater than $A/30$. Comparison of calibration by using tubes and perforated plates as dummy resistances is shown in Fig. 5.36 (KANEMASU et al. 1969).

In calibration described by MONTEITH and BULL (1970), either different distances between wet filter paper and the sensor are used, or the evaporation area is varied by covering the filter paper with cellulose acetate plates with different numbers of holes 1.05 mm in diameter. A linear relationship between square root of Δt and the diffusive path length with brass extension tube was found.

The main advantage of unventilated transpiration porometers is that they are simple instruments, suitable for field measurements. The main disadvantages are the errors caused by temperature variations and consequent possible convection in the cup; and in porometers without an adjustable diffusive path length the transit times may be too long, causing changes in stomatal aperture, or too short, causing errors in reading.

Ventilated Porometers

A substantial change in design of the transpiration porometer was made by SLATYER (1966c) and by BYRNE et al. (1970), with the introduction of internal ventilation. The ventilation of the enclosed leaf surface and the rapid mixing of the enclosed air enable more precise calculation of leaf resistances while maintaining a constant, reproducible low boundary layer resistance. This eliminates errors caused by temperature gradients and the resulting convection which disturbs diffusion in the still air: in an aspirated porometer, diffusion is restricted to within the leaf and the thin boundary layer; a sensor with as short a response time as possible is used and the air is dried by insertion of a compartment containing a desiccant. The aspirated porometer described by these authors samples a very small area of leaf (a few mm²). The leaf surface and the humidity sensor are ventilated by a small fan.

This type of instrument was modified by KAUFMANN (1968c), see Fig. 5.37. The leaf area sampled was made larger (1.2 cm diameter), and the internal volume was 65.2 cm². As in SLATYER's instrument, an Aminco-DUNMORE humidity sensor (Narrow-Range Humidity Sensor Type TH range 5 to 14%, Hygrodynamics Inc., Silver Spring, Maryland), a battery-driven small fan and a removable drying compartment with silica gel were used. The transit time (Δt) for a pre-determined increase in humidity inside the chamber (e.g. from 12 to 14 per cent relative humidity) was measured after fixing the porometer to a leaf surface, starting with the enclosed air dried below the lower humidity limit by the silica gel. A similar device, but with a different internal arrangement, was described by KNIPLING (1968a), TURNER et al. (1969), TURNER and PARLANGE (1970).

Calibration of Ventilated Porometers

Calibration is in terms of r_1 [s cm^{-1}] in conditions of constant and measured air humidity and temperature, with ventilation similar to that caused by the small fan in the chamber. The rate of evaporation from a wet filter paper of known surface area (*i.e.* the area of the opening to the chamber) is measured both from a freely evaporating surface and from the surface covered with an artificial membrane of high diffusive resistance, such as millipore filters or plastic microperforated filters, *e.g.* Nuclepore filter (General Electric), 8 to 10 μm thick with

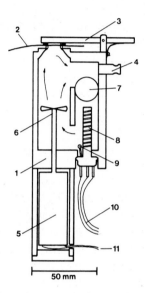

Fig. 5.37 Schematical cross section of a transpiration porometer with fan and drying device (KAUFMANN 1967). 1 — Perspex body, 2 — leaf, 3 — leaf clamp, 4 — opening for inserting needle leaves, 5 — motor, 6 — fan, 7 — silica gel drier (movable vertically to the cross-section plane) which can be inserted into the air circuit when drying the air, 8 — lithium chloride humidity sensor, 9 — thermistor, (8 plus 9: Hydrodynamics narrow range humidity sensing element), 10 — connections to the meter, 11 — connection to the battery.

about 3 700 pores, 10 μm in diameter per cm^2 (KAUFMANN 1967). From the evaporation rates and the absolute air humidities, resistance r_a (free wet filter paper) and $r_a + r_{membrane}$ (wet filter paper covered with the membrane), and hence $r_{membrane}$, can be calculated using Equation 5.25. The wet filter paper surface and the membrane-covered surface can be made into "packages" which can be placed, instead of the leaf blade, across the opening of the porometer or (perhaps better) enclosed in the chambers (SLAVÍK 1973). The transit time (Δt) is measured for wet filter paper, giving the r_a of the instrument, and for wet filter paper covered with a membrane for which $r_{membrane}$ has been previously determined.

BYRNE et al. (1970) described another relatively very simple calibration technique for a ventilated porometer. In place of the leaf, a connector is attached to the
opening. This connector contains a pad of fibreglass filter paper which covers
the opening; the other end is connected by plastic tubing to a surgical syringe.
Moist plastic foam is enclosed in the syringe, so that doses of a known volume
of air saturated with water vapour at room temperature can be "injected" into the
porometer. The volume of saturated air to be added to increase the humidity
inside the porometer from the arbitrarily chosen initial value to the final value is
found by trial and error. At slow rates of injection a motor driven syringe can be
used. The resistance of the instrument itself ($r_{\text{instr.}}$), which should be subtracted
from any reading in order to get the leaf resistance (r_1), is found in the usual
manner by using wet filter paper instead of a leaf:

$$r_{\text{instr.}} = \frac{A \, \Delta t \, \Delta C}{m},$$ (5.40)

where A is the area of leaf sampled [cm^2], Δt time lapse [s], ΔC is the difference
in water vapour density between full saturation at leaf (instrument) temperature
and the mean humidity between the two readings, m is the amount of water
vapour [g] added.
    Possible sources of error in non-steady state porometers are (besides calibration):
(1) Moisture absorbed by the sensitive coating of the sensor itself,
(2) moisture absorbed on − or interchanged with − the inner walls of the cup and
(3) responses time of the humidity sensor in its dynamic and static response (see
e.g. MONTEITH and BULL 1970, MORROW and SLATYER (1971b). MORROW and
SLATYER (1971b) drew the attention to the errors caused by the drift of sensor
reading due to the storage and previous cycle of drying. They result both in long
term drift of calibration and in instrumental drift. They suggest to store the tested
type [unventilated porometer of similar version as described by KANEMASU et al.
(1969)] with closed porometer chamber connected to the silica gel dehydrator and
the instrument drift should be reduced by repeated dehydrations prior to a series
of measurements, by determining this drift and recalibrating periodically.
(4) Changes in stomatal aperture brought about by enclosure of the part of the
leaf in the cup, especially during relatively long exposures (of a few minutes) and
in particular when the leaf is shaded during the procedure.
(5) Difference in temperature between leaf and air and error in measuring both
temperatures. If $T_1 > T_{\text{air}}$, the difference between these two temperatures results
in underestimation of leaf resistance and vice versa. If this difference is e.g. of the
order of $+1\,°C$, the ratio $r_{\text{true}}/r_{\text{apparent}}$ is about 0.8 when the leaf resistance is low
($r_1 = 1$ to $2\,\text{s cm}^{-1}$); it increases with higher resistances. With the temperature
difference being $5\,°C$ the above ratio may be as low as 0.5 (MORROW and SLATYER
1971a).

A great emphasis must be put on the measurement of leaf and bulk air temperature. In some types of porometers isothermal conditions are assumed but may not in fact exist (TURNER and PARLANGE 1970). Leaf temperature is mostly measured by an attached thermistor situated in the opening. The possible errors due to heat conduction are discussed *e.g.* by PERRIER (1971). Air temperature in ventilated porometers is measured with sufficient precision by means of a thermistor situated mostly at the humidity sensor. In unventilated porometers the determination of air temperature (if measured) may be very critical and cause substantial errors.

Steady State Diffusion Porometers

A new type of water vapour diffusion porometer was designed by BEARDSELL *et al.* (1972). It measures the steady state transpiration rate of a leaf (leaves) enclosed in a chamber. The principle is the following: Dry air is blown into the ventilated chamber at a rate (measured) just sufficient to keep the pre-determined air humidity constant. A balance is maintained between the flux of transpired water and the air flow. Errors due to the difference between the static and dynamic response of the sensor and due to adsorption and desorption of water on the inner walls of the chamber are eliminated.

The water flux due to transpiration (q, [g s^{-1}])

$$q = \frac{C_{sat(T)} - C_a}{r} = \frac{fC_a}{A} \quad [\text{g cm}^{-2} \text{ s}^{-1}], \tag{5.41}$$

where $C_{sat(T)}$ and C_a are water vapour concentrations at leaf and air temperatures, respectively, r is total resistance, f is air flux [cm^3 s^{-1}], A is leaf area in the chamber [cm^2]. It follows under isothermal conditions that total resistance

$$r = \left(\frac{100}{(\text{r.h.})} - 1\right)\frac{A}{f} \quad [\text{s cm}^{-1}], \tag{5.42}$$

r.h. is relative humidity, r.h. and A being constant and known, r is (under isothermal conditions) a unique function of the influx of the dry air (f) so that the flow meter can be directly calibrated in terms of r calculated as above, which is a substantial advantage.

The Perspex chamber is big (about 0.8 dm^2), the humidity sensor is a PCRC--11 sulphonated polystyrene electrohumidity sensor (Phys. Chem. Research Corporation, New York, U.S.A.). The null balancing times was about 60 to 10 s.

MEIDNER and SPANNER (1959) constructed a differential transpiration porometer. Two sites close together on a leaf were ventilated with two air jets of the same size, rate (about 40 cm^3 min^{-1}) and temperature, but different humidities (*e.g.* 75 and 85 per cent), controlled by bubbling through saturated solutions of NaCl and KCl, respectively. This causes a difference in the transpiration rates at the two sites, resulting in a temperature difference which was measured with a thermopile (Fig. 5.38). An empirical relationship is obtained between the thermopile output and the leaf resistance as measured with the mass flow resistance porometer. As calibrated in this way, the method does not take cuticular transpiration into consideration. The method can also be used for hypostomatous leaves such as most broad-leaf tree species.

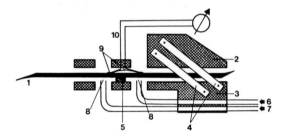

Fig. 5.38 Diagram of the differential transpiration porometer (MEIDNER and SPANNER 1959). 1 — leaf, 2 and 3 — two Perspex blocks with parallel platforms 5 × 10 × 60 mm, 4 — double hinge ensuring parallel position of plates, 5 — soft plastic pad, 6 — air inlet (75% r.h.), 7 — air inlet (85% r.h.), 8 — air spaces, 9 — thermopile, 10 — leads to the galvanometer.

5.6.4.2. Diffusion Porometers Using Other Gases

The diffusive flux of a gas q [g cm^{-2} s^{-1}] across a membrane (such as a leaf lamina) is assumed to be directly proportional to the difference in concentration of the gas $(C_1 - C_2)$ [g cm^{-3}] and inversely proportional to the resistance of the membrane (r) [s cm^{-1}]. Thus

$$q = \frac{C_1 - C_2}{r_{total}} . \tag{5.43}$$

The total diffusive resistance consists of that of the two epidermal layers containing the stomata (r_s), the diffusive resistance of the intercellular spaces in the mesophyll (r_i), and also that of the boundary layers on both surfaces of the leaf (r_a):

$$r_{total} = r_s^{adax} + r_i + r_s^{abax} + r_a^{adax} + r_a^{abax} . \tag{5.44}$$

Gases have been selected for diffusion porometry largely on account of the ease with which they can be analysed. The extent to which they affect the physiology of the leaf and hence cause artifacts in stomatal aperture is largely unknown.

GREGORY and ARMSTRONG (1936) measured the difference between the diffusion of hydrogen through the leaf and diffusion of air in the opposite direction using a chamber sealed to the leaf. As the diffusion of hydrogen is faster, the pressure in the cup falls. This is compensated for by the pressure of hydrogen generated in a simple electrolytic device. The authors measured the electric current necessary to generate the amount of hydrogen needed to maintain constant pressure in the leaf cup. The equipment is rather complex (Fig. 5.39), and the long exposure to hydrogen may affect stomatal apertures.

SPANNER (1953) measured the rate of hydrogen diffusion across the leaf by passing a slow stream of hydrogen over one side of the leaf while measuring the temperature increase (DUFOUR effect) at the other side of the leaf, where hydrogen (diffused through the leaf) mixed with nitrogen of the air, by means of an extremely fine thermocouple. It is not necessary to seal a chamber to the leaf. This method is again suitable for amphistomatous leaves only.

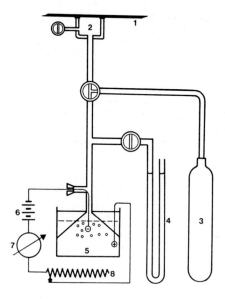

Fig. 5.39 Diagram of the hydrogen diffusion porometer (GREGORY and ARMSTRONG 1936). 1 — leaf, 2 — porometer cup, 3 — H_2 cylinder, 4 — manometer, 5 — hydrogene generating vessel, 6 — battery, 7 — microampermeter, 8 — sliding resistance.

A diffusion porometer for continuous measurement of diffusion rate and hence diffusive resistance of amphistomatous leaves, using nitrous oxide (N_2O) as the diffusing gas, was proposed by SLATYER and JARVIS (1966). This gas can easily be measured with an infrared analyzer, and does not cause any physiological changes in photosynthetic uptake of carbon dioxide or transpiration, according to the tests made by the authors. A concentration difference of nitrous oxide is established across the leaf by adding this gas to the air passing through one of the two compartments of the gas exchange chamber which was designed for the simultaneous measurement of exchanges at the upper and lower surfaces of a leaf. Concentrations of N_2O up to 5 000 ppm were used. The resistance [s cm^{-1}] calculated from the flux of N_2O [cm^3 cm^{-1} s^{-1}] and the known concentration difference [$cm^3 \ cm^{-3}$] comprises not only the internal resistance of the leaf (including the resistance of the stomata) but also the resistance in both boundary layers which must be determined experimentally (see Section 5.5.6). This resistance can be reduced by stirring the air in the chambers, so reducing the thickness of the boundary layers. The authors determined the diffusion coefficient of N_2O in air (D_{N_2O}) by comparing the diffusion of H_2O and N_2O in a model pore membrane. They found the ratio $D_{H_2O} : D_{N_2O} = 1.54$. If D_{H_2O} is $0.28 \ cm^{-2} \ s^{-1}$ at 28 °C, D_{N_2O}, then equals $0.18 \ cm^2 \ s^{-1}$. This procedure makes it possible to calculate approximate absolute values of the diffusion resistance of the leaf mesophyll including the stomata in the adaxial ("upper") and abaxial ("lower") epidermis ($r_1 + r_s^{ad} + r_s^{ab}$) to water vapour.

MORESHET et al. (1967, 1968) designed a porometer using radioactive argon diffusing from a cylindrical chamber across an amphistomatous leaf blade to a sampling tube (after 30 s diffusion), in which it is taken to a crystal scintillation counter. The reference concentration of radioactive argon in the chamber was measured by a flow-through GM cell in an air sample withdrawn with a hypodermic needle. The very short half-life of argon is a disadvantage, since it makes it necessary to work near an atomic reactor.

5.6.4.3 Mass Flow Porometers

The viscous flow of air forced by pressure difference through an amphistomatous leaf blade can be measured by mass flow porometers. The resistance of the leaf lamina to the viscous flow of air depends primarily on the resistance of stomata and the intercellular spaces.

Generally amphistomatous leaves may be measured by mass flow porometers so that the measured resistance to viscous flow of air across the leaf is the sum of the resistances of both epidermises with stomata and of the resistance of the intercellular system of the mesophyll. In hypostomatous leaves, where the chamber (porometer cup) is attached to the abaxial (lower) surface of the leaf, the air flux through the stomata of the lower epidermis plus a lateral air flow through communicating intercellular spaces to the stomata on the same side of the leaf around the chamber (cup) is measured.

It is again disadvantageous that measurement by mass flow porometers involve enclosing part of one side of the leaf in a chamber which may influence stomatal opening. The effect of the cup should be minimized by making the time when the cup is attached as short as possible. The mechanical effect of attaching the porometer chamber does not have any significant influence on the stomatal aperture. However, the chamber should be constructed in such a way that the illumination of the measured part of the leaf is decreased as little as possible by its attachment. It is also a disadvantage that the necessary difference in pressure on the two sides of the leaf blade may cause (mechanically) changes in the shape of the guard cells and so influence stomatal opening. There is also the possibility of dehydration of the mesophyll cells by mass flow through the intercellular spaces.

The theory of viscous flow porometry has been investigated by PENMAN (1942), BANGE (1953), STREBEYKO (1965a,b), WAGGONER (1965), JARVIS et al. (1967), MILTHORPE and PENMAN (1967), JARVIS (1971), and others.

Mass flow through the stomata and the intercellular spaces is a physiological artifact, as normal gas exchange takes place practically exclusively by diffusion.

MEIDNER and SPANNER (1959) assume cube root (for the graminaceous type of stomata) and square root (for eliptical stomata) relationships between diffusive conductivity and viscous conductivity. According to WAGGONER (1965) the diffusive resistance of the stomata r_s is proportional to the pressure drop in a pressure-drop porometer ΔP raised to the yth power

$$\log r_s = by \log \Delta P , \qquad (5.45)$$

where b may be derived directly from the the measured sizes of the stomata.

JARVIS et al. (1967) predict departure from a simple power-type relationship between diffusive viscous resistances for two reasons: (1) molecular slip in connection with stomatal geometry and (2) the fact that the internal resistance (r_i) adds a constant term to viscous and diffusive stomatal resistances. A detailed treatment of this problem is offered by MILTHORPE and PENMAN (1967).

The cup can be readily sealed to the leaf by means of a silicone rubber ring (RASCHKE 1965a) and clamp, or with a cement (e.g. with Terostat). The old LEICK's sealing material (LEICK 1928) is also useful: 2 parts pure calophony are melted in a porcelain dish with 3 parts (by weight) pure vaseline and boiled: 2 parts of white office rubber eraser cut into small pieces are added and mixed and boiled until a yellow-grey mass is obtained after cooling. This has an almost unlimited life when kept in a closed container.

In general there exist three types of mass flow porometers: In flow rate porometers the rate of air flow across the leaf is measured by means of flow meters in terms of $[cm^3\ cm^{-2}\ s^{-1}] = [cm\ s^{-1}]$ (viscous flow). In resistance porometers the resistance of the leaf lamina is compared with known mechanical resistances to viscous flow (e.g. capillaries). The third type are pressure-drop porometers. They determine time necessary for a drop in air pressure in the porometer chamber attached to the leaf caused by "leakage" through the leaf.

Flow Rate Porometers

The so-called soap-bubble porometer described by RASCHKE (1965a) seems to be very suitable for field measurements (Fig. 5.40). It is a cup porometer in which the flow rate is measured with a soap-bubble flow meter (LEVY 1964). Its operation is as follows. The level of the soap solution or detergent solution is raised above the air inlet by pressing a rubber ball. A bubble is formed and this is driven by the air flow through the measuring tube of a burette. The time for the bubble to pass through a pre-determined volume is measured with a stop-watch and the flow rate calculated. An inflated air-bed on which the investigator sits, kneels or even lies can serve as a source of pressure air under field conditions. Pressure control by device (*c*) facilitates the change of pressure, which may be very low (sometimes only 15 mm of water), this being physiologically very advantageous.

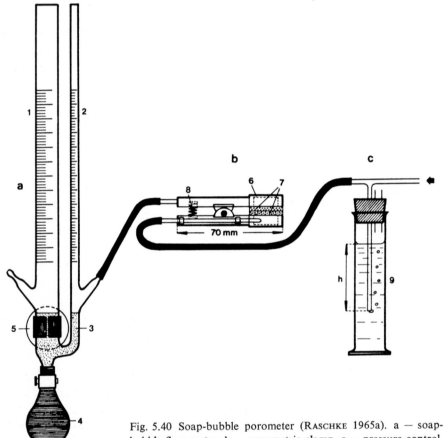

Fig. 5.40 Soap-bubble porometer (RASCHKE 1965a). a — soap-bubble flow meter, b — porometric clamp, c — pressure control, 1 and 2 — burettes suitable for two different flow-rate ranges, 3 — soap solution, 4 — rubber bulb, facilitating the increase of the soap solution level above the inlet orifice, 5 — magnetic holder, 6 — porometer cup, 7 — silicone rubber sealing, 8 — spring, 9 — pressure regulator, h — pressure in mm of H_2O.

Construction and operation of a simple type of flow rate porometer (GLOSER 1962 — unpublished data, SOLÁROVÁ 1965, GLOSER 1967) suitable both for field and laboratory work will be described here as this is applicable to similar instruments. A leaf cup (Fig. 5.41) of a functional area of 0.5 cm^2 is designed as a pincer chamber made from aluminium sheet and duralumin. The edge sealed to the leaf

is formed by a rubber ring (5). The leaf is pressed against the cup with a polyurethane foam ring (7) on the other pincer arm. Air (4) is drawn from the leaf cup
(2) by means of a thin-walled rubber tube (3 mm inner diameter) covered with
a sealed cover slip (3) (this prevents any substantial changes in illumination of the
measured part of the leaf). The complete porometer is shown in Fig. 5.42. The
capillary, with a mercury column about 40 mm long, has an inner diameter
of about 2 mm and is roughly 400 mm long. It is calibrated along its entire length
according to the procedure described on page 196. A decrease in length of the
mercury column of 100 mm corresponds to about 250 mm³ of air under these
conditions. The time [s] for the mercury column to drop by *e.g.* 300 mm (from

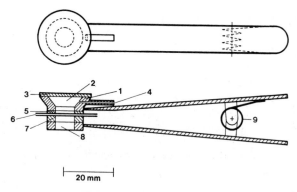

Fig. 5.41 A clamp porometric chamber (SOLÁROVÁ 1965). 1 — upper part, *i.e.* the chamber
proper, 2 — its conical part, 3 — sealed cover glass, 4 — air outlet from chamber to the porometer, 5 — soft silicone rubber annulus, 6 — leaf, 7 — polyurethane foam annulus, 8 — lower
part of the porometric clamp, 9 — spring.

one mark to another) is measured. This drop is chosen on the basis of preliminary
measurements: the measurement is accurate for a short drop provided that the
stomata are fairly closed. The tube must be thoroughly washed before use, first
with a chromic acid mixture, then several times with distilled water, and finally
with pure acetone. It is then dried with air (under reduced pressure). The mercury
should be as pure as possible. The air-tightness of the whole porometer must
be checked first: there should be no drop in the mercury column when it is drawn
up above the upper mark with the hose clamp (4) closed and the leaf cup sealed,
with LEICK sealing material, onto a PVC plate, for example. The parts of the rubber
tubes pressed by the hose clamps must be soft, so that closure is perfect. Clamps
were found to be much more useful than glass taps.
   The measuring procedure is as follows: (a) The mercury column is drawn up
the porometer capillary to the upper mark. This may be done in two ways. The
necessary volume of air is drawn out (with the upper clamp (4) open) with a 10 ml
all-glass syringe (3) (Fig. 5.42), so that the required suction is applied. This

procedure was found to be more useful than the other usual method of with-drawing the air out with a rubber ball: a rubber ball with a uni-directional valve is pressed (the valve allows air to flow out but not in) while the upper clamp (4) is closed. This results in suction. The upper clamp is then carefully released (the lower clamp (2) near the porometric chamber is of course closed) and the mercury is drawn to the required position. The upper clamp (4) is then left closed. If a syringe is used, the upper clamp (4) is not essential; however, care must be taken

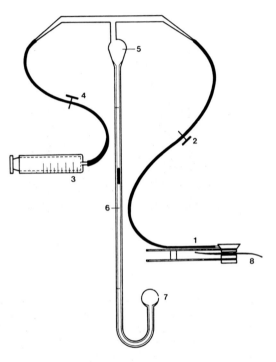

Fig. 5.42 A simple porometer according to GLOSER (1967) modified by SOLÁROVÁ (1965). 1 — porometric clamp, 2 — lower screw, 3 — injection syringe, 4 — upper clamp, 5 — upper safety space, 6 — calibrated capillary with mercury column, 7 — lower safety flask, 8 — leaf.

to ensure that the piston of the syringe is not moved during the measurement. It is essential to measure always at the same initial suction, so that the volume of air in which the suction caused by the weight of the mercury is observed, is always the same, *i.e.* the volume of air above the mercury up to the piston of the injection syringe (or to the clamp) and up to the lower clamp in the path to the porometric chamber.

(b) The leaf cup is attached to the leaf, the lower clamp is opened and the time taken for the upper meniscus of the mercury column to move from the upper mark to the lower one is measured.

BIERHUIZEN *et al.* (1965) described a porometer constructed in such a way that the force by which the cups are pressed together against the leaf can be regulated. The authors used a normal resistance porometer circuit (see later) for laboratory work, and a sphygmomanometer (blood-pressure meter) in the field.

The porometer described by RASCHKE (1965b) is an excellent recording flow rate porometer for long-term laboratory measurements. Saturation of the leaf segment with water is ensured in the chamber used for leaf segments (Fig. 5.43).

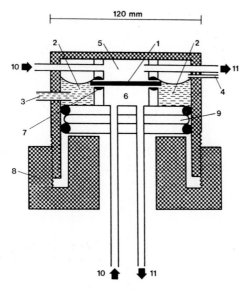

Fig. 5.43 Slightly modified porometer cup for recording mass flow porometer described by RASCHKE (1965b) [GLINKA and MEIDNER 1968]. The cup was designed for leaf discs (1) lying in contact with water (2) in a circular reservoir with water inlet (3) and outlet (4). The two chambers, upper (5) and lower (6) with air inlets and outlets. 7 — leaf disc washers, 8 — screw used to tighten the leaf disc between washers (7), 9 — movable platform carrying lower chamber (6), 10 — inlets and 11 — outlets of the upper and lower chamber for ventilating the chambers when readings are not being taken.

The flow rate is measured by a capillary flow meter and is converted to a pressure difference which is indicated by liquid-filled manometers. The liquid-filled parts of these manometers, operating as cylindrical lenses, concentrate light on silicone photocells. The construction of this very reliable laboratory porometer is rather complex; the reader is referred to the description in the original paper.

If the flow rate is measured, suction of pressure must be maintained during the measurement. If the suction is generated by the weight of a liquid (*e.g.* mercury) column, the level of which falls in a capillary connected to the attached cup during the measurement, it is useful to use a short mercury column.

For calculating the viscous resistance from flow rate data obtained by viscous flow porometers, an equation similar to Eq. 5.25 may be used by substituting the difference in concentration of the measured gas by the difference in pressure across the leaf.

Resistance Porometers

Resistance porometers are a type of mass flow porometer in which the leaf resistance is compared directly with known adjustable resistances (capillary and/or needle valves). The resistance porometer designed by GREGORY and PEARSE (1934)

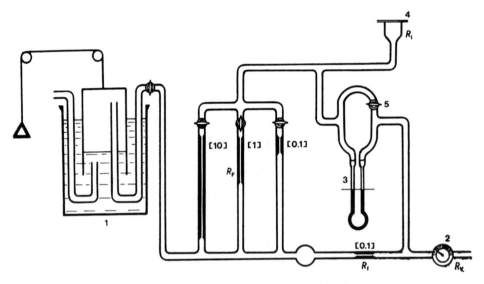

Fig. 5.44 Scheme of the porometer based on Wheatstone bridge (HEATH and RUSSELL 1951). 1 — aspirator used to press air into the device, 2 — calibrated needle valve, 3 — differential manometer, 4 — leaf, 5 — stopcock, R_F — series of three graded capillary resistances, R_f — standard capillary resistance, R_l — measured leaf resistance, R_v — resistance of the needle valve. See the description of the function of the device in the text.

and SPANNER and HEATH (1951) was improved by HEATH and RUSSELL (1951) using the Wheatstone bridge principle (Fig. 5.44). One arm of the bridge was formed by three graduated capillary resistances, of which one, R_F, was chosen on the basis of preliminary pilot experiments and connected in series with the resistance of the leaf to be measured (R_l). The remaining two arms of the bridge were formed by another standard resistance, R_f, in series with an accurately calibrated needle valve (2, resistance R_V). A sensitive differential manometer (3) was used here instead of the zero galvanometer used in an electric Wheatstone bridge. The manometer contained Brodie solution and liquid paraffin, making its sensitivity five times greater than if it were filled with water. The source of over-pressure was a gasometric aspirator (1) filled with liquid paraffin. Since it is neces-sary to maintain a constant pressure only for a very short period of the measure-

ment (relative values are used), the result does not depend on the absolute magnitude of the pressure – a large rubber ball can also be used as the source of pressure.

Measurement is as follows. The cup, with the rim covered with a sealing material, is pressed against a rubber ring attached to the leaf (4), the aspiration (1)
is put into operation, a suitable resistance R_F is selected (higher for larger and

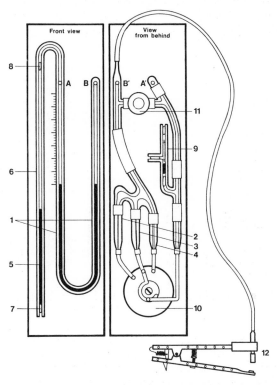

Fig. 5.45 The pocket size porometer according to WEATHERLEY (1966). 1 – manometer
serving as capillary flow meter, 2, 3 and 4 – three alternative standard capillaries switched by
a Plexiglass rotating switch (10), 5 – mercury column falling down in the capillary (6), 7 and
8 – cotton wool stoppers, 9 – mercury cotton-wool gravity operated valve, 10 – rotating
switch, 11 – flat type stopcock, 12 – porometer clip. A and A' and B and B', respectively are
interconnected.

more open stomata or for high stomatal density) and the calibrated needle resistance (2) is set in such a way that the differential manometer (3) shows no pressure
difference when the tap (5) is closed.

Then

$$\log R_1 = \log R_V + \log \frac{R_F}{R_f}. \qquad (5.46)$$

The selected resistances R_F are graded in multiples of 10, making it possible
to reach maximum accuracy within various ranges.

A pocket size resistance porometer was designed by WEATHERLEY (1966) (Fig. 5.45). The flow rate induced by a plug of mercury (5) falling in a tube (6) is measured with a mercury manometer (1) bridged with three interchangeable capillary resistances (2,3,4). The mercury is retained by cotton wool stops (7 and 8) inserted in the ends of the tube (6). The leaf in a porometer cup is connected in series to resistances (2, 3 or 4). When tube (6) and the arms of the mercury manometer (1) are always kept strictly in the same plane, the decrease in pressure can

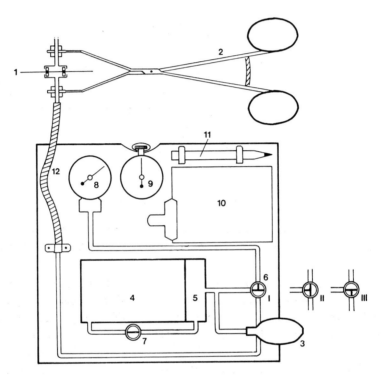

Fig. 5.46 The experimental set for field porometry (SHIMSHI 1967). 1 — leaf, 2 — surgical pincers with porometer cups, 3 — rubber bulb for pressing the air into a double chamber (4 and 5), 6 — three-way stopcock in position I, II and III, 7 — tube connecting two parts of the pressure vessel, 8 — pressure gauge, 9 — watch, 10 — data paper, 11 — pencil.

be changed by changing the angle at which the apparatus is held (normally 45°). In order to move the mercury column (5) back to the top of the tube [to the upper cotton wool stop (8)] in preparation for the next measurement, the apparatus is tilted and the air is pushed out by the returning mercury through the gravity operated mercury-cotton wool valve (9). Capillaries (2,3,4) can be interchanged by a simple Perspex rotating switch (10). The inner volume of connecting tubing is kept as small as possible. The flat-type tap (11) allows the standard resistance (2, 3 or 4) to be short-circuited. The porometer cup is made from a few millimeters of latex rubber tubing (2.5 to 4.5 in diameter) which has been pressed against a hot clean surface so that it is self-adhesive on the leaf.

A portable Wheatstone bridge porometer for field use was described by MORES-HET (1964). A recording version of the resistance porometer was constructed by HEATH and MANSFIELD (1962) and ALLAWAY and MANSFIELD (1969).

When using resistance porometers, the resistance of a capillary to viscous flow of air may be calculated from POISEUILLE's law:

$$q_{air} = \frac{\pi a^4 \, \Delta p}{8 \eta l} \, ,$$ (5.47)

where q_{air} is the viscous flow rate [cm^3 s^{-1}] through a tube of radius a [cm] and length 1 [cm]; Δp is the pressure difference in [g cm^{-1} s^{-2}], η is the coefficient of viscosity of air [g cm^{-1} s^{-1}] or [N m^{-2} s]. Hence the resistance to viscous flow

$$R = \frac{\Delta p}{q_{air}} = \frac{8 \eta l}{\pi a^4} \quad [\text{g cm}^{-4} \text{ s}^{-1}] .$$ (5.48)

Pressure-drop Porometers

In the so-called pressure-drop porometer, the rate of decrease of pressure in a system resulting from the leakage of air through the lamina of a leaf is measured.

A simple pressure-drop porometer was constructed by ALVIM (1965). Air is pumped into an inflatable rubber air reservoir with a rubber bulb pump until a certain pressure, measured by a pressure gauge (of the type used for measuring blood pressure, either a dial gauge or a sphygmometer) is reached. A three-way valve connected the reservois with the porometer cup, which is attached to the leaf. This is then opened so that the leaf is connected to the system. The time taken for a pre-determined pressure drop caused by air leakage through the leaf blade is measured. Other pressure-drop porometers were designed by BIERHUIZEN et al. (1965), MEIDNER (1965) STREBEYKO (1965a), WILLIAMS and SINCLAIR (1969) MOHSIN (1970).

An improved version of ALVIM's porometer, described by SHIMSHI (1967), is shown in Fig. 5.46. Instead of a rubber air reservoir there is a rigid air container (e.g. a glass bottle) into which air is pumped by a rubber bulb pump. By changing the size of the reservoir, the sensitivity of the instrument can be readily altered.

Conductance K [moles s^{-1} cm^{-2} (mm Hg)$^{-1}$] can be calculated from the equation:

$$K = \frac{V \ln (P_2/P_1)}{RTA \, \Delta t} \, ,$$ (5.49)

where V is the volume of the porometer [cm^3], P_1 is initial pressure and P_2 is final pressure [mm Hg], R is the gas constant, T absolute temperature [K], A cup area [cm^2], Δt time for the pressure drop [s] (*e.g.* SHIMSHI 1967, PEASLEE and MOSS 1968).

A recording pressure-drop porometer was designed by JODO (1970): the frequency of the drops in pressure in the line ending in a porometer cup is recorded (Fig. 5.47).

The pressures used in most pressure-drop porometers mainly those using sphygmomanometers are high enough to affect the guard cell configuration (RASCHKE 1965a).

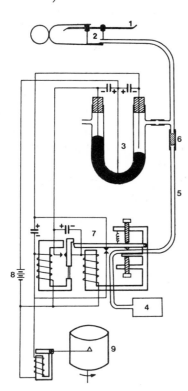

Fig. 5.47 Diagram of the recording pressure-drop porometer (JODO 1970). 1 — leaf, 2 — porometer cup with a leaf clamp, 3 — mercury contact manometer, 4 — membrane pump, 5 — silicone tube, 6 — filter, 7 — electric valve, 8 — battery, 9 — recorder.

5.7 Guttation

Guttation is the secretion of liquid water from the leaf, in the form of solutions containing low concentrations of inorganic and organic compounds, by special structures called hydathodes. Hydathodes are usually located at the margin or the tip of the leaf blades; they consist of thin-walled parenchymatous tissue, epithem, adjacent to the ends of the leaf xylem elements. The epithem secretes water through pores in the epidermis of the leaf surface as a result of hydrostatic pressure in the xylem. The pores are formed by inactive stomata. Guttation is a process serving to equilibrate the water balance of a plant during rapid absorption by roots and minimum or zero transpiration. The plants exhibit guttation especially when they have a rapidly growing and active root system (*e.g.* young plants), when ecological conditions favour activity of the roots (moist, warm and aerated soil), and when transpiration is low and practically zero (night, humid air). It is often difficult to distinguish guttation from condensation of dew, as they frequently occur suimultaneously in natural conditions.

The extent of guttation may be measured in several ways. A simple gravimetric determination is to absorb all drops of guttation water from the leaf surface with weighed strips of filter paper, and then rapidly to reweigh the paper. Errors may arise through water loss by evaporation or by water dripping off the strips. The guttation water may also be measured volumetrically by sucking it up with micropipettes (see TOMAN and SYNAK 1954).

HÖHN and BOY (1956) constructed a guttograph for measuring and recording the guttation rate. This is based on recording number of drops of guttation water. The plants are turned upside down during the measurement, and the dripping guttation water is collected in a vessel. The water drips onto coloured recording paper. The frequency of the drops may also be recorded by a photocell (in a similar way to a fraction collector used for chromatography). ÚLEHLA's method of measuring strips described in Section 3.4.2. may also be used for the measurement of the rate of guttation or total amount of guttated water.

Table 5.4 *Review of methods of stomatal aperture*

Method	What is measured	Equipment
1 **Direct observation** (Section 5.6.3.1)	Number (per mm^2), shape and size (length, width in μm) of stomatal guard cells and apertures	
1.1 Direct microscopy on living material		Microscope with medium or high magnification
1.2 Direct microscopy on fixed epidermis		Microscope with high magnification
1.3 Direct, still and time--lapse microphoto-graphy		Special camera adaptation (expensive) using medium magnification
2 **Indirect microscopy** or microphotography of leaf surface reprints (impressions) (Section 5.6.3.2) 2.1 Using dissolved compounds	Same as 1	Solution of collodion (or metacrylate or cellulose acetate)
2.2 Using silicon rubber negatives		Solution for silicon rubber polymerization
3 **Infiltration methods** (Section 5.6.3.3)	Resistance of stomata plus intercellular spaces to viscous flow of the liquid. Relative values only	
3.1 Infiltration under atmospheric pressure	As in 3	Liquids with different viscosity and surface tension

determination (modified from SLAVÍK *1971).*

Advantages and usefulness	Disadvantages and limitations
Living material. By focusing, the shape of stomatal aperture may be observed. The same stoma may be repeatedly observed	Limited number of stomata observed. Difficult with hairy leaves and very small stomata. Possible effect of external conditions during determination
Very simple for suitable material	Leaves with easily stripped epidermis only. Possible changes in aperture and size due to fixation
As in 1.1	As in 1.1 Effect to changed light conditions very critical, especially in time lapse photography
Simple and rapid technique suitable for use in the field and for quantitative evaluation of surface anatomy. Large number of stomata may be measured	Results depend on the viscosity of the liquid: how far the pore is penetrated, so that the reprints do not necessarily represent the aperture itself. Cannot be repeated on the same spot. Stomatal aperture may be affected by the organic solvent and/or by excessive cooling of the leaf due to its vaporization. Unreliable with apertures of less than about 1 μm
As in 2.1	As in 2.1
Simple and rapid method yielding relative values	Objective evaluation, cannot be repeated on the same leaf. Results are not comparable for different plant species. Useless in leaves with a damaged cuticles (e.g. by pests). Wettability and solubility of leaf surface may interfere

Table 5.4

Method	What is measured	Equipment
3.2 Pressure infiltration	As in 3	Special, inexpensive equipment with small pressure bomb
4 Porometry 4.1 Water vapour diffusion porometry (Section 5.6.4.1)	A sum of boundary layer resistance plus transpiration resistance, *i.e.* diffusive resistance to water vapour of stomata, intercellular spaces and cuticle	Equipment for measuring rate of change of humidity and air and leaf temperature in leaf chamber
4.2 Diffusion porometry (Section 5.6.4.2)	Diffusive resistance of the leaf to model gases which may be recalculated for CO_2 [s cm^{-1}]	Usually leaf chambers on both sides of a leaf are used; the concentration of the gas is measured by elaborate and expensive equipment
4.3 Viscous flow porometry (mass flow) (Section 5.6.4.3)	Viscous flow conductivity (or resistance) of the leaf usually to air	
4.3.1 Viscous flow rate measurements	Viscous flow rate measured, viscous resistance calculated	Leaf chambers, flowmeter
4.3.2 Resistance porometers	Viscous flow resistance directly measured	Laboratory equipment with leaf chambers
4.3.3 Pressure-drop measurements	Viscous conductance (resistance) may be calculated from pressure drop, time and volume of the porometer chamber	Mostly simple devices using leaf chamber(s)
5 Calculation of transpiration resistances and hence stomatal resistance from determination of transpiration rate (Section 5.5.6)	Transpiration rate from a leaf surface may be recalculated to the transpiration resistance values	Equipment for gasometric measurements of transpiration rate using leaf chambers. Leaf and air temperatures must be measured simultaneously

(continued)

Advantages and usefulness	Disadvantages and limitations
Values representing large number of stomata. Useful for field work, and also for linear (needle) leaves and both hypostomatous (or epistomatous) and amphistomatous leaves	As in 3.1
Yields diffusive resistance values for water vapour	Includes the flux of water through the cuticle
A useful laboratory method yielding directly values of diffusive resistance	Expensive laboratory equipment needed, only for amphistomatous leaves. Includes the intercellular space resistance through the leaf
Useful for field measurements or sophisticated laboratory devices, which can be adapted for recording Laboratory methods	Pressure difference (if high) may cause artifact and change stomatal aperture
	The calculation of diffusive resistances is not simple: the relationship of viscous flow conductivity (resistance) to diffusive conductivity (resistance) also depends on the geometry of the stomatal aperture. Includes the intercellular space resistance through the leaf
Useful for field measurements	High pressure difference may cause artifacts
Useful with simultaneous gasometric measurements of CO_2 exchange in controlled conditions	

6

Table Appendix

Table 6.1 *Pertinent physical values, their symbols and units.*

Basic Units	Secondary SI Units	SI Unit and its Symbol	Dimension of the Unit	Acceptable Units and Dimensions	Other Units and their Conversion Factors
Length l		the metre, m	m	$1\ \text{cm} = 0.01\ \text{m}$	1 inch, in $= 25.4\ \text{mm} = 0.0254\ \text{m}$ 1 foot, ft $= 12\ \text{in} = 0.3048\ \text{m}$ 1 yard, yd $= 36\ \text{in} = 0.9144\ \text{m}$ 1 mile $= 1760\ \text{yd} = 1.60934\ \text{km}$
	Area A		m^2	$1\ \text{cm}^2 = 10^{-4}\ \text{m}^2$	$1\ \text{in}^2 = 6.4516\ \text{cm}^2 = 6.4516 \times 10^{-4}\ \text{m}^2$ $1\ \text{ft}^2 = 929.0304\ \text{cm}^2$ $1\ \text{yd}^2 = 8361.2736\ \text{cm}^2$
	Volume V		m^3	$1\ \text{cm}^3 = 10^{-6}\ \text{m}^3$ $1\ \text{dm}^3 = 10^{-3}\ \text{m}^3$ litre, $1\ \text{l} = \text{dm}^3$	$1\ \text{in}^3 = 16.3871\ \text{cm}^3 = 1.63871 \times 10^{-5}\ \text{m}^3$ $1\ \text{ft}^3 = 2.83168 \times 10^{-2}\ \text{m}^3$ 1 pint, pt $= 0.568261\ \text{l}$ 1 quart, qt $= 1.136521$ 1 gallon (USA), USgal $= 3.78541 \times 10^{-3}\ \text{m}^3$ 1 gallon (UK), gal $= 4.54609 \times 10^{-3}\ \text{m}^3$
	Velocity v, u		m s^{-1}	$1\ \text{cm s}^{-1} = 0.01\ \text{m s}^{-1}$ $1\ \text{km h}^{-1} = 0.27\ \text{m s}^{-1}$	$1\ \text{ft/s (fps)} = 0.3048\ \text{m s}^{-1}$ $1\ \text{mile/h (mph)} = 0.44704\ \text{m s}^{-1}$
	Acceleration		m s^{-2}	$1\ \text{cm s}^{-2} = 10^{-2}\ \text{m s}^{-2}$	
Mass m		the kilogram, kg	kg	$1\ \text{g} = 0.001\ \text{kg}$	1 ounce, oz $= 28.3495\ \text{g}$ 1 pound, $1\ \text{lb} = 453.59239\ \text{g}$ 1 ounce fluid, fl oz $= 28.4131\ \text{cm}^3$ 1 quarter $= 12.7006\ \text{kg}$ 1 ton $= 1016.05\ \text{kg}$ $1\ \text{lb/ft}^3 = 16.0185\ \text{kg m}^{-3}$
	Density ϱ		kg m^{-3}	$1\ \text{g cm}^{-3} = 1000\ \text{kg m}^{-3}$	
	Force F	the newton, N	$1\ \text{N} = \text{kg m s}^{-2}$	$1\ \text{dyn} = 1\ \text{g cm s}^{-2} = 10^{-5}\ \text{N}$	
	Pressure p, P	the pascal, Pa	$1\ \text{Pa} = \text{N m}^{-2} = \text{kg m}^{-1}\,\text{s}^{-2}$	$1\ \text{bar} = 1000\ \text{mbar} = 10^5\ \text{N m}^{-2}$	$1\ \text{atm} = 760\ \text{torr} = 1013.25\ \text{mbar} = 101325\ \text{N m}^{-2}$ $1\ \text{torr} = 133.322\ \text{N m}^{-2}$

Physical value	Unit	Symbol / definition	Relations	Conversions
Time t	the second, s	s	day, 1 d = 24 h = 1440 min = 86400 s; hour, 1 h = 60 min = 3600 s; minute, 1 min = 60 s [all in mean solar units]	1 dyn cm^{-2} = 0.1 N m^{-2}; 1 mm Hg = 1.00000014 torr; 1 in Hg = 25.4 mm Hg; 1 cm H$_2$O = 98.0665 N m^{-2} = 0.9806 65 mbar; 1 in H$_2$O (4°C) = 2.491 mbar
Energy	the joule, J	$1\,\mathrm{J} = \mathrm{kg\,m^2\,s^{-2}}$	calorie (international table) $1\,\mathrm{cal_{IT}} = 4.1868\,\mathrm{J}$; calorie (thermochemical) $1\,\mathrm{cal_{th}} = 4.184\,\mathrm{J}$	$1\,\mathrm{erg} = 10^{-7}\,\mathrm{J}$; 1 British thermal unit (Btu) = 1.05506 kJ
Power	the watt, W	$1\,\mathrm{W} = \mathrm{J\,s^{-1}} = \mathrm{kg\,m^2\,s^{-3}}$; $\mathrm{J\,m^{-2}\,s^{-1}} = \mathrm{W\,m^{-2}}$		
Energy flux			$1\,\mathrm{cal\,cm^{-2}\,s^{-1}} = 41.868\,\mathrm{kW\,m^{-2}}$; langley min^{-1}, $1\,\mathrm{ly\,min^{-1}} = 1\,\mathrm{cal\,cm^{-2}\,min^{-1}} = 697.8\,\mathrm{W\,m^{-2}}$	
Electric current I	the ampere, A	A	$1\,\mu\mathrm{A} = 0.000001\,\mathrm{A}$	
Temperature T	the kelvin, K	K	$°\mathrm{C} = [T(°\mathrm{C}) + 273.15]\,\mathrm{K}$	1 degree Fahrenheit, $°\mathrm{F} = 5/9[T(°\mathrm{F}) + 459.67]\,\mathrm{K}$
Amount of substance n	the mole, mol	mol		
Luminous intensity I_v	the candela, cd	cd		

Table 6.2 *Prefixes for units.*

	Prefix	Symbol		Prefix	Symbol
10^{-1}	deci	d	10	deka	da
10^{-2}	centi	c	10^2	hekto	h
10^{-3}	milli	m	10^3	kilo	k
10^{-6}	micro	μ	10^6	mega	M
10^{-9}	nano	n	10^9	giga	G
10^{-12}	piko	p	10^{12}	tera	T
10^{-15}	femto	f			
10^{-18}	atto	a			

Table 6.3 *Some mathematical symbols.*

$=$	equal
$\neq$	non equal
$\equiv$	identically equal
$\hat{=}$	corresponds
$\approx$	approximately equal
$\sim$ or $\propto$	proportional to
∞	infinity

Table 6.4 *Some physical properties of liquid water.*

Property		−5	0	5	10	15	20	25	30
					Temperature [°C]				
Density	[g cm^{-3}]	0.99918	0.99987	0.99999	0.99973	0.99913	0.99823	0.99708	0.99568
Dynamic viscosity	[mg cm^{-1} s^{-1}]	—	17.921	15.188	13.077	11.404	10.050	8.937	8.007
Surface tension	[g s^{-2}]	76.4	75.6	74.8	74.2	73.4	72.7	71.9	71.1
Latent heat of vaporization	[J g^{-1}]	—	2 501	—	2 477	2 465	2 454	2 442	2 430
	[cal g^{-1}]	—	597.3	594.5	591.7	588.9	586.0	583.2	580.4
Specific heat at constant pressure	[J g^{-1} °C^{-1}]	—	4.217	—	4.192	4.186	4.182	4.179	4.178
	[cal g^{-1} °C^{-1}]	—	1.0074	1.0037	1.0013	0.9998	0.9988	0.9983	0.9980
Thermal conductivity	[mJ cm^{-1} s^{-1} °C^{-1}]	—	5.6	—	5.8	5.9	5.9	—	6.1
	[cal cm^{-1} s^{-1} °C^{-1}]	—	1.34	1.37	1.40	1.42	1.44	1.46	1.48

All data at atmospheric pressure 1 013.25 mbar = 760 torr.

Table 6.5 *Some physical properties of water vapour* (*From* VAN WIJK 1963).

Temperature [°C]		0	5	10	15	20	25	30	35	40
Saturation vapour pressure over water e_0	[mbar]	6.108		12.272	17.044	23.373	31.671	42.430		
	[mm Hg]	4.58	6.53	9.20	12.78	17.52	23.75	31.82	42.20	55.30
Density of vapour in water vapour saturated air	[g m^{-3}]	4.847		9.399	12.83	17.30	23.05	30.38		
Diffusivity of water vapour in air D	[cm^2 s^{-1}]	0.224		0.239	0.247	0.255	0.262	0.270		

Table 6.6 *Absolute air humidity* $[\mathrm{mg\,l^{-1} = g\,m^{-3}}]$ *at various relative air humidities* [%] *and air temperatures* [°C].

T [°C]	r.h. 10	20	30	40	50	60	70	80	90	100
−10	0.215	0.431	0.646	0.862	1.077	1.292	1.508	1.723	1.939	2.154
− 9	0.234	0.468	0.702	0.936	1.170	1.404	1.638	1.872	2.106	2.340
− 8	0.254	0.507	0.761	1.015	1.269	1.522	1.776	2.030	2.283	2.537
− 7	0.275	0.550	0.826	1.101	1.376	1.651	1.926	2.202	2.477	2.752
− 6	0.299	0.598	0.896	1.195	1.494	1.793	2.092	2.390	2.689	2.988
− 5	0.324	0.649	0.973	1.298	1.622	1.946	2.271	2.595	2.920	3.244
− 4	0.352	0.704	1.055	1.407	1.759	2.111	2.463	2.814	3.166	3.518
− 3	0.381	0.763	1.144	1.525	1.907	2.288	2.669	3.050	3.432	3.813
− 2	0.413	0.826	1.239	1.652	2.065	2.478	2.891	3.304	3.717	4.130
− 1	0.447	0.894	1.340	1.787	2.234	2.681	3.128	3.574	4.021	4.468
0	0.485	0.969	1.454	1.938	2.423	2.908	3.392	3.877	4.361	4.846
+ 1	0.519	1.038	1.557	2.076	2.596	3.115	3.634	4.153	4.672	5.191
2	0.556	1.111	1.667	2.223	2.779	3.334	3.890	4.446	5.001	5.557
3	0.595	1.189	1.784	2.378	2.973	3.567	4.162	4.756	5.351	5.945
4	0.636	1.272	1.907	2.543	3.179	3.815	4.451	5.086	5.722	6.358
5	0.680	1.359	2.039	2.718	3.398	4.077	4.757	5.436	6.116	6.795
6	0.726	1.451	2.177	2.903	3.629	4.354	5.080	5.806	6.531	7.257
7	0.775	1.549	2.324	3.099	3.874	4.648	5.423	6.198	6.972	7.747
8	0.827	1.653	2.480	3.307	4.134	4.960	5.787	6.614	7.440	8.267
9	0.882	1.763	2.645	3.526	4.408	5.290	6.171	7.053	7.934	8.816
10	0.940	1.879	2.819	3.758	4.698	5.638	6.577	7.517	8.456	9.396
11	1.00	2.00	3.00	4.00	5.01	6.01	7.01	8.01	9.01	10.01
12	1.07	2.13	3.20	4.26	5.33	6.40	7.46	8.53	9.59	10.66
13	1.13	2.27	3.40	4.54	5.67	6.80	7.94	9.07	10.21	11.34
14	1.21	2.41	3.62	4.82	6.03	7.24	8.44	9.65	10.85	12.06
15	1.28	2.56	3.85	5.13	6.41	7.69	8.97	10.26	11.54	12.82
16	1.36	2.73	4.09	5.45	6.82	8.18	9.54	10.90	12.27	13.63
17	1.45	2.89	4.34	5.79	7.24	8.68	10.13	11.58	13.02	14.47
18	1.54	3.07	4.61	6.14	7.68	9.22	10.75	12.29	13.82	15.36
19	1.63	3.26	4.89	6.52	8.15	9.78	11.41	13.04	14.67	16.30
20	1.73	3.46	5.19	6.92	8.65	10.37	12.10	13.83	15.56	17.29
21	1.83	3.67	5.50	7.33	9.17	11.00	12.83	14.66	16.50	18.33
22	1.94	3.88	5.83	7.77	9.71	11.65	13.59	15.54	17.48	19.42
23	2.06	4.11	6.17	8.22	10.28	12.34	14.39	16.45	18.50	20.56
24	2.18	4.35	6.53	8.71	10.89	13.06	15.24	17.42	19.59	21.77
25	2.30	4.61	6.91	9.22	11.52	13.82	16.13	18.43	20.74	23.04
26	2.44	4.87	7.31	9.75	12.19	14.62	17.06	19.50	21.93	24.37
27	2.58	5.15	7.73	10.30	12.88	15.46	18.03	20.61	23.18	25.76
28	2.72	5.45	8.17	10.89	13.62	16.34	19.06	21.78	24.51	27.23
29	2.88	5.75	8.63	11.50	14.38	17.26	20.13	23.01	25.88	28.76
30	3.04	6.07	9.11	12.14	15.18	18.22	21.25	24.29	27.32	30.36

Table 6.6 *(continued)*

$T\,°C$	r.h. 10	20	30	40	50	60	70	80	90	100
31	3.20	6.41	9.61	12.82	16.02	19.22	22.43	25.63	28.84	32.04
32	3.31	6.62	9.92	13.23	16.54	19.85	23.16	26.46	29.77	33.08
33	3.57	7.13	10.70	14.26	17.83	21.39	24.96	28.52	32.09	35.65
34	3.76	7.52	11.27	15.03	18.79	22.55	26.31	30.06	33.82	37.58
35	3.96	7.92	11.88	15.84	19.80	23.76	27.72	31.68	35.64	39.60
36	4.17	8.34	12.51	16.68	20.86	25.03	29.20	33.37	37.54	41.71
37	4.39	8.78	13.18	17.57	21.96	26.35	30.74	35.14	39.53	43.92
38	4.62	9.24	13.87	18.49	23.11	27.73	32.35	36.98	41.60	46.22
39	4.86	9.73	14.59	19.45	24.32	29.18	34.04	38.90	43.77	48.63
40	5.11	10.23	15.34	20.46	25.57	30.68	35.80	40.91	46.03	51.14

Table 6.7 *Absolute air humidity* [mg l^{-1} = g m^{-3}] *at various relative air humidities* [%] *and air temperatures* [°C] *in the range between 40% to 100% relative humidity and between 15.0 °C to 30.0 °C.*

Sequence of the parts of the Table on the next pages:

°C	r.v. [%]					
	40　　　50	60	70	80	90	100
15	1A	2A	3A	4A	5A	6A
19	1B	2B	3B	4B	5B	6B
23	1C	2C	3C	4C	5C	6C
27	1D	2D	3D	4D	5D	6D
30						

1A

T [°C]	r.h. [%]										
	40	41	42	43	44	45	46	47	48	49	50
15.0	5.13	5.26	5.39	5.51	5.64	5.77	5.90	6.03	6.15	6.28	6.41
15.1	5.16	5.29	5.42	5.55	5.68	5.81	5.94	6.06	6.19	6.32	6.45
15.2	5.19	5.32	5.45	5.58	5.71	5.84	5.97	6.10	6.23	6.36	6.49
15.3	5.23	5.36	5.49	5.62	5.75	5.88	6.01	6.14	6.27	6.40	6.53
15.4	5.26	5.39	5.52	5.65	5.78	5.92	6.05	6.18	6.31	6.44	6.57
15.5	5.29	5.42	5.56	5.69	5.82	5.95	6.09	6.22	6.35	6.48	6.62
15.6	5.32	5.46	5.59	5.72	5.86	5.99	6.12	6.26	6.39	6.52	6.66
15.7	5.35	5.49	5.62	5.76	5.89	6.03	6.16	6.29	6.43	6.56	6.70
15.8	5.39	5.52	5.66	5.79	5.93	6.06	6.20	6.33	6.47	6.60	6.74
15.9	5.42	5.55	5.69	5.83	5.96	6.10	6.24	6.37	6.51	6.64	6.78
16.0	5.45	5.59	5.72	5.86	6.00	6.14	6.27	6.41	6.55	6.68	6.82
16.1	5.48	5.62	5.76	5.90	6.04	6.17	6.31	6.45	6.59	6.72	6.86
16.2	5.52	5.66	5.80	5.93	6.07	6.21	6.35	6.49	6.63	6.77	6.90
16.3	5.55	5.69	5.83	5.97	6.11	6.25	6.39	6.53	6.67	6.81	6.95
16.4	5.59	5.73	5.87	6.01	6.15	6.29	6.43	6.57	6.71	6.85	6.99
16.5	5.62	5.76	5.90	6.04	6.18	6.33	6.47	6.61	6.75	6.89	7.03
16.6	5.65	5.80	5.94	6.08	6.22	6.36	6.51	6.65	6.79	6.93	7.07
16.7	5.69	5.83	5.97	6.12	6.26	6.40	6.54	6.69	6.83	6.97	7.11
16.8	5.72	5.87	6.01	6.15	6.30	6.44	6.58	6.73	6.87	7.01	7.16
16.9	5.76	5.90	6.04	6.19	6.33	6.48	6.62	6.77	6.91	7.05	7.20
17.0	5.79	5.94	6.08	6.23	6.37	6.52	6.66	6.81	6.95	7.10	7.24
17.1	5.83	5.97	6.12	6.26	6.41	6.56	6.70	6.85	6.99	7.14	7.28
17.2	5.86	6.01	6.15	6.30	6.45	6.59	6.74	6.89	7.03	7.18	7.33
17.3	5.90	6.04	6.19	6.34	6.49	6.63	6.78	6.93	7.08	7.22	7.37
17.4	5.93	6.08	6.23	6.38	6.52	6.67	6.82	6.97	7.12	7.27	7.42
17.5	5.97	6.12	6.26	6.41	6.56	6.71	6.86	7.01	7.16	7.31	7.46
17.6	6.00	6.15	6.30	6.45	6.60	6.75	6.90	7.05	7.20	7.35	7.50
17.7	6.04	6.19	6.34	6.49	6.64	6.79	6.94	7.09	7.25	7.40	7.55
17.8	6.07	6.22	6.37	6.53	6.68	6.83	6.98	7.14	7.29	7.44	7.59
17.9	6.11	6.26	6.41	6.56	6.72	6.87	7.02	7.18	7.33	7.48	7.64
18.0	6.14	6.29	6.45	6.60	6.76	6.91	7.06	7.22	7.37	7.53	7.68
18.1	6.18	6.33	6.49	6.64	6.80	6.95	7.11	7.26	7.42	7.57	7.73
18.2	6.22	6.37	6.53	6.68	6.84	7.00	7.15	7.31	7.46	7.62	7.77
18.3	6.25	6.41	6.57	6.72	6.88	7.04	7.19	7.35	7.51	7.66	7.82
18.4	6.29	6.45	6.61	6.77	6.92	7.08	7.24	7.40	7.55	7.71	7.87
18.5	6.33	6.49	6.65	6.81	6.96	7.12	7.28	7.44	7.60	7.76	7.92
18.6	6.37	6.53	6.69	6.85	7.01	7.17	7.32	7.48	7.64	7.80	7.96
18.7	6.41	6.57	6.73	6.89	7.05	7.21	7.37	7.53	7.69	7.85	8.01
18.8	6.44	6.61	6.77	6.93	7.09	7.25	7.41	7.57	7.73	7.90	8.06
18.9	6.48	6.64	6.81	6.97	7.13	7.29	7.46	7.62	7.78	7.94	8.10
19.0	6.52	6.68	6.85	7.01	7.17	7.34	7.50	7.66	7.82	7.99	8.15

1B

T [°C]	r.h. [%]										
	40	41	42	43	44	45	46	47	48	49	50
19.0	6.52	6.68	6.85	7.01	7.17	7.34	7.50	7.66	7.82	7.99	8.15
19.1	6.56	6.72	6.89	7.05	7.22	7.38	7.54	7.71	7.87	8.04	8.20
19.2	6.60	6.77	6.93	7.10	7.26	7.43	7.59	7.76	7.92	8.09	8.25
19.3	6.64	6.81	6.97	7.14	7.30	7.47	7.64	7.80	7.97	8.13	8.30
19.4	6.68	6.85	7.01	7.18	7.35	7.52	7.68	7.85	8.02	8.18	8.35
19.5	6.72	6.89	7.06	7.22	7.39	7.56	7.73	7.90	8.06	8.23	8.40
19.6	6.76	6.93	7.10	7.27	7.44	7.61	7.77	7.94	8.11	8.28	8.45
19.7	6.80	6.97	7.14	7.31	7.48	7.65	7.82	7.99	8.16	8.33	8.50
19.8	6.84	7.01	7.18	7.35	7.52	7.70	7.87	8.04	8.21	8.38	8.55
19.9	6.88	7.05	7.22	7.40	7.57	7.74	7.91	8.08	8.26	8.43	8.60
20.0	6.92	7.09	7.27	7.44	7.61	7.79	7.96	8.13	8.30	8.48	8.65
20.1	6.96	7.14	7.31	7.48	7.66	7.83	8.01	8.18	8.35	8.53	8.70
20.2	7.00	7.18	7.35	7.53	7.70	7.88	8.05	8.23	8.40	8.58	8.75
20.3	7.04	7.22	7.40	7.57	7.75	7.93	8.10	8.28	8.45	8.63	8.81
20.4	7.08	7.26	7.44	7.62	7.79	7.97	8.15	8.33	8.50	8.68	8.86
20.5	7.13	7.30	7.48	7.66	7.84	8.02	8.20	8.38	8.55	8.73	8.91
20.6	7.17	7.35	7.53	7.71	7.88	8.06	8.24	8.42	8.60	8.78	8.96
20.7	7.21	7.39	7.57	7.75	7.93	8.11	8.29	8.47	8.65	8.83	9.01
20.8	7.25	7.43	7.61	7.79	7.98	8.16	8.34	8.52	8.70	8.88	9.07
20.9	7.29	7.47	7.66	7.84	8.02	8.20	8.39	8.57	8.75	8.94	9.12
21.0	7.33	7.51	7.70	7.88	8.07	8.25	8.43	8.62	8.80	8.99	9.17
21.1	7.37	7.56	7.74	7.93	8.11	8.30	8.48	8.67	8.85	9.04	9.22
21.2	7.42	7.60	7.79	7.98	8.16	8.35	8.53	8.72	8.91	9.09	9.28
21.3	7.46	7.65	7.84	8.02	8.21	8.40	8.58	8.77	8.96	9.15	9.33
21.4	7.51	7.69	7.88	8.07	8.26	8.45	8.63	8.82	9.01	9.20	9.39
21.5	7.55	7.74	7.93	8.12	8.31	8.50	8.68	8.87	9.06	9.25	9.44
21.6	7.59	7.78	7.97	8.16	8.35	8.54	8.73	8.92	9.11	9.30	9.49
21.7	7.64	7.83	8.02	8.21	8.40	8.59	8.78	8.98	9.17	9.36	9.55
21.8	7.68	7.87	8.07	8.26	8.45	8.64	8.83	9.03	9.22	9.41	9.60
21.9	7.73	7.92	8.11	8.31	8.50	8.69	8.88	9.08	9.27	9.46	9.66
22.0	7.77	7.96	8.16	8.35	8.55	8.74	8.93	9.13	9.32	9.52	9.71
22.1	7.82	8.01	8.21	8.40	8.60	8.79	8.99	9.18	9.38	9.57	9.77
22.2	7.86	8.06	8.25	8.45	8.65	8.84	9.04	9.24	9.43	9.63	9.82
22.3	7.91	8.10	8.30	8.50	8.70	8.89	9.09	9.29	9.49	9.68	9.88
22.4	7.95	8.15	8.35	8.55	8.75	8.94	9.14	9.34	9.54	9.74	9.94
22.5	8.00	8.20	8.40	8.60	8.80	9.00	9.20	9.40	9.60	9.80	10.00
22.6	8.04	8.24	8.44	8.64	8.85	9.05	9.25	9.45	9.65	9.85	10.05
22.7	8.09	8.29	8.49	8.69	8.90	9.10	9.30	9.50	9.70	9.91	10.11
22.8	8.13	8.33	8.54	8.74	8.94	9.15	9.35	9.56	9.76	9.96	10.17
22.9	8.18	8.38	8.59	8.79	8.99	9.20	9.40	9.61	9.81	10.02	10.22
23.0	8.22	8.43	8.63	8.84	9.04	9.25	9.46	9.66	9.87	10.07	10.28

1C

T [°C]	r.h. [%]										
	40	41	42	43	44	45	46	47	48	49	50
23.0	8.22	8.43	8.63	8.84	9.04	9.25	9.46	9.66	9.87	10.07	10.28
23.1	8.27	8.48	8.68	8.89	9.10	9.31	9.51	9.72	9.93	10.13	10.34
23.2	8.32	8.53	8.74	8.94	9.15	9.36	9.57	9.78	9.99	10.19	10.40
23.3	8.37	8.58	8.79	9.00	9.21	9.42	9.63	9.83	10.04	10.25	10.46
23.4	8.42	8.63	8.84	9.05	9.26	9.47	9.68	9.89	10.10	10.31	10.52
23.5	8.47	8.68	8.89	9.10	9.31	9.53	9.74	9.95	10.16	10.37	10.59
23.6	8.51	8.73	8.94	9.16	9.37	9.58	9.79	10.01	10.22	10.43	10.65
23.7	8.56	8.78	8.99	9.21	9.42	9.64	9.85	10.06	10.28	10.49	10.71
23.8	8.61	8.83	9.04	9.26	9.47	9.69	9.91	10.12	10.34	10.55	10.77
23.9	8.66	8.88	9.10	9.31	9.53	9.75	9.96	10.18	10.40	10.61	10.83
24.0	8.71	8.93	9.15	9.36	9.58	9.80	10.02	10.24	10.45	10.67	10.89
24.1	8.76	8.98	9.20	9.42	9.64	9.86	10.08	10.30	10.52	10.73	10.95
24.2	8.81	9.03	9.25	9.47	9.69	9.91	10.13	10.36	10.58	10.80	11.02
24.3	8.86	9.09	9.31	9.53	9.75	9.97	10.19	10.41	10.64	10.86	11.08
24.4	8.91	9.14	9.36	9.58	9.81	10.03	10.25	10.47	10.70	10.92	11.14
24.5	8.97	9.19	9.41	9.64	9.86	10.09	10.31	10.53	10.76	10.98	11.21
24.6	9.02	9.24	9.47	9.69	9.92	10.14	10.37	10.59	10.82	11.04	11.27
24.7	9.07	9.29	9.52	9.75	9.97	10.20	10.43	10.65	10.88	11.11	11.33
24.8	9.12	9.35	9.57	9.80	10.03	10.26	10.48	10.71	10.94	11.17	11.39
24.9	9.17	9.40	9.63	9.86	10.08	10.31	10.54	10.77	11.00	11.23	11.46
25.0	9.22	9.45	9.68	9.91	10.14	10.37	10.60	10.83	11.06	11.29	11.52
25.1	9.27	9.50	9.74	9.97	10.20	10.43	10.66	10.89	11.12	11.36	11.59
25.2	9.33	9.56	9.79	10.02	10.26	10.49	10.72	10.96	11.19	11.42	11.65
25.3	9.38	9.61	9.85	10.08	10.32	10.55	10.78	11.02	11.25	11.49	11.72
25.4	9.43	9.67	9.90	10.14	10.37	10.61	10.85	11.08	11.32	11.55	11.79
25.5	9.49	9.72	9.96	10.20	10.43	10.67	10.91	11.14	11.38	11.62	11.86
25.6	9.54	9.78	10.02	10.25	10.49	10.73	10.97	11.21	11.45	11.68	11.92
25.7	9.59	9.83	10.07	10.31	10.55	10.79	11.03	11.27	11.51	11.75	11.99
25.8	9.64	9.89	10.13	10.37	10.61	10.85	11.09	11.33	11.57	11.82	12.06
25.9	9.70	9.94	10.18	10.43	10.67	10.91	11.15	11.40	11.64	11.88	12.12
26.0	9.75	9.99	10.24	10.48	10.73	10.97	11.21	11.46	11.70	11.95	12.19
26.1	9.81	10.05	10.30	10.54	10.79	11.03	11.28	11.52	11.77	12.01	12.26
26.2	9.86	10.11	10.35	10.60	10.85	11.09	11.34	11.59	11.83	12.08	12.33
26.3	9.92	10.16	10.41	10.66	10.91	11.16	11.40	11.65	11.90	12.15	12.40
26.4	9.97	10.22	10.47	10.72	10.97	11.22	11.47	11.72	11.97	12.22	12.47
26.5	10.03	10.28	10.53	10.78	11.03	11.28	11.53	11.78	12.03	12.28	12.54
26.6	10.08	10.33	10.59	10.84	11.09	11.34	11.59	11.85	12.10	12.35	12.60
26.7	10.14	10.39	10.64	10.90	11.15	11.40	11.66	11.91	12.17	12.42	12.67
26.8	10.19	10.45	10.70	10.96	11.21	11.47	11.72	11.98	12.23	12.49	12.74
26.9	10.25	10.50	10.76	11.02	11.27	11.53	11.79	12.04	12.30	12.55	12.81
27.0	10.30	10.56	10.82	11.07	11.33	11.59	11.85	12.11	12.36	12.62	12.88

1D

T [°C]	r.h. [%]										
	40	41	42	43	44	45	46	47	48	49	50
27.0	10.30	10.56	10.82	11.07	11.33	11.59	11.85	12.11	12.36	12.62	12.88
27.1	10.36	10.62	10.88	11.14	11.40	11.66	11.92	12.18	12.44	12.70	12.95
27.2	10.42	10.68	10.94	11.20	11.46	11.72	11.98	12.25	12.51	12.77	13.03
27.3	10.48	10.74	11.00	11.27	11.53	11.79	12.05	12.32	12.58	12.84	13.10
27.4	10.54	10.80	11.06	11.33	11.59	11.86	12.12	12.38	12.65	12.91	13.18
27.5	10.60	10.86	11.13	11.39	11.66	11.92	12.19	12.45	12.72	12.99	13.25
27.6	10.65	10.92	11.19	11.46	11.72	11.99	12.26	12.52	12.79	13.06	13.32
27.7	10.71	10.98	11.25	11.52	11.79	12.06	12.32	12.59	12.86	13.13	13.40
27.8	10.77	11.04	11.31	11.58	11.85	12.12	12.39	12.66	12.93	13.20	13.47
27.9	10.83	11.10	11.37	11.65	11.92	12.19	12.46	12.73	13.00	13.28	13.55
28.0	10.89	11.16	11.44	11.71	11.98	12.26	12.53	12.80	13.07	13.35	13.62
28.1	10.95	11.23	11.50	11.78	12.05	12.32	12.60	12.87	13.15	13.42	13.70
28.2	11.01	11.29	11.56	11.84	12.12	12.39	12.67	12.94	13.22	13.50	13.77
28.3	11.07	11.35	11.63	11.91	12.18	12.46	12.74	13.02	13.29	13.57	13.85
28.4	11.13	11.41	11.69	11.97	12.25	12.53	12.81	13.09	13.37	13.65	13.92
28.5	11.20	11.48	11.76	12.04	12.32	12.60	12.88	13.16	13.44	13.72	14.00
28.6	11.26	11.54	11.82	12.10	12.38	12.67	12.95	13.23	13.51	13.79	14.08
28.7	11.32	11.60	11.88	12.17	12.45	12.74	13.02	13.30	13.59	13.87	14.15
28.8	11.38	11.66	11.95	12.23	12.52	12.80	13.09	13.37	13.66	13.94	14.23
28.9	11.44	11.73	12.01	12.30	12.59	12.87	13.16	13.45	13.73	14.02	14.30
29.0	11.50	11.79	12.08	12.36	12.65	12.94	13.23	13.52	13.80	14.09	14.38
29.1	11.56	11.85	12.14	12.43	12.72	13.01	13.30	13.59	13.88	14.17	14.46
29.2	11.63	11.92	12.21	12.50	12.79	13.08	13.38	13.67	13.96	14.25	14.54
29.3	11.69	11.99	12.28	12.57	12.86	13.16	13.45	13.74	14.03	14.33	14.62
29.4	11.76	12.05	12.35	12.64	12.93	13.23	13.52	13.82	14.11	14.41	14.70
29.5	11.82	12.12	12.41	12.71	13.00	13.30	13.60	13.89	14.19	14.48	14.78
29.6	11.88	12.18	12.48	12.78	13.07	13.37	13.67	13.97	14.27	14.56	14.86
29.7	11.95	12.25	12.55	12.85	13.15	13.44	13.74	14.04	14.34	14.64	14.94
29.8	12.01	12.31	12.61	12.91	13.22	13.52	13.82	14.12	14.42	14.72	15.02
29.9	12.08	12.38	12.68	12.98	13.29	13.59	13.89	14.19	14.50	14.80	15.10
30.0	12.14	12.44	12.75	13.05	13.36	13.66	13.96	14.27	14.57	14.88	15.18

2A

T [°C]	r.h. [%]										
	50	51	52	53	54	55	56	57	58	59	60
15.0	6.41	6.54	6.67	6.79	6.92	7.05	7.18	7.31	7.43	7.56	7.69
15.1	6.45	6.58	6.71	6.84	6.97	7.10	7.22	7.35	7.48	7.61	7.74
15.2	6.49	6.62	6.75	6.88	7.01	7.14	7.27	7.40	7.53	7.66	7.79
15.3	6.53	6.66	6.79	6.92	7.06	7.19	7.32	7.45	7.58	7.71	7.84
15.4	6.57	6.71	6.84	6.97	7.10	7.23	7.36	7.49	7.62	7.76	7.89
15.5	6.62	6.75	6.88	7.01	7.14	7.28	7.41	7.54	7.67	7.80	7.94
15.6	6.66	6.79	6.92	7.05	7.19	7.32	7.45	7.59	7.72	7.85	7.98
15.7	6.70	6.83	6.96	7.10	7.23	7.37	7.50	7.63	7.77	7.90	8.03
15.8	6.74	6.87	7.01	7.14	7.28	7.41	7.54	7.68	7.81	7.95	8.08
15.9	6.78	6.91	7.05	7.19	7.32	7.46	7.59	7.73	7.86	8.00	8.13
16.0	6.82	6.96	7.09	7.23	7.36	7.50	7.64	7.77	7.91	8.04	8.18
16.1	6.86	7.00	7.14	7.27	7.41	7.55	7.68	7.82	7.96	8.09	8.23
16.2	6.90	7.04	7.18	7.32	7.45	7.59	7.73	7.87	8.01	8.14	8.28
16.3	6.95	7.08	7.22	7.36	7.50	7.64	7.78	7.92	8.05	8.19	8.33
16.4	6.99	7.13	7.27	7.41	7.55	7.68	7.82	7.96	8.10	8.24	8.38
16.5	7.03	7.17	7.31	7.45	7.59	7.73	7.87	8.01	8.15	8.29	8.43
16.6	7.07	7.21	7.35	7.49	7.64	7.78	7.92	8.06	8.20	8.34	8.48
16.7	7.11	7.26	7.40	7.54	7.68	7.82	7.96	8.11	8.25	8.39	8.53
16.8	7.16	7.30	7.44	7.58	7.73	7.87	8.01	8.15	8.30	8.44	8.58
16.9	7.20	7.34	7.48	7.63	7.77	7.91	8.06	8.20	8.34	8.49	8.63
17.0	7.24	7.38	7.53	7.67	7.82	7.96	8.10	8.25	8.39	8.54	8.68
17.1	7.28	7.43	7.57	7.72	7.86	8.01	8.15	8.30	8.44	8.59	8.73
17.2	7.33	7.47	7.62	7.77	7.91	8.06	8.20	8.35	8.50	8.64	8.79
17.3	7.37	7.52	7.67	7.81	7.96	8.11	8.25	8.40	8.55	8.70	8.84
17.4	7.42	7.56	7.71	7.86	8.01	8.16	8.30	8.45	8.60	8.75	8.90
17.5	7.46	7.61	7.76	7.91	8.06	8.21	8.35	8.50	8.65	8.80	8.95
17.6	7.50	7.65	7.80	7.95	8.10	8.25	8.40	8.55	8.70	8.85	9.00
17.7	7.55	7.70	7.85	8.00	8.15	8.30	8.45	8.61	8.76	8.91	9.06
17.8	7.59	7.74	7.90	8.05	8.20	8.35	8.50	8.66	8.81	8.96	9.11
17.9	7.64	7.79	7.94	8.10	8.25	8.40	8.55	8.71	8.86	9.01	9.17
18.0	7.68	7.83	7.99	8.14	8.30	8.45	8.60	8.76	8.91	9.07	9.22
18.1	7.73	7.88	8.04	8.19	8.35	8.50	8.66	8.81	8.97	9.12	9.28
18.2	7.77	7.93	8.09	8.24	8.40	8.55	8.71	8.87	9.02	9.18	9.33
18.3	7.82	7.98	8.13	8.29	8.45	8.61	8.76	8.92	9.08	9.23	9.39
18.4	7.87	8.03	8.18	8.34	8.50	8.66	8.81	8.97	9.13	9.29	9.44
18.5	7.92	8.07	8.23	8.39	8.55	8.71	8.87	9.03	9.18	9.34	9.50
18.6	7.96	8.12	8.28	8.44	8.60	8.76	8.92	9.08	9.24	9.40	9.56
18.7	8.01	8.17	8.33	8.49	8.65	8.81	8.97	9.13	9.29	9.45	9.61
18.8	8.06	8.22	8.38	8.54	8.70	8.86	9.02	9.18	9.35	9.51	9.67
18.9	8.10	8.27	8.43	8.59	8.75	8.91	9.08	9.24	9.40	9.56	9.72
19.0	8.15	8.31	8.48	8.64	8.80	8.97	9.13	9.29	9.45	9.62	9.78

2B

T [°C]	r.h. [%]										
	50	51	52	53	54	55	56	57	58	59	60
19.0	8.15	8.31	8.48	8.64	8.80	8.97	9.13	9.29	9.45	9.62	9.78
19.1	8.20	8.36	8.53	8.69	8.86	9.02	9.18	9.35	9.51	9.68	9.84
19.2	8.25	8.42	8.58	8.74	8.91	9.07	9.24	9.40	9.57	9.73	9.90
19.3	8.30	8.47	8.63	8.80	8.96	9.13	9.29	9.46	9.63	9.79	9.96
19.4	8.35	8.52	8.68	8.85	9.02	9.18	9.35	9.52	9.68	9.85	10.02
19.5	8.40	8.57	8.74	8.90	9.07	9.24	9.41	9.57	9.74	9.91	10.08
19.6	8.45	8.62	8.79	8.96	9.12	9.29	9.46	9.63	9.80	9.97	10.13
19.7	8.50	8.67	8.84	9.01	9.18	9.35	9.52	9.69	9.85	10.02	10.19
19.8	8.55	8.72	8.89	9.06	9.23	9.40	9.57	9.74	9.91	10.08	10.25
19.9	8.60	8.77	8.94	9.11	9.28	9.46	9.63	9.80	9.97	10.14	10.31
20.0	8.65	8.82	8.99	9.17	9.34	9.51	9.68	9.85	10.03	10.20	10.37
20.1	8.70	8.88	9.05	9.22	9.39	9.57	9.74	9.91	10.09	10.26	10.43
20.2	8.75	8.93	9.10	9.28	9.45	9.63	9.80	9.97	10.15	10.32	10.50
20.3	8.81	8.98	9.16	9.33	9.51	9.68	9.86	10.03	10.21	10.38	10.56
20.4	8.86	9.03	9.21	9.39	9.56	9.74	9.92	10.09	10.27	10.45	10.62
20.5	8.91	9.09	9.27	9.44	9.62	9.80	9.98	10.15	10.33	10.51	10.69
20.6	8.96	9.14	9.32	9.50	9.68	9.86	10.03	10.21	10.39	10.57	10.75
20.7	9.01	9.19	9.37	9.55	9.73	9.91	10.09	10.27	10.45	10.63	10.81
20.8	9.07	9.25	9.43	9.61	9.79	9.97	10.15	10.33	10.51	10.69	10.87
20.9	9.12	9.30	9.48	9.66	9.85	10.03	10.21	10.39	10.57	10.76	10.94
21.0	9.17	9.35	9.54	9.72	9.90	10.09	10.27	10.45	10.63	10.82	11.00
21.1	9.22	9.41	9.59	9.78	9.96	10.15	10.33	10.51	10.70	10.88	11.07
21.2	9.28	9.46	9.65	9.83	10.02	10.20	10.39	10.57	10.76	10.95	11.13
21.3	9.33	9.52	9.71	9.89	10.08	10.26	10.45	10.64	10.82	11.01	11.20
21.4	9.39	9.57	9.76	9.95	10.14	10.32	10.51	10.70	10.89	11.07	11.26
21.5	9.44	9.63	9.82	10.01	10.19	10.38	10.57	10.76	10.95	11.14	11.33
21.6	9.49	9.68	9.87	10.06	10.25	10.44	10.63	10.82	11.01	11.20	11.39
21.7	9.55	9.74	9.93	10.12	10.31	10.50	10.69	10.88	11.07	11.26	11.46
21.8	9.60	9.79	9.99	10.18	10.37	10.56	10.75	10.95	11.14	11.33	11.52
21.9	9.66	9.85	10.04	10.24	10.43	10.62	10.81	11.01	11.20	11.39	11.59
22.0	9.71	9.90	10.10	10.29	10.49	10.68	10.87	11.07	11.26	11.46	11.65
22.1	9.77	9.96	10.16	10.35	10.55	10.74	10.94	11.13	11.33	11.52	11.72
22.2	9.82	10.02	10.22	10.41	10.61	10.81	11.00	11.20	11.40	11.59	11.79
22.3	9.88	10.08	10.28	10.47	10.67	10.87	11.07	11.26	11.46	11.66	11.86
22.4	9.94	10.14	10.34	10.53	10.73	10.93	11.13	11.33	11.53	11.73	11.93
22.5	10.00	10.20	10.40	10.60	10.80	11.00	11.20	11.40	11.60	11.80	12.00
22.6	10.05	10.25	10.45	10.66	10.86	11.06	11.26	11.46	11.66	11.86	12.06
22.7	10.11	10.31	10.51	10.72	10.92	11.12	11.32	11.53	11.73	11.93	12.13
22.8	10.17	10.37	10.57	10.78	10.98	11.18	11.39	11.59	11.80	12.00	12.20
22.9	10.22	10.43	10.63	10.84	11.04	11.25	11.45	11.66	11.86	12.07	12.27
23.0	10.28	10.49	10.69	10.90	11.10	11.31	11.52	11.72	11.93	12.13	12.34

2C

T [°C]	r.h. [%]										
	50	51	52	53	54	55	56	57	58	59	60
23.0	10.28	10.49	10.69	10.90	11.10	11.31	11.52	11.72	11.93	12.13	12.34
23.1	10.34	10.55	10.76	10.96	11.17	11.38	11.58	11.79	12.00	12.21	12.41
23.2	10.40	10.61	10.82	11.03	11.24	11.44	11.65	11.86	12.07	12.28	12.48
23.3	10.46	10.67	10.88	11.09	11.30	11.51	11.72	11.93	12.14	12.35	12.56
23.4	10.52	10.73	10.95	11.16	11.37	11.58	11.79	12.00	12.21	12.42	12.63
23.5	10.59	10.80	11.01	11.22	11.43	11.64	11.85	12.07	12.28	12.49	12.70
23.6	10.65	10.86	11.07	11.28	11.50	11.71	11.92	12.13	12.35	12.56	12.77
23.7	10.71	10.92	11.13	11.35	11.56	11.78	11.99	12.20	12.42	12.63	12.84
23.8	10.77	10.98	11.20	11.41	11.63	11.84	12.06	12.27	12.49	12.70	12.92
23.9	10.83	11.05	11.26	11.48	11.69	11.91	12.12	12.34	12.56	12.77	12.99
24.0	10.89	11.11	11.32	11.54	11.76	11.98	12.19	12.41	12.63	12.84	13.06
24.1	10.95	11.17	11.39	11.61	11.83	12.05	12.26	12.48	12.70	12.92	13.14
24.2	11.02	11.24	11.46	11.68	11.89	12.11	12.33	12.55	12.77	12.99	13.21
24.3	11.08	11.30	11.52	11.74	11.96	12.18	12.40	12.63	12.85	13.07	13.29
24.4	11.14	11.36	11.59	11.81	12.03	12.25	12.48	12.70	12.92	13.14	13.36
24.5	11.21	11.43	11.65	11.88	12.10	12.32	12.55	12.77	12.99	13.22	13.44
24.6	11.27	11.49	11.72	11.94	12.17	12.39	12.62	12.84	13.07	13.29	13.52
24.7	11.33	11.56	11.78	12.01	12.24	12.46	12.69	12.91	13.14	13.37	13.59
24.8	11.39	11.62	11.85	12.08	12.30	12.53	12.76	12.99	13.21	13.44	13.67
24.9	11.46	11.69	11.91	12.14	12.37	12.60	12.83	13.06	13.29	13.52	13.74
25.0	11.52	11.75	11.98	12.21	12.44	12.67	12.90	13.13	13.36	13.59	13.82
25.1	11.59	11.82	12.05	12.28	12.51	12.74	12.98	13.21	13.44	13.67	13.90
25.2	11.65	11.89	12.12	12.35	12.58	12.82	13.05	13.28	13.52	13.75	13.98
25.3	11.72	11.96	12.19	12.42	12.66	12.89	13.12	13.36	13.59	13.83	14.06
25.4	11.79	12.02	12.26	12.49	12.73	12.96	13.20	13.43	13.67	13.91	14.14
25.5	11.86	12.09	12.33	12.57	12.80	13.04	13.27	13.51	13.75	13.98	14.22
25.6	11.92	12.16	12.40	12.64	12.87	13.11	13.35	13.59	13.82	14.06	14.30
25.7	11.99	12.23	12.47	12.71	12.95	13.19	13.42	13.66	13.90	14.14	14.38
25.8	12.06	12.30	12.54	12.78	13.02	13.26	13.50	13.74	13.98	14.22	14.46
25.9	12.12	12.37	12.61	12.85	13.09	13.33	13.57	13.82	14.06	14.30	14.54
26.0	12.19	12.43	12.68	12.92	13.16	13.41	13.65	13.89	14.13	14.38	14.62
26.1	12.26	12.50	12.75	12.99	13.24	13.48	13.73	13.97	14.22	14.46	14.70
26.2	12.33	12.57	12.82	13.07	13.31	13.56	13.80	14.05	14.30	14.54	14.79
26.3	12.40	12.65	12.89	13.14	13.39	13.64	13.88	14.13	14.38	14.63	14.87
26.4	12.47	12.72	12.96	13.21	13.46	13.71	13.96	14.21	14.46	14.71	14.96
26.5	12.54	12.79	13.04	13.29	13.54	13.79	14.04	14.29	14.54	14.79	15.04
26.6	12.60	12.86	13.11	13.36	13.61	13.86	14.12	14.37	14.62	14.87	15.12
26.7	12.67	12.93	13.18	13.43	13.69	13.94	14.19	14.45	14.70	14.96	15.21
26.8	12.74	13.00	13.25	13.51	13.76	14.02	14.27	14.53	14.78	15.04	15.29
26.9	12.81	13.07	13.32	13.58	13.84	14.09	14.35	14.61	14.86	15.12	15.38
27.0	12.88	13.14	13.40	13.65	13.91	14.17	14.43	14.69	14.94	15.20	15.46

2D

T [°C]	r.h. [%]										
	50	51	52	53	54	55	56	57	58	59	60
27.0	12.88	13.14	13.40	13.65	13.91	14.17	14.43	14.69	14.94	15.20	15.46
27.1	12.95	13.21	13.47	13.73	13.99	14.25	14.51	14.77	15.03	15.29	15.55
27.2	13.03	13.29	13.55	13.81	14.07	14.33	14.59	14.85	15.11	15.38	15.64
27.3	13.10	13.36	13.63	13.89	14.15	14.41	14.68	14.94	15.20	15.46	15.72
27.4	13.18	13.44	13.70	13.97	14.23	14.49	14.76	15.02	15.29	15.55	15.81
27.5	13.25	13.52	13.78	14.05	14.31	14.58	14.84	15.11	15.37	15.64	15.90
27.6	13.32	13.59	13.86	14.12	14.39	14.66	14.92	15.19	15.46	15.72	15.99
27.7	13.40	13.67	13.93	14.20	14.47	14.74	15.01	15.27	15.54	15.81	16.08
27.8	13.47	13.74	14.01	14.28	14.55	14.82	15.09	15.36	15.63	15.90	16.16
27.9	13.55	13.82	14.09	14.36	14.63	14.90	15.17	15.44	15.71	15.98	16.25
28.0	13.62	13.89	14.16	14.44	14.71	14.98	15.25	15.52	15.80	16.07	16.34
28.1	13.70	13.97	14.24	14.52	14.79	15.06	15.34	15.61	15.89	16.16	16.43
28.2	13.77	14.05	14.32	14.60	14.87	15.15	15.42	15.70	15.97	16.25	16.52
28.3	13.85	14.13	14.40	14.68	14.96	15.23	15.51	15.79	16.06	16.34	16.62
28.4	13.92	14.20	14.48	14.76	15.04	15.32	15.59	15.87	16.15	16.43	16.71
28.5	14.00	14.28	14.56	14.84	15.12	15.40	15.68	15.96	16.24	16.52	16.80
28.6	14.08	14.36	14.64	14.92	15.20	15.48	15.77	16.05	16.33	16.61	16.89
28.7	14.15	14.44	14.72	15.00	15.29	15.57	15.85	16.13	16.42	16.70	16.98
28.8	14.23	14.51	14.80	15.08	15.37	15.65	15.94	16.22	16.51	16.79	17.08
28.9	14.30	14.59	14.88	15.16	15.45	15.74	16.02	16.31	16.60	16.88	17.17
29.0	14.38	14.67	14.96	15.24	15.53	15.82	16.11	16.40	16.68	16.97	17.26
29.1	14.46	14.75	15.04	15.33	15.62	15.91	16.20	16.49	16.78	17.07	17.36
29.2	14.54	14.83	15.12	15.41	15.71	16.00	16.29	16.58	16.87	17.16	17.45
29.3	14.62	14.91	15.21	15.50	15.79	16.08	16.38	16.67	16.96	17.26	17.55
29.4	14.70	14.99	15.29	15.58	15.88	16.17	16.47	16.76	17.06	17.35	17.64
29.5	14.78	15.08	15.37	15.67	15.96	16.26	16.56	16.85	17.15	17.44	17.74
29.6	14.86	15.16	15.46	15.75	16.05	16.35	16.65	16.94	17.24	17.54	17.84
29.7	14.94	15.24	15.54	15.84	16.14	16.44	16.74	17.03	17.33	17.63	17.93
29.8	15.02	15.32	15.62	15.92	16.22	16.52	16.83	17.13	17.43	17.73	18.03
29.9	15.10	15.40	15.71	16.01	16.31	16.61	16.91	17.22	17.52	17.82	18.12
30.0	15.18	15.48	15.79	16.09	16.40	16.70	17.00	17.31	17.61	17.92	18.22

3A

T [°C]	r.h. [%]										
	60	61	62	63	64	65	66	67	68	69	70
15.0	7.69	7.82	7.95	8.07	8.20	8.33	8.46	8.59	8.71	8.84	8.97
15.1	7.74	7.87	8.00	8.13	8.25	8.38	8.51	8.64	8.77	8.90	9.03
15.2	7.79	7.92	8.05	8.18	8.31	8.44	8.57	8.70	8.83	8.95	9.08
15.3	7.84	7.97	8.10	8.23	8.36	8.49	8.62	8.75	8.88	9.01	9.14
15.4	7.89	8.02	8.15	8.28	8.41	8.54	8.67	8.80	8.94	9.07	9.20
15.5	7.94	8.07	8.20	8.33	8.46	8.60	8.73	8.86	8.99	9.12	9.26
15.6	7.98	8.12	8.25	8.38	8.52	8.65	8.78	8.91	9.05	9.18	9.31
15.7	8.03	8.17	8.30	8.43	8.57	8.70	8.84	8.97	9.10	9.24	9.37
15.8	8.08	8.22	8.35	8.49	8.62	8.75	8.89	9.02	9.16	9.29	9.43
15.9	8.13	8.27	8.40	8.54	8.67	8.81	8.94	9.08	9.21	9.35	9.48
16.0	8.18	8.32	8.45	8.59	8.72	8.86	9.00	9.13	9.27	9.40	9.54
16.1	8.23	8.37	8.50	8.64	8.78	8.92	9.05	9.19	9.33	9.46	9.60
16.2	8.28	8.42	8.56	8.69	8.83	8.97	9.11	9.25	9.38	9.52	9.66
16.3	8.33	8.47	8.61	8.75	8.89	9.02	9.16	9.30	9.44	9.58	9.72
16.4	8.38	8.52	8.66	8.80	8.94	9.08	9.22	9.36	9.50	9.64	9.78
16.5	8.43	8.57	8.71	8.85	8.99	9.13	9.27	9.41	9.55	9.70	9.84
16.6	8.48	8.62	8.76	8.90	9.05	9.19	9.33	9.47	9.61	9.75	9.89
16.7	8.53	8.67	8.82	8.96	9.10	9.24	9.38	9.53	9.67	9.81	9.95
16.8	8.58	8.72	8.87	9.01	9.15	9.30	9.44	9.58	9.73	9.87	10.01
16.9	8.63	8.77	8.92	9.06	9.21	9.35	9.50	9.64	9.78	9.93	10.07
17.0	8.68	8.83	8.97	9.12	9.26	9.41	9.55	9.70	9.84	9.99	10.13
17.1	8.73	8.88	9.03	9.17	9.32	9.46	9.61	9.76	9.90	10.05	10.19
17.2	8.79	8.94	9.08	9.23	9.37	9.52	9.67	9.81	9.96	10.11	10.25
17.3	8.84	8.99	9.14	9.28	9.43	9.58	9.73	9.87	10.02	10.17	10.32
17.4	8.90	9.04	9.19	9.34	9.49	9.64	9.79	9.93	10.08	10.23	10.38
17.5	8.95	9.10	9.25	9.40	9.55	9.70	9.84	9.99	10.14	10.29	10.44
17.6	9.00	9.15	9.30	9.45	9.60	9.75	9.90	10.05	10.20	10.35	10.50
17.7	9.06	9.21	9.36	9.51	9.66	9.81	9.96	10.11	10.26	10.41	10.56
17.8	9.11	9.26	9.42	9.57	9.72	9.87	10.02	10.17	10.32	10.48	10.63
17.9	9.17	9.32	9.47	9.62	9.78	9.93	10.09	10.23	10.38	10.54	10.69
18.0	9.22	9.37	9.53	9.68	9.83	9.99	10.14	10.29	10.44	10.60	10.75
18.1	9.28	9.43	9.58	9.74	9.89	10.05	10.20	10.35	10.51	10.66	10.82
18.2	9.33	9.49	9.64	9.80	9.95	10.11	10.26	10.42	10.57	10.73	10.88
18.3	9.39	9.54	9.70	9.86	10.01	10.17	10.32	10.48	10.64	10.79	10.95
18.4	9.44	9.60	9.76	9.92	10.07	10.23	10.39	10.54	10.70	10.86	11.01
18.5	9.50	9.66	9.82	9.97	10.13	10.29	10.45	10.61	10.76	10.92	11.08
18.6	9.56	9.72	9.87	10.03	10.19	10.35	10.51	10.67	10.83	10.99	11.15
18.7	9.61	9.77	9.93	10.09	10.25	10.41	10.57	10.73	10.89	11.05	11.21
18.8	9.67	9.83	9.99	10.15	10.31	10.47	10.63	10.80	10.96	11.12	11.28
18.9	9.72	9.89	10.05	10.21	10.37	10.53	10.70	10.86	11.02	11.18	11.34
19.0	9.78	9.94	10.11	10.27	10.43	10.60	10.76	10.92	11.08	11.25	11.41

3B

T [°C]	r.h. [%]										
	60	61	62	63	64	65	66	67	68	69	70
19.0	9.78	9.94	10.11	10.27	10.43	10.60	10.76	10.92	11.08	11.25	11.41
19.1	9.84	10.00	10.17	10.33	10.50	10.66	10.82	10.99	11.15	11.32	11.48
19.2	9.90	10.06	10.23	10.39	10.56	10.72	10.89	11.05	11.22	11.38	11.55
19.3	9.96	10.12	10.29	10.46	10.62	10.79	10.95	11.12	11.29	11.45	11.62
19.4	10.02	10.18	10.35	10.52	10.68	10.85	11.02	11.19	11.35	11.52	11.69
19.5	10.08	10.24	10.41	10.58	10.75	10.92	11.08	11.25	11.42	11.59	11.76
19.6	10.13	10.30	10.47	10.64	10.81	10.98	11.15	11.32	11.49	11.66	11.82
19.7	10.19	10.36	10.53	10.70	10.87	11.04	11.21	11.38	11.55	11.72	11.89
19.8	10.25	10.42	10.59	10.77	10.94	11.11	11.28	11.45	11.62	11.79	11.96
19.9	10.31	10.48	10.66	10.83	11.00	11.17	11.34	11.52	11.69	11.86	12.03
20.0	10.37	10.54	10.72	10.89	11.06	11.24	11.41	11.58	11.75	11.93	12.10
20.1	10.43	10.61	10.78	10.96	11.13	11.30	11.48	11.65	11.83	12.00	12.17
20.2	10.50	10.67	10.85	11.02	11.20	11.37	11.55	11.72	11.90	12.07	12.25
20.3	10.56	10.74	10.91	11.09	11.26	11.44	11.62	11.79	11.97	12.14	12.32
20.4	10.62	10.80	10.98	11.15	11.33	11.51	11.68	11.86	12.04	12.22	12.39
20.5	10.69	10.86	11.04	11.22	11.40	11.58	11.75	11.93	12.11	12.29	12.47
20.6	10.75	10.93	11.11	11.29	11.46	11.64	11.82	12.00	12.18	12.36	12.54
20.7	10.81	10.99	11.17	11.35	11.53	11.71	11.89	12.07	12.25	12.43	12.61
20.8	10.87	11.06	11.24	11.42	11.60	11.78	11.96	12.14	12.32	12.50	12.68
20.9	10.94	11.12	11.30	11.48	11.67	11.85	12.03	12.21	12.39	12.58	12.76
21.0	11.00	11.18	11.37	11.55	11.73	11.92	12.10	12.28	12.46	12.65	12.83
21.1	11.07	11.25	11.43	11.62	11.80	11.99	12.17	12.35	12.54	12.72	12.91
21.2	11.13	11.32	11.50	11.69	11.87	12.06	12.24	12.43	12.61	12.80	12.98
21.3	11.20	11.38	11.57	11.75	11.94	12.13	12.31	12.50	12.69	12.87	13.06
21.4	11.26	11.45	11.64	11.82	12.01	12.20	12.38	12.57	12.76	12.95	13.13
21.5	11.33	11.51	11.70	11.89	12.08	12.27	12.46	12.65	12.83	13.02	13.21
21.6	11.39	11.58	11.77	11.96	12.15	12.34	12.53	12.72	12.91	13.10	13.29
21.7	11.46	11.65	11.84	12.03	12.22	12.41	12.60	12.79	12.98	13.17	13.36
21.8	11.52	11.71	11.90	12.10	12.29	12.48	12.67	12.86	13.05	13.25	13.44
21.9	11.59	11.78	11.97	12.16	12.36	12.55	12.74	12.94	13.13	13.32	13.51
22.0	11.65	11.84	12.04	12.23	12.43	12.62	12.81	13.01	13.20	13.40	13.59
22.1	11.72	11.91	12.11	12.30	12.50	12.70	12.89	13.09	13.28	13.48	13.67
22.2	11.79	11.98	12.18	12.38	12.57	12.77	12.97	13.16	13.36	13.55	13.75
22.3	11.86	12.05	12.25	12.45	12.65	12.84	13.04	13.24	13.44	13.63	13.83
22.4	11.93	12.12	12.32	12.52	12.72	12.92	13.12	13.32	13.51	13.71	13.91
22.5	12.00	12.20	12.39	12.59	12.79	12.99	13.19	13.39	13.59	13.79	13.99
22.6	12.06	12.27	12.47	12.67	12.87	13.07	13.27	13.47	13.67	13.87	14.07
22.7	12.13	12.34	12.54	12.74	12.94	13.14	13.34	13.55	13.75	13.95	14.15
22.8	12.20	12.41	12.61	12.81	13.01	13.22	13.42	13.62	13.82	14.03	14.23
22.9	12.27	12.48	12.68	12.88	13.09	13.29	13.49	13.70	13.90	14.11	14.31
23.0	12.34	12.55	12.75	12.96	13.16	13.37	13.57	13.78	13.98	14.19	14.39

3C

T [°C]	r.h. [%]										
	60	61	62	63	64	65	66	67	68	69	70
23.0	12.34	12.55	12.75	12.96	13.16	13.37	13.57	13.78	13.98	14.19	14.39
23.1	12.41	12.62	12.83	13.03	13.24	13.44	13.65	13.86	14.06	14.27	14.48
23.2	12.48	12.69	12.90	13.11	13.31	13.52	13.73	13.94	14.15	14.35	14.56
23.3	12.56	12.77	12.97	13.18	13.39	13.60	13.81	14.02	14.23	14.44	14.65
23.4	12.63	12.84	13.05	13.26	13.47	13.68	13.89	14.10	14.31	14.52	14.73
23.5	12.70	12.91	13.12	13.34	13.55	13.76	13.97	14.18	14.39	14.60	14.82
23.6	12.77	12.99	13.20	13.41	13.62	13.84	14.05	14.26	14.47	14.69	14.90
23.7	12.84	13.06	13.27	13.49	13.70	13.92	14.13	14.34	14.56	14.77	14.99
23.8	12.92	13.13	13.35	13.56	13.78	13.99	14.21	14.42	14.64	14.86	15.07
23.9	12.99	13.21	13.42	13.64	13.86	14.07	14.29	14.51	14.72	14.94	15.16
24.0	13.06	13.28	13.50	13.71	13.93	14.15	14.37	14.59	14.80	15.02	15.24
24.1	13.14	13.36	13.58	13.79	14.01	14.23	14.45	14.67	14.89	15.11	15.33
24.2	13.21	13.43	13.65	13.87	14.09	14.32	14.54	14.76	14.98	15.20	15.42
24.3	13.29	13.51	13.73	13.95	14.18	14.40	14.62	14.84	15.06	15.29	15.51
24.4	13.36	13.58	13.81	14.03	14.26	14.48	14.70	14.93	15.15	15.37	15.60
24.5	13.44	13.67	13.89	14.11	14.34	14.56	14.79	15.01	15.24	15.46	15.69
24.6	13.52	13.74	13.97	14.19	14.42	14.65	14.87	15.10	15.32	15.55	15.77
24.7	13.59	13.82	14.05	14.27	14.50	14.73	14.96	15.18	15.41	15.64	15.86
24.8	13.67	13.90	14.13	14.35	14.58	14.81	15.04	15.27	15.50	15.72	15.95
24.9	13.74	13.97	14.20	14.43	14.66	14.89	15.12	15.35	15.58	15.81	16.04
25.0	13.82	14.05	14.28	14.51	14.74	14.98	15.21	15.44	15.67	15.90	16.13
25.1	13.90	14.13	14.37	14.60	14.83	15.06	15.29	15.53	15.76	15.99	16.22
25.2	13.98	14.21	14.45	14.68	14.91	15.15	15.38	15.62	15.85	16.08	16.32
25.3	14.06	14.30	14.53	14.77	15.00	15.24	15.47	15.70	15.94	16.17	16.41
25.4	14.14	14.38	14.61	14.85	15.09	15.32	15.56	15.79	16.03	16.27	16.50
25.5	14.22	14.46	14.70	14.93	15.17	15.41	15.65	15.88	16.12	16.36	16.60
25.6	14.30	14.54	14.78	15.02	15.26	15.49	15.73	15.97	16.21	16.45	16.69
25.7	14.38	14.62	14.86	15.10	15.34	15.58	15.82	16.06	16.30	16.54	16.78
25.8	14.46	14.70	14.94	15.18	15.43	15.67	15.91	16.15	16.39	16.63	16.87
25.9	14.54	14.78	15.03	15.27	15.51	15.75	16.00	16.24	16.48	16.72	16.97
26.0	14.62	14.86	15.11	15.35	15.60	15.84	16.08	16.33	16.57	16.82	17.06
26.1	14.70	14.95	15.20	15.44	15.69	15.93	16.18	16.42	16.67	16.91	17.16
26.2	14.79	15.04	15.28	15.53	15.77	16.02	16.27	16.51	16.76	17.01	17.25
26.3	14.87	15.12	15.37	15.62	15.86	16.11	16.36	16.61	16.86	17.10	17.35
26.4	14.96	15.21	15.45	15.70	15.95	16.20	16.45	16.70	16.95	17.20	17.45
26.5	15.04	15.29	15.54	15.79	16.04	16.29	16.54	16.79	17.04	17.30	17.55
26.6	15.12	15.38	15.63	15.88	16.13	16.38	16.64	16.89	17.14	17.39	17.64
26.7	15.21	15.46	15.71	15.97	16.22	16.47	16.73	16.98	17.23	17.49	17.74
26.8	15.29	15.55	15.80	16.06	16.31	16.56	16.82	17.07	17.33	17.58	17.84
26.9	15.38	15.63	15.89	16.14	16.40	16.66	16.91	17.17	17.42	17.68	17.93
27.0	15.46	15.72	15.97	16.23	16.49	16.75	17.00	17.26	17.52	17.77	18.03

3D

T [°C]	r.h. [%]										
	60	61	62	63	64	65	66	67	68	69	70
27.0	15.46	15.72	15.97	16.23	16.49	16.75	17.00	17.26	17.52	17.77	18.03
27.1	15.55	15.81	16.07	16.32	16.58	16.84	17.10	17.36	17.62	17.88	18.13
27.2	15.64	15.90	16.16	16.42	16.68	16.94	17.20	17.46	17.72	17.98	18.24
27.3	15.72	15.99	16.25	16.51	16.77	17.03	17.29	17.56	17.82	18.08	18.34
27.4	15.81	16.08	16.34	16.60	16.86	17.13	17.39	17.65	17.92	18.18	18.44
27.5	15.90	16.17	16.43	16.69	16.96	17.22	17.49	17.75	18.02	18.28	18.55
27.6	15.99	16.25	16.52	16.79	17.05	17.32	17.58	17.85	18.12	18.38	18.65
27.7	16.08	16.34	16.61	16.88	17.15	17.41	17.68	17.95	18.22	18.48	18.75
27.8	16.16	16.43	16.70	16.97	17.24	17.51	17.78	18.05	18.32	18.59	18.85
27.9	16.25	16.52	16.79	17.06	17.33	17.61	17.88	18.15	18.42	18.69	18.96
28.0	16.34	16.61	16.88	17.16	17.43	17.70	17.97	18.24	18.52	18.79	19.06
28.1	16.43	16.71	16.98	17.25	17.53	17.80	18.07	18.35	18.62	18.89	19.17
28.2	16.52	16.80	17.07	17.35	17.62	17.90	18.17	18.45	18.72	19.00	19.27
28.3	16.62	16.89	17.17	17.45	17.72	18.00	18.28	18.55	18.83	19.11	19.38
28.4	16.71	16.99	17.26	17.54	17.82	18.10	18.38	18.65	18.93	19.21	19.49
28.5	16.80	17.08	17.36	17.64	17.92	18.20	18.48	18.76	19.04	19.32	19.60
28.6	16.89	17.17	17.45	17.74	18.02	18.30	18.58	18.86	19.14	19.42	19.70
28.7	16.98	17.27	17.55	17.83	18.11	18.40	18.68	18.96	19.24	19.53	19.81
28.8	17.08	17.36	17.64	17.93	18.21	18.50	18.78	19.06	19.35	19.63	19.92
28.9	17.17	17.45	17.74	18.03	18.31	18.60	18.88	19.17	19.45	19.74	20.02
29.0	17.26	17.55	17.83	18.12	18.41	18.70	18.98	19.27	19.56	19.84	20.13
29.1	17.36	17.65	17.93	18.22	18.51	18.80	19.09	19.38	19.67	19.95	20.24
29.2	17.45	17.74	18.03	18.32	18.61	18.90	19.19	19.48	19.77	20.06	20.35
29.3	17.55	17.84	18.13	18.42	18.72	19.01	19.30	19.59	19.88	20.17	20.47
29.4	17.64	17.94	18.23	18.52	18.82	19.11	19.40	19.70	19.99	20.29	20.58
29.5	17.74	18.04	18.33	18.63	18.92	19.22	19.51	19.81	20.10	20.40	20.69
29.6	17.84	18.13	18.44	18.73	19.02	19.32	19.62	19.91	20.21	20.51	20.80
29.7	17.93	18.23	18.53	18.83	19.13	19.42	19.72	20.02	20.32	20.62	20.91
29.8	18.03	18.33	18.63	18.93	19.23	19.53	19.83	20.13	20.43	20.73	21.03
29.9	18.12	18.43	18.73	19.03	19.33	19.63	19.93	20.23	20.54	20.84	21.14
30.0	18.22	18.52	18.83	19.13	19.43	19.74	20.04	20.34	20.64	20.95	21.25

4A

T [°C]	r.h. [%]										
	70	71	72	73	74	75	76	77	78	79	80
15.0	8.97	9.10	9.23	9.36	9.49	9.62	9.74	9.87	10.00	10.13	10.26
15.1	9.03	9.16	9.29	9.42	9.55	9.68	9.81	9.94	10.07	10.19	10.32
15.2	9.08	9.21	9.35	9.48	9.61	9.74	9.87	10.00	10.13	10.26	10.39
15.3	9.14	9.27	9.40	9.53	9.67	9.80	9.93	10.06	10.19	10.32	10.45
15.4	9.20	9.33	9.46	9.59	9.73	9.86	9.99	10.12	10.25	10.38	10.52
15.5	9.26	9.39	9.52	9.65	9.79	9.92	10.05	10.18	10.32	10.45	10.58
15.6	9.31	9.45	9.58	9.71	9.85	9.98	10.11	10.24	10.38	10.51	10.64
15.7	9.37	9.50	9.64	9.77	9.91	10.04	10.17	10.31	10.44	10.57	10.71
15.8	9.43	9.56	9.70	9.83	9.96	10.10	10.23	10.37	10.50	10.64	10.77
15.9	9.48	9.62	9.75	9.89	10.02	10.16	10.30	10.43	10.57	10.70	10.84
16.0	9.54	9.68	9.81	9.95	10.08	10.22	10.36	10.49	10.63	10.76	10.90
16.1	9.60	9.74	9.87	10.01	10.15	10.28	10.42	10.56	10.69	10.83	10.97
16.2	9.66	9.80	9.93	10.07	10.21	10.35	10.49	10.62	10.76	10.90	11.04
16.3	9.72	9.86	9.99	10.13	10.27	10.41	10.55	10.69	10.83	10.97	11.10
16.4	9.78	9.92	10.06	10.20	10.33	10.47	10.61	10.75	10.89	11.03	11.17
16.5	9.84	9.98	10.12	10.26	10.40	10.54	10.68	10.82	10.96	11.10	11.24
16.6	9.89	10.04	10.18	10.32	10.46	10.60	10.74	10.88	11.03	11.17	11.31
16.7	9.95	10.10	10.24	10.38	10.52	10.67	10.81	10.95	11.09	11.23	11.38
16.8	10.01	10.16	10.30	10.44	10.59	10.73	10.87	11.01	11.16	11.30	11.44
16.9	10.07	10.22	10.36	10.50	10.65	10.79	10.94	11.08	11.22	11.37	11.51
17.0	10.13	10.28	10.42	10.57	10.71	10.86	11.00	11.15	11.29	11.51	11.65
17.1	10.19	10.34	10.48	10.63	10.78	10.92	11.07	11.21	11.36	11.51	11.65
17.2	10.25	10.40	10.55	10.69	10.84	10.99	11.14	11.28	11.43	11.58	11.72
17.3	10.32	10.46	10.61	10.76	10.91	11.06	11.20	11.35	11.50	11.65	11.79
17.4	10.38	10.53	10.68	10.82	10.97	11.12	11.27	11.42	11.57	11.72	11.86
17.5	10.44	10.59	10.74	10.89	11.04	11.19	11.34	11.49	11.64	11.79	11.94
17.6	10.50	10.65	10.80	10.95	11.10	11.25	11.40	11.56	11.71	11.86	12.01
17.7	10.56	10.72	10.87	11.02	11.17	11.32	11.47	11.62	11.77	11.93	12.08
17.8	10.63	10.78	10.93	11.08	11.24	11.39	11.54	11.69	11.84	12.00	12.15
17.9	10.69	10.84	10.99	11.15	11.30	11.45	11.61	11.76	11.91	12.07	12.22
18.0	10.75	10.90	11.06	11.21	11.37	11.52	11.67	11.83	11.98	12.14	12.29
18.1	10.82	10.97	11.13	11.28	11.44	11.59	11.75	11.90	12.06	12.21	12.37
18.2	10.88	11.04	11.19	11.35	11.51	11.66	11.82	11.97	12.13	12.28	12.44
18.3	10.95	11.11	11.26	11.42	11.58	11.73	11.89	12.05	12.20	12.36	12.52
18.4	11.01	11.17	11.33	11.49	11.64	11.80	11.96	12.12	12.28	12.43	12.59
18.5	11.08	11.24	11.40	11.56	11.71	11.87	12.03	12.19	12.35	12.51	12.67
18.6	11.15	11.31	11.47	11.62	11.78	11.94	12.10	12.26	12.42	12.58	12.74
18.7	11.21	11.37	11.53	11.69	11.85	12.01	12.17	12.33	12.49	12.66	12.82
18.8	11.28	11.44	11.60	11.76	11.92	12.08	12.25	12.41	12.57	12.73	12.89
18.9	11.34	11.51	11.67	11.83	11.99	12.16	12.32	12.48	12.64	12.80	12.97
19.0	11.41	11.57	11.74	11.90	12.06	12.23	12.39	12.55	12.71	12.88	13.04

4B

T [°C]	r.h. [%]										
	70	71	72	73	74	75	76	77	78	79	80
19.0	11.41	11.57	11.74	11.90	12.06	12.23	12.39	12.55	12.71	12.88	13.04
19.1	11.48	11.64	11.81	11.97	12.14	12.30	12.46	12.63	12.79	12.96	13.12
19.2	11.55	11.71	11.88	12.04	12.21	12.37	12.54	12.70	12.87	13.03	13.20
19.3	11.62	11.78	11.95	12.12	12.28	12.45	12.61	12.78	12.95	13.11	13.28
19.4	11.69	11.85	12.02	12.19	12.35	12.52	12.69	12.86	13.02	13.19	13.36
19.5	11.76	11.92	12.09	12.26	12.43	12.60	12.76	12.93	13.10	13.27	13.44
19.6	11.82	11.99	12.16	12.33	12.50	12.67	12.84	13.01	13.18	13.35	13.51
19.7	11.89	12.06	12.23	12.40	12.57	12.74	12.91	13.08	13.25	13.42	13.59
19.8	11.96	12.13	12.30	12.48	12.65	12.82	12.99	13.16	13.33	13.50	13.67
19.9	12.03	12.20	12.38	12.55	12.72	12.89	13.06	13.24	13.41	13.58	13.75
20.0	12.10	12.27	12.45	12.62	12.79	12.97	13.14	13.31	13.48	13.66	13.83
20.1	12.17	12.35	12.52	12.70	12.87	13.04	13.22	13.39	13.57	13.74	13.91
20.2	12.25	12.42	12.60	12.77	12.95	13.12	13.30	13.47	13.65	13.82	14.00
20.3	12.32	12.50	12.67	12.85	13.02	13.20	13.38	13.55	13.73	13.90	14.08
20.4	12.39	12.57	12.75	12.92	13.10	13.28	13.45	13.63	13.81	13.99	14.16
20.5	12.47	12.64	12.82	13.00	13.18	13.36	13.53	13.71	13.89	14.07	14.25
20.6	12.54	12.72	12.90	13.08	13.25	13.43	13.61	13.79	13.97	14.15	14.33
20.7	12.61	12.79	12.97	13.15	13.33	13.51	13.69	13.87	14.05	14.23	14.41
20.8	12.68	12.87	13.05	13.23	13.41	13.59	13.77	13.95	14.13	14.31	14.49
20.9	12.76	12.94	13.12	13.30	13.49	13.67	13.85	14.03	14.21	14.40	14.58
21.0	12.83	13.01	13.20	13.38	13.56	13.75	13.93	14.11	14.29	14.48	14.66
21.1	12.91	13.09	13.27	13.46	13.64	13.83	14.01	14.20	14.38	14.56	14.75
21.2	12.98	13.17	13.35	13.54	13.72	13.91	14.09	14.28	14.47	14.65	14.84
21.3	13.06	13.25	13.43	13.62	13.80	13.99	14.18	14.36	14.55	14.74	14.92
21.4	13.13	13.32	13.51	13.70	13.89	14.07	14.26	14.45	14.64	14.82	15.01
21.5	13.21	13.40	13.59	13.78	13.97	14.16	14.34	14.53	14.72	14.91	15.10
21.6	13.29	13.48	13.67	13.86	14.05	14.24	14.43	14.62	14.81	15.00	15.19
21.7	13.36	13.55	13.75	13.94	14.13	14.32	14.51	14.70	14.89	15.09	15.28
21.8	13.44	13.63	13.82	14.02	14.21	14.40	14.59	14.79	14.98	15.17	15.36
21.9	13.51	13.71	13.90	14.10	14.29	14.48	14.68	14.87	15.06	15.26	15.45
22.0	13.59	13.79	13.98	14.18	14.37	14.57	14.76	14.96	15.15	15.35	15.54
22.1	13.67	13.87	14.06	14.26	14.45	14.65	14.85	15.04	15.24	15.44	15.63
22.2	13.75	13.95	14.14	14.34	14.54	14.74	14.93	15.13	15.33	15.53	15.72
22.3	13.83	14.03	14.23	14.43	14.62	14.82	15.02	15.22	15.42	15.62	15.81
22.4	13.91	14.11	14.31	14.51	14.71	14.91	15.11	15.31	15.51	15.71	15.90
22.5	13.99	14.19	14.39	14.59	14.79	14.99	15.19	15.39	15.59	15.80	16.00
22.6	14.07	14.27	14.47	14.68	14.88	15.08	15.28	15.48	15.68	15.88	16.09
22.7	14.15	14.35	14.56	14.76	14.96	15.16	15.37	15.57	15.77	15.97	16.18
22.8	14.23	14.43	14.64	14.84	15.05	15.25	15.45	16.66	15.86	16.06	16.27
22.9	14.31	14.52	14.72	14.93	15.13	15.34	15.54	15.74	15.95	16.15	16.36
23.0	14.39	14.60	14.80	15.01	15.21	15.42	15.63	15.83	16.04	16.24	16.45

4C

T [°C]	r.h. [%]										
	70	71	72	73	74	75	76	77	78	79	80
23.0	14.39	14.60	14.80	15.01	15.21	15.42	15.63	15.83	16.04	16.24	16.45
23.1	14.48	14.68	14.89	15.10	15.30	15.51	15.72	15.93	16.13	16.34	16.55
23.2	14.56	14.77	14.98	15.19	15.39	15.60	15.81	16.02	16.23	16.44	16.64
23.3	14.65	14.86	15.06	15.27	15.48	15.69	15.90	16.11	16.32	16.53	16.74
23.4	14.73	14.94	15.15	15.36	15.57	15.78	16.00	16.21	16.42	16.63	16.84
23.5	14.82	15.03	15.24	15.45	15.66	15.88	16.09	16.30	16.51	16.72	16.94
23.6	14.90	15.11	15.33	15.54	15.75	15.97	16.18	16.39	16.61	16.82	17.03
23.7	14.99	15.20	15.41	15.63	15.84	16.06	16.27	16.49	16.70	16.92	17.13
23.8	15.07	15.29	15.50	15.72	15.93	16.15	16.36	16.58	16.80	17.01	17.23
23.9	15.16	15.37	15.59	15.81	16.02	16.24	16.46	16.67	16.89	17.11	17.32
24.0	15.24	15.46	15.68	15.89	16.11	16.33	16.55	16.77	16.98	17.20	17.42
24.1	15.33	15.55	15.77	15.99	16.21	16.43	16.64	16.86	17.08	17.30	17.52
24.2	15.42	15.64	15.86	16.08	16.30	16.52	16.74	16.96	17.18	17.40	17.62
24.3	15.51	15.73	15.95	16.17	16.39	16.62	16.84	17.06	17.28	17.50	17.72
24.4	15.60	15.82	16.04	16.26	16.49	16.71	16.93	17.16	17.38	17.60	17.82
24.5	15.69	15.91	16.13	16.36	16.58	16.81	17.03	17.25	17.48	17.70	17.93
24.6	15.77	16.00	16.22	16.45	16.68	16.90	17.13	17.35	17.58	17.80	18.03
24.7	15.86	16.09	16.32	16.54	16.77	17.00	17.22	17.45	17.67	17.90	18.13
24.8	15.95	16.18	16.41	16.64	16.86	17.09	17.32	17.55	17.77	18.00	18.23
24.9	16.04	16.27	16.50	16.73	16.96	17.19	17.41	17.64	17.87	18.10	18.33
25.0	16.13	16.36	16.59	16.82	17.05	17.28	17.51	17.74	17.97	18.20	18.43
25.1	16.22	16.45	16.69	16.92	17.15	17.38	17.61	17.84	18.07	18.31	18.54
25.2	16.32	16.55	16.78	17.01	17.25	17.48	17.71	17.95	18.18	18.41	18.64
25.3	16.41	16.64	16.88	17.11	17.35	17.58	17.81	18.05	18.28	18.52	18.75
25.4	16.50	16.74	16.97	17.21	17.44	17.68	17.92	18.15	18.39	18.62	18.86
25.5	16.60	16.83	17.07	17.31	17.54	17.78	18.02	18.25	18.49	18.73	18.97
25.6	16.69	16.93	17.17	17.40	17.64	17.88	18.12	18.36	18.60	18.83	19.07
25.7	16.78	17.02	17.26	17.50	17.74	17.98	18.22	18.46	18.70	18.94	19.18
25.8	16.87	17.12	17.36	17.60	17.84	18.08	18.32	18.56	18.80	19.05	19.29
25.9	16.97	17.21	17.45	17.70	17.94	18.18	18.42	18.67	18.91	19.15	19.39
26.0	17.06	17.30	17.55	17.79	18.04	18.28	18.52	18.77	19.01	19.26	19.50
26.1	17.16	17.40	17.65	17.89	18.14	18.38	18.63	18.88	19.12	19.37	19.61
26.2	17.25	17.50	17.75	17.99	18.24	18.49	18.74	18.98	19.23	19.48	19.72
26.3	17.35	17.60	17.85	18.10	18.34	18.59	18.84	19.09	19.34	19.59	19.83
26.4	17.45	17.70	17.95	18.20	18.45	18.70	18.95	19.20	19.45	19.69	19.94
26.5	17.55	17.80	18.05	18.30	18.55	18.80	19.05	19.30	19.55	19.80	20.06
26.6	17.64	17.89	18.15	18.40	18.65	18.90	19.16	19.41	19.66	19.91	20.17
26.7	17.74	17.99	18.25	18.50	18.75	19.01	19.26	19.52	19.77	20.02	20.28
26.8	17.84	18.09	18.35	18.60	18.86	19.11	19.37	19.62	19.88	20.13	20.39
26.9	17.93	18.19	18.45	18.70	18.96	19.22	19.47	19.73	19.99	20.24	20.50
27.0	18.03	18.29	18.55	18.80	19.06	19.32	19.58	19.84	20.09	20.35	20.61

4D

T [°C]	r.h. [%]										
	70	71	72	73	74	75	76	77	78	79	80
27.0	18.03	18.29	18.55	18.80	19.06	19.32	19.58	19.84	20.09	20.35	20.61
21.1	18.13	18.39	18.65	18.91	19.17	19.43	19.69	19.95	20.21	20.47	20.73
27.2	18.24	18.50	18.76	19.02	19.28	19.54	19.80	20.06	20.32	20.58	20.84
27.3	18.34	18.60	18.86	19.13	19.39	19.65	19.91	20.17	20.44	20.70	20.96
27.4	18.44	18.71	18.97	19.23	19.50	19.76	20.02	20.29	20.55	20.81	21.08
27.5	18.55	18.81	19.08	19.34	19.61	19.87	20.14	20.40	20.67	20.93	21.20
27.6	18.65	18.91	19.18	19.45	19.71	19.98	20.25	20.51	20.78	21.05	21.31
27.7	18.75	19.02	19.29	19.55	19.82	20.09	20.36	20.63	20.89	21.16	21.43
27.8	18.85	19.12	19.39	19.66	19.93	20.20	20.47	20.74	21.01	21.28	21.55
27.9	18.96	19.23	19.50	19.77	20.04	20.31	20.58	20.85	21.12	21.39	21.66
28.0	19.06	19.33	19.60	19.88	20.15	20.42	20.69	20.96	21.24	21.51	21.78
28.1	19.17	19.44	19.71	19.99	20.26	20.54	20.81	21.08	21.36	21.63	21.90
28.2	19.27	19.55	19.82	20.10	20.38	20.65	20.93	21.20	21.48	21.75	22.03
28.3	19.38	19.66	19.94	20.21	20.49	20.77	21.04	21.32	21.60	21.87	22.15
28.4	19.49	19.77	20.05	20.32	20.60	20.88	21.16	21.44	21.72	21.99	22.27
28.5	19.60	19.88	20.16	20.44	20.72	21.00	21.28	21.56	21.84	22.12	22.40
28.6	19.70	19.98	20.27	20.55	20.83	21.11	21.39	21.67	21.96	22.24	22.52
28.7	19.81	20.09	20.38	20.66	20.94	21.23	21.51	21.79	22.08	22.36	22.64
28.8	19.92	20.20	20.49	20.77	21.06	21.34	21.63	21.91	22.19	22.48	22.76
28.9	20.02	20.31	20.60	20.88	21.17	21.46	21.74	22.03	22.31	22.60	22.89
29.0	20.13	20.42	20.71	20.99	21.28	21.57	21.86	22.15	22.43	22.72	23.01
29.1	20.24	20.53	20.82	21.11	21.40	21.69	21.98	22.27	22.56	22.85	23.14
29.2	20.35	20.65	20.94	21.23	21.52	21.81	22.10	22.39	22.68	22.98	23.27
23.3	20.47	20.76	21.05	21.34	21.64	21.93	22.22	22.52	22.81	23.10	23.39
29.4	20.58	20.87	21.17	21.46	21.76	22.05	22.34	22.64	22.93	23.23	23.52
29.5	20.69	20.99	21.28	21.58	21.87	22.17	22.47	22.76	23.06	23.35	23.65
29.6	20.80	21.10	21.40	21.70	21.99	22.29	22.59	22.89	23.18	23.48	23.78
29.7	20.91	21.21	21.51	21.81	22.11	22.41	22.71	23.01	23.31	23.61	23.91
29.8	21.03	21.33	21.63	21.93	22.23	22.53	22.83	23.13	23.43	23.73	24.03
29.9	21.14	21.44	21.74	22.05	22.35	22.65	22.95	23.26	23.56	23.86	24.16
30.0	21.25	21.55	21.86	22.16	22.47	22.77	23.07	23.38	23.68	23.99	24.29

5A

T [°C]	r.h. [%]										
	80	81	82	83	84	85	86	87	88	89	90
15.0	10.26	10.39	10.52	10.64	10.77	10.90	11.03	11.16	11.28	11.41	11.54
15.1	10.32	10.45	10.58	10.71	10.84	10.97	11.10	11.23	11.36	11.48	11.61
15.2	10.39	10.52	10.65	10.78	10.91	11.04	11.17	11.30	11.43	11.56	11.69
15.3	10.45	10.58	10.71	10.84	10.98	11.11	11.24	11.37	11.50	11.63	11.76
15.4	10.52	10.65	10.78	10.91	11.04	11.17	11.31	11.44	11.57	11.70	11.83
15.5	10.58	10.71	10.85	10.98	11.11	11.24	11.38	11.51	11.64	11.77	11.91
15.6	10.64	10.78	10.91	11.04	11.18	11.31	11.44	11.58	11.71	11.85	11.98
15.7	10.71	10.84	10.98	11.11	11.25	11.38	11.51	11.65	11.78	11.92	12.05
15.8	10.77	10.91	11.04	11.18	11.31	11.45	11.58	11.72	11.85	11.99	12.12
15.9	10.84	10.97	11.11	11.24	11.38	11.52	11.65	11.79	11.93	12.06	12.20
16.0	10.90	11.04	11.17	11.31	11.45	11.59	11.72	11.86	12.00	12.13	12.27
16.1	10.97	11.11	11.24	11.38	11.52	11.66	11.79	11.93	12.07	12.21	12.35
16.2	11.04	11.17	11.31	11.45	11.59	11.73	11.87	12.01	12.14	12.28	22.42
16.3	11.10	11.24	11.38	11.52	11.66	11.80	11.94	12.08	12.22	12.36	12.50
16.4	11.17	11.31	11.45	11.59	11.73	11.87	12.01	12.15	12.29	12.43	12.57
16.5	11.24	11.38	11.52	11.66	11.80	11.94	12.08	12.22	12.36	12.51	12.65
16.6	11.31	11.45	11.59	11.73	11.87	12.01	12.16	12.30	12.44	12.58	12.72
16.7	11.38	11.52	11.66	11.80	11.94	12.09	12.23	12.37	12.51	12.65	12.80
16.8	11.44	11.59	11.73	11.87	12.01	12.16	12.30	12.44	12.59	12.73	12.87
16.9	11.51	11.66	11.80	11.94	12.09	12.23	12.37	12.52	12.66	12.80	12.95
17.0	11.58	11.72	11.87	12.01	12.16	12.30	12.44	12.59	12.73	12.88	13.02
17.1	11.65	11.80	11.94	12.09	12.23	12.38	12.52	12.67	12.81	12.96	13.10
17.2	11.72	11.87	12.01	12.16	12.31	12.45	12.60	12.74	12.89	13.03	13.18
17.3	11.79	11.94	12.09	12.23	12.38	12.53	12.67	12.82	12.97	13.11	13.26
17.4	11.86	12.01	12.16	12.31	12.45	12.60	12.75	12.90	13.05	13.19	13.34
17.5	11.94	12.08	12.23	12.38	12.53	12.68	12.83	12.98	13.12	13.27	13.42
17.6	12.01	12.16	12.31	12.45	12.60	12.75	12.90	13.05	13.20	13.35	13.50
17.7	12.08	12.23	12.38	12.53	12.68	12.83	12.98	13.13	13.28	13.43	13.58
17.8	12.15	12.30	12.45	12.60	12.75	12.90	13.06	13.21	13.36	13.51	13.66
17.9	12.22	12.37	12.52	12.68	12.83	12.98	13.13	13.28	13.44	13.59	13.74
18.0	12.29	12.44	12.60	12.75	12.90	13.06	13.21	13.36	13.51	13.67	13.82
18.1	12.37	12.52	12.67	12.83	12.98	13.14	13.29	13.44	13.60	13.75	13.91
18.2	12.44	12.60	12.75	12.91	13.06	13.22	13.37	13.53	13.68	13.84	13.99
18.3	12.52	12.67	12.83	12.98	13.14	13.30	13.45	13.61	13.76	13.92	14.08
18.4	12.59	12.75	12.90	13.06	13.22	13.38	13.53	13.69	13.85	14.00	14.16
18.5	12.67	12.82	12.98	13.14	13.30	13.46	13.61	13.77	13.93	14.09	14.25
18.6	12.74	12.90	13.06	13.22	13.38	13.54	13.69	13.85	14.01	14.17	14.33
18.7	12.82	12.98	13.14	13.30	13.46	13.62	13.78	13.94	14.10	14.26	14.42
18.8	12.89	13.05	13.21	13.37	13.53	13.70	13.86	14.02	14.18	14.34	14.50
18.9	12.97	13.13	13.29	13.45	13.61	13.78	13.94	14.10	14.26	14.42	14.59
19.0	13.04	13.20	13.37	13.53	13.69	13.86	14.02	14.18	14.34	14.51	14.67

5B

T [°C]	r.h. [%]										
	80	81	82	83	84	85	86	87	88	89	90
19.0	13.04	13.20	13.37	13.53	13.69	13.86	14.02	14.18	14.34	14.51	14.67
19.1	13.12	13.28	13.45	13.61	13.78	13.94	14.10	14.27	14.43	14.60	14.76
19.2	13.20	13.36	13.53	13.69	13.86	14.02	14.19	14.35	14.52	14.68	14.85
19.3	13.28	13.44	13.61	13.78	13.94	14.11	14.27	14.44	14.61	14.77	14.94
19.4	13.36	13.52	13.69	13.86	14.02	14.19	14.36	14.53	14.69	14.86	15.03
19.5	13.44	13.60	13.77	13.94	14.11	14.28	14.44	14.61	14.78	14.95	15.12
19.6	13.51	13.68	13.85	14.02	14.19	14.36	14.53	14.70	14.87	15.04	15.20
19.7	13.59	13.76	13.93	14.10	14.27	14.44	14.61	14.78	14.95	15.12	15.29
19.8	13.67	13.84	14.01	14.19	14.36	14.53	14.70	14.87	15.04	15.21	15.38
19.9	13.75	13.92	14.10	14.27	14.44	14.61	14.78	14.96	15.13	15.30	15.47
20.0	13.83	14.00	14.18	14.35	14.52	14.70	14.87	15.04	15.21	15.39	15.56
20.1	13.91	14.09	14.26	14.44	14.61	14.78	14.96	15.13	15.31	15.48	15.65
20.2	14.00	14.17	14.35	14.52	14.70	14.87	15.05	15.22	15.40	15.57	15.75
20.3	14.08	14.26	14.43	14.61	14.78	14.96	15.14	15.31	15.49	15.67	15.84
20.4	14.16	14.34	14.52	14.69	14.87	15.05	15.23	15.40	15.58	15.76	15.94
20.5	14.25	14.42	14.60	14.78	14.96	15.14	15.32	15.50	15.67	15.85	16.03
20.6	14.33	14.51	14.69	14.87	15.05	15.23	15.41	15.59	15.77	15.94	16.12
20.7	14.41	14.59	14.77	14.95	15.13	15.32	15.50	15.68	15.86	16.04	16.22
20.8	14.49	14.68	14.86	15.04	15.22	15.40	15.59	15.77	15.95	16.13	16.31
20.9	14.58	14.76	14.94	15.13	15.31	15.49	15.67	15.86	16.04	16.22	16.41
21.0	14.66	14.84	15.03	15.21	15.40	15.58	15.76	15.95	16.13	16.32	16.50
21.1	14.75	14.93	15.12	15.30	15.49	15.67	15.86	16.04	16.23	16.41	16.60
21.2	14.84	15.02	15.21	15.39	15.58	15.77	15.95	16.14	16.32	16.51	16.70
21.3	14.92	15.11	15.30	15.49	15.67	15.86	16.05	16.23	16.42	16.61	16.79
21.4	15.01	15.20	15.39	15.58	15.76	15.95	16.14	16.33	16.52	16.70	16.89
21.5	15.10	15.29	15.48	15.67	15.86	16.05	16.23	16.42	16.61	16.80	16.99
21.6	15.19	15.38	15.57	15.76	15.95	16.14	16.33	16.52	16.71	16.90	17.09
21.7	15.28	15.47	15.66	15.85	16.04	16.23	16.42	16.61	16.80	17.00	17.19
21.8	15.36	15.56	15.75	15.94	16.13	16.32	16.52	16.71	16.90	17.09	17.28
21.9	15.45	15.65	15.84	16.03	16.22	16.42	16.61	16.80	17.00	17.19	17.38
22.0	15.54	15.73	15.93	16.12	16.32	16.51	16.70	16.90	17.09	17.29	17.48
22.1	15.63	15.83	16.02	16.22	16.41	16.61	16.80	17.00	17.19	17.39	17.58
22.2	15.72	15.92	16.11	16.31	16.51	16.70	16.90	17.10	17.29	17.49	17.68
22.3	15.81	16.01	16.21	16.41	16.60	16.80	17.00	17.19	17.39	17.59	17.79
22.4	15.90	16.10	16.30	16.50	16.70	16.90	17.09	17.29	17.49	17.69	17.89
22.5	16.00	16.19	16.39	16.59	16.79	16.99	17.19	17.39	17.59	17.79	17.99
22.6	16.09	16.29	16.49	16.69	16.89	17.09	17.29	17.49	17.69	17.89	18.09
22.7	16.18	16.38	16.58	16.78	16.98	17.19	17.39	17.59	17.79	17.99	18.19
22.8	16.27	16.47	16.67	16.88	17.08	17.28	17.49	17.69	17.89	18.09	18.30
22.9	16.36	16.56	16.77	16.97	17.18	17.38	17.58	17.79	17.99	18.19	18.40
23.0	16.45	16.66	16.86	17.07	17.27	17.48	17.68	17.89	18.09	18.30	18.50

5C

T [°C]	r.h. [%]										
	80	81	82	83	84	85	86	87	88	89	90
23.0	16.45	16.66	16.86	17.07	17.27	17.48	17.68	17.89	18.09	18.30	18.50
23.1	16.55	16.75	16.96	17.17	17.37	17.58	17.78	17.99	18.20	18.40	18.61
23.2	16.64	16.85	17.06	17.27	17.47	17.68	17.89	18.10	18.30	18.51	18.72
23.3	16.74	16.95	17.16	17.37	17.58	17.78	17.99	18.20	18.41	18.62	18.83
23.4	16.84	17.05	17.26	17.47	17.68	17.89	18.10	18.31	18.52	18.73	18.94
23.5	16.94	17.15	17.36	17.57	17.78	17.99	18.20	18.41	18.62	18.83	19.05
23.6	17.03	17.24	17.46	17.67	17.88	18.09	18.31	18.52	18.73	18.94	19.15
23.7	17.13	17.34	17.56	17.77	17.98	18.20	18.41	18.62	18.84	19.05	19.26
23.8	17.23	17.44	17.66	17.87	18.08	18.30	18.51	18.73	18.94	19.16	19.37
23.9	17.32	17.54	17.76	17.97	18.19	18.40	18.62	18.83	19.05	19.27	19.48
24.0	17.42	17.64	17.85	18.07	18.29	18.51	18.72	18.94	19.16	19.37	19.59
24.1	17.52	17.74	17.96	18.18	18.40	18.61	18.83	19.05	19.27	19.49	19.71
24.2	17.62	17.84	18.06	18.28	18.50	18.72	18.94	19.16	19.38	19.60	19.82
24.3	17.72	17.94	18.17	18.39	18.61	18.83	19.05	19.27	19.49	19.71	19.94
24.4	17.82	18.05	18.27	18.49	18.71	18.94	19.16	19.38	19.61	19.83	20.05
24.5	17.93	18.15	18.37	18.60	18.82	19.05	19.27	19.49	19.72	19.94	20.17
24.6	18.03	18.25	18.48	18.70	18.93	19.15	19.38	19.60	19.83	20.06	20.28
24.7	18.13	18.35	18.58	18.81	19.03	19.26	19.49	19.72	19.94	20.17	20.40
24.8	18.23	18.46	18.68	18.91	19.14	19.37	19.60	19.83	20.05	20.28	20.51
24.9	18.33	18.56	18.79	19.02	19.25	19.48	19.71	19.94	20.17	20.40	20.63
25.0	18.43	18.66	18.89	19.12	19.35	19.59	19.82	20.05	20.28	20.51	20.74
25.1	18.54	18.77	19.00	19.23	19.47	19.70	19.93	20.16	20.40	20.63	20.86
25.2	18.64	18.88	19.11	19.34	19.58	19.81	20.04	20.28	20.51	20.75	20.98
25.3	18.75	18.99	19.22	19.46	19.69	19.92	20.16	20.39	20.63	20.86	21.10
25.4	18.86	19.09	19.33	19.57	19.80	20.04	20.27	20.51	20.74	20.98	21.22
25.5	18.97	19.20	19.44	19.68	19.91	20.15	20.39	20.62	20.86	21.10	21.34
25.6	19.07	19.31	19.55	19.79	20.03	20.26	20.50	20.74	20.98	21.22	21.45
25.7	19.18	19.42	19.66	19.90	20.14	20.38	20.62	20.86	21.09	21.33	21.57
25.8	19.29	19.53	19.77	20.01	20.25	20.49	20.73	20.97	21.21	21.45	21.69
25.9	19.39	19.64	19.88	20.12	20.36	20.60	20.84	21.09	21.33	21.57	21.81
26.0	19.50	19.74	19.99	20.23	20.47	20.72	20.96	21.20	21.44	21.69	21.93
26.1	19.61	19.86	20.10	20.34	20.59	20.83	21.08	21.32	21.57	21.81	22.06
26.2	19.72	19.97	20.21	20.46	20.71	20.95	21.20	21.44	21.69	21.93	22.18
26.3	19.83	20.08	20.33	20.58	20.82	21.07	21.32	21.56	21.81	22.06	22.31
26.4	19.94	20.19	20.44	20.69	20.94	21.19	21.44	21.68	21.93	22.18	22.43
26.5	20.06	20.31	20.56	20.81	21.06	21.31	21.56	21.81	22.06	22.31	22.56
26.6	20.17	20.42	20.67	20.92	21.17	21.42	21.67	21.93	22.18	22.43	22.68
26.7	20.28	20.53	20.78	21.04	21.29	21.54	21.79	22.05	22.30	22.55	22.81
26.8	20.39	20.64	20.90	21.15	21.41	21.66	21.91	22.17	22.42	22.68	22.93
26.9	20.50	20.76	21.01	21.27	21.52	21.78	22.03	22.29	22.54	22.80	23.06
27.0	20.61	20.87	21.12	21.38	21.64	21.90	22.15	22.41	22.67	22.92	23.18

5D

T [°C]	r.h. [%]										
	80	81	82	83	84	85	86	87	88	89	90
27.0	20.61	20.87	21.12	21.38	21.64	21.90	22.15	22.41	22.67	22.92	23.18
27.1	20.73	20.99	21.24	21.50	21.76	22.02	22.28	22.54	22.80	23.05	23.31
27.2	20.84	21.10	21.36	21.63	21.89	22.15	22.41	22.67	22.93	23.19	23.45
27.3	20.96	21.22	21.49	21.75	22.01	22.27	22.53	22.79	23.06	23.32	23.58
27.4	21.08	21.34	21.61	21.87	22.13	22.40	22.66	22.92	23.19	23.45	23.71
27.5	21.20	21.46	21.73	21.99	22.26	22.52	22.79	23.05	23.32	23.58	23.85
27.6	21.31	21.58	21.85	22.11	22.38	22.65	22.91	23.18	23.45	23.71	23.98
27.7	21.43	21.70	21.97	22.23	22.50	22.77	23.04	23.31	23.58	23.84	24.11
27.8	21.55	21.82	22.09	22.36	22.63	22.90	23.17	23.44	23.70	23.97	24.24
27.9	21.66	21.93	22.21	22.48	22.75	23.02	23.29	23.56	23.83	24.11	24.38
28.0	21.78	22.05	22.33	22.60	22.87	23.15	23.42	23.69	23.96	24.24	24.51
28.1	21.90	22.18	22.45	22.73	23.00	23.28	23.55	23.82	24.10	24.37	24.65
28.2	22.03	22.30	22.58	22.85	23.13	23.41	23.68	23.96	24.23	24.51	24.78
28.3	22.15	22.43	22.70	22.98	23.26	23.54	23.81	24.09	24.37	24.64	24.92
28.4	22.27	22.55	22.83	23.11	23.39	23.67	23.94	24.22	24.50	24.78	25.06
28.5	22.40	22.68	22.96	23.24	23.52	23.80	24.08	24.36	24.64	24.92	25.20
28.6	22.52	22.80	23.08	23.36	23.64	23.93	24.21	24.49	24.77	25.05	25.33
28.7	22.64	22.92	23.21	23.49	23.77	24.06	24.34	24.62	24.90	25.19	25.47
28.8	22.76	23.05	23.33	23.62	23.90	24.19	24.47	24.75	25.04	25.32	25.61
28.9	22.89	23.17	23.46	23.74	24.03	24.32	24.60	24.89	25.17	25.46	25.74
29.0	23.01	23.30	23.58	23.87	24.16	24.45	24.73	25.02	25.31	25.59	25.88
29.1	23.14	23.43	23.72	24.00	24.29	24.58	24.87	25.16	25.45	25.74	26.02
29.2	23.27	23.56	23.85	24.14	24.43	24.72	25.01	25.30	25.59	25.88	26.17
29.3	23.39	23.69	23.98	24.27	24.56	24.85	25.15	25.44	25.73	26.02	26.31
29.4	23.52	23.82	24.11	24.40	24.70	24.99	25.28	25.58	25.87	26.16	26.46
29.5	23.65	23.95	24.24	24.54	24.83	25.13	25.42	25.72	26.01	26.31	26.60
29.6	23.78	24.08	24.37	24.67	24.96	25.26	25.56	25.85	26.15	26.45	26.74
29.7	23.91	24.20	24.50	24.80	25.10	25.40	25.70	25.99	26.29	26.59	26.89
29.8	24.03	24.33	24.63	24.93	25.23	25.53	25.83	26.13	26.43	26.73	27.03
29.9	24.16	24.46	24.77	25.07	25.37	25.67	25.97	26.27	26.57	26.88	27.18
30.0	24.29	24.59	24.90	25.20	25.50	25.81	26.11	26.41	26.71	27.02	27.32

6A

T [°C]	r.h. [%]										
	90	91	92	93	94	95	96	97	98	99	100
15.0	11.54	11.67	11.80	11.92	12.05	12.18	12.31	12.44	12.56	12.69	12.82
15.1	11.61	11.74	11.87	12.00	12.13	12.26	12.39	12.52	12.64	12.77	12.90
15.2	11.69	11.82	11.95	12.08	12.20	12.33	12.46	12.59	12.72	12.85	12.98
15.3	11.76	11.89	12.02	12.15	12.28	12.41	12.54	12.67	12.80	12.93	13.06
15.4	11.83	11.96	12.09	12.23	12.36	12.49	12.62	12.75	12.88	13.01	13.14
15.5	11.91	12.04	12.17	12.30	12.43	12.57	12.70	12.83	12.96	13.09	13.23
15.6	11.98	12.11	12.24	12.38	12.51	12.64	12.78	12.91	13.04	13.17	13.31
15.7	12.05	12.19	12.32	12.45	12.59	12.72	12.85	12.99	13.12	13.25	13.39
15.8	12.12	12.26	12.39	12.53	12.66	12.80	12.93	13.07	13.20	13.33	13.47
15.9	12.20	12.33	12.47	12.60	12.74	12.87	13.01	13.14	13.28	13.41	13.55
16.0	12.27	12.41	12.54	12.68	12.81	12.95	13.09	13.22	13.36	13.49	13.63
16.1	12.35	12.48	12.62	12.76	12.89	13.03	13.17	13.30	13.44	13.58	13.71
16.2	12.42	12.56	12.70	12.83	12.97	13.11	13.25	13.39	13.52	13.66	13.80
16.3	12.50	12.63	12.77	12.91	13.05	13.19	13.33	13.47	13.61	13.74	13.88
16.4	12.57	12.71	12.85	12.99	13.13	13.27	13.41	13.55	13.69	13.83	13.97
16.5	12.65	12.79	12.93	13.07	13.21	13.35	13.49	13.63	13.77	13.91	14.05
16.6	12.72	12.86	13.00	13.14	13.29	13.43	13.57	13.71	13.85	13.99	14.13
16.7	12.80	12.94	13.08	13.22	13.36	13.51	13.65	13.79	13.93	14.08	14.22
16.8	12.87	13.01	13.16	13.30	13.44	13.59	13.73	13.87	14.02	14.16	14.30
16.9	12.95	13.09	13.23	13.38	13.52	13.67	13.81	13.95	14.10	14.24	14.39
17.0	13.02	13.17	13.31	13.46	13.60	13.75	13.89	14.04	14.18	14.33	14.47
17.1	13.10	13.25	13.39	13.54	13.68	13.83	13.98	14.12	14.27	14.41	14.56
17.2	13.18	13.33	13.47	13.62	13.77	13.91	14.06	14.21	14.35	14.50	14.65
17.3	13.26	13.41	13.56	13.70	13.85	14.00	14.15	14.29	14.44	14.59	14.74
17.4	13.34	13.49	13.64	13.79	13.93	14.08	14.23	14.38	14.53	14.68	14.83
17.5	13.42	13.57	13.72	13.87	14.02	14.17	14.32	14.47	14.62	14.77	14.92
17.6	13.50	13.65	13.80	13.95	14.10	14.25	14.40	14.55	14.70	14.85	15.00
17.7	13.58	13.73	13.88	14.03	14.19	14.34	14.49	14.64	14.79	14.94	15.09
17.8	13.66	13.81	13.96	14.12	14.27	14.42	14.57	14.73	14.88	15.03	15.18
17.9	13.74	13.89	14.05	14.20	14.35	14.51	14.66	14.81	14.97	15.12	15.27
18.0	13.82	13.97	14.13	14.28	14.44	14.59	14.74	14.90	15.05	15.21	15.36
18.1	13.91	14.06	14.22	14.37	14.53	14.68	14.83	14.99	15.14	15.30	15.45
18.2	13.99	14.15	14.30	14.46	14.61	14.77	14.93	15.08	15.24	15.39	15.55
18.3	14.08	14.23	14.39	14.55	14.70	14.86	15.02	15.17	15.33	15.49	15.64
18.4	14.16	14.32	14.48	14.63	14.79	14.95	15.11	15.26	15.42	15.58	15.74
18.5	14.25	14.40	14.56	14.72	14.88	15.04	15.20	15.36	15.51	15.67	15.83
18.6	14.33	14.49	14.65	14.81	14.97	15.13	15.29	15.45	15.61	15.77	15.92
18.7	14.42	14.58	14.74	14.90	15.06	15.22	15.38	15.54	15.70	15.86	16.02
18.8	14.50	14.66	14.82	14.98	15.15	15.31	15.47	15.63	15.79	15.95	16.11
18.9	14.59	14.75	14.91	15.07	15.23	15.40	15.56	15.72	15.88	16.04	16.21
19.0	14.67	14.83	15.00	15.16	15.32	15.49	15.65	15.81	15.97	16.14	16.30

6B

T [°C]	r.h. [%]										
	90	91	92	93	94	95	96	97	98	99	100
19.0	14.67	14.83	15.00	15.16	15.32	15.49	15.65	15.81	15.97	16.14	16.30
19.1	14.76	14.92	15.09	15.25	15.42	15.58	15.74	15.91	16.07	16.24	16.40
19.2	14.85	15.01	15.18	15.34	15.51	15.67	15.84	16.00	16.17	16.33	16.50
19.3	14.94	15.10	15.27	15.44	15.60	15.77	15.93	16.10	16.27	16.43	16.60
19.4	15.03	15.19	15.36	15.53	15.69	15.86	16.03	16.20	16.36	16.53	16.70
19.5	15.12	15.28	15.45	15.62	15.79	15.96	16.12	16.29	16.46	16.63	16.80
19.6	15.20	15.37	15.54	15.71	15.88	16.05	16.22	16.39	16.56	16.73	16.89
19.7	15.29	15.46	15.63	15.80	15.97	16.14	16.31	16.48	16.65	16.82	16.99
19.8	15.38	15.55	15.72	15.90	16.07	16.24	16.41	16.58	16.75	16.92	17.09
19.9	15.47	15.64	15.82	15.99	16.16	16.33	16.50	16.68	16.85	17.02	17.19
20.0	15.56	15.73	15.91	16.08	16.25	16.43	16.60	16.77	16.94	17.12	17.29
20.1	15.65	15.83	16.00	16.18	16.35	16.52	16.70	16.87	17.05	17.22	17.39
20.2	15.75	15.92	16.10	16.27	16.45	16.62	16.80	16.97	17.15	17.32	17.50
20.3	15.84	16.02	16.19	16.37	16.55	16.72	16.90	17.07	17.25	17.43	17.60
20.4	15.94	16.11	16.29	16.47	16.64	16.82	17.00	17.18	17.35	17.53	17.71
20.5	16.03	16.21	16.39	16.56	16.74	16.92	17.10	17.28	17.45	17.63	17.81
20.6	16.12	16.30	16.48	16.66	16.84	17.02	17.20	17.38	17.56	17.74	17.91
20.7	16.22	16.40	16.58	16.76	16.94	17.12	17.30	17.48	17.66	17.84	18.02
20.8	16.31	16.49	16.67	16.86	17.04	17.22	17.40	17.58	17.76	17.94	18.12
20.9	16.41	16.59	16.77	16.95	17.13	17.32	17.50	17.68	17.86	18.04	18.23
21.0	16.50	16.68	16.87	17.05	17.23	17.42	17.60	17.78	17.96	18.15	18.33
21.1	16.60	16.78	16.97	17.15	17.33	17.52	17.70	17.89	18.07	18.26	18.44
21.2	16.70	16.88	17.07	17.25	17.44	17.62	17.81	17.99	18.18	18.36	18.55
21.3	16.79	16.98	17.17	17.35	17.54	17.73	17.91	18.10	18.28	18.47	18.66
21.4	16.89	17.08	17.27	17.45	17.64	17.83	18.02	18.20	18.39	18.58	18.77
21.5	16.99	17.18	17.37	17.56	17.74	17.93	18.12	18.31	18.50	18.69	18.88
21.6	17.09	17.28	17.47	17.66	17.85	18.04	18.23	18.42	18.61	18.79	18.98
21.7	17.19	17.38	17.57	17.76	17.95	18.14	18.33	18.52	18.71	18.90	19.09
21.8	17.28	17.48	17.67	17.86	18.05	18.24	18.44	18.63	18.82	19.01	19.20
21.9	17.38	17.58	17.77	17.96	18.15	18.35	18.54	18.73	18.93	19.12	19.31
22.0	17.48	17.67	17.87	18.06	18.26	18.45	18.64	18.84	19.03	19.23	19.42
22.1	17.58	17.78	17.97	18.17	18.36	18.56	18.75	18.95	19.14	19.34	19.53
22.2	17.68	17.88	18.08	18.27	18.47	18.67	18.86	19.06	19.26	19.45	19.65
22.3	17.79	17.98	18.18	18.38	18.58	18.77	18.97	19.17	19.37	19.56	19.76
22.4	17.89	18.09	18.29	18.48	18.68	18.88	19.08	19.28	19.48	19.68	19.88
22.5	17.99	18.19	18.39	18.59	18.79	18.99	19.19	19.39	19.59	19.79	19.99
22.6	18.09	18.29	18.49	18.70	18.90	19.10	19.30	19.50	19.70	19.90	20.10
22.7	18.19	18.40	18.60	18.80	19.00	19.21	19.41	19.61	19.81	20.02	20.22
22.8	18.30	18.50	18.70	18.91	19.11	19.31	19.52	19.72	19.93	20.13	20.33
22.9	18.40	18.60	18.81	19.01	19.22	19.42	19.63	19.83	20.04	20.24	20.45
23.0	18.50	18.71	18.91	19.12	19.32	19.53	19.74	19.94	20.15	20.35	20.56

6C

T [°C]	r.h. [%]										
	90	91	92	93	94	95	96	97	98	99	100
23.0	18.50	18.71	18.91	19.12	19.32	19.53	19.74	19.94	20.15	20.35	20.56
23.1	18.61	18.82	19.02	19.23	19.44	19.65	19.85	20.06	20.27	20.47	20.68
23.2	18.72	18.93	19.14	19.34	19.55	19.76	19.97	20.18	20.39	20.59	20.80
23.3	18.83	19.04	19.25	19.46	19.67	19.88	20.09	20.29	20.50	20.71	20.92
23.4	18.94	19.15	19.36	19.57	19.78	19.99	20.20	20.41	20.62	20.83	21.04
23.5	19.05	19.26	19.47	19.68	19.89	20.11	20.32	20.53	20.74	20.95	21.17
23.6	19.15	19.37	19.58	19.79	20.01	20.22	20.43	20.65	20.86	21.07	21.29
23.7	19.26	19.48	19.69	19.91	20.12	20.34	20.55	20.76	20.98	21.19	21.41
23.8	19.37	19.59	19.80	20.02	20.23	20.45	20.67	20.88	21.10	21.31	21.53
23.9	19.48	19.70	19.92	20.13	20.35	20.57	20.78	21.00	21.22	21.43	21.65
24.0	19.59	19.81	20.03	20.24	20.46	20.68	20.90	21.12	21.33	21.55	21.77
24.1	19.71	19.92	20.14	20.36	20.58	20.80	21.02	21.24	21.46	21.68	21.90
24.2	19.82	20.04	20.26	20.48	20.70	20.92	21.14	21.36	21.58	21.80	22.02
24.3	19.94	20.16	20.38	20.60	20.82	21.04	21.27	21.49	21.71	21.93	22.15
24.4	20.05	20.27	20.50	20.72	20.94	21.16	21.39	21.61	21.83	22.06	22.28
24.5	20.17	20.39	20.61	20.84	21.06	21.29	21.51	21.73	21.96	22.18	22.41
24.6	20.28	20.51	20.73	20.96	21.18	21.41	21.63	21.86	22.08	22.31	22.53
24.7	20.40	20.62	20.85	21.07	21.30	21.53	21.75	21.98	22.21	22.43	22.66
24.8	20.51	20.74	20.97	21.19	21.42	21.65	21.88	22.10	22.33	22.56	22.79
24.9	20.63	20.85	21.08	21.31	21.54	21.77	22.00	22.23	22.46	22.68	22.91
25.0	20.74	20.97	21.20	21.43	21.66	21.89	22.12	22.35	22.58	22.81	23.04
25.1	20.86	21.09	21.32	21.55	21.79	22.02	22.25	22.48	22.71	22.94	23.17
25.2	20.98	21.21	21.44	21.68	21.91	22.14	22.38	22.61	22.84	23.07	23.31
25.3	21.10	21.33	21.57	21.80	22.03	22.27	22.50	22.74	22.97	23.21	23.44
25.4	21.22	21.45	21.69	21.92	22.16	22.39	22.63	22.87	23.10	23.34	23.57
25.5	21.34	21.57	21.81	22.05	22.28	22.52	22.76	22.99	23.23	23.47	23.71
25.6	21.45	21.69	21.93	22.17	22.41	22.65	22.88	23.12	23.36	23.60	23.84
25.7	21.57	21.81	22.05	22.29	22.53	22.77	23.01	23.25	23.49	23.73	23.97
25.8	21.69	21.93	22.17	22.42	22.66	22.90	23.14	23.38	23.62	23.86	24.10
25.9	21.81	22.05	22.30	22.54	22.78	23.02	23.27	23.51	23.75	23.99	24.24
26.0	21.93	22.17	22.42	22.66	22.91	23.15	23.39	23.64	23.88	24.13	24.37
26.1	22.06	22.30	22.55	22.79	23.04	23.28	23.53	23.77	24.02	24.26	24.51
26.2	22.18	22.43	22.67	22.92	23.17	23.41	23.66	23.91	24.15	24.40	24.65
26.3	22.31	22.55	22.80	23.05	23.30	23.55	23.79	24.04	24.29	24.54	24.79
26.4	22.43	22.68	22.93	23.18	23.43	23.68	23.93	24.18	24.43	24.68	24.93
26.5	22.56	22.81	23.06	23.31	23.56	23.81	24.06	24.31	24.56	24.81	25.07
26.6	22.68	22.93	23.19	23.44	23.69	23.94	24.19	24.45	24.70	24.95	25.20
26.7	22.81	23.06	23.31	23.57	23.82	24.07	24.33	24.58	24.84	25.09	25.34
26.8	22.93	23.19	23.44	23.70	23.95	24.21	24.46	24.72	24.97	25.23	25.48
26.9	23.06	23.31	23.57	23.83	24.08	24.34	24.60	24.85	25.11	25.36	25.62
27.0	23.18	23.44	23.70	23.95	24.21	24.47	24.73	24.99	25.24	25.50	25.76

6D

T [°C]	r.h. [%]										
	90	91	92	93	94	95	96	97	98	99	100
27.0	23.18	23.44	23.70	23.95	24.21	24.47	24.73	24.99	25.24	25.50	25.76
27.1	23.31	23.57	23.83	24.09	24.35	24.61	24.87	25.13	25.39	25.65	25.91
27.2	23.45	23.71	23.97	24.23	24.49	24.75	25.01	25.27	25.53	25.79	26.05
27.3	23.58	23.84	24.10	24.37	24.63	24.89	25.15	25.41	25.68	25.94	26.20
27.4	23.71	23.98	24.24	24.50	24.77	25.03	25.29	25.56	25.82	26.08	26.35
27.5	23.85	24.11	24.38	24.64	24.91	25.17	25.44	25.70	25.97	26.23	26.50
27.6	23.98	24.24	24.51	24.78	25.04	25.31	25.58	25.84	26.11	26.38	26.64
27.7	24.11	24.38	24.65	24.91	25.18	25.45	25.72	25.99	26.25	26.52	26.79
27.8	24.24	24.51	24.78	25.05	25.32	25.59	25.86	26.13	26.40	26.67	26.94
27.9	24.38	24.65	24.92	25.19	25.46	25.73	26.00	26.27	26.54	26.81	27.08
28.0	24.51	24.78	25.05	25.33	25.60	25.87	26.14	26.41	26.69	26.96	27.23
28.1	24.65	24.92	25.19	25.47	25.74	26.02	26.29	26.56	26.84	27.11	27.38
28.2	24.78	25.06	25.33	25.61	25.89	26.16	26.44	26.71	26.99	27.26	27.54
28.3	24.92	25.20	25.48	25.75	26.03	26.31	26.58	26.86	27.14	27.41	27.69
28.4	25.06	25.34	25.62	25.89	26.17	26.45	26.73	27.01	27.29	27.56	27.84
28.5	25.20	25.48	25.76	26.04	26.32	26.60	26.88	27.16	27.44	27.72	28.00
28.6	25.33	25.61	25.90	26.18	26.46	26.74	27.02	27.30	27.59	27.87	28.15
28.7	25.47	25.75	26.04	26.32	26.60	26.89	27.17	27.45	27.74	28.02	28.30
28.8	25.61	25.89	26.18	26.46	26.75	27.03	27.32	27.60	27.88	28.17	28.45
28.9	25.74	26.03	26.32	26.60	26.89	27.18	27.46	27.75	28.03	28.32	28.61
29.0	25.88	26.17	26.46	26.74	27.03	27.32	27.61	27.90	28.18	28.47	28.76
29.1	26.02	26.31	26.60	26.89	27.18	27.47	27.76	28.05	28.34	28.63	28.92
29.2	26.17	26.46	26.75	27.04	27.33	27.62	27.92	28.21	28.50	28.79	29.08
29.3	26.31	26.61	26.90	27.19	27.48	27.78	28.07	28.36	28.65	28.95	29.24
29.4	26.46	26.75	27.05	27.34	27.63	27.93	28.22	28.52	28.81	29.11	29.40
29.5	26.60	26.90	27.19	27.49	27.78	28.08	28.38	28.67	28.97	29.26	29.56
29.6	26.74	27.04	27.34	27.64	27.93	28.23	28.53	28.83	29.13	29.42	29.72
29.7	26.89	27.19	27.49	27.79	28.09	28.38	28.68	28.98	29.28	29.58	29.88
29.8	27.03	27.33	27.63	27.93	28.24	28.54	28.84	29.14	29.44	29.74	30.04
29.9	27.18	27.48	27.78	28.08	28.39	28.69	28.99	29.29	29.60	29.90	30.20
30.0	27.32	27.62	27.93	28.23	28.54	28.84	29.14	29.45	29.75	30.06	30.36

7

References

ABELE, J. E.: The physical background to freezing point osmometry and its medical-biological applications. — Amer. J. med. Electronics **2**: 32—41, 1963.

ABOU RAYA, M. A.: Apparatus for measuring transpiration rate at different air speeds. — Proc. egyptian Acad. Sci. **10**: 12—16, 1955.

AITCHISON, G. D., BUTLER, P. F., GURR, C. G.: Techniques associated with the use of gypsum block soil moisture meters. — Aust. J. appl. Sci. **2**: 56—75, 1951.

ALCOCK, M. B.: An improved electronic instrument for estimation of pasture yield. — Nature **203**: 1309—1310, 1964.

ALJIBURY, F. K., TOMLINSON, W. M., HOUSTON, C. E.: Tensiometers, automatic timing for sprinkler control. — Calif. Agr. **19**: 2—4, 1965.

ALLAWAY, W. G., MANSFIELD, T. A.: Automated system for following stomatal behavior of plants in growth cabinets. — Can. J. Bot. **47**: 1995—1998, 1969.

ALVIM, P. de T.: A new type of porometer for measuring stomatal opening and its use in irrigation studies. — In: ECKARDT, F. E. (ed.): Methodology of Plant Eco-Physiology. (Arid Zone Res. **25**). Pp. 325—329. UNESCO, Paris 1965.

ALVIM, P. de T., HAVIS, J. R.: An improved infiltration series for studying stomatal opening as illustrated with coffee. — Plant Physiol. **29**: 97—98, 1954.

VAN ANDEL, O. M.: Determination of the osmotic value of exudation sap by means of the thermoelectric method of Baldes and Johnson. — Proc. kon. nederl. Akad. Wetensch. **55**: 40—48, 1952.

VAN ANDEL, O. M.: The influence of salts on the exudation of tomato plants. — Acta bot. neerl. **2**: 445—521, 1953.

ANDERSON, A. B. C.: A method of determining the soil moisture content based on the variation of the electrical capacitance of the soil at low frequency, with moisture content. — Soil Sci. **56**: 29—41, 1943.

ANDERSON, A. B. C., EDLEFSEN, N. E.: The electrical capacity of the 2-electrode plaster of paris block as an indicator of soil moisture. — Soil Sci. **54**: 35—46, 1942.

ANDERSON, W. P., AIKMAN, D. P., MEIRI, A: Excised root exudation — as standing gradient — Proc. roy. Soc. London **174**: 445—458, 1970.

ANDERSON, W. P., COLLINS, J. C.: The exudation from excised maize roots bathed in sulphate media. — J. exp. Bot. **20**: 72—80, 1969.

ANDERSON, W. P., HOUSE, C. R.: A correlation between structure and function in the root of *Zea mays*. — J. exp. Bot. **18**: 544—555, 1967.

ANDERSON, W. P., REILLY, E. J.: A study of the exudation of excised maize roots after removal of the epidermis and outer cortex. — J. exp. Bot. **19**: 19—30, 1968.

ANDERSSON, N. E., HERTZ, C. H.: Positive Spitzenentladung als Hygrometer geringer Trägheit. — Z. angew. Phys. **7**: 361—366, 1955.

ANDERSSON, N. E., HERTZ C. H., RUFELT, H.: A new fast recording hygrometer for plant transpiration measurements. — Physiol. Plant. **7**: 753—767, 1954.

ANTOSZEWSKI, R., LIS, E. K.: Phosphorus-32 beta gauge for measuring the water content of fleshy fruits. — Biol. Plant. **15**: 119—122, 1973.

APPERLOO, M.: Bepaling watergehalte van spinazie via de dielektrische constante. Voorlopig onderzoek naar de mogelijkheit van snelle en nauwkeurige vaststelling. — Electronica **9**: 81—83, 1956.

ARCICHOVSKIJ, V.: Untersuchungen über die Saugkraft der Pflanzen. I. Über die Methoden der Saugkraftmessungen. — Planta **14**: 517—527, 1931.

ARCICHOVSKIJ, V., ARCICHOVSKAJA, N.: Untersuchungen über die Saugkraft der Pflanzen: II. Die gravimetrische Methode der Saugkraftmessungen an den Blättern. — Planta **14**: 528—532, 1931.

ARCICHOVSKIJ, V., KISSELEW, N., KRASSULIN, N., MENJINSKAJA, E., OSSIPOV, A.: Untersuchungen über die Saugkraft der Pflanzen. III. Die Saugkraft der Bäume. Saugkraftmessungen nach der Potometer-Methode. — Planta **14**: 533—544, 1931.

ARCICHOVSKIJ, V., OSSIPOV, A.: Untersuchungen über die Saugkraft der Pflanzen. IV. Saugkraftmessungen nach der Schlierenmethode. — Planta **14**: 545—551, 1931.

ARISZ, W. H., HELDER, R. J., VAN NIE, R.: Analysis of the exudation process in tomato plants. — J. exp. Bot. **2**: 257—297, 1951.

ARLAND, A.: Das Problem des Wasserhaushaltes bei landwirtschaftlichen Kulturpflanzen in kritisch-experimenteller Betrachtung. I. u. II. — Wiss. Arch. Landwirtsch., Abt. A **1**: 1—160, **2**: 423—433, 1929.

ARLAND, A.: Die Transpirationsintensität der Pflanzen als Grundlage bei der Ermittlung optimaler acker- und pflanzenbaulicher Kulturmassnahmen. — Abh. sächs. Akad. Wiss. Leipzig, Math. Naturwiss. Kl. **44**: 1—80, 1952.

ARNOLD, A. J.: A measured volume dispenser for watering plants. — J. agr. Eng. Res. **15**: 323—324, 1970.

ARVIDSSON, I.: Austrocknungs- und Dürreresistenzverhältnisse einiger Repräsentanten öländischer Pflanzenvereine nebst Bemerkungen über Wasserabsorption durch oberirdische Organe. — Oikos **1** (suppl.): 1—181, 1951.

ARVIDSSON, I.: Plants as dew collectors. — Int. Union Geod. Geophys. Gen. Assembly Toronto 1957, **2**: 481—484, 1958.

ASHBY, E., WOLF, R.: A critical examination of the gravimetric method of determining suction force. — Ann. Bot. (N. S.) **11**: 261—268, 1947.

ASHTON, F. M.: Effects of a series of cycles of alternating low and high soil water contents on the rate of apparent photosynthesis in sugar cane. — Plant Physiol. **31**: 266—274, 1956.

Aspirations-Psychrometer-Tafeln. — Akademie-Verlag, Berlin 1955.

AVERY, D. J.: The supply of air to leaves in assimilation chambers. — J. exp. Bot. **17**: 655 to 677, 1966.

BAILEY, L. F., ROTHACHER, J. S., CUMMINGS, W. H.: A critical study of the cobalt chloride method of measuring transpiration. — Plant Physiol. **27**: 562—574, 1952.

BAKER, D. A., MILBURN, J. A.: Lateral movement of inorganic solutes in plants. — Nature **205**: 306—307, 1965.

BAKER, D. A., WEATHERLEY, P. E.: Water and solute transport by exuding root systems of *Ricinus communis*. — J. exp. Bot. **20**: 485—496, 1969.

BALDES, E. J.: Theory of the thermo-electric measurement of osmotic pressure. — Biodynamica **46**: 1—8, 1939.

BALDES, E. J., JOHNSON, A. F.: The thermo-electric osmometer; its construction and use. — Biodynamica **47**: 1—11 S, 1939.

BANGE, G. G. J.: On the quantitative explanation of stomatal transpiration. — Acta bot. neerl. **2**: 255—297, 1953.

BARANETZKY, I.: Eine Mitteilung über die Periodizität des Blutens bei krautartigen Pflanzen und deren Ursachen. — Bot. Zeit. **31**: 65, 1873.

BARGER, G.: A microscopical method of determining molecular weights. — J. chem. Soc. **85**: 286—324, 1904.

Barrs, H. D.: Heat of respiration as a possible cause of error in the estimation by psychrometric methods of water potential in plant tissue. — Nature 203: 1136—1137, 1964.

Barrs, H. D.: Comparison of water potentials in leaves as measured by two types of thermocouple psychrometer. — Aust. J. biol. Sci. 18: 36—52, 1965a.

Barrs, H. D.: Psychrometric measurement of leaf water potential: lack of error attributable to leaf permeability. — Science 149: 63—65, 1965b.

Barrs, H. D.: Determination of water deficits in plant tissues. In: Kozlowski, T. T. (ed.): Water Deficits and Plant Growth. Vol. 1. Pp. 235—368. Academic Press, New York—London 1968.

Barrs, H. D., Freeman, B., Blackwell, J., Ceccato, R. D.: Comparisons of leaf water potential and xylem water potential in tomato plants. — Aust. J. biol. Sci. 23: 485—487, 1970.

Barrs, H. D., Kramer, P. J.: Water potential increase in sliced leaf tissue as a cause of error in vapor phase determinations of water potential. — Plant Physiol. 44: 959—864, 1969.

Barrs, H. D., Slatyer, R. O.: Experience with three vapour methods for measuring water potential in plants. — In: Eckardt, F. E. (ed.): Methodology of Plant Eco-Physiology. (Arid Zone Res. 25). Pp. 369—384. UNESCO, Paris 1965.

Barrs, H. D., Weatherley, P. E.: A re-examination of the relative turgidity technique for estimating water deficits in leaves. — Aust. J. biol. Sci. 15: 413—428, 1962.

Batyuk, V. P., Bratachevskiĭ, Yu. A.: Novyĭ metod vyznachennya vologosti roslyn u pol'o-vykh umovykh. [A new method of determining hydration of plants in field conditions.] — Dopov. ukr. Akad. s.-g. Nauk 1958 (4): 14—16, 1958.

Baumbach, S.: Vergleichsmessungen mit verschiedenen Verdunstungsmessern unter definierter Versuchsbedingungen. — Ber. deut. Wetterdienstes US Zone 5: 211—216, 1952.

Baumgartner, A.: Thermoelektrische Untersuchungen über die Geschwindigkeit des Transpirationsstromes. — Z. Bot. 28: 81—136, 1934—5.

van Bavel, C. H. M.: Measurement of soil moisture content by the neutron method. — US Dept. Agr. 41/24: 1—29, 1958.

van Bavel, C. H. M.: Neutron measurement of surface soil moisture. — J. geophys. Res. 66: 4193—4198, 1961.

van Bavel, C. H. M., Nakayama, F. S., Ehrler, W. L.: Measuring transpiration resistance of leaves. — Plant Physiol. 40: 535—540, 1965.

van Bavel, C. H. M., Nielson, D. R., Davidson, J. M.: Calibration and characteristics of two neutron moisture probes. — Proc. Soil Sci. Soc. Amer. 25: 329—334, 1961.

van Bavel, C. H. M., Nixon, P. R., Hauser, V. I.: Soil moisture measurement with the neutron method. — US Dept. Agr. Res. Serv. 41/70: 1—39, 1963.

van Bavel, C. H. M., Underwood, D. N., Swanson, R. W. A.: Soil moisture measurement by neutron moderation. — Soil Sci. 82: 29—41, 1956.

Bearce, B. C., Kohl, H. C. Jr.: Measuring osmotic pressure of sap within live cells by means of a visual melting point apparatus. — Plant Physiol. 46: 515—519, 1970.

Beardsell, M. F., Jarvis, P. G., Davidson, B.: A null-balance diffusion porometer suitable for use with leaves of many shapes. — J. appl. Ecol. 23: 677—685, 1972.

Beck, W. A.: Osmotic pressure, osmotic value and suction tension. — Plant Physiol. 3: 413—440, 1928.

Begg, J. E., Bierhuizen, J. F., Lemon, E. R., Misra, D. K., Slatyer, R. O., Stern, W. R.: Diurnal energy and water exchanges in bulrush millet in an area of high solar radiation. — Agr. Meteorol. 1: 294—312, 1964.

Begg, J. E., Turner, N. C.: Water potential gradients in field tobacco. — Plant Physiol. 46: 343—346, 1970.

Belcher, D. J., Cuykendall, T. R., Sack, H. S.: The measurement of soil moisture and density by neutron and gamma-ray scattering. — Tech. Rep. 127 CAA, USA, 1950.

BERGER-LANDEFELDT, U.: Zur schnellen Registrierung von Temperatur und Dampfdruck bzw. Taupunkt in und über Pflanzenbeständen. — Ber. deut. bot. Ges. **67**: 357—365, 1954.

BERGNER, P. E. E.: Tracer theory: A review. — Isotop. Radiat. Technol. **3**: 245—262, 1966.

BIELORAI, H.: Beta-ray gauging technique for measuring leaf water content changes of *Citrus* seedlings as affected by the moisture status in the soil. — J. exp. Bot. **19**: 489—495, 1968.

BIERHUIZEN, J. F., SLATYER, R. O.: An apparatus for the continuous and simultaneous measurement of photosynthesis and transpiration under controlled environmental conditions. — CSIRO Div. Land. Res. reg. Surv. tech. Paper **24**: 1—16, 1964.

BIERHUIZEN, J. F., SLATYER, R. O., ROSE, C. W.: A porometer for laboratory and field operation. — J. exp. Bot. **16**: 182—191, 1965.

BJÖRKMAN, O., HOLMGREN, P.: Photosynthetic adaptation to light intensity in plants native to shaded and exposed habitats. — Physiol. Plant. **19**: 854—859, 1966.

BLOODWORTH, M. E., PAGE, J. B.: Use of thermistors for the measurement of soil moisture and temperature. — Proc. Soil Sci. Soc. Amer. **21**: 11—15, 1957.

BLOODWORTH, M. E., PAGE, J. B., COWLEY, W. R.: A thermoelectric method for determining the rate of water movement in plants. — Proc. Soil Sci. Soc. Amer. **19**: 411—414, 1955.

BLOODWORTH, M. E., PAGE, J. B., COWLEY, W. R.: Some applications of the thermoelectric method for measuring water flow rates in plants. — Agron. J. **48**: 222—228, 1956.

BLUM, A., SULLIVAN, C. Y., EASTIN, J. D.: On the pressure chamber technique for estimating leaf water potential in sorghum. — Agron. J. **65**: 337—338, 1973.

BLUM, G.: Osmotischer Wert, Saugkraft, Turgor. Protoplasmatologia. — In: HEILBRUNN, L. V., WEBER, F.: Handbuch der Protoplasmaforschung, Vol. II C 7a. Pp. 1—102, 1958.

BODMAN, G. B., DAY, P. R.: Freezing point of a group of California soils and their extracted clays. — Soil Sci. **55**: 225—246, 1943.

BOLGARINA, V. P., EÏDUS, L. Kh.: Luchevoï metod opredeleniya zelenoï massy rastenii na kornyu. [Radiometric method for determining fresh mass of plants on the root.] — Biofizika **1**: 653—656, 1956.

DE BOODT, M., DE LEENHEER, L., MORTIER, P.: De praktische bruikbaarheid in-de bodenkunde van de methode van de neutrondiffusie. — Med. Landbouwhogeschool (Gent) **28**: 85—132, 1963.

BORNKAMM, R.: Standortsbedingungen und Wasserhaushalt von Trespen-Halbtrockenrasen (*Mesobromion*) im oberen Leinegebiet. — Flora **146**: 23—67, 1958.

BOSS, G.: Die Brauchbarkeit des Piche-Evaporimeter bei Verdunstungsmessungen. — Ber. deut. Wetterdienstes US Zone **5**: 194—202, 1952.

BOUGET, S. J., ELRICK, D. E., TANNER, C. B.: Electrical resistance units for moisture measurements: their moisture hysteresis, uniformity and sensitivity. — Soil Sci. **86**: 298—304, 1958.

BOURDEAU, P. F., SCHOPMEYER, C. S.: Oleoresin exudation pressure in slash pine: its measurement, heritability, and relation to oleoresin yield. — In: THIMANN, K. V. (ed.): The Physiology of Forest Trees. Pp. 313—319. Reinhold Press, New York 1958.

BOUYOUCOS, G. J.: The alcohol method for determining moisture content of soil. — Soil Sci. **32**: 173—179, 1931.

BOUYOUCOS, G. J.: Nylon electrical resistance unit for continuous measurements of soil moisture in the field. — Soil Sci. **67**: 319—330, 1949.

BOUYOUCOS, G. J.: More durable plaster of paris moisture blocks. — Soil Sci. **76**: 447—451, 1953.

BOUYOUCOS, G. J.: Cylindrical versus rectangular plaster of paris blocks for measuring soil moisture. — Agron. J. **56**: 92—93, 1964.

BOUYOUCOS, G. J., McCOLL, M. M.: The freezing point method as a new means of measuring the concentration of the soil solution directly in the soil. — Mich. agr. exp. Sta., Tech. Bull. 24, 1915.

368   REFERENCES

BOUYOUCOS, G. J., McCOLL, M. M.: Further studies on the freezing point lowering of soils. — Mich. agr. exp. Sta., Tech. Bull. 31, 1916.

BOUYOUCOS, G. J., McCOLL, M. M.: Determining the absolute salt content of soils by the freezing point method. — J. agr. Res. 15: 331—336, 1918.

BOUYOUCOS, G. J., MICK, A. H.: An electrical resistance method for the continuous measurement of soil moisture under field conditions. — Mich. agr. exp. Sta., Tech. Bull. 172, 1940.

BOUYOUCOS, G. J., MICK, A. H.: A fabric absorption unit for continuous measurement of soil moisture in the field. — Soil Sci. 66: 217—232, 1948.

BOX, J. E. Jr.: Design and calibration of a thermocouple psychrometer which uses the Peltier effects. — In: International Symposium on Humidity and Moisture. — Vol. 11. Pp. 101—121. Reinhold Publ. Co., New York 1965a.

BOX, J. E. Jr.: Measurement of water stress in cotton plant leaf discs with a thermocouple psychrometer. — Agron. J. 57: 367—370, 1965b.

BOX, J. E. Jr., LEMON, E. R.: Preliminary field investigations of electrical resistance - moisture stress relations in cotton and grain sorghum plants. — Proc. Soil Sci. Soc. Amer. 22: 193—196, 1958.

BOYER, J. S.: Isopiestic technique: measurement of accurate leaf water potentials. — Science 154: 1459—1460, 1966.

BOYER, J. S.: Leaf water potentials measured with a pressure chamber. — Plant Physiol. 42: 133—137, 1967a.

BOYER, J. S.: Matric potentials of leaves. — Plant Physiol. 42: 213—217, 1967b.

BOYER, J. S.: Relationship of water potential to growth of leaves. — Plant Physiol. 43: 1056 to 1062, 1968.

BOYER, J. S.: Measurement of the water status of plants. — Annu. Rev. Plant Physiol. 20: 351—364, 1969.

BOYER, J. S.: Leaf enlargement and metabolic rates in corn, soybean, and sunflower at various leaf water potentials. — Plant Physiol. 46: 233—235, 1970.

BOYER, J. S., GHORASHY, S. R.: Rapid field measurement of laef water potential in soybean. — Agron. J. 63: 344—345, 1971.

BOYER, J. S., KNIPLING, E. B.: Isopiestic technique for measuring leaf water potentials with a thermocouple psychrometer. — Proc. nat. Acad. Sci. USA 54: 1044—1051, 1965.

BOYSEN-JENSEN, P.: Über neue Apparate zur Messung der Kohlensäure-Assimilation, der Respiration, der Öffnungsweite der Spaltöffnungen und der Beleuchtungsstärke. — Planta 5: 456—472, 1928.

BRACH, E. J., MASON, W. J.: A stable multivibrator for measuring impedance of plant leaves and stems. — Can. J. Bot. 43: 995—997, 1965.

BRANTON, D., JACOBSON, L.: Iron transport in pea plants. — Plant Physiol. 37: 539—545, 1962.

BRASTAD, W. A., BORCHARDT, L. F.: Electric hygrometer of small dimensions. — Rev. sci. Instr. 24: 1143—1144, 1953.

BRAUN, H. J.: Eine Methode für die Untersuchung des Wasserverbrauches der Holzpflanzen. I. Das Prinzip der Methode und ihre Brauchbarkeit. — Forstw. Centralblatt 89: 189—194, 1970.

BRAUN, H. J.: Eine Methode für die Untersuchung des Wasserverbrauches der Holzpflanzen. II. Ergebnisse der Testversuche. — Forstw. Centralblatt 90: 319—328, 1971.

BRAUN, H. J., SCHMIDT, P.: Methodische Versuche zur direkten Messung des absoluten Wasserverbrauches der Holzpflanzen. — Z. Pflanzenphysiol. 66: 337—342, 1972.

BREAZEALE, E. L., McGEORGE, W. T., BREAZEALE, J. F.: Moisture absorption by plants from an atmosphere of high humidity. — Plant Physiol. 25: 413—419, 1950.

BREAZEALE, E. L., McGEORGE, W. T., BREAZEALE, J. F.: Water absorption and transpiration by leaves. — Soil Sci. 72: 239—244, 1951.

BRIGGS, G. E.: Some aspects of free space in plant tissues. — New Phytol. **56**: 305—324, 1957.

BRIGGS, G. E., ROBERTSON, R. N.: Apparent free space. — Annu. Rev. Plant Physiol. **8**: 11—30, 1957.

BRIGGS, L. J., SHANTZ, H. L.: The wilting coefficient for different plants and its indirect determination. — Bot. Gaz. **53**: 20—37, 1912a.

BRIGGS, L. J., SHANTZ, H. L.: The relative wilting coefficients for different plants. — Bot. Gaz. **53**: 229—235, 1912b.

BRIGGS, L. J., SHANTZ, H. L.: Die relativen Welkungskoeffizienten verschiedener Pflanzen. — Flora **105**: 224—230, 1913.

BRIX, H.: Errors in measurement of leaf water potential of some woody plants with the Schardakow dye method. — Forest. Branch, Can. Dep. Forest. Publ. **1164**: 5—11, 1966.

BRIX, H., KRAMER, P. J.: Measurement of diffusion pressure deficits of plant tissues with a thermocouple method. — Manuscript. Dep. Bot., Duke Univ., Durham, N.C. 1962.

BROGÅRDH, T., JOHNSSON, A.: A flexible apparatus for continuous recording of water uptake by plants. — Med. biol. Engin. **1973**: 286—292, 1973a.

BROGÅRDH, T., JOHNSSON, A.: Oscillatory transpiration and water uptake of *Avena* plants. II. Effects of deformation of xylem vessels. — Physiol. Plant. **28**: 341—345, 1973b.

BROUWER, R.: Water absorption by the roots of *Vicia faba* at various transpiration strengths. I. Analysis of the uptake and the factors determining it. — Proc. kon. ned. Akad. Wetensch., Ser. C **56**: 106—115, 1953.

BROUWER, R.: Water movement across the root. — In: FOGG, G. E. (ed.): The State and Movement of Water in Living Organisms. Pp. 131—149. Cambridge Univ. Press 1965.

BROWN, R. W.: Leaf sampler for relative water content measurements: design and application. — Agron. J. **61**: 644—647, 1969.

BROWN, R. W.: Measurement of water potential with thermocouple psychrometers: Construction and applications. — USDA Forest Res. Paper INT-80, pp. 1—27, 1970.

BROWN, R. W., VAN HAVEREN, B. P. (eds.): Psychrometry in Water Relations Research. — Utah Agr. Exp. Sta., Logan 1973 (in press).

BROYER, T. C.: The movement of material into plants. I. Osmosis and the movement of water in plants. — Bot. Rev. **13**: 1—58, 1947.

BROYER, T. C., FURNSTAL, A. H.: A press for recovery of fluids from plant tissues. — Plant Physiol. **16**: 419—421, 1941.

BUCKINGHAM, E.: Studies on the movement of soil moisture. — US Dep. Agr. Bur. Soils Bull. 38, 1907.

BUFFEL, K.: New techniques for comparative permeability studies on the oat coleoptile with reference to the mechanism of auxin action. — Nederl. Akad. Wetersch. **14**: 1—46, 1952.

BURLEY, J. W., NWOKE, F. I. O., LEISTER, G. L., POHAM, R. A.: The relationship of xylem maturation to the absorption and translocation of P^{32}. — Amer. J. Bot. **57**: 436—442, 1970.

BURSTRÖM, H.: Growth and water absorption of *Helianthus* tuber tissue. — Physiol. Plant. **6**: 685—691, 1953.

BURSTRÖM, H.: Root surface development, sucrose inversion and free space. — Physiol. Plant. **10**: 741—751, 1957.

BURSTRÖM, H.: Definition and determination of water saturation. — Protoplasma **61**: 294—301, 1966.

BUSCALIONI, L., POLLACCI, G.: L'applicazione delle pellicole di collodio allo studio di alcuni processi fisiologici della piante ed in particolar modo alla traspirazione. — Atti Ist. Bot. Univ. Pavia 7: 83—95, 1901a.

BUSCALIONI, L., POLLACCI, G.: Ulteriori ricerche sull' applicazione delle pellicole di collodio allo studio di alcuni processi fisiologici delle piante ed in particolar modo della traspirazione vegetale. — Atti. Ist. Bot. Univ. Pavia 7: 127—170, 1901b.

BUSCHBOM, U.: Zur Methodik kontinuierlicher Wassergehalt-Bestimmungen an Blättern mittels β-Strahlenabsorption. — Planta **95**: 146—166, 1970.

BUTTERY, B. R., BOATMAN, S. G.: Manometric measurement of turgor pressures in laticiferous phloem tissues. — J. exp. Bot. **17**: 283—296, 1966.

BYRNE, G. F., ROSE, C. W., SLATYER, R. O.: An aspirated diffusion prorometer. — Agr. Meteorol. **7**: 39—44, 1970.

ČABART, J.: Gamaskopická methoda stanovení půdní vlhkosti a její zpřesňování. [Gammascopic method for soil moisture determination.] — Sborník ČSAZV, Lesnictví **4**: 659—672, 1958.

ČABART, J., VÁLEK, B.: Vertikální způsob stanovení půdní vlhkosti gamaskopickou methodou. [Vertical arrangement in gammascopic determination of soil moisture.] — Sborník ČSAZV, Lesnictví **5**: 695—704, 1959.

CAILLOUX, M.: Mesure quantitative de l'eau absorbée par un seul poil radiculaire. — Ann. ACFAS **10**: 83, 1944.

CAILLOUX, M.: L'absorption de l'eau par les poils radiculaires. — Bull. Soc. fr. Physiol. vég. **12**: 147—156, 1966.

CAILLOUX, M.: Metabolism and the absorption of water by root hairs. — In: KOLEK, J. (ed.): Structure and Function of Primary Root Tissues. (Proc. Symp. Tatranská Lomnica Sept. 7—10, 1971). Pp. 315—322, Veda, Bratislava 1974.

CALISSENDORFF, C.: An *in situ* leaf and soil water psychrometer having low temperature sensitivity. — M. Sc. Thesis, Washington State Univ., Pullman, Washington 1970.

CAMPBELL, A. G., PHILLIPS, D. S. M., O'REILLY, E. D.: An electronic instrument for pasture yield estimation. — J. brit. Grassland Soc. **17**: 89, 1962.

CAMPBELL, E. C., CAMPBELL, G. S., BARLOW, W. K.: A dewpoint hygrometer for water potential measurement. — Agr. Meteorol. **12**: 113—121, 1973.

CAMPBELL, G. S., GARDNER, W. H.: Psychrometric measurement of soil water potential: temperature and bulk density effects. — Proc. Soil Sci. Soc. Amer. **35**: 8—12, 1971.

CAMPBELL, G. S., TRULL, J. W., GARDNER, W. H.: A welding technique for Peltier thermocouple psychrometers. — Proc. Soil Sci. Soc. Amer. **32**: 887—889, 1968.

CAMPBELL, G. S., ZOLLINGER, W. D., TAYLOR, S. A.: Sample changer for thermocouple psychrometers: construction and some applications. — Agron. J. **58**: 315—318, 1966.

CANNELL, G. H., ASHBELL, C. W.: Prefabrication of mold and construction of cylindrical electrode type resistance unit. — Soil Sci. **97**: 108—112, 1964.

CARDER, A. C.: Atmometer assemblies, a comparison. — Can. J. Plant Sci. **40**: 700—706, 1960.

CARR, D. J., GAFF, D. F.: The role of the cell-wall water in the water relations of leaves. — In: Plant-Water Relationships in Arid and Semi-Arid Conditions. Proc. Madrid Symp. (Arid Zone Res. 16). Pp. 117—125. UNESCO Paris.

CARR, M. K. V.: The internal water status of the tea plant (*Camellia sinensis*): Some results illustrating the use of the pressure chamber technique. — Agr. Meteorol. **9**: 447—460, 1971/2.

CARY, J. W., FISHER, H. D.: Plant moisture stress: A portable freezing-point meter compared with the psychrometer. — Agron. J. **61**: 302—305, 1969.

CARY, J. W., FISHER, H. D.: Plant water potential gradients measured in the field by freezing point. — Physiol. Plant. **24**: 397—402, 1971.

ČATSKÝ, J.: The role played by growth in the determination of water deficit in plants. — Biol. Plant. **1**: 277—286, 1959.

ČATSKÝ, J.: Determination of water deficit in disks cut out from leaf blades. — Biol. Plant. **2**: 76—78, 1960.

ČATSKÝ, J.: Water saturation deficit in the wilting plant. The preference of young leaves and the translocation of water from old into young leaves. — Biol. Plant. **4**: 306—314, 1962.

ČATSKÝ, J.: Water saturation deficit and its development in young and old leaves. — In: RUTTER, A. J. and WHITEHEAD, F. H. (ed.): The Water Relations of Plants. (Proc. Symp. London 1961). Pp. 101—112. Blackwell Sci. Publ., Oxford 1963.

ČATSKÝ, J.: Leaf-disk method for determining water saturation deficit. — In: ECKARDT, F. E. (ed.): Methodology of Plant Eco-Physiology. (Arid Zone Res. 25). Pp. 353—360, UNESCO, Paris 1965.

ČATSKÝ, J.: Temperature effects in the measurement of water saturation deficit (relative water content) in tobacco and kale. — Biol. Plant. 11: 381—383, 1969.

ČERMÁK, J., DEML, M., PENKA, M.: A new method of sap flow rate determination in tree. — Biol. Plant. 15: 171—178, 1973.

CETL, I.: Návrh jednoduché metody ke zjištění odolnosti rostlin vůči suchu. [A simple method for drought resistance determination.] — Českosl. Biol. 2: 361—369, 1953.

CETL, I.: Odolnost polních plodin vůči suchu a možnosti jejího zvýšení. [Drought resistance in crops and its possible increase.] — Rozpravy ČSAV, Praha, řada MPV 67: 1—108, 1957.

CETL, I., PENKA, M.: Intensita transpirace a schopnost udržovat vodu u jarního a ozimého ječmene při odstupňovaných dávkách minerálních živin. [Transpiration rate and water-holding capacity in spring and winter barley with various nutrition levels.] — Českosl. Biol. 7: 81—86, 1958.

CHARBONNIÈRE, R.: Le permittivité des matériaux et son application au dosage de l'humidité.— In: Techniques d'Étude des Facteurs Physiques de la Biosphère. INRA, Paris 1970.

CHEESMAN, J. H., ROBERTS, E. C., TIFFANY, L. H.: Effects of nitrogen level and osmotic pressure of the nutrient solution on incidence of Puccinia graminis and Helminthosporium sativum infection in Merion Kentucky Bluegrass. — Agron. J. 57: 599—602, 1965.

CHESHEVA, Z. P.: Metody opredeleniya svyazannoï vody. [Methods of bound water determination.] — Izv. gos. nauch.-issl. Inst. koll. Chim. 1934: 2, 1934.

CHILDS, E. C.: A note on electrical methods of determining soil moisture. — Soil Sci. 55: 219—223, 1943.

CLARKSON, D. T., SANDERSON, J.: Relationship between the anatomy of cereal roots and the absorption of nutrients and water. — Agr. Res. Council Letcombe Lab. Annu. Rep. Pp. 16—25. 1970.

CLAUSEN, J. J., KOZLOWSKI, T. T.: Use of the relative turgidity technique for measurement of water stresses in gymnosperm leaves. — Can. J. Bot. 43: 305—316, 1965.

CLOSS, R. L.: The heat pulse method for measuring rate of sap flow in a plant stem. — New Zeal. J. Sci. 1: 281—288, 1958.

CLOSS, R. L., JONES, R. L.: The construction and installation of gypsum soil moisture meters. — New Zeal. J. Sci. Technol. B 37: 150—157, 1955.

COHEN, O. P., TADMOR, N. H.: A comparison of neutron moderation and gravimetric sampling for soil-moisture determination with emphasis on the cost factor. — Agr. Meteorol. 3: 97—102, 1966.

COLEMAN, E. A., HANAWALT, W. B., BURCK, C. R.: Some improvements in tensiometer design. — J. Amer. Soc. Agron. 38: 455—458, 1946.

COLEMAN, E. A., HENDRIX, T. M.: The fibreglass electrical soil moisture instrument. — Soil Sci. 67: 425—438, 1949.

COLLANDER, R.: The permeability of Nitella cells to non-electrolytes. — Physiol. Plant. 7: 420—445, 1954.

COLLINS, B. C.: A standing dew meter. — Meteorol. Mag. 90: 114—117, 1961.

COLLINS, J. C., HOUSE, C. R.: The exchange of sodium ions in the root of Zea mays. — J. exp. Bot. 20: 497—506, 1969.

COLLINS, J. C., LINSTEAD, P. J.: Effect of calcium on the potassium flux. — Planta 84: 353—357, 1969.

COMPTON, C.: Water deficit in *Citrus*. — Proc. Amer. Soc. hort. Sci. **34**: 91—95, 1936.

CONDON, B. N., MILLAR, B. D., NEWHOUSE, I., BRYAN, A. J.: A semi-automatic measuring system for Spanner type thermocouple psychometer. — Sci. Instr. (J. Phys. E) **4**: 575—579, 1971.

COPE, F., TRICKETT, E. S.: Soils and Ferts. 28: 1965 Cit. according to NEWBOULD *et al.* 1968.

CORNEJO, T. A., VAADIA, Y.: Estudio preliminar de los metodos para determinar la condición del aqua en las hojas de las plantas y su relación con condición del aqua en e suelo. [Preliminary study of two methods for determining the condition of water in plant leaves and its relation with the condition of water in the soil.] — Agronomia (Perú) **27**: 211—219, 1960.

COURTIN, G. M., BLISS, L. C.: A hydrostatic lysimeter to measure evapotranspiration under remote field conditions. — Arctic alpine Res. **3**: 81—89, 1971.

COWAN, I. R.: Transport of water in the soil-plant-atmosphere system. — J. appl. Ecol. **2**: 221—239, 1965.

COWAN, I. R., MILTHORPE, F. L.: Plant factors influencing the water status of plant tissues. — In: KOZLOWSKI, T. T. (ed.): Water Deficits and Plant Growth. Vol. 1. Pp. 137—193. Academic Press, New York—London 1968.

CRAFTS, A. S., BROYER, T. C.: Migration of salts and water into xylem of the roots of higher plants. — Amer. J. Bot. **25**: 529—535, 1938.

CRAFTS, A. S., CURRIER, H. B., STOCKING, C. R.: Water in the Physiology of Plants. — Waltham, Mass. 1949.

CRONEY, D., COLEMAN, J. D., CURRIER, E. W.: The electrical resistance method of measuring soil moisture. — Brit. J. appl. Phys. **2**: 85—91, 1951.

CROSS, N. L.: A humidity control system for a small chamber. — J. sci. Instr. (J. Phys. E.) Ser. 2, **1**: 65—68, 1968.

CURRIER, H. B.: Cryoscopy of small amounts of expressed tissue sap. — Plant Physiol. **19**: 544—550, 1944.

CURTIS, O. F., SCHOFIELD, H. T.: A comparison of osmotic concentration of supplying and receiving tissues and its bearing on the Münch hypothesis of the translocation mechanism. — Amer. J. Bot. **20**: 502—512, 1933.

CZERSKI, J.: Gasometric method of volume determination of intercellular spaces in plant tissues. — Acta Soc. Bot. Pol. **33**: 247—262, 1964.

CZERSKI, J.: Gasometric method of water deficit measurement in leaves. — Biol. Plant. **10**: 275—283, 1968.

CZERSKI, J.: An easily operated apparatus to register the amount of the aqueous solution absorbed by a plant root system. — Acta Soc. Bot. Pol. **41**: 187—196, 1972.

DAINTY, J.: Water relations of plant cells. — In: PRESTON, R. D. (ed.): Advances in Botanical Research **1**: 279—326, 1963.

DAINTY, J., GINZBURG, B. Z.: The measurement of hydraulic conductivity (osmotic permeability to water) of internodal Characean cells by means of transcellular osmosis. — Biochem. biophys. Acta **79**: 102—111, 1964.

DAINTY, J., HOPE, A. B.: The water permeability of cells of *Chara australis* R. BR. — Aust. J. biol. Sci. **12**: 136—145, 1959.

DALE, J. E.: The use of the infiltration method in the study of the behaviour of the stomata of upland cotton. — Empire Cotton Growing Rev. **35**: 254—259, 1958.

DALTON, F. N., RAWLINS, S. L.: Design criteria for Peltier-effect thermocouple psychometer. — Soil Sci. **105**: 12—17, 1968.

DAMAGNEZ, J.: Conditions d'utilisation de la souche à neutrons pour la détermination d'humidité sans le sol. Influence de la densité apparente et de la sature. — In: Radioisotopes Soil Plant. Pp. 159—169. Int. at. Energy Agency. Wien 1962.

DARWIN, F.: Observations on stomata. — Phil. Trans. roy. Soc. London Ser. B 190: 531—621, 1898.

DARWIN, F., PERTZ, D. T. M.: On a new method of estimating the aperture of stomata. — Proc. roy. Soc. London B 84: 136—154, 1911.

DAUM, C. R.: A method for determining water transport in trees. — Ecology 48: 425—431, 1967.

DAVIDSON, J. M., BIGGAR, J. W., NIELSON, D. R.: Gamma-radiation attenuation for measuring bulk density and transient water flow in porous media. — J. geophys. Res. 68: 4777—4783, 1963.

DAVIS, R. M. Jr.: The refractometer reading of muskmelon leaf sap in relation to growing conditions. — Proc. Amer. Soc. hort. Sci. 83: 599—604, 1963.

DAVSON, H., DANIELI, J. F.: The Permeability of Natural Membranes. — Cambridge Univ. Press, 1943.

DECKER, J. P., GAYLOR, W. G., COLE, F. D.: Measuring transpiration of undisturbed tamarisk shrubs. — Plant Physiol. 37: 393—397, 1962.

DECKER, J. P., WETZEL, B. F.: A method for measuring transpiration of intact plants under controlled light, humidity and temperature. — Forest. Sci. 3: 350—354, 1957.

DECKER, J. P., WIEN, J. D.: Transpirational surges in Tamarix and Eucalyptus as measured with an infrared gas analyzer. — Plant Physiol. 35: 340—343, 1960.

DETLING, J. K., KLIKOFF, L. G.: Comparison of two field techniques for determination of water potential in two halophytes. — Amer. Middland Naturalist 85: 235—238, 1971.

DIEM, M.: Feuchtmessung mit Hilfe thermoelektrischer Psychrometer. — Arch. Meteorol. Geophys. Bioklimatol., Ser. B 5: 59—65, 1953.

DIETRICH, M.: Die Transpiration der Schatten- und Sonnenpflanzen in ihrer Beziehung zum Standort. — Jahrb. wiss. Bot. 65: 98—194, 1926.

DIXON, H. H.: A thermoelectric method of cryoscopy. — Sci. Proc. roy. Dublin Soc. 13: 49—62, 1911.

DIXON, H. H.: Transpiration and the ascent of sap in plants. — McMillan, London 1914.

DIXON, H. H., ATKINS, W. R. G.: On osmotic pressure in plants and on a thermoelectric method of determining freezing points. — Sci. Proc. roy. Dublin Soc. 12: 275—311, 1910.

DJAVANCHIR, A.: Mise au point d'une chambre de transpiration pour mesurer le résistance stomatique. — Oecol. Plant. 5: 301—318, 1970.

DOWNEY, L. A., MILLER, J. W.: Rapid measurements of relative turgidity in maize (Zea mays L.). — New Phytol. 70: 555—560, 1971.

VAN DUIN, R. H. A., DE VRIES, D. A.: A recording apparatus for measuring thermal conductivity and some results obtained with it in soil. — Netherlands. J. agr. Sci. 2: 168—175, 1954.

DUMANSKIĬ, A. V., VOĬTSEKHOVSKIĬ, R. V.: Metody opredeleniya gidrofilnosti dispersnykh sistem. [Methods of determining hydrophilous features of disperse coloids.] — Koloid. Zh. 6: 413—425, 1948.

DUNIWAY, J. M.: Comparison of pressure chamber and thermocouple psychrometer determination of leaf water status in tomato. — Plant Physiol. 48: 106—107, 1971.

DUNMORE, F. W.: An electric hygrometer and its application to radiometeorography. — Bull. amer. meteorol. Soc. 19: 225—243, 1938a.

DUNMORE, F. W.: An electric hygrometer and its application to radio meteorography. — J. Res. nat. Bur. Stand. 20: 723—744, 1938b.

ECKARDT, F. E.: Eco-physiological measuring techniques applied to research on water relations of plants in arid and semi-arid regions. — In: Plant-Water Relationships in Arid and Semi-Arid Conditions. Reviews of Research. (Arid Zone Res. 15). Pp. 139—171. UNESCO, Paris 1960.

EDLEFSEN, N. E.: Some thermodynamics of aspects of the use of soil moisture by plants. — Trans. Amer. Geophys. Union Trans. 1941, Part III 22: 917—940, 1941.

EHLIG, C. F.: Measurement of energy status of water in plants with a thermocouple psychrometer. — Plant Physiol. **37**: 288—290, 1962.

EHLIG, C. F., GARDNER, W. R.: Relationship between transpiration and the internal water relations of plants. — Agron. J. **56**: 127—130, 1964.

EHRLER, W. L., VAN BAVEL, C. H. M., NAKAYAMA, F. S.: Transpiration, water absorption, and internal water balance of cotton plants as affected by light and changes in saturation deficit. — Plant. Physiol. **41**: 71—74, 1966.

EHRLER, W. L., NAKAYAMA, F. S., VAN BAVEL, C. H. M.: Cyclic changes in water balance and transpiration of cotton leaves in a steady environment. — Physiol. Plant. **18**: 766—775, 1965.

EHWALD, R.: A method for continual measurement of water uptake by roots using a registring polarimeter. — In: KOLEK, J. (ed.): Structure and Function of Primary Root Tissues. (Proc. Symp. Tatranská Lomnica Sept. 7—10, 1971). Pp. 323—327, Veda, Bratislava 1974.

EIDMANN, F. E.: Untersuchungen über die Wurzelatmung und Transpiration unserer Holzarten. — Schriftenreihe H. Göring Akad. dent. Forstwirsch. **5**, 1943.

ELKINS, C. B. Jr., WILLIAMS, G. G.: Still and time-lapse photography of plant stomata. — Crop Sci. **2**: 164—166, 1962.

EMMERT, E. M., BALL, F. K.: The effect of soil moisture on the availability of nitrate, phosphate and potassium to the tomato plant. — Soil Sci. **35**: 295—306, 1933.

ENGLAND, C. B.: Changes in fibre-glass soil moisture electrical resistance elements in long-term instalation. — Proc. Soil Sci. Soc. Amer. **29**: 229—231, 1965.

ERNEST, E. C. M.: Suction-pressure gradients and the measurement of suction pressure. — Ann. Bot. **45**: 717—731, 1931.

ERNEST, E. C. M.: Studies in the suction pressure of plant cells II. — Ann. Bot. **48**: 293—305, 1934a.

ERNEST, E. C. M.: The water relations of plant cells. — Linnean Soc. (London) Bot. **49**: 495, 1934b.

EVANS, E. C., III., VAUGHAN, B. E.: New methods for effecting watertight seals on corn roots. — Plant Physiol. **41**: 1077—1078, 1966.

EVENARI, M., RICHTER, R.: Physiological-ecological investigations in the wilderness of Judaea. — J. Linnean Soc. (London), Bot. **51**: 333—381, 1937.

FALK, S. O.: A microwave hygrometer for measuring plant transpiration. — Z. Pflanzenphysiol. **55**: 31—37, 1966a.

FALK, S. O.: Effect on transpiration and water uptake by rapid changes in the osmotic potential of the nutrient solution. — Physiol. Plant. **19**: 602—617, 1966b.

FALK, S. O., HERTZ, C. H., VIRGIN, H. I.: On the relation between turgor pressures and tissue rigidity. I. Experiments on resonance frequency and tissue rigidity. — Physiol. Plant. **11**: 802—817, 1958.

FAWCET, R. G., COLLIS, S. E.: A filter-paper method for determining the moisture characteristics of soil. — Aust. J. exp. Agr., Animal Husbandry **7**: 162—167, 1967.

FEDOROVSKIĬ, D. V.: Zavisimost koefficienta zavyadaniya ot vida rasteniĭ. [Dependence of wilting point on plant species.] — Pochvovedenie **1948** (10): 612—621, 1948.

FEDOTOV, V. D., MIFTAKHUTDINOVA, F. G., MURTAZIN, Sh. F.: Issledovaniye protonnoi relaksatsii v zhivikh rastitelnykh tkanyakh metodom spinovogo ekha. [The spin echo NMR study of proton relaxation in living plant tissues.] — Biofizika **14**: 873—882, 1969.

FERGUSON, H., GARDNER, W. H.: Water content measurement in soil columns by gamma ray absorption. — Proc. Soil Sci. Soc. Amer. **26**: 11—14, 1962.

FERGUSON, J. D., STREET, H. E.: The promotion and inhibition of excised root growth by various sugars and sugar alcohols. — Ann. Bot. **22**: 513—523, 1958.

FIECHTER, A., VETSCH, U.: Die Wasserbestimmung in Zellmaterial von *Saccharomyces cerevisiae* nach Karl Fischer. — Experientia **13**: 72, 1957

FILIPPOV, L. A.: K voprosu o kriticheskoi dlya rasteniya vlazhnosti pochvy v svyazi s sutoch-nymi izmeneniyami placha khlopchatnika. [Critical soil moisture and daily course of bleeding in cotton plant.] — Dokl. Akad. Nauk SSSR **106**: 145—147, 1956.

FISCHER, K.: Neues Verfahren zur massanalytischen Bestimmung des Wassergehaltes von Flüssigkeiten und festen Körpern. — Angew. Chem. **48**: 394—396, 1935.

FISCUS, E. F.: *In situ* measurement of root-water potential. — Plant Physiol. **50**: 191—193, 1972.

FISHER, H. D.: An inexpensive method of determining plant moisture stress using freezing point depression. — Soil Sci. **113**: 383—385, 1972.

FLEISCHHAUER-BINZ, E.: Die Messung von Bodensaugkräften mit Tensiometern. — Planta **37**: 565—594, 1949.

FLETCHER, J. E.: Dielectric methods for determining soil moisture. — Proc. Soil Sci. Soc. Amer. **4**: 84—88, 1939.

FLETCHER, J. E., ROBINSON, M. E.: A capacitance meter for estimating forage weight. — J. Range Management **9**: 96, 1956.

FRANK, A. B., HARRIS, D. G.: Measurement of leaf water potential with a pressure chamber. — Agron. J. **65**: 334—335, 1973.

FRANKOVÁ, T., KOLEK, J.: The effect of mannitol on the oxygen consumption in plant tissues. — Biol. Plant. **9**: 81—85, 1967.

FROESCHEL, P.: Das Druckstomatometer, ein neuer pflanzenphysiologischer Apparat zur Messung der Apertur der Stomata. — Cellule **56**: 63—70, 1953.

FROESCHEL, P., CHAPMAN, P.: A new method of measuring the size of the stomata apertures. — Cellule **54**: 233—250, 1951.

FRY, K. E., WALKER, R. B.: A pressure-infiltration method for estimating stomatal opening in conifers. — Ecology **48**: 155—157, 1967.

FUKUDA, Y.: Über die Hydratur der Pflanzen und eine empirische Formel der Verdunstung und Transpiration. — Pflanzenforschung 19, Fischer, Jena 1935.

FUKUDA, Y.: The functional relationship between the suction intensity of protoplasma and the retention capacity of vacuolar sap on the plant hydrature. — Jap. J. Bot. **16**: 181—209, 1958.

FULLER, E. N., SCHETTLER, P. D., GIDDINGS, J. C.: A new method for prediction of binary gas-phase diffusion coefficients. — Ind. eng. Chem. **58**: 18—27, 1966.

FURR, J. R., REEVE, J. O.: Range of soil moisture percentages through which plant undergo permanent wilting in some soils from semi-arid unirrigated areas. — J. agr. Res. **71**: 149—170, 1945.

GAASTRA, P.: Photosynthesis of crop plants as influenced by light, carbon dioxide, temperature, and stomatal diffusion resistance. — Med. Landbouwhogeschool (Wageningen) **59**: 1—68, 1959.

GAFF, D. F., CARR, D. J.: The quantity of water in the cell wall and its significance. — Aust. J. biol. Sci. **14**: 299—311, 1961.

GAFF, D. F., CARR, D. J.: An examination of the refractometric method for determining the water potential of plant tissues. — Ann. Bot. (N. S.) **28**: 351—368, 1964.

GALE, J., POLJAKOFF-MAYBER, A., KAHANE, I.: The gas diffusion porometer technique and its application to the measurement of leaf mesophyll resistance. — Israel J. Bot. **16**: 187 to 204, 1967.

GARDNER, R.: A method of measuring the capillary tension of soil moisture over a wide moisture range. — Soil Sci. **43**: 277—283, 1937.

GARDNER, W., KIRKHAM, D.: Determination of soil moisture by neutron scattering. — Soil. Sci. **73**: 391—401, 1952.

GARDNER, W. R.: Dynamic aspects of water availability to plants. — Soil Sci. **89**: 63—73, 1960a.

GARDNER, W. R.: Measurement of capillary conductivity and diffusivity with a tensiometer. — Seventh Int. Congr. Soil Sci. Trans. (Madison, Wisc.) 1: 300—304, 1960b.

GARDNER, W. R.: Relation of root distribution to water uptake and availability. — Agron. J. 56: 41—45, 1964.

GARDNER, W. R.: Dynamic aspects of soil-water availability to plants. — Ann. Rev. Plant Physiol. 16: 323—342, 1965.

GARDNER, W. R., EHLIG, C. F.: Physical aspects of the internal water relations of plant leaves. — Plant Physiol. 40: 705—710, 1965.

GARDNER, W. R., NIEMAN, R. H.: Lower limit of water availability to plants. — Science 143: 1460—1462, 1964.

GÄUMANN, E., JAAG, O.: Untersuchungen über die pflanzliche Transpiration. — Ber. schweiz. bot. Ges. 45: 411—518, 1936.

GEIGER, R.: Bericht über ein von G. Hofmann entwickeltes Gerät zur Registrierung der Bodenluftfeuchtigkeit. — Ber. deut. Wetterdienstes US Zone 32: 24—27, 1952.

GENKEL', P. A., MARGOLINA, K. P.: [The determination of the capacity of plants to withstand dehydration.] (In Russ.) — Dokl. Akad. Nauk SSSR 86: 849—852, 1952.

GESSNER, F.: Die Wasseraufnahme durch Blätter und Samen. — In: RUHLAND, W. (ed.): Handbuch der Pflanzenphysiologie 3: 213—246, Springer-Verlag, Berlin—Göttingen—Heidelberg 1956.

GEZALYAN, M. G.: O napryazhennosti vodnogo rezhima neizolirovannykh list'ev i novom metode ego issledovaniya. [Water stress in attached leaves and a new method for studying it.] — Dokl. Akad. Nauk Arm. SSR 42: 236—240, 1966.

GEZALYAN, M. G.: Issledovanie ovodnennosti list'ev na osnove opredeleniya dielektricheskoĭ pronitsaemosti vody. [Study of water content in leaves measured by dielectric permeability of water.] — Thesis, Akad. Nauk Arm. SSR, Erevan 1968.

GINSBURG, H., GINZBURG, B. Z.: Radial water and solute flows in roots of *Zea mays*. I. Water flow. — J. exp. Bot. 21: 580—592, 1970a.

GINSBURG, H., GINZBURG, B. Z.: Radial water and solute flows in roots of *Zea mays*. II. Ion fluxes across root cortex. — J. exp. Bot. 21: 593—604, 1970b.

GLINKA, Z., KATCHANSKY, M. Y.: The effect of water potential on the CO_2 compensation point of maize and sunflower leaf tissue. — Israel J. Bot. 19: 533—541, 1970.

GLINKA, Z., MEIDNER, H.: The measurement of stomatal responses to stimuli in leaves and leaf discs. — J. exp. Bot. 19: 152—166, 1968.

GLOSER, J.: Some problems of the determination of stomatal aperture by the microrelief method. — Biol. Plant. 9: 28—33, 1967.

GLOVER, J.: A method for the continuous measurement of transpiration of single leaves under natural conditions. — Ann. Bot. (N. S.) 5: 25—34, 1941.

GLUBRECHT, H., NIEMANN, E. G., RUNDFELDT, H.: Dichtebestimmung mit Gammastrahlen an biologischen Objekten. — Atompraxis 5: 237—239, 1959.

GOEDEWAGEN, M. A. J. cited after TROUGHTON, A.: The underground organs of herbage grasses. — Commonwealth agr. Bur. Bucks, 1957.

GOODE, J. E., HEGARTY, T. W.: Measurement of water potential of leaves by methods involving immersion in sucrose solutions. — Nature 276: 109—110, 1965.

GORTNER, R. A., GORTNER, W. A.: The cryoscopic method for the determination of "bound water". — J. gen. Physiol. 17: 327—339, 1933.

GOSSE, G., DE PARCEVAUX, S.: Application de l'absorption des ondes centimétrique à la mesure de l'humidité des végétaux. — In: Techniques d'Étude des Facteurs Physiques de la Biosphère. Pp. 359—369. INRA, Paris 1970.

GRADMANN, H.: Untersuchungen über die Wasserverhältnisse des Bodens als Grundlage des Pflanzenwachstums. — Jahrb. wiss. Bot. 69: 1—100, 1928.

GRADMANN, H.: Über die Messung von Bodensaugwerten. — Jahrb. wiss. Bot. **80**: 92—111, 1934.

GREEN, P. B.: Simultaneous measurement of growth rate and turgor pressure in growing *Nitella* cells. — Amer. Soc. Plant Physiol. Annu. Meeting, College Park, Maryland, 1966.

GREEN, P. B., STANTON, F. W.: Turgor pressure: direct manometric measurement in single cell of *Nitella*. — Science **155**: 1675—1676, 1967.

GREENE, M. T., MARVIN, J. W.: The water content of maple stems. I. Application of the Karl Fischer method for the analyses of water in maple bark and wood. — Plant Physiol. **33**: 169—173, 1958.

GREGORY, F. G.: A convenient method for attaching potometers. — Ann. Bot. (N.S.) **2**: 253—255, 1938.

GREGORY, F. G., ARMSTRONG, J. I.: The diffusion porometer. — Proc. roy. Soc. London, Ser. B, **121**: 27—42, 1936.

GREGORY, F. G., PEARSE, H. L.: The resistance porometer and its application to the study of stomatal movement. — Proc. roy. Soc. London, Ser. B, **114**: 477—493, 1934.

GREGORY, F. G., WOODFORD, E. K.: An apparatus for the study of the oxygen, salt and water uptake of various zones of the root, with some preliminary results with *Vicia faba*. — Ann. Bot. (N.S.) **3**: 147—154, 1939.

GRIEVE, B. J., WENT, F. W.: An electric hygrometer apparatus for measuring water-vapour loss from plants in the field. — In: ECKARDT, F. E. (ed.): Methodology of Plant Eco-Physiology. (Arid Zone Res. 25). Pp. 247—257, UNESCO Paris 1965.

GROENEWEGEN, H., MILLS, J. A.: Uptake of mannitol into the shoots of intact barley plants. — Aust. J. biol. Sci. **13**: 1—5, 1960.

GROOT, J.: The use of silicone rubber plastic for replicating leaf surfaces. — Acta bot. neerl. **18**: 703—708, 1969.

GROSSENBACHER, K. A.: Autonomic cycle of rate of exudation of plants. — Amer. J. Bot. **26**: 107—109, 1939.

GURR, C. G.: Use of gamma rays in measuring water content and permeability in unsaturated columns of soil. — Soil Sci. **94**: 224—229, 1962.

GUSEV, N. A.: Nekotorye metody issledovaniya vodnogo rezhima rasteniĭ. [Some methods in studying water relations in plants.] — Vsesoyuz. bot. Obshch. Akad. Nauk SSSR, Leningrad **1960**: 1—60, 1960.

GUSEV, N. A.: O charakteristike sostoyaniya vody v rasteniyakh. [Characteristics in plant water status.] — Fiziol. Rast. **9**: 432—437, 1962.

GUSEV, N. A., SEDYKH, N. V.: Sovremennye metody issledovaniya sostoyaniya vody i moleku-lyarnykh mekhanismov vodoobmena rastitelnoĭ kletki. [Modern methods of studying water status and molecular mechanisms of water exchange in plant cells.] — Selskokh. Biol. **6**: 930—939, 1971.

HABERLANDT, G.: Die Lichtsinnesorgane der Laubblätter. — Leipzig 1905.

HACK, H. R. B.: An improved method of soil moisture control with observations on tomato growth and water uptake. — J. exp. Bot. **22**: 323—336, 1971.

HAINES, F. M.: The absorption of water by leaves in fogged air. — J. exp. Bot. **4**: 106—107, 1953.

HAISE, H. R.: How to measure the moisture in the soil. — US Dept. Agr. Yearbook 1955: 362—371, 1955.

HAISE, H. R., KELLEY, O. J.: Relation of soil moisture tension to heat transfer and electrical resistance in plaster of paris blocks. — Soil Sci. **61**: 411—422, 1946.

HAISE, H. R., KELLEY, O. J.: Causes of diurnal fluctuations of tensiometers. — Soil Sci. **70**: 301—313, 1950.

HALADA, J., ZACH, B.: Konduktometrický přístroj na zjišťování obsahu vody v půdě. [Conductometer for soil moisture determination.] — Sborník ČSAZV (Praha) A, **27**: 107—112, 1954.

HALKET, A. C.: On various methods for determining osmotic pressures. — New Phytol. **12**: 164, 1913.

HALMA, F. F.: Some phases in the water relations of *Citrus*. — Proc. Amer. Soc. hort. Sci. **31**: 108—109, 1934.

HAMMEL, H. T.: Measurement of turgor pressure and its gradient in the phloem of oak. — Plant Physiol. **43**: 1042—1048, 1968.

HAMNER, C. L., CARLSON, R. F., TUKEY, H. B.: Improvement in keeping quality of succulent plants and cut flowers by treatment under water in partial vacuum. — Science **102**: 332—333, 1945.

HANCOCK, C. K., BURDICK, R. L.: Rapid determination of water in wet soils. — Soil Sci. **83**: 197—205, 1957.

HANCOCK, C. K., HUDGINS, C. M.: Determination of water in soils by an indirect conductivity method. — Anal. Chem. **26**: 1738—1740, 1954.

HANKS, R. J., BOWERS, S. A.: Neutron meters access tube influences soil temperature. — Proc. Soil Sci. Soc. Amer. **24**: 62—63, 1960.

HANKS, R. J., SHAWCROFT, R. W.: An economical lysimeter for evaporation studies. — Proc. Soil Sci. Soc. Amer. **57**: 634—635, 1965.

HANSEN, H. C.: The water-retaining power of the soil. — J. Ecol. **14**: 111—119, 1926.

HARGITAY, B., KUHN, W., WIRZ, H.: Eine mikrokryoskopische Methode für sehr kleine Lösungsmengen — Experientia **7**: 276, 1951.

HARMS, W. R., McGREGOR, W. H. D.: A method for measuring the water balance of pine needles. — Ecology **43**: 531—532, 1962.

HARRIS, J. A.: An extension to 5.99° of tables to determine the osmotic pressure of expressed vegetable sap from the depression on the freezing point. — Amer. J. Bot. **2**: 418—419, 1915.

HARRIS, J. A., GORTNER, R. A.: Notes on the calculation of the osmotic pressure of expressed vegetable saps from the depression of the freezing point, with a table for the values of P for Δ 0.001° to Δ 2.999°. — Amer. J. Bot. **1**: 75—78, 1914.

HÄRTEL, H.: Über die Quellbarkeit pflanzlicher Membranen. — Phyton (Austria) **3**: 69—83, 1951.

HÄRTEL, O.: Über die Möglichkeit der Anwendung der plasmometrischen Methode Höflers auf nichtzylindrische Zellen. — Protoplasma **57**: 354—370, 1963.

HATAKEYAMA, I.: Studies on the freezing of living and dead tissues of plants, with special reference to the colloidally bound water in living state. — Mem. Coll. Sci., Univ Kyoto, Ser. B., **28** (3): 401—429, 1961.

HATSCHECK, E.: A direct demonstration of bound water in gelatin gel. — Trans. Faraday Soc. **32**: 787, 1936.

HAYWARD, H. E., BLAIR, W. M., SKALING, P. E.: Device for measuring entry of water into roots. — Bot. Gaz. **104**: 152—160, 1942.

HEATH, O. V. S.: Studies on the stomatal behaviour. V. The role of carbon dioxide in the light response of stomata. Part I. Investigation of the cause of abnormally wide stomatal opening within porometer cups. — J. exp. Bot. **1**: 29—62, 1950.

HEATH, O. V. S., MANSFIELD, T. A.: A recording porometer with detachable cups operating on four separate leaves. — Proc. roy. Soc. London B **156**: 1—13, 1962.

HEATH, O. V. S., RUSSELL, J.: The Wheatstone bridge porometer. — J. exp. Bot. **2**: 111—116, 1951.

HEIGEL, K.: Ergebnisse von Verdunstungsmessungen mit Piche-Evaporimetern, ihre Abhängigkeit von einigen meteorologischen Faktoren und von verschiedenen Standorten. — Meteorol. Rundschau **10**: 101—107, 1957.

HEINRICH, G.: Fusarinsäure in ihrer Wirkung auf die Wasserpermeabilität des Protoplasmas. — Protoplasma **55**: 320—356, 1962.

HELLMUTH, E. O., GRIEVE, B. J.: Measurement of water potential of leaves with particular reference to the Shardakow method. — Flora **159**: 147—167, 1970.

HENDERSON, F. Y.: The preparation of three-colour strips for transpiration measurements. — Ann. Bot. **50**: 321—324, 1936.

HERRICK, E. M.: A three-wire thermocouple system for use in cryoscopic investigations. — Amer. J. Bot. **21**: 673—687, 1934.

HESSE, W.: Meteorologische Einflüsse bei der Pflanzentranspiration. — Ann. Meteorol. **5**: 7—12, 1952.

HESSE, W.: Messung der Pflanzentranspiration mit Kleinlysimetern. — Z. Acker. Pflanzenbau **98**: 107—118, 1954a.

HESSE, W.: Ergebnisse von Pflanzentranspirationsmessungen mit Kleinlysimetern in Zusammenhang mit meteorologischen Einflüssen. — Angew. Meteorol. **2**: 65—82, 1954b.

HEWLETT, J. D., DOUGLASS, J. E., CLUTTER, J. L.: Instrumental and soil moisture variance using the neutron-scattering method. — Soil Sci. **97**: 19—24, 1964.

HEWLETT, J. D., KRAMER, P. J.: The measurement of water deficits in broadleaf plants. — Protoplasma **57**: 381—391, 1963.

HIGASHI, K.: Studies on Bound Water. — Monograph Series of the Research Institute of Applied Electricity. — No 5, pp. 9—35, 1955.

HILL, A. V.: A thermal method of measuring the vapour pressure of an aqueous solution. — Proc. roy. Soc. London A **127**: 9—19, 1930.

HINSCH, N., NIEMANN, E. G.: Ein kombiniertes Gerät zur Messung von Dichte und Wassergehalt im lebenden Baumstamm. — Atompraxis **10**: 372—376, 1964.

HINSON, W. H., KITCHING, R.: A readily constructed transistorized instrument for electrical resistance measurements in biological research. — J. appl. Ecol. **1**: 301—305, 1964.

HIRST, J. M.: A method for recording the formation and persistence of water deposits on plant shoots. — Quart. J. roy. meteorol. Soc. **80**: 227—231, 1954.

HLUCHOVSKÝ, B., SRB, V.: O použitelnosti Molischovy infiltrační metody ke zjišťování otevřenosti průduchů. [Use of Molisch's infiltration method for the determination of stomata aperture.] — Českosl. Biol. **7**: 376—380, 1958.

HOFFMAN, G. J., HERKELRATH, W. N.: Design features of intact leaf thermocouple psychrometres for measuring water potential. — Trans. amer. Soc. agr. Eng. **11**: 631—634, 1968.

HOFFMAN, G. J., HERKELRATH, W. N., AUSTIN, R. S.: Simultaneous cycling of Peltier thermocouple psychrometers for rapid water potential measurements. — Agron. J. **61**: 597—601, 1969.

HOFFMAN, G. J., RAWLINS, S. L.: Silver foil psychrometer for measuring leaf water potential *in situ*. — Science **177**: 802—804, 1972.

HOFFMAN, G. J., SPLINTER, W. E.: Instrumentation for measuring water potential of an intact plant-soil system. — Trans. amer. Soc. agr. Eng. **11**: 38—42, 1968a.

HOFFMAN, G. J., SPLINTER, W. E.: Water potential measurements of an intact plant-soil system. — Agron. J. **60**: 408—413, 1968b.

HÖFLER, K.: Die plasmolytisch-volumetrische Methode und ihre Anwendbarkeit zur Messung des osmotischen Wertes lebender Pflanzenzellen. — Ber. deut. bot. Ges. **35**: 706—726, 1917.

HÖFLER, K.: Eine plasmolytisch-volumetrische Methode zur Bestimmung des osmotischen Wertes von Pflanzenzelle. — Denkschrift Wien. Akad. mathem. - naturwiss. Kl. **95**: 99—115, 1918a.

HÖFLER, K.: Permeabilitätbestimmung nach der plasmometrischen Methode. — Ber. deut. bot. Ges. **36**: 414—422, 1918b.

HÖFLER, K.: Über Eintritts- und Rückgangsgeschwindigkeit der Plasmolyse und eine Methode zur Bestimmung der Wasserpermeabilität des Protoplasten. — Jahrb. wiss. Bot. **73**: 300, 1930.

HÖFLER, K., MIGSCH, H., ROTTENBURG, W.: Über die Austrockungsresistenz landwirtschaftlicher Kulturpflanzen. — Forschungsdienst 12: 50—51, 1941.

HOFMANN, G.: Die Thermodynamik der Taubildung. — Ber. deut. Wetterdienstes 3: 1—45, 1955.

HOFMANN, G.: Dew measurement by thermodynamical means. — Int. Union Geod. Geophys. gen. Assembly Toronto 2: 443—445, 1958.

HÖHN, K., BOY, H.: Wasseraufnahme und Wasserverbrauch von Avena-Sprossen bei fehlender Transpiration. — Beitr. Biol. Pfl. 34: 67—82, 1956.

HOLMES, J. W.: Calibration and field use of the neutron scattering method of measuring soil water content. — Aust. J. appl. Sci. 7: 45—58, 1956a.

HOLMES, J. W.: Measuring soil water content and evaporation by the neutron scattering method. — Netherlands J. agr. Sci. 4: 30—34, 1956b.

HOLMES, J. W., JENKINSON, A. F.: Techniques for using the neutron moisture meter. — J. agr. Eng. Res. 4: 100—109, 1959.

HOLMES, J. W., TURNER, K. G.: The measurement of water content of soils by neutron scattering: A portable apparatus for field use. — J. agr. Engin. Res. 3: 199—204, 1958.

VAN DEN HONERT, T. H.: Water transport in plants as a catenary process. — Disc. Faraday Soc. 1948(3): 146—153, 1948.

HOPMANS, P. A. M.: Rhythms in stomatal opening of bean leaves. — Meded. Landbouwhogeschool Wageningen 71—3: 1—86, 1971.

HOUSE, C. R., FINDLAY, N.: Mechanism of fluid exudation from isolated maize roots. — Nature 221: 649—650, 1966a.

HOUSE, C. R., FINDLAY, N.: Analysis of transient changes in fluid exudation from isolated maize roots. — J. exp. Bot. 17: 627—640, 1966b.

HOUSE, C. R., JARVIS, P.: Effect of temperature on the radial exchange of labelled water in maize roots. — J. exp. Bot. 19: 31—40, 1968.

HSIAO, T. C., ACEVEDO, E., HENDERSON, D. W.: Maize leaf elongation: continuous measurements and close dependence on plant water status. — Science 168: 590—591, 1970.

HSIEH, J. J. C.: A technique for controlling soil water content in the vicinity of root hairs and its application to soil-water-plant studies. — Ph. D. Thesis Washington State Univ. Pullman, Washington 1962.

HSIEH, J. J. C., HUNGATE, F. P.: Temperature compensated Peltier psychrometer for measuring plant and soil water potentials. — Soil Sci. 110: 253—257, 1970.

HUBER, B.: Zur Methodik der Transpirationsbestimmung am Standort. — Ber. deut. bot. Ges. 45: 611—618, 1927.

HUBER, B.: Weitere quantitative Untersuchungen über das Wasserleitungssystem der Pflanzen. — Jahrb. wiss. Bot. 67: 877—959, 1928.

HUBER, B.: Beobachtung und Messung pflanzlicher Saftströme. — Ber. deut. bot. Ges. 50: 89—109, 1932.

HUBER, B.: Die Gefässleitung. — In: RUHLAND, W. (ed.): Handbuch der Pflanzenphysiologie Vol. 3. Pp. 541—592. Springer-Verlag Berlin—Göttingen—Heidelberg 1956.

HUBER, B., HÖFLER, K.: Die Wasserpermeabilität des Protoplasmas. — Jahrb. wiss. Bot. 73: 351—511, 1930.

HUBER, B., MILLER, R.: Methoden zur Wasserdampf- und Transpirationsregistrierung im laufenden Luftstrom. — Ber. deut. bot. Ges. 67: 223—234, 1954.

HUBER, B., SCHMIDT, E.: Eine Kompensationsmethode zur thermoelektrischen Messung langsamer Saftströme. — Ber. deut. bot. Ges. 55: 514—529, 1937.

HUBER, B., ZIEGLER, H.: Atmung und Wasserhaushalt. In: RUHLAND, W. (ed.): Handbuch der Pflanzenphysiologie. — Vol. 12/2. Pp. 150—167. Springer Verlag, Berlin—Göttingen—Heidelberg 1960.

HÜBNER, G.: Zum Wassertransport in Vicia faba. — Flora 148: 549—594, 1960.

Hu Ju, G., Kramer, P. J.: Radial salt transport in corn roots. — Plant Physiol. **42**: 985—990, 1967.

Hunter, A. S., Kelley, O. J.: A new technique for studying the absorption of moisture and nutrients from soil by plant roots. — Soil Sci. **62**: 441—450, 1946.

Hyde, F. J., Lawrence, J. T.: Electronic assessment of pasture growth. — Electronic Eng. **36**: 666—670, 1964.

Hygen, G.: Studies in plant transpiration I. — Physiol. Plant. **4**: 57—183, 1951.

Hygen, G.: Studies in plant transpiration II. — Physiol. Plant. **6**: 106—133, 1953.

Idle, D. B.: Scanning electron microscopy of leaf surface replicas and the measurement of stomatal aperture. — Ann. Bot. **33**: 75—76, 1969a.

Idle, D. B.: The calculation of transpiration rate and diffusion resistance of a single leaf from micrometeorological information subject to errors of measurement. — Ann. Bot. **34**: 159—176, 1969b.

Iljin, W. S.: Der Einfluss des Wassermangels auf die Kohlenstoffassimilation durch die Pflanzen. — Flora N. F. **116**: 360—378, 1923.

Impens, I.: Leaf wetness, diffusion resistance and transpiration rates of bean leaves (*Phaseolus vulgaris* L.) through comparison of "wet" and "dry" leaf temperatures. — Oecol. Plant. **1**: 327—334, 1966.

Impens, I., Stewart, D. W., Allen, L. H. Jr., Lemon, E. R.: Diffusive resistance at, and transpiration rates from leaves *in situ* within the vegetative canopy of a corn crop. — Plant Physiol. **42**: 99—104, 1967.

Ingvalson, R. D., Oster, J. D., Rawlins, S. L., Hoffman, G. J.: Measurement of water potential in soil with a combined thermocouple psychrometer and salinity sensor. — Proc. Soil Sci. Soc. Amer. **34**: 570—574, 1970.

International Society of Soil Science, Soil Physics Terminology, Bulletin No 23, 2, 1963.

Ishmukhametova, N. N., Khokhlova, L. P.: Sravnitel'noe izuchenie sostayaniya vody lisťev ozimykh rasteniĭ diel'kometricheskim metodom. [Water status in leaves of winter plants studied by measurement of dielectric constant]. — Fiziol. Rast. **18**: 169—176, 1971.

Ishmukhametova, N. N., Rybkina, G. V.: Izuchenie sostoyaniya vody v khloroplastekh bobov metodom izmereniya dielektricheskoĭ postoyannoĭ na sverkhvysokikh chastotakh. [Study of the water state in bean chloroplasts by the determination of dielectric constant at UHF.] — Fiziol. Rast. **19**: 385—389, 1972.

Ishmukhametova, N. N., Startseva, A. V.: Izuchenie sostoyaniya vody v steblyakh gorokha metodom dielektricheskikh izmereniĭ na S.V.Ch. [Water status in pea stalks as studied by the method of UHF dielectric measurement.] — Dokl. Akad. Nauk SSSR **200**: 996—998, 1971.

Ittner, E.: Der Tagesgang der Geschwindigkeit des Transpirationsstromes im Stamme einer 75-jährigen Fichte. — Oecol. Plant. **3**: 177—183, 1968.

Ivanov, L. A.: O metode opredeleniya ispareniya rasteniĭ v estestvennykh usloviyakh ikh proizrastaniya. [On a method determining transpiration rate of plants in natural conditions.] — Lesn. Zh. **48**: 1—7, 1918.

Ivanov, L. A.: Zur Methodik der Transpirationsbestimmung am Standort. — Ber. deut. bot. Ges. **46**: 306—310, 1928.

Ivanov, L. A., Silina, A. A., Tsel'niker, Yu. L.: O metode bystrogo vzveshivaniya dlya opredeleniya transpiratsii v estestvennykh usloviyakh. [Cut-shoot method for transpiration measurement in natural conditions]. — Bot. Zh. **35**: 171—185, 1950.

Ivanov, O. V.: Ustanovka dlya odnovremennoĭ registratsii postupleniya vody v kornevuyu sistemu rasteniya, transpiratsii i izmeneniya vesa zelenoĭ massy. [Device for simultaneous recording of water uptake by plant root system, transpiration and fresh weight changes.] — Sb. Tr. agron. Fiz. **9**: 180—187, 1962.

IVANOV, O. V.: Avtomaticheskiĭ terrapotometer. [Automatical terra-potometer.] — Bot. Zh. **48**: 688—693, 1963.

IVANOV, O. V.: Gravitron — avtomaticheskaya ustanovka dlya dlitelnoĭ nepreryvnoĭ registratsii izmeneniĭ biomassy rasteniya s odnovremennym uchetom kolichestva vody postupayushcheĭ v kornevuyu sistemu. [Gravitron, an automatic device for continuous recording the plant biomass with simultaneous determination of water uptake by root system.] — Bot. Zh. **50**: 517—522, 1965.

JACKSON, J. E., WEATHERLEY, P. E.: The effect of hydrostatic pressure gradients on the movement of potassium across the root cortex. — J. exp. Bot. **13**: 128—143, 1962a.

JACKSON, J. E., WEATHERLEY, P. E.: The effect of hydrostatic pressure gradients on the movement of sodium and calcium across the root cortex. — J. exp. Bot. **13**: 404—413, 1962b.

JACKSON, W. T.: Flooding injury studied by approach-graft and split root system techniques. — Amer. J. Bot. **43**: 496—502, 1956.

JACKSON, W. T.: Use of carbowaxes (polyethylene glycols) as osmotic agents. — Plant Physiol. **37**: 513—519, 1962.

JACOB, H. P.: A probe for the photoelectric measurement of soil moisture. — Z. Landeskultur **3**: 328—332, 1962.

JANÁČ, J., ČATSKÝ, J., JARVIS, P. G. *et al.*: Infra-red gas analysers and other physical analysers. — In: ŠESTÁK, Z., ČATSKÝ, J., JARVIS, P. G. (ed.): Plant Photosynthetic Production. Manual of Methods. Pp. 111—197. — Dr W. Junk N. V. Publishers, The Hague 1971.

JANES, B. E.: Adjustment mechanisms of plants subjected to varied osmotic pressures of nutrient solution. — Soil Sci. **101**: 180—188, 1966.

JARVIS, P., HOUSE, C. R.: The radial exchange of labelled water in maize roots. — J. exp. Bot. **18**: 695—706, 1967.

JARVIS, P., HOUSE, C. R.: The radial exchange of labelled water in isolated steles of maize roots. — J. exp. Bot. **20**: 507—515, 1969.

JARVIS, P., HOUSE, C. R.: Evidence for symplasmic ion transport in maize roots. — J. exp. Bot. **21**: 83—90, 1970.

JARVIS, P. G.: The estimation of resistances to carbon dioxide transfer. — In: ŠESTÁK, Z., ČATSKÝ, J., JARVIS, P. G. (ed.): Plant Photosynthetic Production. Manual of Methods. Pp. 566—631. Dr W. Junk N. V. Publishers, The Hague 1971.

JARVIS, P. G., ČATSKÝ, J. *et al.*: General principles of gasometric methods and the main aspects of installation design. — In: ŠESTÁK, Z., ČATSKÝ, J., JARVIS, P. G. (ed.): Plant Photosynthetic Production. Pp. 49—110. — Dr. W. Junk N. V. Publishers, The Hague 1971.

JARVIS, P. G., JARVIS, M. S.: The water relations of tree seedlings. IV. Some aspects of the tissue water relations and drought resistance. — Physiol. Plant. **16**: 501—516, 1963.

JARVIS, P. G., JARVIS, M. S.: The water relation of tree seedlings. V. Growth and root respiration in relation to osmotic potential at the root medium. — In: SLAVÍK, B. (ed.): Water Stress in Plants. Pp. 167—182. — Dr W. Junk N. V. Publishers, The Hague 1965.

JARVIS, P. G., ROSE, C. W., BEGG, J. E.: An experimental and theoretical comparison of viscous and diffusive resistances to gas flow through amphistomatous leaves. — Agr. Meteorol. **4**: 103—117, 1967.

JARVIS, P. G., SLATYER, R. O.: A controlled-environment chamber for studies of gas exchange by each surface of a leaf. — CSIRO Aust., Div. Land Res. Tech. Paper **29**: 1—16, 1966a.

JARVIS, P. G., SLATYER, R. O.: Calibration of β-gauges for determining leaf water status. — Science **153**: 78—79, 1966b.

JARVIS, P. G., SLATYER, R. O.: The role of the mesophyll cell wall in leaf transpiration. — Physiol. Plant. **90**: 303—322, 1970.

JENNINGS, G., MONTEITH, J. L.: A sensitive recording dew-balance. — Quart. J. roy. meteorol. Soc. **80**: 222—226, 1954.

JENSEN G.: Relationship between water and nitrate in excised tomato root systems. — Physiol. Plant. **15**: 791—803, 1962.

JENSEN, R. D., TAYLOR, S. A., WIEBE, H. H.: Negative transport and resistance to water flow through plants. — Plant Physiol. **36**: 633—638, 1961.

JODO, S.: [Stomatal movement and water relations in crops. I. Performance test on a newly improved recording porometer.] In Jap. — Proc. Crop Sci. Soc. Japan **39**: 431—439, 1970.

JOHNSON, C. M., STOUT, P. R., PEARSON, G. A.: Constant-temperature baths for use in freezing point measurements. — Plant Physiol. **26**: 196—197, 1951.

JOHNSSON, A.: Oscillatory transpiration and water uptake of *Avena* plants. I. Preliminary observations. — Physiol. Plant. **28**: 40—50, 1973.

JONES, F. E., WEXLER, A.: A barium chloride film hygrometer element. — J. geophys. Res. **65**: 2087—2095, 1960.

JONES, H. G.: Estimation of plant water status with the beta-gauge. — Agr. Meteorol. **11**: 345—355, 1973.

JONES, R. J.: Electronic pasture sampler. — CSIRO Aust. Div. trop. Pastures Annu. Rep. 1963 (p. 20), 1964 (p. 26), 1965 (p. 39), 1966 (p. 44).

JONES, R. J., HAYDOCK, K. P.: Yield estimation of tropical and temperature pasture species using an electronic capacitance meter. — J. agr. Sci. **75**: 27—36, 1970.

JORDAN, W. R., RITCHIE, J. T.: Influence of soil water stress on evaporation, root absorption, and internal water status of cotton. — Plant Physiol. **48**: 783—788, 1971.

KAINDL, K., HUBMER, H.: Die Bestimmung der Pflanzenmasse mittels Strahlungsabsorption. — Atompraxis **6**: 316—319, 1960.

KALRA, Y. P.: Application of split-root technique in orthophosphate absorption experiments. — J. agr. Sci. **77**: 77—81, 1971.

KAMIYA, N., TAZAWA, M.: Studies on water permeability of a single plant cell by means of transcellular osmosis. — Protoplasma **46**: 394—422, 1956.

KAMIYA, N., TAZAWA, M., TAKATA, T.: Water permeability of the cell wall in *Nitella*. — Plant Cell Physiol. **3**: 285—292, 1962.

KAMIYA, N., TAZAWA, M., TAKATA, T.: The relation of turgor pressure to cell volume in *Nitella flexilis* with special reference to mechanical properties of the cell wall. — Protoplasma **57**: 501—521, 1963.

KANEMASU, E. T., THURTELL, G. W., TANNER, C. B.: Design, calibration and field use of a stomatal diffusion porometer. — Plant Physiol. **44**: 881—885, 1969.

KATO, I.: [Studies on the transpiration and evapotranspiration amount by the chamber method.] In Jap. — Tokai—Kinki nat. agr. exp. Sta. Jap. **1967**: 1—14, 1967.

KATO, I., NAITO, Y., TANIGUCHI, R., KAMOTA, F.: [Studies on the method for measuring amounts of evapotranspiration. I. A method of measuring humidity using transpiration chambers.] In Jap. — Proc. Crop Sci. Soc. Jap. **28**: 286—288, 1960.

KAUFMANN, M. R.: Evaluation of the pressure chamber technique for estimating plant water potential of forest tree species. — Forest Sci. **14**: 369—374, 1968a.

KAUFMANN, M. R.: Evaluation of the pressure chamber method for measurement of water stress in *Citrus*. — Proc. Amer. Soc. hortic. Sci. **93**: 186—190, 1968b.

KAUFMANN, M. R.: Water relations of pine seedlings in relation to root and shoot growth. — Plant Physiol. **43**: 281—288, 1968c.

KAUSCH, W.: Saugkraft und Wassernachleitung in Boden als physiologische Faktoren. Unter besonderer Berücksichtigung des Tensiometers. — Planta **45**: 217—253, 1955.

KAWATA, Shin-ichiro, LAI, K. L.: On the correlation between the endodermis differentiation and the water absorption in the crown roots of rice plant. — Proc. Crop Sci. Soc. Jap. **37**: 631—640, 1968.

Kay, B. D., Low, P. F.: Measurement of the total suction of soils by a thermistor psychrometer. — Proc. Soil Sci. Soc. Amer. **34**: 373—376, 1970.

Kazaryan, V. O., Gezalyan, M. G.: Ob ovodnennosti neizolirovannykh listev i ee izmenenii v zavisimosti ot korneobespechennosti rasteniĭ. [Water content in attached leaves and its changes with root amount.] — Dokl. Akad. Nauk Arm. SSR **46**: 195—199, 1968.

Kelley, O. J.: A rapid method of calibrating various instruments for measuring soil moisture *in situ*. — Soil Sci. **58**: 433—440, 1944.

Kemper, W. D.: Estimation of osmotic stress in soil water from the electrical resistance of finely porous ceramic units. — Soil Sci. **87**: 345—349, 1959.

Kennedy, J. S., Booth, C. O.: Water relations of leaves from woody and herbaceous plants. — Nature **181**: 1271—1272, 1958.

Kennedy, J. S., Mittler, T. E.: A method for obtaining phloem sap via the mouth-parts of aphids. — Nature **171**: 528, 1953.

Kenny, P., McGruddy, P. J.: A circuit for a self-timing stomatal diffusion porometer. — Agr. Meteorol. **10**: 393—399, 1972.

Ketel, D. H., Dirkse, W. G., Ringoet, A.: Water uptake from foliar-applied drops and its further distribution in the oat leaf. — Acta bot. neerl. **21**: 155—165, 1972.

Kijne, J. W., Taylor, S. A.: A constant temperature bath controlled with a precision of 0.001° centigrade. — Proc. Soil Sci. Soc. Amer. **27**: 110, 1963.

Killian, C.: Le déficit de saturation hydrique chez les plantes sahariennes. — Rev. gén. Bot. **54**: 81—101, 1947.

Kitching, R.: A precision portable electrical resistance bridge incorporating a centre zero null detector. — J. agr. Engin. Res. **10**: 264—266, 1965.

Klausing, O.: Erfahrungen mit einem neuen Piche-Atmographen. Ein Beitrag zum Verdunstungsproblem. — Meteorol. Rundschau **10**: 158—162, 1957.

Klemm, M.: Untersuchungen über die Geschwindigkeit des Transpirationsstromes. 1. Mitt. Die Transpirationsstromgeschwindigkeit von Pappeln bei Dürrebelastung. — Flora **152**: 580—589, 1962.

Klemm, M., Klemm, W.: Die Verwendung von Radioisotopen zur kontinuierlichen Bestimmung des Tagesverlaufes der Transpirationsstromgeschwindigkeit bei Bäumen. — Flora **154**: 89—93, 1964.

Klemm, W.: Eine neue Methode zur Bestimmung der Holzfeuchterverlaufes in wachsenden Bäumen mit Hilfe von Gammastrahlen. — Flora **147**: 465—470, 1959.

Klemm, W.: Die Durchstrahlungsmethode zur Untersuchung des Wasserhaushaltes von Waldbäumen. — Vakutronik-Inf. **1965**(4): 18—19, 1965.

Klemm, W.: Die Entwicklung der Durchstrahlungsmethode zur Untersuchung des Wasserhaushaltes von Bäumen. — Isotopenpraxis **2**: 262—267, 1966.

Klepper, B.: Diurnal pattern of water potential in woody plants. — Plant Physiol. **43**: 1931—1934, 1968.

Klepper, B., Barrs, H. D.: Effects of salt secretion on psychrometric determinations of water potential of cotton leaves. — Plant Physiol. **43**: 1138—1140, 1968.

Klepper, B., Browning, V. D., Taylor, H. M.: Stem diameter in relation to plant water status. — Plant Physiol. **48**: 683—685, 1971.

Klepper, B., Ceccato, R. D.: Determinations of leaf and fruit water potential with a pressure chamber. — Hort. Res. (Edinb.) **9**: 1—7, 1969.

Klepper, B., Greenway, H.: Effects of water stress on phosphorus transport to the xylem. — Planta **80**: 142—146, 1968.

Kleshnin, G. V., Strogonov, B. P., Shul'gin, I. A.: Novyĭ metod opredeleniya transpiratsii. [New method of determining transpiration.] — Fiziol. Rast. **1**: 188—192, 1954.

KLINE, J. R., MARTIN, J. R., JORDAN, C. F., KORANDA, J. J.: Measurement of transpiration in tropical trees with tritiated water. — Ecology 51: 1068—1970.

KLOZ, J.: Nový typ přístroje na měření vody spotřebované rostlinami. [New device for measuring water consumption in plants]. — Českosl. Biol. 2: 174—181, 1953.

KLUTE, A., PETERS, D. B.: A recording tensiometer with a short response time. — Proc. Soil Sci. Soc. Amer. 26: 87—88, 1962.

KLUTE, A., RICHARDS, L. A.: Effect of temperature on relative vapour pressure of water in soil: Apparatus and preliminary measurements. — Soil Sci. 93: 391—396, 1962.

KNIGHT, R. C.: A convenient modification of the porometer. — New Phytol. 14: 214—216 1915.,

KNIGHT, R. O.: The Plant in Relation to Water. — Dover Publications, Inc., New York 1965.

KNIPLING, E. B.: Comparison of the dye method with the thermocouple psychrometer method for measuring leaf water potentials. — Plant Physiol. 40 (suppl.) xxxv—xxxvi, 1965

KNIPLING, E. B.: Measurement of leaf water potential by the dye method. — Ecology 48: 1038—1041, 1967a.

KNIPLING, E. B.: Effect of leaf aging on water deficit — water potential relationships of dogwood leaves growing in two environments. — Physiol. Plant. 20: 65—72, 1967b.

KNIPLING, E. B.: A hygrometer for measuring the transpiration resistance of leaves. — US Army Cold Regions Res. Engin. Lab., Hannover, New Hampshire 1968a.

KNIPLING, E. B.: A pressure chamber for measuring leaf water stress. — Informal Memorandum, US Army Cold Regions Res. Engin. Lab., Hannover, New Hampshire 1968b.

KNIPLING, E. B., KRAMER, P. J.: Comparison of the dye method with the thermocouple psychrometer for measuring leaf water potentials. — Plant Physiol. 42: 1315—1320, 1967.

KOCH, W.: Der Tagesgang der Produktivität der Transpiration. — Planta 48: 418—452, 1957.

KOCH, W., KLEIN, E., WALZ, H.: Neuartige Gaswechsel-Messanlage für Pflanzen in Laboratorium und Freiland. — Siemens-Zeitschrift 42: 392—404, 1968.

KOCH, W., LANGE, O. L., SCHULZE, E.-D.: Ecophysiological investigations on wild and cultivated plants in the Negev Desert. I. Methods: A mobile laboratory for measuring carbon dioxide and water vapour exchange. — Oecologia 8: 296—309, 1971.

KOCH, W., WALZ, H.: Kleinklimaanlage der Messung des pflanzlichen Gaswechsel. Ein neuartiges Verfahren zur Feuchteregelung und Transpirationsmessung. — Naturwiss. 54: 321—322, 1967.

KOCHERINA, E. I.: K voprosu o vlazhnosti zavyadaniya rasteniĭ. [On the problem of wilting point]. — Pochvovedenie 1948(1): 20—24, 1948.

KOLLER, D., SAMISH, Y.: A null-point compensating system for simultaneous and continuous measurement of net photosynthesis and transpiration by controlled gas-stream analysis. — Bot. Gaz. 125: 81—88, 1964.

KOMAROVA, N. A.: Vytesnenie pochvennykh rastvorov metodom zameshchaniya zhidkostyami i ispol'zovanie metoda v pochvennykh issledovaniyakh. [Removal of soil solution by liquids and its use in soil studies]. — Tr. pochv. Inst. V. V. Dokuchaeva 51: 5—97, 1956.

KORVEN, H. C., TAYLOR, J. A.: The Peltier effect and its use for determining relative activity of soil water. — Can. J. Soil Sci. 39: 76—85, 1959.

KOSMIN, P. L., FILIPPOV, P. L.: Ispolzovanie sverkhvysokikh chastot dlya beskontaktnogo opredeleniya vlazhnosti lista i steblya. [Use of hyperfrequency for contactless determination of water content in leaf and stem]. — Fiziol. Rast. 15: 177—181, 1968.

KOVÁŘ, L.: On the influence of surface active substances upon the imbibition of plant tissues. — Protoplasma 8: 585—627, 1930.

KOZINKA, V.: Die Gewinnung von Pressaft für kryoskopische Bestimmungen des osmotischen Wertes bei Pflanzen. [Experimenteller Beitrag zum Studium der Wirkung der Abtötung durch Chloroformdämpfe auf die Gefrierpunktserniedrigung des Pressaftes]. — Biológia (Bratislava) 15: 567—583, 1960.

KOZINKA, V.: The effect of high concentrations of growth substances on water uptake. — Biol. Plant. **8**: 235—245, 1966.

KOZINKA, V.: Inhibition of water uptake by high concentratiosn of auxin-like substances. — Biol. Plant. **12**: 180—190, 1970.

KOZINKA, V., KLENOVSKÁ, S.: The uptake of mannitol by higher plants. — Biol. Plant. **7**: 285—292, 1965.

KOZINKA, V., NIŽŇÁNSKY, A.: Biometric analysis of the relationship between the osmotic pressure of the cell sap and its refractive index. — Biol. Plant. **5**: 77—84, 1963.

KOZLOWSKI, T. T.: Water Metabolism in Plants. Biological Monographs. — Harper and Row, New York 1964.

KRAMER, P. J.: The absorption of water by root systems of plants. — Amer. J. Bot. **19**: 148—164, 1932.

KRAMER, P. J.: Bound water. — In: RUHLAND, W. (ed.): Handbuch der Pflanzenphysiologie Vol. 1. Pp. 223—242. Springer-Verlag, Berlin—Göttingen—Heidelberg 1955.

KRAMER, P. J.: Outer space in plants. Some possible implications of the concept. — Science **125**: 633—635, 1957.

KRAMER, P. J.: Measurement of water potential and osmotic potential of leaf tissue with a thermocouple psychrometer. — Dep. Bot., Duke Univ., Durham, N. C. Mimeographed 1967.

KRAMER, P. J.: Plant and Soil Water Relationships. A Modern Synthesis. — McGraw-Hill Book Co., New York 1969.

KRAMER, P. J., BRIX, H.: Measurement of water stress in plants. — In: ECKARDT, F. E. (ed.): Methodology of Plant Eco-Physiology. (Arid Zone Res. 25). Pp. 343—351. UNESCO, Paris 1965.

KRAMER, P. J., KNIPLING, E. B., MILLER, L. N.: Terminology of cell-water relations. — Science **153**: 889—890, 1966.

KREČMER, V.: Rosa jako činitel meteorologický, půdní, fysiologický, ekologický a rosa v lesnictví. [Dew as meteorological, soil, physiological, ecological factors in woodland]. — Lesnická Práce **30**: 340—374, 1951.

KREEB, K.: Hydratur und Ertrag. — Ber. deut. bot. Ges. **70**: 121—136, 1957.

KREEB, K.: Die Bedeutung der Hydratur für die Kontrolle der Wasserversorgung bei Kulturpflanzen. Pp. 1—83. — Diss. Stuttgart—Hohenheim, 1958.

KREEB, K.: Über die gravimetrische Methode zur Bestimmung des Saugspannung und das Problem des negativen Turgors. I. Mitt. — Planta **55**: 274—282, 1960.

KREEB, K.: Zur Frage des negativen Turgors bei mediterranen Hartlaubpflanzen unter natürlichen Bedingungen. — Planta **56**: 479—489, 1961.

KREEB, K.: Untersuchungen zum Wasserhaushalt der Pflanzen unter extrem ariden Bedingungen. — Planta **59**: 442—458, 1963.

KREEB, K.: Die Bedeutung des Quellungswassers der Zelle bei der kryoskopischen Bestimmung des osmotischen Wertes. — Ber. deut. bot. Ges. **78**: 159—166, 1965a.

KREEB, K.: Untersuchungen zu den osmotischen Zustandgrössen. I. Mitt. Ein tragbares elektronisches Mikrokryoskop für ökophysiologische Arbeiten. — Planta **65**: 269—279, 1965b.

KREEB, K.: Untersuchungen zu den osmotischen Zustandgrössen. II. Mitt. Eine elektronische Methode zur Messung der Saugspannung (NTC-Methode). — Planta **66**: 156—164, 1965c.

KREEB, K.: Determination of the internal water balance (hydrature) in the field by measuring suction force and refractive index. — In: ECKARDT, F. E. (ed.): Methodology of Plant Eco-Physiology. (Arid Zone Res. 25). Pp. 385—391, UNESCO, Paris 1965d.

KREEB, K.: Die Registrierung des Wasserzustandes über die elektrische Leitfähigkeit der Blätter. — Ber. deut. bot. Ges. **79**: 150—162, 1966.

KREEB, K.: Entgegnung an R. O. Slatyer. — Z. Pflanzenphysiol. **56**: 95—97, 1967.

KREEB, K., BORCHARD, W.: Thermodynamische Betrachtungen zum Wasserhaushalt der Pflanze. — Z. Pflanzenphysiol. **56**: 186—202, 1967.

KREEB, K., ÖNAL, M.: Über die gravimetrische Methode zur Bestimmung der Saugspannung und das Problem des negativen Turgors. II. Mitt. Die Berücksichtigung von Atmungsverlusten während der Messungen. — Planta **56**: 409—415, 1961.

KRYUCHKOV, P. A.: Metody vydeleniya pochvennykh rastvorov. [Methods in obtaining soil solutions]. — Sovr. metody issled. fiziko-khimich. svoĭstv pochv. — 1, 1947.

KÜHN, W.: Zur Wassergehaltsbestimmung an lebenden Bäumen mittels schneller Neutronen und niederenergetischer Gammastrahlung. — Kerntechnik **3**: 382—385, 1961.

KUPERMAN, I. A., BOCHKOV, G. A., LABZUN, P. K.: Prostoe ustroĭstvo dlya opredeleniya pogloshcheniya vody rasteniem. [A simple device for the determination of water absorption by plants]. — Fiziol. Rast. **15**: 380—382, 1968.

KÜSTER, H. J.: Ein Verfahren zur kurzfristigen Messung der Wasseraufnahme bei Potometerversuchen. — Ber. deut. bot. Ges. **69**: 67—74, 1956.

KÜSTER, H. J.: Zur Methodik der Bestimmung der Wasserbilanz im Potometerversuch. — Z. Bot. **46**: 67—74, 1958.

KUZNETSOV, S. V.: Ispoĺzovanie yaderno-magnitnogo rezonansa. [Use of the nuclear-magnetic resonance]. — Vestnik seĺskokhoz. Nauki **6**: 117—123, 1961.

KVĚT, J., MARṢHALL, J. K.: Assessment of leaf area and other assimilating plant surfaces. — In: ŠESTÁK, Z., ČATSKÝ, J., JARVIS, P. G. (ed.): Plant Photosynthetic Production. Manual of Methods. Pp. 517—555. — Dr. W. Junk N. V. Publishers, The Hague 1971.

LACHENMAYER, J.: Transpiration und Wasserabsorption intakter Pflanzen nach vorausgegangener Verdunkelung bei Konstanz der Lichtintensität und der übrigen Aussenfaktoren. — Jahrb. wiss. Bot. **76**: 765—827, 1932.

LADEFOGED, K.: A method for measuring the water consumption of larger intact trees. — Physiol. Plant. **13**: 648—658, 1960.

LAGERWERFF, J. V., OGATA, G., EAGLE, H. E.: Control of osmotic pressure of culture solutions with polyethylene glycol. — Science **133**: 1486—1487, 1961.

LAMBERT, J. R., SCHILFGAARDE, J. VAN: A method of determining the water potential of intact plants. — Soil Sci. **100**: 1—9, 1965.

LANDOLT-BÖRNSTEIN Physikalisch-chemische Tabellen. — P. 26, 1966.

LANDSBERG, J. J., LUDLOW, M. M.: A technique for determining resistance to mass transfer through the boundary layers of plants with complex structure. — J. appl. Ecol. **7**: 187—192, 1970.

LANG, A. R. G.: Osmotic coefficients and water potentials of sodium chloride solutions from 0 to 40°C. — Aust. J. Chem. **20**: 2017—2023, 1967.

LANG, A. R. G.: Psychrometric measurement of soil water potential *in situ* under cotton plants. — Soil Sci. **106**: 460—464, 1968.

LANG, A. R. G., BARRS, H. D.: An apparatus for measuring water potentials in the xylem of intact plants. — Aust. J. biol. Sci. **18**: 487—497, 1965.

LANG, A. R. G., KLEPPER, B., CUMMING, M. J.: Leaf water balance during oscillation of stomatal aperture. — Plant Physiol. **44**: 826—830, 1969.

LANG, A. R. G., TRICKETT, E. S.: Automatic scanning of Spanner and droplet psychrometers having outputs up to 30 μV. — J. sci. Instr. **42**: 777—782, 1965.

LARCHER, W.: Schnellmethode zur Unterscheidung lebender von toten Zellen mit Hilfe der Eigenfluoreszenz pflanzlicher Zellsäfte. — Mikroskopie **8**: 299—302, 1953.

LARCHER, W.: Frosttrocknis an der Waldgrenze und in der alpinen Zwergstrauchheide auf dem Patscherkofel. — Veröff. Ferdinandeum, Innsbruck **37**: 49—81, 1957.

LARCHER, W.: Transpiration and photosynthesis of detached leaves and shoots of *Quercus pubescens* and *Quercus ilex* during desiccation under standard conditions. — Bull. Res. Counc. Israel **8D**: 213—224, 1960.

LARKUM, A. W. D.: Some observations with a new potometer on the absorption of water by young barley plants. — J. exp. Bot. **20**: 25—33, 1969.

LATIES, G. G., BUDD, K.: The development of differential permeability in isolated steles of corn roots. — Proc. nat. Acad. Sci. USA **52**: 462—469, 1964.

LAWLOR, D. W.: Plant growth in polyethylene glycol solutions in relation to the osmotic potential of the root medium and the leaf water balance. — J. exp. Bot. **20**: 895—911, 1969.

LAWLOR, D. W.: Absorption of polyethylene glycols by plants and their effects on plant growth. — New Phytol. **69**: 501—513, 1970.

LAWLOR, D. W.: An automatic multichannel thermocouple psychrometer based on an operational amplifier. — J. appl. Ecol. **9**: 581—588, 1972.

LAWLOR, D. W.: Growth and water absorption of wheat with parts of the roots at different water potentials. — New Phytol. **72**: 297—305, 1973.

LEBEDEV, G. V.: Skorost obmena vody v nabukhshikh semenakh rastenii. [Rate of water exchange in swollen seeds]. — Dokl. Akad. Nauk SSSR **128**: 632—634, 1959.

LEBEDEV, G. V., CHUCHKIN, V. G., SABININA, E. D., BRYUKVIN, V. G.: Pribor dlya nepreryvnoi zapisi processa pogloshcheniya vody rasteniyami. [An apparatus for continuous recording of water uptake by plants]. — Fiziol. Rast. **2**: 1110—1114, 1964.

LEBEDEV, G. V., SOLOVEV, A. K.: Pribor dlya izucheniya skorosti pogloshcheniya vody rasteniyami. [An apparatus of studying the rate of water uptake by plants]. — Fiziol. Rast. **2**: 752—756, 1964.

LEE, C. Y., WILKE, C. R.: Measurements of vapor diffusion coefficient. — Ind. eng. Chem. **46**: 2381—2387, 1954.

LEE, R., GATES, D. M.: Diffusion resistance in leaves as related to their stomatal anatomy and microstructure. — Amer. J. Bot. **51**: 963—975, 1964.

LEICK, E.: Ein neues Universal-Doppel-Porometer. — Ber. deut. bot. Ges. **45**: 43—59, 1928.

LEICK, E.: Bestimmung der Transpiration und Evaporation mit Rücksicht auf die Bedürfnisse der Ökologie. — In: ABDERHALDEN, E. (ed.): Handbuch der biologischen Arbeitsmethoden. Vol. XI/4. Pp. 1573—1735, Urban & Schwarzenberg, Wien 1939.

LEISTER, G. L., POPHAM, R. A., BURLEY, J. W.: An isolation unit for supplying radioisotopes to specific segments of intact root. — Amer. J. Bot. **57**: 436—442, 1970.

LE JEUNE, G., ARNOULD, G.: The determination of moisture in soils and solid substances by means of an apparatus based on the changes in the dielectric constants. — Compt. rend. Acad. Sci. (Paris) **246**: 1217—1219, 1958.

LEMÉE, G., GONZALEZ, P.: Comparaison de méthodes de mesure du potentiel hydrique (tension de succion, DPD) dans les feuilles par équilibre osmotique et par équilibre de pression de vapeur. — In: ECKARDT, P. E. (ed.): Methodology of Plant Eco-Physiology. (Arid Zone Res. 25). Pp. 361—368. UNESCO, Paris 1965.

LEMÉE, G., LAISNÉ, G.: La méthode réfractiométrique de mesure de la suction. — Rev. gén. Bot. **58**: 336—347, 1951.

LEONARD, R. A., LOW, P. F.: A self-adjusting null-point tensiometer. — Proc. Soil Sci. Soc. Amer. **26**: 123—125, 1962.

LESHEM, B.: Toxic effect of carbowaxes (polyethylene glycols) on *Pinus halepensis* MILL. seedlings. — Plant Soil **24**: 322—342, 1966.

LESHEM, Y., THAINE, R.: A note on the measurement of stomatal aperture. — New Phytol. **68**: 1047—1049, 1969.

LEVITT, J.: The osmotic equivalent and osmotic potential difference of plant cell. — Physiol. Plant. **4**: 446—448, 1951.

LEVITT, J.: The Hardiness of Plants. — Acad. Press, New York 1956.

LEVITT, J.: Osmotic pressure measuring devices. — In: NEWMAN, D. W. (ed.): Instrumentation in Experimental Biology. Pp. 405—427. MacMillan, New York 1964.

LEVITT, J.: The measurement of drought resistance. — In: ECKARDT, F. E. (ed.): Methodology of Plant Eco-physiology. — Arid Zone Res. Vol. 25. Pp. 407—412. UNESCO, Paris 1965.

LEVITT, J., SCARTH, G. W., GIBBS, R. D.: Water permeability of isolated protoplasts in relation to volume changes. — Protoplasma 26: 237—248, 1936.

LEVY, A.: The accuracy of the bubble meter method for gas flow measurements. — J. sci. Instr. 41: 449—453, 1964.

LEYTON, L.: Continuous recording of sap flow rates in tree stems. — Proc. 14th IUFRO-Congress, Munich, Pt. 1, Sect. 11, pp. 240—251, 1967.

LEYTON, L.: Problems and techniques in measuring transpiration from trees. — In: LUCKWILL, L. C., CUTTING, C. V. (eds.) Physiology of Tree Crops. — Pp. 101—112. Academic Press, London—New York 1970.

LIENEWEG, F.: Absolute und relative Feuchtebestimmungen mit dem Lithiumchlorid-Feuchtemesser. — Siemens-Zeitschrift 29: 212—218, 1955.

LIGON, J. T.: Evaluation of the gamma transmission method for determining soil water balance and evapotranspiration. — Trans. Amer. Soc. Agr. Eng. 12: 121—126, 1969.

LINDNER, H.: An instrument for measuring the water content and temperature of soil by means of thermistors. — Albrecht-Thaer Arch. 8: 79—87, 1964.

LIVINGSTON, B. E.: A method for controlling plant moisture. — Plant World 11: 39—40, 1908.

LIVINGSTON, B. E.: Porous clay cones for the auto-irrigation of potted plants. — Plant World 21: 202—208, 1918.

LIVINGSTON, B. E., BAKKE, A. L.: The transpiring power of plant foliage as measured by the method of standardized hygrometric paper. — Carnegie Inst. Wash. Publ. 13: 1914.

LIVINGSTON, B. E., KOKETSU, R.: The water-supplying power of the soil as related to the wilting of plants. — Soil Sci. 9: 469—485, 1920.

LIVINGSTON, B. E., SHREVE, F. B.: Improvements in the method for determining the transpiring power of plant surfaces by hygrometric paper. — Plant World 19: 287—309, 1916.

LJUNGGREN, K.: A review of the use of radioisotope tracers for evaluating parameters pertaining to the flow of material in plant and natural systems. — Isotop. Radiat. Technol. 5: 3—24, 1967.

LLOYD, F. E.: The physiology of stomata. — Carnegie Inst. Wash. Publ. 82: 1—142, 1908.

LOCKHART, J. A.: A new method for the determination of osmotic pressure. — Amer. J. Bot. 46: 704—708, 1959.

LOMAS, J., SHASHOUA, Y.: The performance of three types of leaf-wetness recorders. — Agr. Meteorol. 7: 159—166, 1970.

LOPUSHINSKY, W.: Effect of water movement on ion movement into the xylem of tomato roots. — Plant Physiol. 39: 494—501, 1964.

LOPUSHINSKY, W.: A portable apparatus for estimating stomatal aperture in conifers. — US Dept. Agr. Forest Serv., Pacific Northwest Forest and Range Exp. Stat., Portland, Oregon, 7 pp., 1969.

LOPUSHINSKY, W., KLOCK, G.: Construction details of ceramic bulb thermocouple psychrometers. — Northwest Sci. Soc. Proc. 1970a.

LOPUSHINSKY, W., KLOCK, G.: Construction details of ceramic bulb psychrometers. — USDA Forest Service, Forest Hydrology Lab., Wenatchee, Washington, 1970b.

LOUGUET, Ph.: Sur une méthode d'étude du mouvement des stomates utilisant la diffusion de l'hydrogène à travers les feuilles. — In: ECKARDT, F. E. (ed.): Methodology of Plant Eco-Physiology. (Arid Zone Res. 25). Pp. 307—316, UNESCO, Paris 1965.

LØVLIE, A., ZEUTHEN, E.: A thermostatic bath showing temperature fluctuations about 2—3. 10^{-4}°C. — Compt. rend. Trav. Labor. Carlsberg, Copenhague 32: 381—392, 1962.

LUCK, W.: Feuchtigkeit. — R. Oldenburg, München—Wien 1964.

LUFT, K. F., KESSELER, G., ZÖRNER, K. H.: Nichtdispersive Ultrarot-Gasanalyse mit dem UNOR. — Chemie-Ingenieur-Technik **39**: 937—945, 1967.

LUNDEGÅRDH, H.: The translocation of salts and water through wheat roots. — Physiol. Plant. **3**: 103—151, 1950.

LUNDEGÅRDH, H.: The transport of water in wood. — Ark. Bot., Ser. 2 **2**: 89—119, 1954.

LUNDEGÅRDH, H.: Pflanzenphysiologie. — VEB Gustav Fischer Verlag, Jena 1960.

LÜTTGE, U., LATIES, G. G.: Dual mechanisms of ion absorption in relation to long distance transport in plants. — Plant Physiol. **41**: 1531—1539, 1966.

LÜTTGE, U., WEIGL, J.: Der Ionentransport in intakten und entrindeten Wurzeln. — Ber. deut. bot. Ges. **77**: 63—70, 1964.

LYON, C. J.: Analysis of osmotic relations by extending the simplified method. — Plant Physiol. **11**: 167—172, 1936.

LYON, C. J.: Improvements of the simplified method for osmotic measurements. — Plant Physiol. **15**: 561—562, 1940.

MACKLON, A. E. S., WEATHERLEY, P. E.: A vapour-pressure instrument for the measurement of leaf and soil water potential. — J. exp. Bot. **16**: 261—270, 1965.

MÄDE, A.: Zur Methodik der relativen Taumessung. — Wiss. Z. M. Luther-Univ. (Halle—Wittemberg) ·**5**: 483—512, 1956.

MAERTENS, C., MORIZET, J., STUDER, R.: Modalités d'utilisation en agronomie d'un humidimètre à ralentissement de neutrons. — Ann. agron. (Paris) **16**: 5—23, 1965.

MANOHAR, M. S.: Measurement of water potential of intact plant tissues. I. Design of a microthermocouple psychrometer. — J. exp. Bot. **17**: 44—50, 1966a.

MANOHAR, M. S.: Measurement of water potential of intact plant tissues. II. Factors affecting the precision of the thermocouple psychrometer technique. — J. exp. Bot. **17**: 51—56, 1966b.

MANOHAR, M. S.: Effect of "osmotic" systems on germination of peas (*Pisum sativum* L.). — Planta **71**: 81—86, 1966c.

MANOHAR, M. S.: Effect of the excision of leaf tissues on the measurement of their water potential with thermocouple psychrometer. — Experientia **22**: 368, 1966d.

MANOHAR, M. S.: Which water potential? Differences between isopiestic thermocouple psychrometer measurements of intact and excised plant materials. — Biol. Plant. **13**: 247—256, 1971.

MANOHAR, M. S., HEYDECKER, W.: Water requirements for seed germination. — Annu. Rep. Nottingham Univ. School Agr. **1964**: 55—62, 1965.

MARKOV, M. V.: Polevoi sposob opredeleniya skorosti aktivnoi podachi vody kornem po kolichestvu vytekayushcheï pasoki. [Field method for determination of rate of active root water supply by means of exudate amount]. — Bot. Zh. **45**: 1502—1503, 1960.

MARR, A. G., VAADIA, Y.: Rapid cryoscopic technique for measuring osmotic properties of drop size samples. — Plant Physiol. **36**: 677—680, 1961.

MARSHALL, D. C.: Measurement of sap flow in conifers by heat transport. — Plant Physiol. **33**: 385—396, 1958.

MARSHALL, D. C.: The freezing of plant tissue. — Aust. J. biol. Sci. **14**: 368—390, 1961.

MARSHALL, D. C.: Sap flow analysis charts. — New Zeal. J. Sci. Technol. **5**: 521—530, 1962.

MASON, T. G., PHILLIS, E.: Studies on foliar hydration in the cotton plant. II. Preliminary observations using the disc-culture. — Ann. Bot. (N.S.) **6**: 455—468, 1942.

MAXIMOV, N. A., PETINOV, N. S.: Opredelenie sosushcheï sily lisťev metodom kompensacii s pomoshchu refraktometra. [The determination of leaf suction force by the compensation method using refractometer]. — Dokl. Akad. Nauk SSSR **62**: 537—540, 1948.

McGuinness, J. L., Dreibelbis, F. R., Harrold, L. L.: Soil moisture measurements with the neutron meter method supplement weighing lysimeters. — Proc. Soil Sci. Soc. Amer. **25**: 339—342, 1961.

McIlroy, I. C.: A sensitive temperature and humidity probe. — Aust. J. agr. Res. **6**: 196 to 199, 1955.

McQueen, I. S., Miller, R. F.: Calibration and evaluation of wide-range gravimetric method for measuring moisture stress. — Soil Sci. **106**: 225—231, 1968.

Mederski, H. J.: Determination of internal water status of plants by beta ray gauging. — Soil Sci. **92**: 143—146, 1961.

Mederski, H. J.: Plant water balance determination by beta gauging technique. — Agron. Abstr. **56**: 126, 1964.

Mederski, H. J., Alles, W.: Beta gauging leaf water status: Influence of changing leaf characteristics. — Plant Physiol. **43**: 470—472, 1968.

Mees, G. C., Weatherley, P. E.: The mechanism of water absorption by roots. I. Preliminary studies on the effects of hydrostatic pressure gradients. — Proc. roy. Soc. London B **147**: 367—380, 1957a.

Mees, G. C., Weatherley, P. E.: The mechanism of water absorption by roots. II. The role of hydrostatic pressure gradients across the cortex. — Proc. roy. Soc. London B **147**: 381—391, 1957b.

Meidner, H.: Measurement of water intake from the atmosphere by leaves. — New Phytol. **53**: 423—426, 1954.

Meidner, H.: A simple porometer for measuring the resistance to air flow offered by stomata. — School Sci. Rev. **47**: 149—151, 1965.

Meidner, H.: A critical study of sensor element diffusion porometers. — J. exp. Bot. **21**: 1060—1066, 1970.

Meidner, H., Mansfield, T. A.: Physiology of Stomata. — McGraw-Hill, London 1968.

Meidner, H., Spanner, D. C.: The differential transpiration porometer. — J. exp. Bot. **10**: 190—205, 1959.

Merill, S. D., Dalton, F. N., Helkelrath, W. N., Hoffman, G. J., Ingvalson, R. D., Oster, J. D., Rawlins, S. L.: Details of construction of a multipurpose thermocouple psychrometer. — US Dep. Agr., US Salinity Lab. Res. Rep. **115**: 1—9, 1968.

Merriam, R. A.: Moisture sampling in wildland soils with a neutron probe. — Iowa State Coll. J. Sci. **34**: 641—647, 1960.

Meyer, B. S.: The water relations of plant cells. — Bot. Rev. **4**: 531—548, 1938.

Meyer, B. S.: A critical evaluation of the terminology of diffusion phenomena. — Plant Physiol. **20**: 142—164, 1945.

Meyer, B. S., Wallace, A. M.: A comparison of two methods of determining the diffusion pressure deficit of potato tissues. — Amer. J. Bot. **28**: 838—843, 1941.

Michael, G.: Eine Methode zur Bestimmung der Spaltöffnungsweite von Koniferen. — Flora **159** A: 559—561, 1969.

Michel, B. E.: Solute potentials of sucrose solutions. — Plant Physiol. **50**: 196—198, 1972.

Michel, B. E., ElSharkawi, H. M.: Investigation of plant water relations with divided root systems of soybean. — Plant Physiol. **46**: 728—731, 1970.

Michler, R., Steubing, L.: Thermoelektrische Messung der Geschwindigkeit des Transpirationsstromes in krautigen Pflanzen. — Flora B **157**: 477—486, 1968.

Milburn, J. A.: The conduction of sap. I. Water conduction and cavitation in water stressed leaves. — Planta **69**: 34—42, 1966.

Milburn, J. A., Johnson, R. P. C.: The conduction of sap. II. Detection of vibrations produced by sap cavitation in *Ricinus* xylem. — Planta **69**: 43—52, 1966.

MILBURN, J. A., WEATHERLEY, P. E.: The influence of temperature on the process of water uptake by detached leaves and leaf discs. — New Phytol. **70**: 929—938, 1971.

MILLAR, A. A., DUYSEN, M. E., WILKINSON, G. E.: Internal water balance of barley under soil moisture stress. — Plant Physiol. **43**: 968—972, 1968.

MILLAR, A. A., LANG, A. R. G., GARDNER, W. R.: Four-terminal Peltier type thermocouple psychrometer for measuring water potential in nonisothermal systems. — Agron. J. **62**: 705—708, 1970.

MILLAR, B. D.: Relative turgidity of leaves: temperature effects in measurement. — Science **154**: 512—513, 1966.

MILLAR, B. D.: Improved thermocouple psychrometer for the measurement of plant and soil water potential. I. Thermocouple psychrometry and an improved instrument design. — J. exp. Bot. **22**: 875—890, 1971a.

MILLAR, B. D.: Improved thermocouple psychrometer for the measurement of plant and soil water potential. II. Operation and calibration. — J. exp. Bot. **22**: 891—905, 1971b.

MILLAR, B. D.: Multiple water bath with precise temperature control. — Agr. Engin. **52**: 456—457, 1971c.

MILLER, A. T. Jr.: Studies on tissue water. I. The determination of blood water by the distillation method. — J. biol. Chem. **143**: 65—73, 1942.

MILLER, A. T. Jr.: Studies on tissue water. II. A macromodification of the distillation method for the determination of tissue water. — J. biol. Chem. **149**: 153—155, 1943.

MILLER, P. C., GATES, D. M.: Transpiration in plants. — Amer. Mid. Naturalist **77**: 77—85, 1967.

MILLER, R. D.: A technique for measuring soil moisture tension in rapidly changing systems. — Soil Sci. **72**: 291—301, 1951.

MILTHORPE, F. L.: The significance of the measurement made by the cobalt chloride paper method. — J. exp. Bot. **6**: 17—19, 1955.

MILTHORPE, F. L.: Plant factors involved in transpiration. — In: Plant-Water Relationships in Arid and Semi-Arid Conditions. Proc. Madrid Symp. (Arid Zone Res. 16). Pp. 107—115. UNESCO, Paris 1961.

MILTHORPE, F. L., PENMAN, H. L.: The diffusive conductivity of the stomata of wheat leaves. — J. exp. Bot. **18**: 422—457, 1967.

MINCHIN, F. R., BAKER, D. A.: Water dependent and water independent fluxes of potassium in exuding root systems of *Ricinus communis*. — Planta **89**: 212—223, 1969.

MINCHIN, F. R., BAKER, D. A.: A mathematical analysis of water and solute transport across the root of *Ricinus communis*. — Planta **94**: 16—26, 1970.

MITCHELL, J., SMITH, D. M.: Aquametry. Application of the Karl Fischer reagent to quantitative analysis involving water. — Intersci. Publ., New York 1948.

MITTLER, T. E.: Studies on the feeding and nutrition of *Tuberolachnus salignus* (GMELIN) (*Homoptera, Aphidae*). II. The nitrogen and sugar composition of invested phloem sap and excreted honeydew. — J. exp. Biol. **35**: 74—83, 1958.

MOHSIN, M. A.: A field porometer for use on paddy plants. — Riso **19**: 169—174, 1970.

MOINAT, A. D.: An auto-irrigator for growing plants in the laboratory. — Plant Physiol. **18**: 280—287, 1943.

MOLISCH, H.: Das Offen- und Geschlossensein der Spaltöffnungen veranschaulicht durch eine neue Methode (Infiltrationsmethode). — Z. Bot. **4**: 106—112, 1912.

MOLOTKOVSKIĬ, G. Kh.: Izuchenie sostoyaniya ustits metodom culluloidnykh otpechatkov. [Stomata aperture studied by celluloid reprint method]. — Dokl. Akad. Nauk SSSR **3**: 27—28, 1935.

MOLZ, F. J.: A study of suction force by the simplified method. I. Effect of external factors. — Amer. J. Bot. **13**: 433—463, 1926.

MOMIN, A. U.: A new simple method of estimating the moisture content of the soil *in situ*. — Ind. J. agr. Sci. **17**: 81—88, 1947.

MONTEITH, J. L.: Dew: facts and fallacies. — In: RUTTER, A. J., WHITEHEAD, F. H. (eds.): The Water Relations of Plants. Pp. 37—56. Blackwell Sci. Publ., Oxford 1963.

MONTEITH, J. L., BULL, T. A.: A diffusive resistance porometer for field use. II. Theory, calibration and performance. — J. appl. Ecol. **7**: 623—638, 1970.

MONTEITH, J. L., OWEN, P. C.: A thermocouple method for measuring relative humidity in the range 95—100%. — J. sci. Instr. **35**: 443—446, 1958.

MONTEITH, J. L., SZEICZ, G., WAGGONER, P. E.: The measurement and control of stomatal resistance in the field. — J. appl. Ecol. **2**: 345—355, 1965.

MONTFORT, C.: Die Wasserbilanz in Nährlösung, Salzlösung und Hochmoorwasser. — Z. Bot. **14**: 97—172, 1922.

MONTFORT, C., HAHN, H.: Atmung und Assimilation als dynamische Kennzeichen abgestufter Trockenresistenz bei Farnen und höheren Pflanzen. — Planta **38**: 503—515, 1950.

MORESHET, S.: A portable Wheatstone bridge porometer for field measurements of stomatal resistance. — Israel J. agr. Res. **14**: 27—30, 1964.

MORESHET, S., STANHILL, G.: The measurement of stomatal diffusion resistance in the laboratory and in the field. — 3rd Annu. Rep. Volcani Inst. (Rehovot). Pp. 29—31. 1967.

MORESHET, S., STANHILL, G., KOLLER, D.: A new method of measuring stomatal diffusion resistance. — Israel J. Bot. **16**: 50, 1967.

MORESHET, S., STANHILL, G., KOLLER, D.: A radioactive tracer technique for the direct measurement of the diffusion resistance of stomata. — J. exp. Bot. **19**: 460—467, 1968.

MORESHET, S., YOCUM, C. S.: A condensation type porometer for field use. — Plant Physiol. **49**: 944—949, 1972.

MORROW, P. A., SLATYER, R. O.: Leaf temperature effects on measurements of diffusive resistance to water vapor transfer. — Plant Physic. **47**: 559—561, 1971a.

MORROW, P. A., SLATYER, R. O.: Leaf resistance measurements with diffusion porometers: Precautions in calibration and use. — Agr. Meteorol. **8**: 223—233, 1971b.

MORSE, H. N.: The osmotic pressure of aqueous solutions. — Carnegie Inst. Wash. Publ. No 198, 1914.

MORTIER, P., DE BOODT, M.: Determination of soil moisture by neutron scattering. — Neth. J. agr. Sci. **4**: 111—113, 1956.

MOSEBACH, G.: Ein Mikroverfahren zur kryoskopischen Untersuchungen saftreicher Gewebe. — Ber. deut. bot. Ges. **58**: 29—40, 1940.

MOSS, P.: Some aspects of the cation status of soil moisture. — Plant Soil **18**: 99—113, 1963.

MOURAVIEFF, I.: Tension de succion et déficit de saturation hydrique du tapis végétal des pelouses sèches de la région de Grasse (Alpes Maritimes). — Soc. Bot. Fr. Bull. **106**: 306—309, 1959.

MOURAVIEFF, I.: Déficit de saturation subléthal et caractères de la plasmolyse chez quelques plantes des régions méditerranéennes au cours de la saison sèche. — Bull. mens. Soc. Linnéenne (Lyon) **32**: 90—95, 1963.

MOURAVIEFF, I.: Sur les propriétés optiques des feuilles. Influence de l'état d'hydratation de la feuille sur la transmission de la lumière. — Bull. mens. Soc. Linnéenne (Lyon) **33**: 365—372, 1964.

MOZINGO, H. N., KLEIN, P., ZEEVI, Y., LEWIS, E. R.: Venus's flytrap observations by scanning electron microscopy. — Amer. J. Bot. **57**: 593—598, 1970.

MURR, L. E.: Measurement of the relative dielectric constant for plants: With particular reference to grasses. — Proc. Pennsylv. Acad. Sci. **39**: 192—201, 1965.

MYERS, G. M. P.: The water permeability of unplasmolyzed tissues. — J. exp. Bot. **2**: 129—144, 1951.

NADEL, M.: Sur la mesure de l'ouverture des stomates. Critique de la methode de fixation des stomates par l'alcool. — Palest. J. Bot. (Rehovot) 3: 21—64, 1940.

NAKAYAMA, F. S., EHRLER, W. L.: Beta ray gauging technique for measuring leaf water content changes and moisture status of plants. — Plant Physiol. 39: 95—98, 1964.

NAMKEN, L. N., BARTHOLIC, J. F., RUNKLES, J. R.: Water stress and stem radial contraction of cotton plants (Gossypium hirsutum L.) under field conditions. — Agron. J. 63: 623—627, 1971.

NAMKEN, L. N., LEMON, E. R.: Field studies of internal moisture relations of the corn plant. — Agron. J. 53: 643—646, 1960.

NÁTR, L., ŠPIDLA, J.: Application of the leaf-disk method to the determination of photosynthesis in cereals. — Biol. Plant. 3: 245—251, 1961.

NĚMEC, B.: Příspěvky k fysiologii a morfologii rostlinné buňky. [Contributions to plant cell physiology and morphology]. — Věstník král. Spol. Nauk Tř. mat. 60: 6—18, 1899.

NESTEROV, V. G., NIKOLAEV, L. V.: Nepreryvnoe izmerenie vlazhnosti lisťev betaskopicheskim metodom. [Continuous measurement of leaf water content with a betascopic method]. — Dokl. TSKhA (Moskva) 133: 465—467, 1968.

NESTEROV, V. G., NIKOLAEV, L. V.: Primenenie metodov infrakrasnoï spektroskopii i betaspektroskopii pri opredelenii soderzhaniya vody i sukhogo veshchestva v zhivykh list'yakh. [Use of infra-red spectroscopical and betaspectroscopical methods for determining content of water and dry matter in living leaves]. — Dokl. TSKhA (Moskva) 154: 359—363, 1969.

NEUMANN, H. H. and THURTELL, G. W.: A Peltier cooled thermocouple dewpoint hygrometer for in situ measurement of water potentials. — In: BROWN, R.W., VAN HAVEREN, B. P. (eds.): Psychrometry in Water Relations Research. — Utah Agric. Exp. Sta, Logan 1973 (in press).

NEUWIRTH, G.: Pflanzensoziologische und ökologische Untersuchungen an Hängen des Lindbusches, der Harslebener Berge und des Steinholzes. — Wiss. Z. M. Luther-Univ., Halle—Wittenberg, Math. Nat. 7: 101—124, 1958.

NEWBOULD, P., MERCER, E. R., LAY, P. M.: Estimations of changes in the water status of soils under field conditions from the attenuation of β-radiation by water held in absorbent nylon pads. — Exp. Agr. 4: 167—177, 1968.

NEWTON, R., MARTIN, W. M.: Physico-chemical studies on the nature of drought resistance in crop plants. — Canad. J. Res. 3: 336—427, 1930.

NIKITIN, V. A.: Bystryï sposob opredeleniya elektroprovodnosti rastiteľnykh tkaneï. [Rapid method for determining of electrical conductivity of plant tissues]. — Fiziol. Rast. 13: 373—376, 1966.

NORDHAUSEN, M.: Weitere Beiträge zum Saftsteigeproblem. — Jahrb. wiss. Bot. 60: 307—353, 1921.

NORTH, C.: A technique for measuring structural features of plant epidermis using cellulose acetate films. — Nature 178: 1186—1187, 1956.

NOY-MEIR, I., GINZBURG, B. Z.: An analysis of the water potential isotherm in plant tissue. II. Comparative studies on leaves of different types. — Aust. J. biol. Sci. 22: 35—52, 1969.

OCHI, H.: Autecological study of mosses in respect to water economy. I. On the minimum hydrability within which mosses are able to survive. — Bot. Mag. (Tokyo) 65: 112—118, 1952.

OGURA, Y.: On the measurement of freezing point and osmotic pressure of tobacco leaf. — Proc. Crop Sci. Soc. Jap. 40: 51—57, 1971.

OKUNTSOV, M. M. A., TARASOVA, E. N.: O sostayanii vody v rasteniyakh. [Water status in plants]. — Dokl. Akad. Nauk SSSR 83: 315—317, 1952.

O'LEARY, J. W.: Root-pressure exudation in woody plants. — Bot. Gaz. 126: 108—115, 1965.

VAN OORSCHOT, J. P. L., BELKSMA, M.: An assembly for the continuous recording of CO_2 exchange and transpiration of whole plants. — Weed Res. 1: 245—257, 1961.

OPPENHEIMER, H. R.: Kritische Betrachtungen zu den Saugkraftmessungen von Ursprung und Blum. — Ber. deut. bot. Ges. **48**: 130—140, 1930.

OPPENHEIMER, H. R.: Zur Kenntnis der hochsommerlichen Wasserbilanz mediterraner Gehölze. — Ber. deut. bot. Ges. **50a**: 185—245, 1932.

OPPENHEIMER, H. R.: Untersuchungen zur Kritik der Saugkraft. — Planta **18**: 525—548, 1933.

OPPENHEIMER, H. R.: Critical remarks on the value of Lloyd's alcohol fixation method for measuring stomatal aperture. — Palest. J. Bot. hort. Sci. **1**: 43—47, 1935.

OPPENHEIMER, H. R.: Critique expérimentale de deux méthodes employées en vue d'établir le déficit de saturation hydrique (DSH) des feuilles. — 8th int. bot. Congr., Rep. Comm. Sect. III, pp. 218—220, Paris 1954.

OPPENHEIMER, H. R.: Zur Kenntnis kritischer Wassersättigungsdefizite in Blättern und ihrer Bestimmung. — Planta **60**: 51—69, 1963.

OPPENHEIMER, H. R., ENGELBERG, N.: Mesure du degré d'ouverture des stomates de conifères. Méthodes anciennes et modernes. — In: ECKARDT, F. E. (ed.): Methodology of Plant Eco-Physiology. (Arid Zone Res. UNESCO 25). Pp. 317—323. UNESCO, Paris 1965.

OPPENHEIMER, H. R., HALEVY, A. H.: Anabiosis of *Ceterach officinarum* LAM. et DC. — Bull. Res. Counc. Israel D **11**: 127—147, 1962.

OPPENHEIMER, H. R., JACOBY, B.: Usefulness of autofluorescence tests as criterion of life in plant tissues. — Protoplasma **53**: 220—226, 1961.

OPPENHEIMER, H. R., LESHEM, B.: Critical thresholds of dehydration in leaves of *Nerium oleander* L. — Protoplasma **61**: 302—321, 1966.

OPPENHEIMER, H. R., SHOMER-ILAN, A.: A contribution to the knowledge of drought resistance of Mediterranean pine trees. — Mitt. florist. - soziol. Arbeitsgem. Stolzenau N. F. **10**: 42—55, 1963.

ORDIN, L., APPLEWHITE, T. H. BONNER, J.: Auxin induced water uptake by *Avena* coleoptile sections. — Plant Physiol. **31**: 44—53, 1956.

ORDIN, L., GAIRON, S.: Diffusion of tritiated water into roots as influenced by water status of tissue. — Plant Physiol. **36**: 331—335, 1961.

ORR, J. S., GILLESPIE, F. C.: Occupancy principle for radioactive tracers in steady state biological systems. — Science **162**: 138—139, 1968.

OSBORNE, T. S., BACON, J. A.: Two improved and inexpensive systems for moisture stabilisation in seeds or other tissues. — Plant Physiol. **36**: 309—312, 1961.

OSTER, J. D., RAWLINS, S. L., INGVALSON, R. D.: Independent measurement of matric and osmotic potential of soil water. — Proc. Soil Sci. Soc. Amer. **33**: 188—192, 1969.

PACARDO, E. P.: Effect of temperature and the role of growth on water uptake of leaf tissues. — Ph.D. Thesis. Dep. Bot., Univ. Aberdeen 1971.

PAINTER, L. I.: Method of subjecting growing plants to a continuous soil moisture stress. — Agron. J. **58**: 459—460, 1966.

PÁLFI, G., JUHÁSZ, J.: The theoretical basis and practical application of a new method of selection for determining water deficiency in plants. — Plant Soil **34**: 503—507, 1971.

PARKER, J.: Moisture retention in leaves of conifers of the Northern Rocky Mountains. — Bot. Gaz. **113**: 210—216, 1951.

PARLANGE, J. Y., WAGGONER, P. E.: Stomatal dimensions and resistance to diffusion. — Plant Physiol. **46**: 337—342, 1970.

PAZOUREK, J., in: NĚMEC, B. (ed.): Botanická mikrotechnika. [Botanical Microtechniques]. — Pp. 404—407. Nakl. ČSAV, Praha 1962.

PEASLEE, D. E., MOSS, D. N.: Stomatal conductivities in K-deficient leaves of maize (*Zea mays* L.). — Crop Sci. **8**: 427—430, 1968.

PECK, A. J.: Theory of the Spanner psychrometer. — I. The thermocouple. — Agr. Meteorol. **5**: 433—447, 1968.

PECK, A. J.: Theory of the Spanner psychrometer. II. Sample effects and equilibration. — Agr. Meteorol. **6**: 111—124, 1969.

PECK, A. J., RABBIDGE, R. M.: Soil-water potential: direct measurement by a new technique. — Science **151**: 1385—1386, 1966a.

PECK, A. J., RABBIDGE, R. M.: Note on an instrument for measuring water potentials particularly in soils. — Conf. Instrumentation Plant Environment Measurements, Aust. CSIRO, Aspendale 1966, p. 20, Soc. Instr. Technol., Melbourne 1966b.

PELTON, W. L.: Evaporation from atmometers and pans. — Can. J. Plant Sci. **44**: 397—404, 1964.

PENKA, M.: Hodnocení půdní vody s biologického hlediska metodou vysychacích křivek. [Soil water evaluation by means of drying curves]. — Československ. Biol. **5**: 105—116, 1956.

PENMAN, H. L.: Theory of porometers used in the study of stomatal movements in leaves. — Proc. roy. Soc. London B **130**: 416—434, 1942.

PENMAN, H. L., SCHOFIELD, R. K.: Some physical aspects of assimilation and transpiration. — Symp. Soc. exp. Biol. **5**: 115—129, 1951.

PEREIRA, H. C.: A cylindrical gypsum block for moisture studies in deep soils. — J. Soil Sci. **2**: 212—223, 1951.

PERRIER, A.: Leaf temperature measurement. — In: ŠESTÁK, Z., ČATSKÝ, J., JARVIS, P. C. (ed.): Plant Photosynthetic Production. Manual of Methods. Pp. 632—671. Dr W. Junk, The Hague 1971.

PERRIER, E. R., JOHNSTON, W. R.: Distribution of thermal neutrons in a soil-water system. — Soil Sci. **93**: 104—112, 1962.

PERRIER, E. R., MARSH, A. W.: Performance characteristics of various electrical resistance units and gypsum materials. — Soil Sci. **86**: 140—147, 1958.

PERSON, H.: Über elektrische Messungen der Bodenfeuchte. — Ber. deut. Wetterdienstes US Zone **5**: 22—24, 1952.

PETTERSSON, S.: Artificially induced water and sulfate transport through sunflower roots. — Physiol. Plant. **19**: 581—601, 1966.

PETTERSSON, S.: A labile-bound component of phosphate in free space of sunflower plant roots. — Physiol. Plant. **24**: 485—490, 1971.

PEYNADO, A., YOUNG, R. H.: Moisture changes in intact citrus leaves monitored by a β gauge technique. — Proc. Amer. Soc. hort. Sci. **92**: 211—220, 1968.

PHARIS, R. P.: Comparative drought resistance of five conifers and foliage moisture content as a viability index. — Ecology **47**: 211—221, 1966.

PHARIS, R. P., FERRELL, W. K.: Differences in drought resistance between coastal and inland sources of Douglas fir. — Can. J. Bot. **44**: 1651—1659, 1966.

PHENE, C. J., HOFFMAN, G. J., RAWLINS, S. L.: Measuring soil matric potential *in situ* by sensing heat dissipation within a porous body: I. Theory and sensor construction. — Proc. Soil Sci. Soc. Amer. **35**: 27—33, 1971.

PHILIP, J. R.: The physical principles of soil water movement during the irrigation cycle. — Proc. int. Congr. Irrig. Drain. **8**: 124—154, 1957.

Physico-chemical Tables. I. and II., (Fysikálně chemické tabulky) Praha 1953.

PICKARD, W. F.: A heat pulse method of measuring water flux in woody plant stems. — Mathem. Biosciences **16**: 247—262, 1973.

PICKARD, W. F., PUCCIA, C. J.: A theory of the steady-state heat step method of measuring water flux in woody plant stems. — Mathem. Biosciences **14**: 1—15, 1972.

PIERPOINT, G.: Measuring surface soil moisture with the neutron depth probe and a surface shield. — Soil Sci. **101**: 189—192, 1966.

PIERPOINT, G.: Direct measurement of internal moisture deficits in trees. — Forest. Chronicle **1967**: 145—148, 1967.

PISEK, A.: Der Wasserhaushalt der Meso- und Hygrophyten. — In: RUHLAND, W. (ed.): Handbuch der Pflanzenphysiologie. Vol. 3. Pp. 825—853. Springer-Verlag, Berlin—Göttingen—Heidelberg 1956.

PISEK, A., BERGER, E.: Kutikuläre Transpiration und Trockenresistenz isolierter Blätter und Sprosse. — Planta 28: 124—155, 1938.

PISEK, A., HEIZMANN, K.: Messungen von Bodenwassergehalt und Welkungsprozentsatz mit der Gipsblockmethode. — Ber. deut. bot. Ges. 74: 465—478, 1962.

PISEK, A., LARCHER, W.: Zusammenhang zwischen Austrockungsresistenz und Frosthärte bei Immergrünen. — Protoplasma 44: 30—46, 1954.

PISEK, A., WINKLER, E.: Die Schliessbewegung der Stomata bei ökologisch verschiedenen Pflanzentypen in Abhängigkeit vom Sättigungszustand der Blätter und vom Licht. — Planta 42: 253—278, 1953.

PLAUT, Z., REINHOLD, L.: The effect of water stress on the movement of ^{14}C-sucrose and tritiated water within the supply leaf of young bean plants. — Aust. J. biol. Sci. 20: 297—307, 1967.

POLSTER, H.: Kritische Bemerkungen zur Arlands "Kurzfristige Vorprüfung von Kulturmassnahmen nach einem neuen Verfahren". — Deut. Landwirtschaft 2: 96—99, 1951.

POLSTER, H., FUCHS, S.: Der Einfluss intermittierender Belichtung auf die Transpiration und Assimilation von Fichte und Lärche bei Dürrebelastung. — Biol. Zentralbl. 79: 465—480, 1960.

POLSTER, H., WEISE, G., NEUWIRTH, G.: Fortschritte in der Verwendung von URAS-Geräten zu Laboratoriums- und Freilandmessungen des CO_2-Stoffwechsels und der Transpiration. — Biol. Beitr. 1: 105—118, 1961.

POSPÍŠILOVÁ, J.: Water potential — water saturation deficit relationship during dehydration and resaturation of leaves. — Biol. Plant. 15: 290—293, 1973.

POSPÍŠILOVÁ, J.: Water potential, water saturation deficit and their relationship in leaves of different insertion levels. — Biol. Plant. 16: 140—143, 1974.

POST, K., SEELEY, J. G.: Automatic watering of greenhouse crops. — Cornell Univ. Agr. exp. Sta. Bull. 793, 1943.

PRAGER, D. J., BOWMAN, R. L.: Freezing point depression: a new method for measuring ultramicroquantities of fluids. — Science 142: 237—239, 1963.

PRÁT, S.: Plasmolysis and permeability II. — Preslia 2: 90—97, 1922.

PRÁT, S.: Fysikální a chemická fysiologie. [Physical and Chemical Physiology]. — In: NĚMEC, B., PRÁT, S., KOŘÍNEK, J.: Učebnice anatomie a fysiologie rostlin pro farmaceuty a přírodopisce. [Textbook of Plant Anatomy and Physiology]. — Melantrich, Praha 1945.

PRINGSHEIM, E. G.: Untersuchungen über Turgordehnung und Membranbeschaffenheit. — Jahrb. wiss. Bot. 74: 749—796, 1931.

PURITCH, G. S., TURNER, J. A.: Effects of pressure increase and release on temperature within a pressure chamber used to estimate plant water potential. — J. exp. Bot. 24: 342—348, 1973.

RAHMAN, ABD EL, HAHIDY, EL M. N.: Observations on the water output of the desert vegetation along Suez Road. — Egypt. J. Bot. 1: 19—38, 1958.

RAMSAY, J. A., BROWN, R. H. J.: Simplified apparatus and procedure for freezing-point determinations upon small volumes of fluid. — J. sci. Instr. 32: 372—375, 1955.

RASCHKE, K.: Ein Verfahren der Transpirationsbestimmung an Blattpunkten, Blatteilen und ganzen Blättern in situ. — Naturwissenschaften 41: 308, 1954.

RASCHKE, K.: Über die physikalischen Beziehungen zwischen Wärmeübergangszahl, Strahlungsaustausch, Temperatur und Transpiration eines Blattes. — Planta 48: 200—238, 1956.

RASCHKE, K.: Über den Einfluss der Diffusionswiderstände auf die Transpiration und Temperatur eines Blattes. — Flora 146: 546—578, 1958.

RASCHKE, K.: Das Seifenblasenporometer (zur Messung der Stomaweite an amphistomatischen Blättern). — Planta **66**: 113—120, 1965a.

RASCHKE, K.: Eignung und Konstruktion registrierender Porometer für das Studium der Schliesszellenphysiologie. — Planta **67**: 225—241, 1965b.

RAST, J.: Zwei neue Mikromethoden der Molekulargewichtsbestimmung. — Abderhalden, E. (ed.): Handbuch der biologischen Arbeitsmethoden. Vol. III A. P. 743. Urban & Schwarzenberg, Wien 1924.

RAWLINS, S. L.: Systematic error in leaf water potential measurements with a thermocouple psychrometer. — Science **146**: 644—646, 1964.

RAWLINS, S. L.: Theory for thermocouple psychrometers used to measure water potential in soil and plant samples. — Agr. Meteorol. **3**: 293—310, 1966.

RAWLINS, S. L.: Some new methods for measuring the components of water potential. — Soil Sci. **112**: 8—16, 1971.

RAWLINS, S. L., DALTON, F. N.: Psychrometric measurement of soil water potential without precise temperature control. — Proc. Soil Sci. Soc. Amer. **31**: 297—301, 1967.

RAY, P. M.: On the theory of osmotic water movement. — Plant Physiol. **35**: 783—795, 1960.

READ, D. W. L., FLECK, S. V., PELTON, W. L.: Self-irrigating greenhouse pots. — Agron. J. **54**: 467—468, 1962.

REDSHAW, A. J., MEIDNER, H.: A thermal method for estimating continuously the rate of flow of sap through an intact plant. — Z. Pflanzenphysiol. **62**: 405—416, 1970.

REEVES, R. E.: The estimation of primary carbinol groups in carbohydrates. — J. Amer. chem. Soc. **63**: 1476, 1941.

REGINATO, R. J., van BAVEL, C. H. M.: Soil water measurement with gamma attenuation. — Proc. Soil Sci. Soc. Amer. **28**: 721—735, 1964.

REHDER, H.: Versuche zur Bestimmung der Saugkraft mit der Schardakow-Methode. — In: RÜBEL, E., LÜDI, W.: Bericht ü. d. Geob. Forschungsinst. Rübel, Zürich **1958**: 91—110, 1959.

REHDER, H.: Saugkraftmessungen an mediterranen Immergrünen mit der Schardakow-Methode. — Ber. deut. bot. Ges. **74**: 84—92, 1961.

REHDER, H., KREEB, K.: Vergleichende Untersuchungen zur Bestimmung der Blattsaugspannung mit der gravimetrischen und der Schardakow-Methode. — Ber. deut. bot. Ges. **74**: 95—98, 1961.

REITMEIER, R. F., RICHARDS, L. A.: Reliability of the pressure membrane method for extraction of soil solution. — Soil Sci. **57**: 119, 1944.

RENNER, O.: Theoretisches und Experimentelles zur Kohäsionstheorie der Wasserbewegung. — Jahrb. wiss. Bot. **56**: 617—667, 1915.

RICE, R.: A fast-response, field tensiometer system. — Trans. Amer. Soc. agr. Eng. **12**: 48—50, 1969.

RICHARDS, B. G.: Thermistor hygrometer for determining the free energy of moisture in unsaturated soils. — Nature **208**: 608—609, 1965a.

RICHARDS, B. G.: A thermistor hygrometer for the direct measurement of the free energy of soil moisture. — CSIRO Soil Mechanics Section, Melbourne, Tech. Rep. **5**: 1—11, 1965b.

RICHARDS, L. A.: A pressure-membrane extraction apparatus for soil solution. — Soil Sci. **51**: 377—386, 1941.

RICHARDS, L. A.: Soil moisture tensiometer: materials and construction. — Soil Sci. **53**: 241—248, 1942.

RICHARDS, L. A.: Pressure membrane apparatus, construction and use. — Agr. Engin. **28**: 451—454, 1947.

RICHARDS, L. A.: Porous plate apparatus for measuring moisture retention and transmission by soil. — Soil Sci. **66**: 105—110, 1948.

RICHARDS, L. A.: Methods of measuring soil moisture tension. — Soil Sci. **68**: 95—112, 1949.

RICHARDS, L. A.: A soil salinity sensor of improved design. — Proc. Soil Sci. Soc. Amer. **30**: 333—337, 1966.

RICHARDS, L. A., BLOOD, H. L.: Some improvements in auto-irrigator apparatus. — J. agr. Res. **49**: 115—121, 1934.

RICHARDS, L. A., CAMPBELL, R. B.: Use of thermistors for measuring the freezing point of solution and soils. — Soil Sci. **65**: 429—436, 1948.

RICHARDS, L. A., CAMPBELL, R. B.: The freezing point of moisture in soil cores. — Proc. Soil Sci. Soc. Amer. **13**: 70—94, 1949.

RICHARDS, L. A., GARDNER, W.: Tensiometers for measuring the capillary tension of soil water. — J. Amer. Soc. Agron. **28**: 352—358, 1936.

RICHARDS, L. A., LOOMIS, W. E.: Limitations of auto-irrigators for controlling soil moisture under growing plants. — Plant Physiol. **17**: 223—235, 1942.

RICHARDS, L. A., OGATA, G.: Thermocouple for vapour pressure measurement in biological and soil systems at high humidity. — Science **128**: 1089—1090, 1958.

RICHARDS, L. A., OGATA, G.: Psychrometric measurements of soil samples equilibrated on pressure membranes. — Proc. Soil Sci. Soc. Amer. **25**: 456—459, 1961.

RICHARDS, L. A., RUSSELL, M. B., NEAL, O. R.: Further developments on apparatus for field moisture studies. — Proc. Soil Sci. Soc. Amer. **2**: 55—64, 1937.

RICHARDS, L. A., WADLEIGH, C. H.: Soil water and plant growth. — In: SHAW, B. T. (ed.): Soil Physical Conditions and Plant Growth. Pp. 73—251. Academic Press, New York 1952.

RICHARDS, L. A., WEAVER, L. R.: Moisture retention by some irrigated soil as related to soil moisture tension. — J. agr. Res. **69**: 215—235, 1944.

RICHARDS, S. J., LAMB, J. Jr.: Field measurements of capillary tension. — J. Amer. Soc. Agron. **29**: 772—780, 1937.

RICHTER, H., HALBWACHS, G., HOLZNER, W.: Saugspannungsmessungen in der Krone eines Mammutbaumes (*Sequoiadendron giganteum*). — Flora **161**: 401—420, 1972.

RICHTER, H., ROTTENBURG, W.: Leitfähigkeitsmessung zur Endpunktanzeige bei der Saug-spannungsbestimmung nach Scholander. — Flora **160**: 440—443, 1971.

RICHTER, K. H.: Messungen mit dem Piche-Evaporimeter. — Techn. Mitt. Instrumenten-wesen deut. Wetterdienstes NF **4**: 90—106, 1958.

RIJTEMA, P. E.: The calculation of non-parallelism of gamma access tubes, using soil sampling data. — J. Hydrol. **9**: 206—212, 1969.

RITCHIE, G. A., HINCKLEY, T. M.: Evidence for error in pressure-bomb estimates of stem xylem potentials. — Ecology **52**: 534—536, 1971.

ROBERTS, B. R.: Light as a source of error in estimates of water potential by vapor equilibra-tion. — Plant Physiol. **44**: 937—938, 1969.

ROBINSON, R. A., STOKES, R. H.: Tables of osmotic and activity coefficients of electrolytes in aqueous solutions at 25 °C. — Faraday Soc. Trans. **45**: 612—624, 1949.

ROBINSON, R. A., STOKES, R. H.: Electrolyte Solutions. — Butterworth Sci. Publ., London (2nd Ed.) 1959.

RODE, A. A.: Pochvennaya vlaga. [Soil moisture]. — Izd. Akad. Nauk SSSR, Moskva 1952.

ROEPKE, R. R.: The thermoelectric method of measuring vapour pressure. — J. phys. Chem. **46**: 359—366, 1942.

ROEPKE, R. R., BALDES, E. J.: A critical study of the thermoelectric method of measuring vapor pressure. — J. biol. Chem. **126**: 349—360, 1938.

ROGERS, W. S. A.: A soil moisture meter. — J. agr. Sci. **25**: 326—343, 1935.

ROLSTON, D. E., HORTON, M. L.: Two beta sources compared for evaluating water status of plants. — Agron. J. **60**: 333—336, 1968.

DE ROO, H. C.: Leaf water potentials of sorghum and corn, estimated with the pressure bomb. — Agron. J. **61**: 969—970, 1969.

DE ROO, H. C.: Leaf water potentials of tobacco, estimated with the pressure bomb. — Tobacco Sci. **14**: 105—106, 1970.

ROOK, D. A., HELLMERS, H., HESKETH, J. D.: Stomata and cuticular surfaces of *Pinus radiata* needles as seen with a scanning electron microscope. — J. Arizona Acad. Sci. **6**: 222—225, 1971.

ROSENE, H. F.: Distribution of the velocities of absorption of water in the onion root. — Plant Physiol. **12**: 1—19, 1937.

ROSENE, H. F.: Comparison of rates of water intake in continuous regions of intact and isolated roots. — Plant Physiol. **16**: 19—38, 1941.

ROSENE, H. F.: Water absorption of root hairs. — Plant Physiol. **18**: 588—607, 1943.

ROTH, R.: Ein Vergleich berechneter und gemessener Tauwerte. — Meteorol. Rundschau **13**: 162—165, 1960.

ROTH, R.: Konstruktive und thermodynamische Eigenschaften des Piche-Evaporimeters. — Arch. Meteorol. Geophys. Bioklimatol., Ser. B **11**: 108—125, 1961.

ROWSE, H. R., MONTEITH, J. L.: A fifty-channel digital recorder for thermocouple psychrometers. — J. sci. Instr. (J. Phys. Eng.) **2**: 397—400, 1969.

RUCH, F., BOSSHARD, U., SAURER, W.: Determination of distribution of water in wheat grains by interference microscopy. — Nature **197**: 1318—1319, 1963.

RÜSCH, J.: Das Verhältnis von Transpiration und Assimilation als physiologische Kenngrösse, untersucht an Pappelklonen. — Züchter **29**: 348—354, 1959.

RUSSELL, R. S., SANDERSON, J.: Nutrient uptake by different parts of the intact roots of plants. — J. exp. Bot. **18**: 491—508, 1967.

RUTTER, A. J., SANDS, K.: The relation of leaf water deficit to soil moisture tension in *Pinus sylvestris* L. I. The effect of soil moisture on diurnal changes in water balance. — New Phytol. **57**: 50—65, 1958.

RYCHNOVSKÁ, M.: Water relations of some steppe plants investigated by means of the reversibility of the water saturation deficit. — In: SLAVÍK, B. (ed.): Water Stress in Plants. Pp. 108—115. Dr W. Junk N. V. Publishers, The Hague 1965.

RYCHNOVSKÁ-SOUDKOVÁ, M.: Study of the reversibility of the water saturation deficit as one of the methods of causal phytogeography. — Biol. Plant. **5**: 173—178, 1963.

RYCHNOVSKÁ, M., KVĚT, J.: Contribution to the ecology of steppe vegetation of the Tihany peninsula. III. Estimation of drought resistance based on the saturation of water deficit. — Ann. biol. (Tihany) **32**: 289—296, 1965.

RYCZKOWSKI, M.: Observations on the osmotic value of the central vacuole sap in *Haemanthus Katharinae* BAK. ovule. — Bull. Acad. pol. Sci. Cl. II. **8**: 145—180, 1960.

RYHINER, A. H., PANKOW, J.: Soil moisture measurement by the gamma transmission method. — J. Hydrol. **9**: 194—205, 1969.

SABININ, D. A.: On the root systems as an osmotic apparatus. — Bull. Inst. Rech. Biol. Univ. Perm. **4** (Suppl.): 1—136, 1925.

SADDLER, H. D. W., PITMAN, M. G.: An apparatus for the measurement of sap flow in unexcised leafy shoots. — J. exp. Bot. **21**: 1048—1059, 1970.

SAMPSON, J.: A method of replicating dry or moist surfaces for examination by light microscopy. — Nature **191**: 932—933, 1961.

SAMUILOV, F. D., NIKIFOROV, E. A., NIKIFOROVA, V. I.: Issledovanie sostoyaniya vody v tkanyakh rastenii metodom yadernogo spinovogo ekha. [The water status in plant tissues investigated by the method of nuclear-spin-echo]. — Dokl. Akad. Nauk SSSR **196**: 723 to 726, 1971.

SANCHEZ-DIAZ, M. F., HESKETH, J. D., KRAMER, P. J.: Wax filaments on sorghum leaves as seen with a scanning electron microscope. — J. Arizona Acad. Sci. **7**: 6—7, 1972.

SÁNCHEZ-DÍAZ, M. F.,KRAMER, P. J.: Turgor differences and water stress in maize and sorghum leaves during drought and recovery. — J. exp. Bot. **24**: 511—515, 1973.

SANDSTRÖM, B.: Ion adsorption in roots lacking epidermis. — Physiol. Plant. **3**: 469—505, 1950.

SAUSSURE, I.: Recherches chimiques sur la végetation. — Paris 1804.

SCARTH, G. W., LOEWY, A., SHAW, M.: Use of the infrared total absorption method for estimating the time course of photosynthesis and transpiration. — Can. J. Res. C **26**: 94—107, 1948.

SCHEUMANN, W.: Spezialpressatz für kleinste Probemengen. — Landwirtsch. Versuchs-Untersuchungswesen **10**: 169—171, 1964.

SCHLÄFLI, A.: Über die Eignung der Refraktometer- und der Schardakowmethode zur Messung osmotischer Zustandsgrössen. — Protoplasma **58**: 75—95, 1964.

SCHLÄFLI, A.: Untersuchungen über den Refraktometerwert und das spezifische Gewicht des Zellsaftes. — Mitt. der Thurgauischen Naturforsch. Ges.-H. **39**: 39—47, 1966.

SCHLOESING, I.: Sur l'analyse des principes solubles de la terre végétable. — Compt. rend. Acad. Sci. Paris **63**, 1866.

SCHOFIELD, R. K.: The pF of the water in soil. — Trans. 3rd int. Congr. Soil Sci. **2**: 37—48, 1935.

SCHOLANDER, P. F., BRADSTREET, E. D., HAMMEL, H. T., HEMMINGSEN, E. A.: Sap concentration in halophytes and other plants. — Plant Physiol. **41**: 529—532, 1966.

SCHOLANDER, P. F., HAMMEL, H. T., BRADSTREET, E. D., HEMMINGSEN, E. A.: Sap pressure in vascular plants. — Science **148**: 339—346, 1965.

SCHOLANDER, P. F., HAMMEL, H. T., HEMINGSEN, E. A., BRADSTREET, E. D.: Hydrostatic pressure and osmotic potential in leaves of mangroves and some other plants. — Proc. nat. Acad. Sci. USA **52**: 119—125, 1964.

SCHÖNHERR, J., BUKOVAC, M. J.: Penetration of stomata by liquids. Dependence on surface tension, wetability, and stomatal morphology. — Plant Physiol. **49**: 813—819, 1972.

SCHORN, M.: Untersuchungen über die Verwendbarkeit der Alkoholfixierungs- und der Infiltrationsmethode zur Messung von Spaltöffnungsweiten. — Jahrb. wiss. Bot. **71**: 783—840, 1929.

SCHRETZENMAYR, M.: Wasserhaushaltuntersuchungen mit der Schardakow-Methode. — Naturwissenschaften **53**: 160—161, 1966.

SCHRETZENMAYR, M.: Untersuchungen zum Messzeitpunkt bei der Schardakow-Methode. — Naturwissenschaften **54**: 94, 1967.

SCHRÖDER, D.: Über den Verlauf des Welkens und die Lebensfähigkeit der Laubblätter. — Diss. Göttingen: 1—110, 1909.

SCHRÖDER, G., Über die Austrocknungsfähigkeit der Pflanzen. — Unters. bot. Inst. Tübingen **2**: 1—52, 1886.

SCOTT, D.: An instrument measuring dew deposition. — Ecology **43**: 341—342, 1962.

SEDYKH, N. V., ISHMUKHAMETOVA, N. N.: Nekotorye novye aspekty primeneniya dielektricheskoï spektroskopii dlya issledovaniya voprosov vodnogo rezhima rastenii. [Some new aspects of the use of dielectric spectroscopy for studying plant water relations.] — Fiziol. Rast. **17**: 945—949, 1970.

ŠESTÁK, Z., ČATSKÝ, J., JARVIS, P. G. (eds.): Plant Photosynthetic Production. Manual of Methods. — Dr W. Junk N. V. Publishers, The Hague 1971.

ŠETLÍK, I., ŠESTÁK, Z.: Use of leaf tissue samples in ventilated chambers for long-term measurements of photosynthesis. — In: ŠESTÁK, Z. ČATSKÝ, J., JARVIS, P. G. (eds.): Plant Photosynthetic Production. Manual of Methods. Pp. 316—342. — Dr W. Junk N. V. Publishers, The Hague 1971.

SEYBOLD, A.: Die pflanzliche Transpiration. I. — Erg. Biol. **5**: 29—165, 1929.

SEYBOLD, A.: Die pflanzliche Transpiration. II. — Erg. Biol. 6: 559—674, 1930.

SHARDAKOV, V. S.: Opredeleniye sosushcheï sily rastitelnykh tkaneï metodom struek. [Determination of suction force in plant tissues by "smear" method]. — Izv. Akad. Nauk SSSR, Otd. Mat. est. Nauk 1938: 1279—1310, 1938.

SHARDAKOV, V. S.: Novyï polevoï metod opredeleniya sosushcheï sily rasteniï. [New field method of suction force determination]. — Dokl. Akad. Nauk SSSR 60: 160—172, 1948.

SHAW, B. T., BAVER, L. D.: An electrothermal method for following moisture changes of the soil in situ. — Proc. Soil Sci. Soc. Amer. 4: 78—83, 1939.

SHEPHERD, W.: Diffusion pressure deficit and turgidity relationships of detached leaves of Trifolium repens L. — Aust. J. agr. Res. 15: 746—751, 1964.

SHERIFF, D. W.: A new apparatus for the measurement of sap flux in small shoots with the magnetohydrodynamic technique. — J. exp. Bot. 23: 1086—1095, 1972.

SHIMSHI, D.: Some aspects of stomatal behaviour, as observed by means of an improved pressure-drop porometer. — Israel J. Bot. 16: 19—28, 1967.

SHIMSHI, D., LIVNE, A.: The estimation of the osmotic potential of plant sap by refractometry and conductometry: A field method. — Ann. Bot. (N. S.) 31: 505—511, 1967.

SHMUELI, E.: Irrigation studies in the Jordan Valley. I. Physiological activity of the banana in relation to soil moisture. — Bull. Res. Counc. Israel 3: 228—247, 1953.

SHMUELI, E., COHEN, O. P.: A critique of Walter's hydrature concept and of his evaluation of water status measurement. — Israel J. Bot. 13: 199—207, 1964.

SHMUELI, E., COHEN, O. P.: In support of the critique on the hydrature concept. — Israel J. Bot. 16: 45—47, 1957.

SHMUK, A, A.: Pochvennyï rastvor. Emulsionnyï method vydeleniya. [Soil solution. Emulsion method of extraction]. — Zh. opyt. Agron. 22, 1921—1923.

SHREVE, E.: Factors governing seasonal changes in transpiration of Encelia farinosa. — Bot. Gaz. 77: 432—439, 1924.

SHULL, C. A.: Absorption of water and the forces involved. — J. Soc. Agron. 22: 459—471, 1930.

SIBIRSKY, W.: Die Bestimmung der Bodenfeuchtigkeit nach der Carbidmethode. — Trans. 3rd Int. Congr. Soil Sci. 1: 10—13, 1935.

SIERP, H., BREWIG, A.: Quantitative Untersuchungen über die Wasserabsorptionszone der Wurzeln. — Jahrb. wiss. Bot. 82, 99—122, 1936.

SIRI, W. E.: Isotopic Tracers and Nuclear Radiations. — McGraw-Hill, New York 1949.

SÍTAŘ, J.: Studium vodního hospodářství rostlin metodou kořenového mostu. [Studies of water relations of plants by root-bridge method]. — Bull. výzk. Úst. zelinář. ČSAZV (Olomouc) 1: 17—27, 1957.

SIVADJIAN, J.: Hygrophotographic method for depicting soil moisture. — Soil Sci. 83: 109 to 112, 1957.

SIVADJIAN, J.: Hygrophotographic measurement of soil moisture following a protracted drought. — Soil Sci. 90: 369—379, 1960.

SKEIB, B.: Ein Messverfahren zur Bestimmung der Wärmekapazität des Erdbodens mittels konstant beheizter Messkörper. — Meteorol. Z. 4: 32—39, 1950.

SKIDMORE, E. L., STONE, J. F.: Physiological role in regulating transpiration rate of the cotton plant. — Agron. J. 56: 405—410, 1964.

SLATYER, R. O.: Absorption of water from atmosphere of different humidity and its transport through plants. — Aust. J. biol. Sci. 9: 552—558, 1956.

SLATYER, R. O.: The significance of the permanent wilting percentage in studies of plant and soil water relations. — Bot. Rev. 23: 586—636, 1957.

SLATYER, R. O.: The measurement of diffusion pressure deficit in plants by a method of vapour equilibration. — Aust. J. biol. Sci. 11: 349—365, 1958.

SLATYER, R. O.: Absorption of water by plants. — Bot. Rev. **26**: 331—392, 1960.

SLATYER, R. O.: An underlying cause of measurement discrepancies in determinations of osmotic characteristics in plant cells and tissues. — Protoplasma **62**: 34—43, 1966a.

SLATYER, R. O.: *In situ* measurements of stomatal resistance. — Proc. Conf. Instrumentation for Plant Environment Measurements. Soc. Inst. Technol., Aust. pp. 5—6, 1966b.

SLATYER, R. O.: Plant-Water Relationships. — Academic Press, London—New York 1967a.

SLATYER, R. O.: Terminology for cell and tissue water relations. — Z. Pflanzenphysiol. **56**: 91—94 and 469—470, 1967b.

SLATYER, R. O.: Effect of errors in measuring leaf temperature and ambient gas concentration on calculated resistances to CO_2 and water vapor exchanges in plant leaves. — Plant Physiol. **47**: 269—274, 1971.

SLATYER, R. O., BARRS, H. D.: Modifications to the relative turgidity technique with notes on its significance as an index of the internal water status of leaves. — In: ECKARDT, F. E. (ed.): Methodology of Plant. Eco-physiology. (Arid Zone Res. 25). Pp. 331—342, UNESCO Paris 1965.

SLATYER, R. O., BIERHUIZEN, J. F.: A differential psychrometer for continuous measurements of transpiration. — Plant Physiol. **39**: 1051—1956, 1964.

SLATYER, R. O., GARDNER, W. R.: Overall aspects of water movement in plants and soils. — In: The State and Movement of Water in Living Organisms. (XIXth Symp. Soc. exp. Biol.). Pp. 113—129. Cambridge Univ. Press, 1965.

SLATYER, R. O., JARVIS, P. G.: Gaseous-diffusion porometer for continuous measurement of diffusive resistance of leaves. — Science **151**: 574—576, 1966.

SLATYER, R. O., McILROY, I. C.: Practical Microclimatology. With Special Reference to the Water Factor in Soil-Plant-Atmosphere Relationships. — UNESCO CSIRO, Australia 1961.

SLATYER, R. O., TAYLOR, S. A.: Terminology in plant- and soil-water relations. — Nature **187**: 922—924, 1960.

SLAVÍK, B.: Osmotické hodnoty dřevin jako indikátor vhodnosti pro stanoviště. — [Osmotic values of trees as an indicator for the habitat]. — Českoslov. Biol. **1**: 125—133, 1952.

SLAVÍK, B.: Rostliny a stanovištní vlhkost. [Plants and water factor of the habitat]. — In: KLIKA, J., NOVÁK, V., GREGOR, A. (eds.): Praktikum fytocenologie, ekologie, půdoznalství a klimatologie. [Manual of Methods of Phytocoenology, Ecology, Soil Science and Climatology]. Pp. 277—326. Nakl. ČSAV, Praha 1954.

SLAVÍK, B.: K dynamice vodního deficitu rostlin. [Dynamics of water deficit in plants]. — Preslia **27**: 124—133, 1955.

SLAVÍK, B.: The influence of water deficit on transpiration. — Physiol. Plant. **11**: 524—536, 1958a.

SLAVÍK, B.: Grafické stanovení intensity průduchové a kutikulární složky transpirace rostlin. [Graphic determination of stomatal and cuticular components of transpiration rate]. — Českoslov. Biol. **7**: 347—352, 1958b.

SLAVÍK, B.: The relation of the refractive index of plant cell sap to its osmotic pressure. — Biol. Plant. **1**. 48—53, 1959.

SLAVÍK, B.: Ecophysiological significance of the height gradient of the equivalent evaporation power of the atmosphere. — Biol. Plant. **2**: 313—324, 1960.

SLAVÍK, B.: Relationship between the osmotic potential of cell sap and the saturation water deficit during wilting of leaf tissue. — Biol. Plant. **5**: 258—264, 1963a.

SLAVÍK, B.: The distribution pattern of transpiration rate, water saturation deficit, stomata number and size, photosynthetic and respiration rate in the area of tobacco leaf blade. — Biol. Plant **5**: 143—153, 1963b.

SLAVÍK, B.: Determination of stomatal aperture. — In: ŠESTÁK, Z., ČATSKÝ, J., JARVIS, P. G. (eds.): Plant Photosynthetic Production. Manual of Methods. Pp. 556—565. — Dr W. Junk N. V. Publishers, The Hague 1971.

SLAVÍK, B.: Transpiration resistances in leaves of maize grown in humid and dry air. — In: SLATYER, R. O. (ed.): Plant Response to Climatic Factors. Pp. 267—269. UNESCO, Paris 1973.

SLAVÍK, B., ČATSKÝ, J.: Differentially measuring infrared analyser with an air-conditioned exposure chamber for photosynthetic rate measurements. — Biol. Plant. **5**: 135—142, 1963.

SLAVÍK B., ČATSKÝ, J.: Colorimetric determination of CO_2 exchange in field and laboratory. — In: ECKARDT, F. E. (ed.): Methodology of Plant Eco-Physiology. (Arid Zone Res. 25.) Pp. 291—298. UNESCO, Paris 1965.

SLAVÍKOVÁ, J.: Eine ökologische Methode zur Wurzelsaugkraftmessung. — Preslia **35**: 241—242, 1963a.

SLAVÍKOVÁ, J.: A critical evaluation of the determination of the root suction force. — Acta Univ. carol. Biol. **1963**: 26—41, 1963b.

SLAVÍKOVÁ, J.: Horizontaler Gradient der Saugkraft eines Wurzelastes und sein Zusammenhang mit dem Wassertransport in Wurzeln. — Acta Horti bot. Univ. carol. **1963**: 73—79, 1963c.

SLAVÍKOVÁ, J.: Compensation of root suction force within a single root system. — Biol. Plant. **9**: 20—27, 1967.

SMITH, R. C.: Time course of exudation from excised corn root segments of different stages of development. — Plant Physiol. **45**: 571—575, 1970.

SOLÁROVÁ, J.: Stomata reactivity in leaves at different insertion level during wilting. — In: SLAVÍK, B. (ed.): Water Stress in Plants. Pp. 147—154. Dr W. Junk N. V. Publishers, The Hague 1965.

SOLÁROVÁ, J.: Unzuverlässlichkeit des Abdruckverfahrens bei der Bestimmung des Spaltöffnungszustandes. — Photosynthetica **2**: 178—183, 1968.

SPANNER, D. C.: The Peltier effect and its use in the measurement of suction pressure. — J. exp. Bot. **2**: 145—168, 1951.

SPANNER, D. C.: On a new method for measuring the stomatal aperture of leaves. — J. exp. Bot. **4**: 283—295, 1953.

SPANNER, D. C., HEATH, O. V. S.: Experimental studies of the relation between carbon assimilation and stomatal movement. II. The use of the resistance porometer in estimating stomatal aperture and diffusive resistance. Part II. Some sources of error in the use of the resistance porometer and some modifications of its design. — Ann. Bot. (N.S.) **15**: 321 to 331, 1951.

SPAUSZUS, S.: Eine Schnellbestimmung der Bodenfeuchtigkeit durch Messung der Dielektrizitätskonstanten. — Z. Pflanzenernähr. Düng. Bodenk. **70**: 23—26, 1955.

SPEIDEL, B.: Untersuchungen zur Physiologie des Blütens bei höheren Pflanzen. — Planta **30**: 67—112, 1939.

SPENCER-GREGORY, H., ROURKE, E.: Hygrometry. — Crosby Lockwood and Son, Ltd., London 1957.

SPOMER, G. G.: Preliminary studies of a method for monitoring moisture tensions in woody stems. — Plant Physiol. **39**: (Suppl.): XLII, 1964.

SPOMER, L. A.: Evaluation of edge injection errors in the floating leaf disk method of measuring leaf tissue water deficit. — Plant Physiol. **49**: 1027—1028, 1972a.

SPOMER, L. A.: A simple inexpensive water bath for precise temperature control. — Agron. J. **64**: 837—838, 1972b.

SPOMER, L. A., LANGHANS, R. W.: Evaluation of pressure bomb and dye method measurements of tissue water potential in greenhouse *Chrysanthemum*. — Hort. Sci. **7**: 412—414, 1972.

SPURNÝ, M.: Ssavé napětí tkáně *Opuntia phaeacanta*, studováno metodou auxografickou, objemoměrnou a váhovou. [Suction tension of *Opuntia phaeacanta* tissue as studied by auxographic, volumetric and gravimetric methods.] Spisy přírodov. Fak. Masaryk. Univ., (Brno) Ser. K5 (9—10): 269—306, 1951.

SREENIVASAYA, M., SRIMATHI, R. A.: Micropress for obtaining press sap from plant tissues. — Indian J. exp. Bot. **5**: 127—128, 1967.

STADELMANN, E.: Mathematische Analyse experimenteller Ergebnisse: Gewinnung der Permeabilitätskonstanten, Stoffaufnahme- und -abgabewerte. — In: RUHLAND, W. (ed.): Handbuch der Pflanzenphysiologie. Vol. 2. Pp. 139—194. Springer Verlag, Berlin—Göttingen—Heidelberg 1956.

STADELMANN, E.: Vergleich und Umrechnung von Permeabilitätskonstanten für Wasser. — Protoplasma **57**: 660—718, 1963.

STAHL, E.: Einige Versuche über Transpiration und Assimilation. — Bot. Zeit. **52**: 117—146, 1894.

STÅLFELT, M. G.: Neuere Methoden zur Ermittlung des Öffnungszustandes der Stomata. — In: ABDERHALDEN, E. (ed.): Handbuch der biologischen Arbeitsmethoden. Vol. XI/4/1. Pp. 167—192. Urban & Schwarzenberg, Berlin—Wien 1939.

STÅLFELT, M. G.: The stomata as a hydrophotic regulator of the water deficit of the plant. — Physiol. Plant. **8**: 572—593, 1955.

STÅLFELT, M. G.: The effect of the water deficit on the stomatal movements in a carbon dioxide-free atmosphere. — Physiol. Plant. **14**: 826—843, 1961.

STANHILL, G.: The effect of differences in soil moisture status on plant growth. A rewiev and analysis of soil moisture regime experiments. — Soil Sci. **84**: 205—215, 1957.

STAPLE, W. J., LEHANE, J. J.: Variability in soil moisture sampling. — Can. J. Soil Sci. **42**: 156—164, 1962.

STEFANOFF, B.: Studien über den Zustand und die Schwankungen des Wassergehaltes in den Blättern und Zweigen einiger Holzpflanzen. — Forstwiss. Centralbl. **75**: 784—799, 1931.

STEIGER, W. R.: A new development in the measurement of high relative humidities. — Science **114**: 152—153, 1951.

STEIN, E.: Bemerkungen zu der Arbeit von Molisch: „Das Offen- und Geschlossensein der Spaltöffnungen veranschaulicht durch eine neue Methode." — Ber. deut. bot. Ges. **30**: 66—68, 1912.

STEPANOV, L. N.: Metod i ustroĭstvo dlya opredeleniya turgornogo davleniya. [Method and device for determining turgor pressure]. — Vest. sel'skokhoz. Nauki **14**(3): 114—116, 1969.

STEUBING, L.: Beiträge zur Tauwasseraufnahme höherer Pflanzen. — Biol. Zentralbl. **68**: 253—259, 1949.

STEUBING, L.: Studien über den Taufall als Vegetationsfaktor. — Ber. deut. bot. Ges. **68**: 55—70, 1955.

STEWARD, G. L., TAYLOR, S. A.: Field experience with the neutron scattering method of measuring soil moisture. — Soil Sci. **83**: 151—158, 1957.

STICE, N. W., BOOHER, L. J.: Plastic tube irrigators with electric control. — Calif. Agr. **19**: 4—5, 1965.

STIGTER, C. J.: Leaf diffusion resistance to water vapour and its direct measurement. I. Introduction and review concerning relevant factors and methods. — Meded. Landbouwhogeschool (Wageningen) **72**(3): 1—47, 1972.

STILES, W.: The suction pressure of the plant cell. — Biochem. J. **16**: 727—728, 1922.

STILES, W.: Studies on toxic action. I. Preliminary observations on the swelling and shrinkage of plant tissues in solutions of toxic substances. — New Phytol. **17**: 349, 1930.

STILES, W.: A diffusive resistance porometer for field use. I. Construction. — J. appl. Ecol. **7**: 617—622, 1970.

STOCKER, O.: Das Wasserhaushalt ägyptischer Wüsten- und Salzpflanzen. — Bot. Abhandl. (Jena) **13**: 1—200, 1928.

STOCKER, O.: Vizsgálatok különbözö termöhelyen nött növények vízhiányának nagyságáról. [Über die Höhe des Wasserdefizites bei Pflanzen verschiedener Standorte.] — Erdészeti Kisérletek (Sopron) **31**: 63—76, 104—114, 1929a.

STOCKER, O.: Eine Feldmethode zur Bestimmung der momentanen Transpirations- und Evaporationsgrösse. — Ber deut. bot. Ges. **47**: 126—136, 1929b.

STOCKER, O.: Das Wasserdefizit von Gefässpflanzen in verschiedener Klimazonen. — Planta **7**: 382—387, 1929c.

STOCKER, O.: Über die Messung der Bodensaugkräften und ihren Verhältnis zu den Wurzelsaugkräften. — Z. Bot. **23**: 27—56, 1930.

STOCKER, O.: Messmethoden der Transpiration. — In: RUHLAND, W. (ed.): Handbuch der Pflanzenphysiologie. Vol. 3. Pp. 293—311. Springer Verlag, Berlin—Göttingen—Heidelberg 1956a.

STOCKER, O.: Die Abhängigkeit der Transpiration von den Umweltfaktoren. — In: RUHLAND, W. (ed.): Handbuch der Pflanzenphysiologie. Vol. 3. Pp. 436—488. Springer Verlag, Berlin—Göttingen—Heidelberg 1956b.

STOCKER, O.: Physiological and morphological changes in plants due to water deficiency. — In: Plant Water Relationships in Arid and Semi-arid Conditions. Reviews of Research. (Arid Zone Res. 15). Pp. 63—104, UNESCO, Paris 1960.

STOCKER, O., KAUSCH, W.: Bodenfeuchte und Tensionmetermessung. — Ber. deut. Wetterdienstes **32**: 16—18, 1952.

STOCKING, C. R.: The calculation of tension in *Cucurbita pepo*. — Amer. J. Bot. **32**: 126—134, 1945.

STOCKING, C. R.: Root pressure. — In: RUHLAND, W. (ed.): Handbuch der Pflanzenphysiologie. Vol. 3. Pp. 583—586. Springer Verlag, Berlin—Göttingen—Heidelberg 1956.

STOECKELER, J. H., AAMODT, E. E.: Use of tensiometers in regulating watering in forest nurseries. — Plant Physiol. **15**: 589—608, 1940.

STOKER, R., WEATHERLEY, P. E.: The influence of the root system on the relationship between the rate of transpiration and depression of leaf water potential. — New Phytol. **70**: 547 to 554, 1971.

STOLZY, L. H., CAHOON, G. A.: A field calibrated portable neutron rate meter for measuring soil moisture in *Citrus* orchards. — Proc. Soil Sci. Soc. Amer. **19**: 419—423, 1955.

STONE, E. C.: Dew as an ecological factor. I. A review of the literature. — Ecology **38**: 407 to 413, 1957a.

STONE, E. C.: Dew as an ecological factor. II. The effect of artificial dew on the survival of *Pinus ponderosa* and associated species. — Ecology **38**: 414—422, 1957b.

STONE, E. C.: The role of dew in pine survival in soils below the wilting point and its measurement. — In: Eckardt, F. E. (ed.): Methodology of Plant Eco-Physiology. (Arid Zone Res. 25). Pp. 421—427. UNESCO, Paris 1965.

STONE, E. C., FOWELLS, H. A.: Survival value of dew under laboratory conditions with *Pinus ponderosa*. — Forest Sci. **1**: 183—188, 1955.

STONE, E. C., WENT, F. W., YOUNG, C. L.: Water absorption from the atmosphere by plants growing in dry soil. — Science **111**: 546—548, 1950.

STONE, J. F., KIRKHAM, D., READ, A. A.: Soil moisture determination by a portable neutron scattering moisture meter. — Proc. Soil Sci. Soc. Amer. **19**: 419—423, 1955.

STREBEYKO, P.: Air passage capacity in leaves. — Acta Soc. Bot. Pol. **34**: 191—199, 1965a.

STREBEYKO, P.: The theory of porometer. — Physiol. Plant. **18**: 725—729, 1965b.

STREBEYKO, P., KARWOWSKA, Z.: Zawartość wody w liściach roślin uprawnych (Badania metodyczne). [Water content of the leaves of crop plants (Methodical studies)]. — Rocz. Nauk roln. **78***A*: 565—578, 1958.

STROGONOV, B. P., LAPINA, L. P.: O vozmozhnom sposobe razdelnogo izucheniya toksicheskogo i osmoticheskogo deĭstviya soleĭ na rasteniya. [A possible method for studying separately the toxic and osmotic action of salts on plants]. — Fiziol. Rast. **11**: 674—680, 1964.

STRUŽKA, V.: Meteorologické přístroje a měření v terénu. [Meteorological Instruments in Field Measurements]. — St. pedag. Nakl., Praha 1956.

STUART, D. M., HADDOCK, J. L.: Inhibition of water uptake in sugar beet roots by ammonia. — Plant Physiol. **43**: 345—350, 1968.

SUSPLUGAS, J., LALAURIE, M., PRIVAT, G.: Technique d'obtention des sucs végétaux. — In: ECKARDT, F. E. (ed.): Methodology of Plant Eco-physiology. (Arid Zone Res. 25). Pp. 393—397. UNESCO, Paris 1965.

SUZUKI, T., KANEKO, M., TORIKATA, H.: [Studies on the water balance in Satsuma orange trees. II. On the estimation methods of water saturation deficit (W.S.D.) of leaves, and the effects of soil moisture and fertilizer supply on the W.S.D. and the apparent photosynthetic rate of leaves]. In Jap. — J. jap. Soc. hort. Sci. **38**: 1—8, 1969.

SWANSON, R. H.: An instrument for detecting sap movement in woody plants. — Rocky Mountain Forest and Range Exp. Stat., Fort Collins, Colorado. Stat. Paper No. **68**; 1—16, 1962.

TAKAOKI, T.: Improvement of cobalt chloride method for determination of bound water in plants. — J. Sci. Hiroshima Univ., Ser. B. Div. 2 (Bot.) **9**: 199—207, 1962.

TAKECHI, O.: Determination of leaf water content by beta ray transmission.— J. agr. Meteorol. **25**: 229—232, 1970.

TAKECHI, O., FURUDOI, Y., ETOH, R.: [Water content and beta ray transmission of various leaves]. In Jap. — J. agr. Meteorol. **26**: 1—3, 1970.

TANNER, C. B., ABRAMS, E., ZUBRISKI, J. C.: Gypsum moisture block calibration based on electrical conductivity in distilled water. — Proc. Soil Sci. Soc. Amer. **13**: 62—65, 1949.

TANNER, C. B., HANKS, J. P.: Moisture hysteresis in gypsum moisture blocks. — Proc. Soil Sci. Soc. Amer. **16**: 48—51, 1952.

TANNER, C. B., SUOMI, V. E.: Lithium chloride dewcell properties and use for dew-point and vapour-pressure gradient measurements. — Trans. amer. geophys. Union **37**: 413—420, 1956.

TARŁOWSKI, J.: Nowa metoda jednoczesnego oznaczania transpiracji i absorpcji wody przez rośliny w kulturach wodnych. [New method for simultaneous determination of transpiration and water absorption of plants in water cultures]. — I. Acta Soc. Bot. Pol. **24**: 705—722 1955, II. Acta Soc. Bot. Pol. **25**: 401—408, 1956, III. Acta Soc. Bot. Pol. **30**: 73—88, 1961.

TAYLOR, S. A.: Field determination of soil moisture. — Agr. Engin. **36**: 657—659, 1955.

TAYLOR, S. A.: Water condition and flow in the soil-plant-atmosphere system. — Amer. Soc. Agron. Spec. Publ. **5**: 81, 1964.

TAYLOR, S. A.: Measuring soil-water potential. — In: ECKARDT, F. E. (ed.): Methodology of Plant Eco-Physiology. (Arid Zone Res. 25). Pp. 149—157. UNESCO, Paris 1965.

TAYLOR, S. A., EVANS, D. D., KEMPER, W. D.: Evaluating soil water. — Utah Agr. exp. Stat. Bull. **426**: 1—67, 1961.

TAYLOR, S. A., SLATYER, R. O.: Water-soil-plant relations terminology. — Proc. int. Congr. Soil Sci. **1**: 394—403, 1961.

TAYLOR, S. A., SLATYER, R. O.: Proposals for a unified terminology in studies of plant-soil-water relationships. — In: Plant Water Relationships in Arid and Semi-arid Conditions. Proc. Madrid Symp. (Arid Zone Res. 16). Pp. 339—349. UNESCO, Paris 1962.

TAZAWA, M.: Neue Methode zur Messung des osmotischen Wertes einer Zelle. — Protoplasma **48**: 342—359, 1957.

TAZAWA, M., KIYOSAWA, K.: Water movement in a plant cell on application of hydrostatic pressure. — Annu. Rep. Biol. Works, Sci. Fac. Univ. Osaka **18**: 57—70, 1970.

TEELE, R. P., SCHUHMAN, S.: A potentiometer for measuring voltages of 10 microvolts to an accuracy of 0.01 microvolt. — J. Res. nat. Bur. Stand. A **22**: 431, 1939.

TEPE, W., LEIDENFROST, E.: Ein Messverfahren zur Bestimmung der Wasserversorgung der Pflanzen. — Z. Acker- Pflanzenbau (I) **115**: 223—230, 1962, (II) **117**: 395—402, 1963.

Thimann, K. V., Loos, G. M., Samuel, E. W.: Penetration of mannitol in higher plants. — Plant Physiol. **38** (suppl.): xxxiii, 1963.

Thoday, D.: On turgescence and the absorption of water by the cells of plants. — New Phytol. **17**: 108—113, 1918.

Thomas, J. R., Namken, L. N., Oerther, G. F., Brown, R. G.: Estimating leaf water content by reflectance measurements. — Agron. J. **63**: 845—847, 1971.

Thomas, M. D., Hill, G. R.: The continuous measurement of photosynthesis, respiration and transpiration of alfalfa and wheat growing under field conditions. — Plant Physiol. **12**: 285—307, 1937.

Thöni, H.: Eine modifizierte Kapillarmethode zur Messung osmotischer Werte von Pflanzensäften und Lösungen. — Experientia **21**: 112, 1965.

Thren, R.: Jahreszeitliche Schwankungen des osmotischen Wertes verschiedener ökologischer Typen in der Umgebung von Heidelberg. — Z. Bot. **26**: 449—526, 1934.

Tinklin, R.: Note on the determination of leaf water-potential. — New Phytol. **66**: 85—88, 1967.

Tobiessen, P.: An inexpensive pressure chamber for xylem water tension measurements. — Ecology **50**: 726—728, 1969.

Toman, M., Synak, J.: Gutace jako indikátor fytotoxicity. [Guttation as indicator of phytotoxicity]. — Biológia (Bratislava) **9**: 99—107, 1954.

Tóth, L., Felföldy, L.: Ozmotikus érték meghatározása új plazmometrikus módszerrel. [Determination of osmotic value by a new plasmometric method]. — Ann. Inst. Biol. (Tihany) Hung. Acad. Sci. **23**: 199—207, 1955.

Tranquillini, W.: Blattemperatur, Evaporation und Photosynthese bei verschiedener Durchströmung der Assimilationsküvette. Mit einem Beitrag zur Kenntnis der Verdunstung in 2000 m Seehöhe. — Ber. deut. bot. Ges. **77**: 204—218, 1964.

Trip, P., Krotkov, G., Nelson, C. D.: Metabolism of mannitol in higher plants. — Amer. J. Bot. **51**: 828—835, 1964.

Troughton, J., Donaldson, L. A.: Probing Plant Structure. A scanning microscope study of some anatomical features in plants and the relationship of these structures to physiological processes. — Chapman & Hall, London 1972.

Trusell, F., Diehl, H.: Efficiency of chemical desiccants. — Anal. Chem. **35**: 674—677, 1963.

Tsel'niker, Yu. L.: [The velocity of water loss from isolated tree leaves and their resistance to dehydration.] In Russ. — Trudy Inst. Lesa Akad. Nauk SSSR **27**: 6—28, 1955.

Tumanow, J. J.: Welken und Dürreresistenz. — Wiss. Arch. Landwirtsch., Abt. *A* **3**: 389—419, 1930.

Turner, N. C., Parlange, J. Y.: Analysis of operation and calibration of a ventilated diffusion porometer. — Plant Physiol. **46**: 175—177, 1970.

Turner, N. C., Pedersen, F. C. C., Wright, W. H.: An aspirated diffusion porometer for field use. — Spec. Bull. Soils Conn. agr. Exp. Sta., XXIX, 1—7, 1969.

Tyree, M. T.: An alternative explanation for the apparently active water exudation in excised roots. — J. exp. Bot. **24**: 33—37, 1973.

Tyree, M. T., Benis, M., Dainty, J.: The water relations of hemlock (*Tsuga canadensis*). III. The temperature dependence of water exchange in a pressure bomb. — Can. J. Bot. **51**: 1537—1543, 1973a.

Tyree, M. T., Dainty, J.: The water relations of hemlock (*Tsuga canadensis*). II. The kinetics of water exchange between the symplast and apoplast. — Can. J. Bot. **51**: 1481—1489, 1973.

Tyree, M. T., Dainty, J., Benis, M.: The water relations of hemlock (*Tsuga canadensis*). I. Some equilibrium water relations as measured by the pressure-bomb technique. — Can. J. Bot. **51**: 1471—1480, 1973b.

TYREE, M. T., HAMMEL, H. T.: The measurement of the turgor pressure and the water relations of plants by the pressure-bomb technique. — J. exp. Bot. **23**: 267—282, 1972.

TYREE, M. T., ZIMMERMANN, M. H.: The theory and practice of measuring transport coefficients and sap flow in the xylem of red maple stems (*Acer rubrum*). — J. exp. Bot. **22**: 1—18, 1971.

TYURINA, M. M.: Kompleksnaya kharakteristika vodouderzhivayushcheĭ sposobnosti rastitelnykh tkaneĭ [Complex charakteristics of the water holding capacity of plant tissues] — Bot. Zh. **57**: 509—521, 1972.

UHLIG, S.: Bestimmung des Verdunstungsanspruches der Luft mit Piche-Evaporimetern. — Mitt. deut. Wetterdienstes **2**: 1—24, 1955.

UHLÍŘ, P.: Methody měření rosy. [Methods of measuring of dew]. — Meteorol. Zprávy (Praha) **2**: 44—49, 1948.

UHLÍŘ, P.: Einige für die agrarmeteorologische Forschung entwickelte Messgeräte. — Int. J. Bioclim. Biometeorol. **1**, 1957.

UHLÍŘ, P., STRUŽKA, V.: Měření doby ovlhnutí listů rostlin. [Measurement of the wetting period of plants]. — Příroda (Praha) **3**: 47—49, 1948.

UHLÍŘ, P., UHLÍŘ, J.: Rosa a doba orosení. [Dew and wetting period]. — Sborník Vys. školy zem. (Praha) **1959**: 175—185, 1959.

ÚLEHLA, J.: The effect of decreasing soil moisture on the water relations of sunflower seedlings. — Biol. Plant. **3**: 312—319, 1961.

ÚLEHLA, J.: Kořenové krvácení bobu a vojtěšky v polních podmínkách. [Root bleading of bean and alfalfa in field conditions]. — Vědecké Práce VÚRV (Praha) **1963**: 37—49, 1963a.

ÚLEHLA, J.: Changes in sap exudation of maize and occurence of lags in exudation during the growing season. — Biol. Plant. **5**: 190—197, 1963b.

ÚLEHLA, J., ZICHOVÁ, L., BAŇOCH, Z.: Influence of soil heterogeneity on water regime and yields of lucerne, sugar beet and wheat. — Biol. Plant. **7**: 449—462, 1965.

ÚLEHLA, V.: Die Quellungsgeschwindigkeit der Zellkolloide als gemeinschaftlicher Faktor in Plasmolyse, Plasmoptyse und ähnlichen Veränderung des Zellvolumens. — Planta **2**: 618—639, 1926.

ÚLEHLA, V.: Die Regulation der Wasserstoffionen-Konzentration durch Sukkulente-Gewebe. — Protoplasma **3**: 469—506, 1928.

UNDERWOOD, D. N., VAN BAVEL, C. H. M., SWANSON, R. W. A.: A portable slow neutron flux meter for measuring soil moisture. — Soil Sci. **77**: 339—340, 1954.

UNGER, K.: Angew. Meteorol. **1**: 280—283, 1953.

UNGER, K.: Zur Anwendung der Durchstrahlungsmethode für phänometrische Messungen mit Hilfe von radioaktiven Strahlungsquellen. — Z. angew. Meteorol. **3**: 115—118, 1958.

UNGER, K.: Die Anwendung radioaktiver Strahlungsquellen zur berührungslosen Massenbestimmung von Pflanzenbeständen und Einzelpflanzen am natürlichen Standort. — Züchter **29**: 289—293, 1959.

UNGER, K.: Experimentelle Untersuchungen mit Hilfe von ionisierenden Strahlen und Neutronen zur Bestimmung der witterungsbedingten Wachstumsintensität von Kulturpflanzen am natürlichen Standort. — Akademie Verlag. Ges. Geest & Portig, Leipzig 1963.

UNGER, K.: Methodische Betrachtungen zur modellmässigen Darstellung des Wasserhaushaltes von Kulturpflanzen. — In: UNGER, K., CLAUS, S., GRAU, I.: Modell zum Wasserhaushalt von Kulturpflanzen. — Studia biophys. **11**: 79—122, 1968.

URSPRUNG, A.: Zur Kenntnis der Saugkraft. VII. Eine neue vereinfachte Methode zur Messung der Saugkraft. — Ber. deut. bot. Ges. **41**: 338, 1923.

URSPRUNG, A.: Über das Eindringen von Wasser und anderer Flüssigkeit in Intrazellularen. — Beih. bot. Centralbl. **41**: 15—40, 1925.

URSPRUNG, A.: Die Messung der osmotischen Zustandgrössen pflanzlicher Zellen und Gewebe. — In: ABDERHALDEN, E. (ed.): Handbuch der biologischen Arbeitsmethoden. Vol. XI/4/2. Pp. 1109—1572. Urban & Schwarzenberg, Berlin—Wien 1939.

URSPRUNG, A., BLUM, G.: Zur Methode der Saugkraftmessung. — Ber. deut. bot. Ges. **34**: 525—549, 1916.

URSPRUNG, A., BLUM, G.: Eine Methode zur Messung des Wand- und Turgordruckes der Zelle, nebst Anwendungen. — Jahrb. wiss. Bot. **63**: 1—110, 1924.

URSPRUNG, A., BLUM, G.: Eine neue Methode zur Messung der Saugkraft von Hartlaub. — Jahrb. wiss. Bot. **67**: 334—348, 1927.

URSPRUNG, A., BLUM, G.: Zwei neue Saugkraft-Messmethoden. I. Die Kapillarmethode zur Messung der statischen Saugkraft der Flüssigkeiten, Quellkörper und Böden. II. Die Hebelmethode zur Messung der Saugkraft von Hartlaub und anderen schwierigen Objekten. — Jahrb. wiss. Bot. **72**: 254, 1930.

VAADIA, Y.: Autonomic diurnal fluctuations in rate of exudation and root pressure of decapitated sunflower plants. — Physiol. Plant. **13**: 701—717, 1960.

VAADIA, Y., RANEY, F. C., HAGAN, R. M.: Plant water deficits and physiological processes. — Annu. Rev. Plant Physiol. **12**: 265—292, 1961.

VAADIA, Y., WAISEL, Y.: Water absorption by the aerial organs of plants. — Physiol. Plant. **16**: 44—51, 1963.

VÁCLAVÍK, J.: Vliv vnějších faktorů na stav průduchů. [Influence of external factors on stomata aperture]. — Thesis. Biol. Fac. Charles Univ., Praha 1955.

VÁCLAVÍK, J.: The maintaining of constant soil moisture levels (lower than maximum capillary capacity) in pot experiments. — Biol. Plant. **8**: 80—85, 1966.

VÁŠA, J.: Stanovení půdních hydrokonstant. [Determination of soil hydroconstants]. — Vodní Hospodářství (Praha) **1959**: 3, 1959.

VASSILJEV, I. M.: Über den Wasserhaushalt von Pflanzen der Sandwüste im südöstlichen Kara-Kum. — Planta **14**: 225—309, 1931.

VEIHMEYER, F. J.: Soil moisture. — In: RUHLAND, W. (ed.): Handbuch der Pflanzenphysiologie Vol. 3. Pp. 64—123, Springer Verlag, Berlin—Göttingen—Heidelberg 1956.

VEIHMEYER, F. J., EDLEFSEN. N. E., HENDRICKSON, A. H.: Use of tensiometers in measuring availability of water to plants. — Plant Physiol. **18**: 16—78, 1943.

VERETENIKOV, A. V.: Primenenie potometra dlya opredeleniya aktivnoĭ poverkhnosti kornevykh okonchaniĭ sosny. [Application of a potometer for the determination of the active surface of root tips in pine trees.] — Fiziol. Rast. **4**: 566—569, 1957.

VIEWEG, G. H., ZIEGLER, H.: Thermoelektrische Registrierung der Geschwindigkeit des Transpirationsstromes I. — Ber. deut. bot. Ges. **73**: 211—226, 1960.

VILKNER, H.: Über die Erdbodenfeuchtigkeit und ihre Messung mittels elektrischer Kapazität. — Z. Meteorol. **11**: 49—58, 1957.

VIRGIN, H. I.: A new method for the determination of the turgor of plant tissues. — Physiol. Plant. **8**: 954—962, 1955.

VITON, P.: Hygromètre à point rosée. — In: Techniques d'Étude des Facteurs Physiques de la Biosphère. Pp. 203—207. INRA, Paris 1970.

VOTCHAL, A. E.: [Methods for continuous estimation of the rate of transpiration of plants.] — Dokl. Akad. Nauk SSSR **29**: 422—425, 1940.

DE VRIES, D. A.: A non stationary method for determining thermal conductivity of soil *in situ*. — Soil Sci. **73**: 83—89, 1952.

DE VRIES, D. A., PECK, A. J.: On the cylindrical probe method of measuring thermal conductivity with special reference to soils. I. Extension of theory and discussion of probe characteristics. — Aust. J. Phys. **11**: 255—271, 1958.

WÄCHTERSHÄUSER, H.: Die Verdunstungsmengen des Piche-Evaporimeters in ihrer korrelativen Abhängigkeit von Sättigungsdefizit und Wind. — Angew. Bot. **28**: 192—203, 1954.

WADLEIGH, C. H., GAUCH, H. G., MAGISTAD, O. C.: Growth and rubber accumulation in guayule as conditioned by soil salinity and irrigation regime. — U.S. Dept. Agr. Tech. Bull. 925, 1946.

WAGGONER, P. E.: Calibration of a porometer in terms of diffusive resistance. — Agr. Meteorol. **2**: 317—329, 1965.

WAGNER, G. H.: Use of porous ceramic cups to sample soil water within the profile. — Soil Sci. **94**: 379—386, 1962.

WAISEL, Y.: Dew absorption by plants of arid zones. — Bull. Res. Counc. Israel **6D**: 180—189, 1958.

WAISEL, Y.: Effect of phenylmercuric acetate on stomatal movement and transpiration of excised *Betula papyrifera* MARSH. leaves. — Plant Physiol. **44**: 685—690, 1969.

WAISTER, P. D.: Equipment for measuring water strees in leaves. — Univ. Nottingham Dep. Hort. Misc. Publ. 15, 1963a.

WAISTER, P. D.: An improved thermocouple for assessing leaf water potential by vapour pressure measurement. — Israel J. Bot. **12**: 192—196, 1963b.

WAISTER, P. D.: Precision of thermocouple psychrometers for measuring leaf water potential. — Nature **205**: 922—923, 1965.

WALLIHAN, E. F.: The use of tensiometers for soil moisture measurement in ecological research. — Ecology **20**: 403—412, 1939.

WALLIHAN, E. F.: Studies of the dielectric method of measuring soil moisture. — Proc. Soil Sci. Soc. Amer. **10**: 39—40, 1946.

WALLIHAN, E. F.: Modification and use of an electric hygrometer for estimating relative stomatal apertures. — Plant Physiol. **39**: 89—90, 1964.

WALLIN, J. R.: Agrometeorological aspects of dew. — Agr. Meteorol. **4**: 85—102, 1967.

WALTER, H.: Über die Pressaftgewinnung für kryoskopische Messungen des osmotischen Wertes bei Pflanzen. — Ber. deut. bot. Ges. **46**: 539—549, 1928.

WALTER, H.: Die kryoskopische Bestimmung des osmotischen Wertes bei Pflanzen. — In: ABDERHALDEN, E. (ed.): Handbuch der biologischen Arbeitsmethoden. XI/4. Pp. 533—571. Urban & Schwarzenberg, Berlin—Wien 1931a.

WALTER, H.: Die Hydratur der Pflanze und ihre physiologisch-ökologische Bedeutung. (Untersuchungen über den osmotischen Wert.) — G. Fischer Verlag, Jena 1931b.

WALTER, H.: Tabellen zur Berechnung des osmotischen Werter von Pflanzenpressäften, Zuckerlösungen und einigen Salzlösungen. — Ber. deut. bot. Ges. **54**: 328—339, 1936.

WALTER, H.: The water economy and the hydrature of plants. — Annu. Rev. Plant Physiol. **6**: 239—252, 1955.

WALTER, H.: Grundlagen der Pflanzenverbreitung. I. Standortslehre (Einführung in die Phytologie III). — E. Ulmer, Stuttgart. First Ed. p. 525, 1951. Second Ed. p. 566, 1960.

WALTER, H.: Zur Klärung des spezifischen Wasserzustandes im Plasma und in der Zellwand bei der höheren Pflanze und seine Bestimmung. I. Allgemeines. II. Methodisches. — Ber. deut. bot. Ges. **76**: 40—53, 54—71, 1963.

WALTER, H.: Zur Klärung des spezifischen Wassserzustandes im Plasma, Teil. III. Ökophysiologische Betrachtungen. — Ber. deut. bot. Ges. **78**: 104—114, 1965.

WALTER, H.: Remarks on the critique of the hydrature concept, by E. SHMUELI and O. P. COHEN. — Israel J. Bot. **15**: 35—36, 1966.

WALTER, H.: Critique of the hydrature concept by Shmueli and Cohen. — Israel J. Bot. **16**: 48, 1967.

WALTER, H., ELLENBERG, H.: Ökologische Pflanzengeographie. — Forschr. Bot. **20**: 100—117, 1958.

WALTER, H., KREEB, K.: Die Hydratation und Hydratur des Protoplasmas der Pflanze und ihre öko-physiologische Bedeutung. — Protoplasmatologia IIC 6. — Springer Verlag, Wien—New York 1970.

WALTER, H., THREN, R.: Die Berechnung des osmotischen Wertes auf Grund von kryoskopischen Messungen und der Vergleich mit Saugkraftbestimmungen. — Jahrb. wiss. Bot. **80**: 20—35, 1934.

WALTER, H., WEISMANN, O.: Über die Gefrierpunkte und osmotischen Werte lebender und toter pflanzlicher Gewebe. — Jahrb. wiss. Bot. **82**: 273—310, 1935.

WARING, R. H., CLEARY, B. D.: Plant moisture stress: evaluation by pressure bomb. — Science **155**: 1248—1254, 1967.

WARING, R. H., HERMANN, R. K.: A modified Piche evaporimeter. — Ecology **47**: 308—310, 1966.

WARREN WILSON, J. (with an appendix by ROSE, C. W.): The components of leaf water potential. I. Osmotic and matric potentials. — Aust. J. biol. Sci. **20**: 329—347, 1967.

WARTIOVAARA, V.: The permeability of *Tolypellopsis* cells for heavy water and methyl alcohol. — Acta bot. fenn. **34**: 1—21, 1944.

WEAST, R. C. (ed.): Handbook of Chemistry and Physics. 49 Ed. — The Chemical Rubber Co., Cleveland, Ohio 1968/69.

WEATHERLEY, P. E.: Note on the diurnal fluctuations in water content of floating leaf disks. — New Phytol. **46**: 276—278, 1947.

WEATHERLEY, P. E.: Studies in the water relations of the cotton plant. I. The field measurement of water deficits in leaves. — New Phytol. **49**: 81—97, 1950.

WEATHERLEY, P. E.: Studies in the water relations of the cotton plant. II. Diurnal and seasonal fluctuations and environmental factors. — New Phytol. **50**: 36—51, 1951.

WEATHERLEY, P. E.: Preliminary investigations into the uptake of sugars by floating leaf disks. — New Phytol. **53**: 204—216, 1954.

WEATHERLEY, P. E.: A new microosmometer. — J. exp. Bot. **11**: 258—268, 1960.

WEATHERLEY, P. E.: Some investigations on water deficits and transpiration under controlled conditions. — In: SLAVÍK, B. (ed.): Water Stress in Plants. Pp. 63—71. Dr W. Junk N. V. Publishers, The Hague 1965a.

WEATHERLEY, P. E.: The state and movement of water in the leaf. — In: FOGG, G. E. (ed.): The State and Movement of Water in Living Organisms. (Symp. Soc. exp. Biol. 19). Pp. 157—184. Cambridge Univ. Press, Cambridge 1965b.

WEATHERLEY, P. E.: Contribution to the general discussion. — In: SLAVÍK, B. (ed.): Water Stress in Plants. Pp. 305—307. Dr W. Junk N. V. Publishers, The Hague 1965c.

WEATHERLEY, P. E.: A porometer for use in the field. — New Phytol. **65**: 376—387, 1966.

WEATHERLEY, P. E., SLATYER, R. O.: Relationship between relative turgidity and diffusion pressure deficit in leaves. — Nature **179**: 1085—1086, 1957.

WEAVER, H. A., JAMISON, V. C.: Limitations in the use of electrical resistance soil moisture units. — Agron. J. **43**: 602—605, 1951.

WEBSTER, R.: The measurement of soil water tension in the field. — New Phytol. **65**: 249—258, 1966.

WEEKS, L. V., STOLZY, L. H.: The use of portable neutron equipment measure the quantity of water in large soil columns. — Proc. Soil Sci. Amer. **22**: 201—203, 1958.

WEHRLE, PH.: Instruments et procédès d'observation météorologiques. — Pp. 337. Office national météorologique, Paris 1942.

WEINBERGER, P., ROMERO, M., OLIVA, M.: Ein methodischer Beitrag zur Bestimmung des subletalen (kritischen) Wassersättigungsdefizits. — Flora **161**: 555—561, 1972.

WELTEN, M.: Physiologisch-ökologische Untersuchungen über den Wasserhaushalt der Pflanzen mit besonderer Berücksichtigung der Wasserabgabewiderstände. — Planta **20**: 45—165, 1933.

WENT, F. W.: Plant growth under controlled conditions. III. Correlation between various physiological processes and growth in the tomato plant. — Amer. J. Bot. **31**: 597—618, 1944.

WENZL, H.: Die Bestimmung des Spaltöffnungszustandes nach dem Abdruckverfahren. — Jahrb. wiss. Bot. **88**: 89—122, 1939.

WERNER, H. O.: Influence of atmospheric and soil moisture conditions on diurnal variations in relative turgidity of potato leaves. — Bull. Univ. Nebraska Col. Agr. Exp. St. **176**: 1—39, 1954.

WERNER, O.: Die Maispflanze auf einem trockenharten Wurzelfaden voll wachsend. — Biol. gen. **7**: 689—708, 1931.

WERNER, O.: Wachstum und Wasserbilanz der Pflanze im Spiegel ihrer Gewichtsänderungen. — Biol. gen. **9**: 199—230, 1933.

WERNER, O.: Neue Wege zur Verbesserung des Vegetationsgefässes. — Bodenkultur **2**: 181—189, 1948.

VAN DER WESTHUIZEN, M.: On the possibility of measuring soil moisture with high frequency electro-magnetic waves. — South Afr. J. agr. Sci. **7**: 589—590, 1964.

WEXLER, A.: Electric hygrometers. — Circ. U.S. nat. Bur. Stand. No. 586, 1957.

WEXLER, A.: Humidity and Moisture. Measurement and Control in Science and Industry. 4. vol. — Reinhold, New York 1965.

WHITEHEAD, F. H., HOOD, J. R. S.: A method of maintaining fractions of field capacity in pot experiments. — New Phytol. **65**: 240—244, 1966.

WHITEMAN, P. C.: The use of relative turgidity as an index of diffusion pressure deficit. — Aust. J. Sci. **24**: 288, 1961.

WHITEMAN, P. C., WILSON, G. L.: Estimation of diffusion pressure deficit by correlation with relative turgidity and beta-radiation absorption. — Aust. J. biol. Sci. **16**: 140—146, 1963.

WIEBE, H. H.: Matric potential of several plant tissues and biocolloids. — Plant Physiol. **41**: 1439—1442, 1966.

WIEBE, H. H., BROWN, R. W., DANIEL, T. W., CAMPBELL, E. C.: Water potential measurements in trees. — BioScience **20**: 225—226, 1970.

WIEBE, H. H., CAMPBELL, G. S., GARDNER, W. H., RAWLINS, S. L., CARY, J. W., BROWN, R. W.: Measurement of plant and soil water status. — Bull. 484, pp. 1—71, Utah Agric. Exp. Sta. 1971.

WIEBE, H. H., KRAMER, P. J.: Translocation of radioactive isotopes, from various regions of roots of barley seedlings. — Plant Physiol. **29**: 342—348, 1954.

VAN WIJK, W. R. (ed.): Physics of Plant Environment. — North Holland Publ. Co., Amsterdam 1963.

WILLIAMS, C. N., SINCLAIR, R.: A sensitive porometer for field use. — J. exp. Bot. **19**: 460 to 467, 1968.

WILSON, C. C.: The porometer method for the continuous estimations of dimensions of stomata. — Plant Physiol. **22**: 582—589, 1947.

WILSON, J. D.: A double-walled pot for the auto-irrigation of plants. — Bull. Torrey bot. Club **56**: 139—153, 1929.

WOLF, D. D., PEARCE, R. B., CARLSON, G. E., LEE, D. R.: Measuring photosynthesis of attached leaves with air sealed chambers. — Crop Sci. **9**: 24—27, 1969.

WOLF, J.: Über das Herstellen mikroskopischer Preparate der Oberflächen verschiedener Objekte mit Hilfe der Adhesionsmethode. — Z. wiss. Mikroskopie **56**: 181—201, 1939.

WOOLEY, J. T.: Radial exchange of labelled water in intact maize roots. — Plant Physiol. **40**: 711—717, 1965.

WORMER, TH. M., OCHS, R.: Humidité du sol, ouverture des stomates et transpiration du palmier à huile et de l'arachide. — Oléagineaux **14**: 571—580, 1959.

WYLIE, R. G.: A new tool for transpiration experiments. — CSIRO (Aust.) annu. Rep. 1962—1963: 32—33, 1963.

YAMADA, Y., TAMAI, S., MIYAGUCHI, T.: The measurement of the thickness of leaves using S^{35}. — Proc. 2nd Jap. Conf. Radioisotopes 1958: 1692—1700, 1961.

YANG, S. J., DE JONG, E.: Measurement of internal water stress in wheat plants. — Canad. J. Sci. 48: 89—95, 1968.

YASTREBOV, M. T.: Vidoizmenenyĭ kapilyarnyĭ metod opredeleniya osmoticheskogo davleniya zhidkosteĭ i ego primenenie v biologii. [Modified capillary method for determining osmotic pressure and its application in biology]. — Tr. Inst. Fiziol. Rast. Akad Nauk SSSR 8: 404—411, 1954.

YEMM, E. W., WILLIS, A. J.: Stomatal movements and changes of carbohydrate in leaves of *Chrysanthemum maximum*. — New Phytol. 53: 373—396, 1954.

YODA, S.: Effect of auxin and gibberellin on osmotic value of pea stem sections. — Plant Cell Physiol. 2: 435—442, 1961.

YOUKER, R. E., DREIBELBIS, F. R.: An improved soil moisture measuring unit for hydrologic studies. — Trans. Amer. Geophys. Union 32: 447—449, 1951.

ZADRAŽIL, K., DAREBNÍK, Z.: Zkušenosti s odporovou metodou měření půdní vlhkosti absorpčními sádrovými bloky. [Resistance method for measuring soil moisture with plaster blocks]. — Sborník ČSAZV — Lesnictví (Praha) 4: 897—906, 1958.

ZELITCH, I.: The control and mechanism of stomatal movement. — In: ZELITCH, I. (ed.): Stomata and Water Relations in Plants. — Conn. Agr. Sta. Bull. 664: 18—42, 1963.

ZIERLER, K. L.: Basic aspects of kinetic theory as applied to tracer-distribution studies. — In: KNISELEY, R. M., TAUXE, W. N. (eds.): Dynamic clinical studies with radioisotopes. — US At. En. Com., Germantown 1964.

ZUR, B.: Osmotic control of the matric soil-water potential. II. Soil-plant system. — Soil Sci. 103: 30—38, 1967.

Author index

A

AAMODT, E. E. see STOECKELER, J. H.
ABDERHALDEN, E. *388, 405, 409, 411*
ABELE, J. E. 74, *364*
ABOU RAYA, M. A. 274, *364*
ABRAMS, E. see TANNER, C. B.
ACEVEDO, E. see HSIAO, T. C.
AIKMAN, D. P. see ANDERSON, W. P.
AITCHISON, G. D., BUTLER, P. F., GURR, C. G. 160, *364*
ALCOCK, M. B. 130, *364*
ALJIBURY, F. K., TOMLINSON, W. M., HOUSTON, C. E. 190, *364*
ALLAWAY, W. G., MANSFIELD, T. A. 323, *364*
ALLEN, L. H. JR. see IMPENS, I.
ALLES, W. see MEDERSKI, H. J.
ALVIM, P. de T. 324, *364*
ALVIM, P. de T., HAVIS, J. R. 299, *364*
VAN ANDEL, O. M. 99, 100, 107, 213, 214, *364*
ANDERSON, A. B. C. 161, *364*
ANDERSON, A. B. C., EDLEFSEN, N. E. 161, 177, *364*
ANDERSON, W. P., AIKMAN, D. P., MEIRI, A. 218, *364*
ANDERSON, W. P., COLLINS, J. C. 213, *364*
ANDERSON, W. P., HOUSE, C. R. 213, 218, *364*
ANDERSON, W. P., REILLY, E. J. 213, 217, 218, *364*
ANDERSSON, N. E., HERTZ, C. H. 276, 277, *364*
ANDERSSON, N. E., HERTZ, C. H., RUFELT, H. 244, 253, 277, *364*
ANTOSZEWSKI, R., LIS, E. K. 124, *364*
APPERLOO, M. 129, *364*
APPLEWHITE, T. H. see ORDIN, L.
ARCICHOVSKAJA, N. see ARCICHOVSKIJ, V.
ARCICHOVSKIJ, V. 24, *365*
ARCICHOVSKIJ, V., ARCICHOVSKAJA, N. 29, 73, *365*

ARCICHOVSKIJ, V., KISSELEW, N., KRASSULIN, N., MENJINSKAJA, E., OSSIPOV, A. 29, 72, *365*
ARCICHOVSKIJ, V., OSSIPOV, A. 25, 72, *365*
ARISZ, W. H., HELDER, R. J., VAN NIE, R. 218, *365*
ARLAND, A. 285, *365*
ARMSTRONG, J. I. see GREGORY, F. G.
ARNOLD, A. J. 190, *365*
ARNOULD, G. see LE JEUNE, G.
ARVIDSSON, I. 153, 155, 248, *365*
ASHBELL, C. W. see CANNELL, G. H.
ASHBY, E., WOLF, R. 25, 26, 72, *365*
ASHTON, F. M. 162, *365*
ASLYNG 157
ATKINS, W. R. G. see DIXON, H. H.
AUSTIN, R. S. see HOFFMAN, G. J.
AVERY, D. J. 263, *365*

B

BACON, J. A. see OSBORNE, T. S.
BAILEY, L. F., ROTHACHER, J. S., CUMMINGS, W. H. 290, *365*
BAKER, D. A. 213
BAKER, D. A., MILBURN, J. A. 201, *365*
BAKER, D. A., WEATHERLEY, P. E. 213, *365*
BAKER, D. A. see MINCHIN, F. R.
BAKKE, A. L. see LIVINGSTON, B. E.
BALDES, E. J. 99, *365*
BALDES, E. J., JOHNSON, A. F. 99, *365*
BALDES, E. J. see ROEPKE, R. R.
BALL, F. K. see EMMERT, E. M.
BANGE, G. G. J. 296, 315, *365*
BAŇOCH, Z. see ÚLEHLA, J.
BARANETZKY, I. 215, *365*
BARGER, G. 98, 107, *365*
BARLOW, W. K. see CAMPBELL, E. C.
BARRS, H. D. 54, 55, 62, 66, 67, 68, 136, 137, 147, *366*
BARRS, H. D., FREEMAN, B., BLACKWELL, J., CECCATO, R. D. 74, *366*

M

Subject Index

A

abaxial leaf surface   287
absolute air humidity see humidity of air, absolute
absolute water capacity   159
absorber thickness   123
absorbing surface of roots   159, 186
absorption power   7
absorption of radioactive radiation   123 to 128
absorption of water and transpiration, simultaneous measurement of   208 to 212
− − − by aerial parts   246 to 248
− − − by complete root system   193 to 201
− − − by root hairs   205, 206
− − − by roots   159, *193 to 212*
− − − by roots, determination of   193 to 212
− − − in parts of a single root   201
absorption rate of water from soil   212
acrylic plastics   262
activity of water, relative   9
adaxial leaf surface   287
adsorption of water vapour   66
Advanced Instruments Inc.   94
age of leaf   126, 128, 137, 139, 142, 150
air, boundary layer of   237, *286 to 289*
− , dielectric properties of   279
air humidity see humidity of air
− − , control of   242
− − , effective (measurement of transpiration rate)   260
air jacket 87 to 89
algae   75, 106, 221
alumel   65
γ-aluminium oxide   245
aluminium oxide hygrometer   279, 280
American Instrument Co.   94
Aminco (DUNMORE) humidity sensor 257, 258, 303, 304, 305, 308, 309
Ammeter, Microvolt   56

ammonia   269, 270
ammonium carbonate   217
Amoeba   80
amphistomatous leaves   287, 294, 301, 306, 314
amplifier for thermistor psychrometry and cryoscopy   60, 67, 91
anaerobiosis   141
angle, contact   247
anhydrone   245
antitranspirants   288
"Anwelkmethode"   285
aphid-stylet technique   80
Apiezon wax "W"   46
apparent free space   131, 132
apricot trees   81
Aqmel   277, 279
aquatic plants   12
araldite   43
arc-welder   44
Argentol   131
argon, radioactive   301
ash trees   223
Asparagus officinalis   116, 136
Aspirations-Psychrometer-Tafeln   241, *365*
aspiration of the thermocouple see ventilation
atmograph see evaporimeter
atmospheric pressure   261, 270
Auriema Ltd.   279
Aushartungsvermögen   153
autofluorescence   152
autoirrigators   190
automatization of psychrometric techniques   55, 68
autoradiography   222
auxanographic method   18
availability of soil water   157 to 191
− − − − , dynamic   158 to 160, 186 to 190
− − − − , static 156, 157, *164 to 185*
− − − − , static, determination of 164 to 185

SUBJECT INDEX 449

— — of intact plants 62 to 65
— — of roots 159, 191 to 193
— — of water vapour 11, 237
— — , relation to relative air humidity 11
— — , units of 12
— — , values of 12
— resaturation deficit (w. r. d.) 151, 152, 154, 155
— saturation deficit 109, 136 to 156, 187
— — — and absorption of beta radiation 125 to 127
— — — and leaf reflectance 131
— — — , calculation of 149
— — — , errors in determination 149
— — — , harmful 151 to 156
— — — , lethal 151
— — — , sublethal 151, 152
— status 1, 3
"water supplying power of soil" 186
water transfer 16
— transport in roots 201
— — , concentration of 286, 288, 289, 292, 302, 311
— vapour, physical properties of (table) 336
— — pressure (tension) 10, 97, 239, 240
— — — (tension), partial 238, 240, 243
— — — , relative 9, 10
— — — , saturation 238, 240
— — -saturated atmosphere 216
— — tension, relative 9, 10
watering of plants 131
wax films 247
weeds 290
weighing of plants 135
— of samples 35
welding of thermocouples 44, 48
Wescor Inc. 57, 62, 65, 68, 102, 167
wet bulb depression 38, 239
— — temperature 65, 239, 240
— junction (thermocouple psychrometer, air humidity) 242
wet-loop psychrometer 38, 39, 50 to 51, 58, 65 to 68, 103, 164
wetting of leaf surface 247, 248
wheat 122, 189, 290
Wheatstone bridge porometer 320, 321, 323
WILD evaporimeter 248
wilting 6
"wilting method" 285

wilting, permanent 186, 187
— , symptoms of 188
wind speed 239
wood 224
— , ring porous 223
— , scattered porous 223
woody plants 136, 231

X

xerophytes 111
xylem 324
— , active (living) 232, 233, 234
— sap 70, 80, 85, 111, 213 to 218
— — exudation 213 to 218
— — volume 213 to 218
— , velocity of water movement in 223 to 231
— vessels 17, 63, 70, 80, 85, 109, 111
— , volume flux of water in 231 to 235
xylene 299, 300
x-y plotter 68

Y

Yellow Springs Instrument Co. 280
yew 116

Z

zero turgor (wilting) 187
— — methods (osmotic potential) 79, 106
zinc chloride flux (soldering) 43

γ see psychrometric constant
ΔT see freezing point depression
Θ see soil water content, volumetric
μ_w see chemical activity of water
π see osmotic pressure
ϱ see density of water
σ see reflection coefficient
τ see matric tension
Φ see hydraulic potential
Ψ_m see matric potential
Ψ_p see pressure turgor potential
Ψ_s see osmotic potential
Ψ_w see (total) water potential

Trees: Structure and Function

By **Martin H. Zimmermann,**
Charles Bullard Professor
of Forestry and Director
of Harvard Forest,
Harvard University

Claud L. Brown,
Professor of Forestry and
Botany, University
of Georgia

with a chapter by
Melvin T. Tyree,
Botany School, University
of Cambridge, England

111 figs. XIV, 336 pp. 1971
Cloth DM 72,—; US $27.80
ISBN 3-540-05367-0

Prices are subject to change
without notice

Trees have the distinction of being the largest and oldest living organisms on earth. Although the herbaceous habit has made unprecedented evolutionary gains, trees are still the most conspicuous plants covering the habitable land surface of the earth. Trees have always been of much interest to botanists, and many of the early investigations concerning the structure and function of plants were conducted with trees. Not only do trees perform all the cellular activities of most unicellular and herbaceous plants . . . they perform a good deal more.

This book is devoted largely to those aspects of tree physiology which are peculiar to tall woody plants. Throughout the book the emphasis is on function as it relates to structure. The authors describe how trees grow, develop and operate, not how these functions are modified in different types of environment. The approach is functional rather than ecological . . . how trees work, not how they behave in various habitats.

The text contains many literature citations without being encyclopedic.

Springer-Verlag
Berlin Heidelberg New York
München Johannesburg London Madrid
New Delhi Paris Rio de Janeiro Sydney
Tokyo Utrecht Wien